# THE
# Ocean
# Engineering
# HANDBOOK

# The Electrical Engineering Handbook Series

*Series Editor*
**Richard C. Dorf**
University of California, Davis

## Titles Included in the Series

*The Avionics Handbook*, Cary R. Spitzer

*The Biomedical Engineering Handbook, 2nd Edition*, Joseph D. Bronzino

*The Circuits and Filters Handbook*, Wai-Kai Chen

*The Communications Handbook*, Jerry D. Gibson

*The Control Handbook*, William S. Levine

*The Digital Signal Processing Handbook*, Vijay K. Madisetti & Douglas Williams

*The Electrical Engineering Handbook, 2nd Edition*, Richard C. Dorf

*The Electric Power Engineering Handbook*, Leo L. Grigsby

*The Electronics Handbook*, Jerry C. Whitaker

*The Engineering Handbook*, Richard C. Dorf

*The Handbook of Formulas and Tables for Signal Processing*, Alexander D. Poularikas

*The Industrial Electronics Handbook*, J. David Irwin

*The Measurement, Instrumentation, and Sensors Handbook*, John G. Webster

*The Mechanical Systems Design Handbook*, Osita D.I. Nwokah

*The RF and Microwave Handbook*, Mike Golio

*The Mobile Communications Handbook, 2nd Edition*, Jerry D. Gibson

*The Ocean Engineering Handbook*, Ferial El-Hawary

*The Technology Management Handbook*, Richard C. Dorf

*The Transforms and Applications Handbook, 2nd Edition*, Alexander D. Poularikas

*The VLSI Handbook*, Wai-Kai Chen

*The Mechatronics Handbook*, Robert H. Bishop

## Forthcoming Titles

*The Communications Handbook, 2nd Edition*, Jerry D. Gibson

*The Circuits and Filters Handbook, 2nd Edition*, Wai-Kai Chen

# THE
# Ocean
# Engineering
# HANDBOOK

### EDITOR-IN-CHIEF

## FERIAL EL-HAWARY

B.H. Engineering Systems Ltd.
Halifax, Nova Scotia, Canada

**CRC Press**
Taylor & Francis Group
Boca Raton London New York

CRC Press is an imprint of the
Taylor & Francis Group, an **informa** business

CRC Press
Taylor & Francis Group
6000 Broken Sound Parkway NW, Suite 300
Boca Raton, FL 33487-2742

First issued in paperback 2019

© 2001 by Taylor and Francis Group, LLC
CRC Press is an imprint of Taylor & Francis Group, an Informa business

No claim to original U.S. Government works

ISBN-13: 978-0-8493-8598-8 (hbk)
ISBN-13: 978-0-367-39769-2 (pbk)

**Visit the Taylor & Francis Web site at**
**http://www.taylorandfrancis.com**

**and the CRC Press Web site at**
**http://www.crcpress.com**

# Dedication

Dedicated To My Family For Our Love of The Oceans

# Preface

*The Ocean Engineering Handbook* is designed to offer the reader a reasonably comprehensive coverage of a number of important areas involving the theory and practice of oceanic and coastal engineering and technology. Of course, one cannot expect to become an expert on a subject as vast and complicated as oceanic engineering from one book, no matter how large. Throughout the book, references are given to more detailed and specialized works on each of the topics treated.

A major challenge in designing this handbook was the breadth and diversity of the subject area. Oceanic technology is remarkably varied. Oceanic engineering applications range from fiber optic applications to position control of ocean-going vessels. Oceanic system theory ranges from marine hydrodynamics to modeling of waves in the oceans. *The Ocean Engineering Handbook* thoroughly covers position control theory and implementation.

The book is organized in six major sections: Marine Hydrodynamics and Vehicle Control, Modeling Considerations, Position Control Systems for Offshore Vessels, Computational Intelligence in Ocean Engineering, Fiber Optics in Oceanographic Applications, and Current Measurement Technology. It is designed to be used as a traditional handbook where one would be able to find the answer to a question about a topic by consulting an article dealing with that topic. The handbook can also be used in different ways as it offers snapshots of the present state-of-the-art in various subjects.

The first chapter of the handbook is on Marine Hydrodynamics and Vehicle Control and is edited by Dr. Zoran Vukic of the University of Zagreb in Croatia. The first section is entitled Anatomy of Sea Level Variability: An Example from the Adriatic and is by M. Orlic. This is followed by Guidance and Control Systems for Marine Vehicles by Z. Vukic and B. Borovic. Section 3 is Sea Ambient Noise: An Example from the Middle Adriatic by D. Matika. S. Krueger introduces Basic Shipboard Instrumentation and Fixed Automatic Stations for Monitoring in the Baltic Sea in Section 4.

Chapter 2 consists of four sections, the first of which is by Advisory Editor/Contributor Hisaaki Maeda from the University of Tokyo who offers a discussion of Marine Hydrodynamics and Dynamics of a Floating Structure. The second is on Mathematical Modeling of Ocean Waves and is by Advisory Editor/Contributor Dr. M. Rahman of DalTech, Dalhousie University in Halifax. Sections 3 and 4 are on Systems Approaches to Heave Compensation in Marine Dynamic Applications and Approaches to Marine Seismic Extraction and are written by Dr. Ferial El-Hawary.

The third chapter is on Position Control Systems for Offshore Vessels written by a Scandinavian team led by Advisory Editors Asgeir J. Sorensen and Thor I. Fossen. Six sections discuss fundamental issues and new design approaches in this significant area. In the first section, J. P. Strand and A. J. Sorensen discuss Marine Positioning Systems. This is followed by a section on Mathematical Modeling of Marine Vessels by A. J. Sorensen. Section 3 is entitled Position and Velocity Observer Design and is written by T. I. Fossen and J. P. Strand. Section 4 by A. J. Sorensen deals with the Design of Controllers for Positioning of Marine Vessels. In section 5, T. I. Fossen and J. P. Strand offer a treatment of Weather Optimal Positioning Systems. The last section deals with methods for thrust control and is by A. J. Sorensen, J. P. Strand, and T. I. Fossen.

The fourth chapter of the handbook deals with the applications of computational intelligence in the ocean environment. Under the leadership of the imminent scholar Dr. C. H. Chen, four teams of contributors address issues in this important emerging area. These begin with an article entitled A Multivariable Online Intelligent Autopilot Design Study written by a team from the University of Plymouth, Devon, U.K. In the second section, Dr. Ray Gosine and the team from Memorial University of Newfoundland offer a detailed discussion of multi-robot cooperation under human supervision. Dr. Donna Kocak, a member of the technical staff at the Harbor Branch Oceanographic Institution offers an excellent introduction and an update on Computer Vision in Ocean Engineering.

Dr. Frank Caimi of the Harbor Branch Oceanographic Institution is Advisory Editor of the fifth chapter of the handbook, which treats Fiber Optics in Oceanographic Applications. This chapter contains four sections and begins with an overview of fiber optics contributed by the advisory editor. This is followed by a discussion of the Basics of Fiber Optic Strain Sensors by Barry Grossman and Syed Murshid. Section three of this chapter is contributed by Tony Dandridge and Clay Kirkendall of the Naval Research Laboratory in Washington, D.C. and covers Fiber Optic Acoustic Sensors. Fiber Optic Telemetry in Ocean Cables is discussed in Section 4 by George Wilkins, Naval Research Laboratory, Washington, D.C.

The final chapter is devoted to current measurement techniques and is offered by Albert J. Williams 3rd of the Woods Hole Oceanographic Institute in Massachusetts.

I invite you to explore with us the many facets of oceans engineering.

# Editor-in-Chief

Ferial El-Hawary, P. Eng., F. MTS, F. EIC, F. IEEE received her M.Sc. from the University of Alberta, Canada in Electrical Engineering and her Ph.D. from Memorial University of Newfoundland in Oceans Engineering. She has published widely, and has made numerous presentations on underwater applications of advanced signal processing and estimation techniques.

She is the cofounder and president of BH Engineering Systems Limited (BHES) of Halifax, and the Modeling and Signal Analysis Research Laboratory in the Faculty of Engineering at the Technical University of Nova Scotia (TUNS), which is now DalTech, a part of Dalhousie University. At BHES, Dr. El-Hawary's activities involve technology transfer from the academic sphere to industry through consulting and the offering of advanced professional development courses for industry.

Dr. El-Hawary has supervised a number of graduate students at TUNS on the application of digital system concepts to underwater dynamic motion estimation and marine seismic methods, and also on the environmental impact of electric power generation. Her research work has been supported by the Natural Sciences and Engineering Research Council of Canada (NSERC) grants.

Dr. El-Hawary has been involved on a worldwide basis in ocean activities both technically and administratively as a member of the IEEE Oceanic Engineering Society Board of Directors, serving as vice-president international and past chairman of the membership development committee. She has been instrumental in promoting the society at the national and international level and, in particular, in organizing oceans conferences held in Canada and outside of North America (Europe). She was also guest editor of a special series of issues of the *IEEE Journal of Oceanic Engineering* dedicated to advanced applications of control and signal processing in the oceans environment.

Her awards and professional memberships include the Institute of Electrical and Electronic Engineers (IEEE) Third Millennium Medal given in recognition and appreciation of her valued services and outstanding contributions (Vancouver, B.C., June 2000); the 1999 Institute of Electrical and Electronics Engineers/Regional Activity Board (IEEE/RAB) achievement award for achievement in promoting IEEE interest in the oceanic engineering community in Atlantic Canada (Halifax, N.S., May 2000); Fellow of the Institute of Electrical and Electronics Engineers/Oceanic Engineering Society (IEEE/OES), for contributions to application of digital system concepts to underwater dynamic motion estimation and marine seismic methods (Seattle, WA, October 1999); Fellow of the Engineering Institute of Canada (EIC), (Ottawa, Ontario, May 1997); and IEEE/Oceanic Engineering Society distinguished service award for outstanding leadership in expanding horizons of oceanic engineering society and promoting oceans conferences beyond the U.S.A. and Canada (Halifax, N.S., October 1997).

# Contributors

**Bruno Borović**
Brodarski Institute
Zagreb, Croatia

**R. S. Burns**
University of Plymouth
Devon, United Kingdom

**Frank M. Caimi**
Florida Institute of Technology
Melbourne, Florida
Harbor Branch Oceanographic
  Institution
Fort Pierce, Florida

**C. H. Chen**
University of Massachusetts
Dartmouth, Massachusetts

**P. J. Craven**
Racal Research Limited
Reading, United Kingdom

**Tony Dandridge**
Naval Research Laboratory
Washington, D.C.

**Ferial El-Hawary**
BH Engineering Systems, Ltd.
Halifax, Nova Scotia, Canada

**Thor I. Fossen**
The Norwegian University of Science
  and Technology
Trondheim, Norway

**R. Gosine**
Memorial University of
  Newfoundland
St. John's, Newfoundland, Canada

**Barry G. Grossman**
Florida Institute of Technology
Melbourne, Florida

**R. Hale**
Memorial University of
  Newfoundland
St. John's, Newfoundland, Canada

**F. Hwang**
Memorial University of
  Newfoundland
St. John's, Newfoundland, Canada

**J. King**
Memorial University of
  Newfoundland
St. John's, Newfoundland, Canada

**Clay Kirkendall**
Naval Research Laboratory
Washington, D.C.

**Donna M. Kocak**
eMerge Interactive, Inc.
Sebastian, Florida
Harbor Branch Oceanographic
  Institution
Fort Pierce, Florida

**Siegfried Krueger**
Baltic Sea Research Institute
Rostock-Warnemuende, Germany

**Hisaaki Maeda**
University of Tokyo
Tokyo, Japan

**Dario Matika**
University of Zagreb
Zagreb, Croatia

**Syed H. Murshid**
Florida Institute of Technology
Melbourne, Florida

**Mirko Orlić**
University of Zagreb
Zagreb, Croatia

**Matiur Rahman**
Dalhousie University
Halifax, Nova Scotia, Canada

**M. Rokonuzzaman**
Memorial University of
  Newfoundland
St. John's, Newfoundland, Canada

**Marit RONæss**
The Norwegian University of Science
  and Technology
Trondheim, Norway

**J. Seshadri**
Memorial University of
  Newfoundland
St. John's, Newfoundland, Canada

**Asgeir J. Sørensen**
The Norwegian University of Science
  and Technology
Trondheim, Norway

**Jann Peter Strand**
ABB Industri AS
Oslo, Norway

**R. Sutton**
University of Plymouth
Devon, United Kingdom

**Zoran Vukić**
University of Zagreb
Zagreb, Croatia

**George Wilkins**
Kailua-Kona, Hawaii

**Albert J. Williams 3rd**
Woods Hole Oceanographic
  Institution
Woods Hole, Massachusetts

# Contents

# 1

# Marine Hydrodynamics and Vehicle Control

Mirko Orlić
*University of Zagreb*

Zoran Vukić
*University of Zagreb*

Bruno Borović
*Brodarski Institute*

Dario Matika
*University of Zagreb*

Siegfried Krueger
*Baltic Sea Research Institute*

## 1.1   Anatomy of Sea Level Variability—An Example from the Adriatic

*Mirko Orlić*

On February, 1 1986 at 01 h EMT an hourly sea level height that surpassed corresponding long-term average by 96 cm was recorded at the Bakar tide-gauge station located on the Croatian coast of the Adriatic Sea (Fig. 1.1). This is one of the largest elevations measured at the station during almost 50 years of continuous operation. As is well known, such episodes bring about the flooding of the north Adriatic coast, with the city of Venice being particularly vulnerable [1]. On the positive side, high (but not too high) sea levels may be beneficial for the operation of ports and, in particular, the maneuverability of the large-draught ships in the port of Bakar improves considerably when the sea level is at its maximum [2]. Consequently, an understanding of sea level variability is not only challenging from the scientific point of view, but is highly applicable as well—particularly having in mind expected rise of sea level during the next century.

The tide-gauge station at Bakar is of the stilling-well type: the pen, recording on a cylindrical drum rotated by clockwork, is driven by a float confined to the well that communicates with the sea through

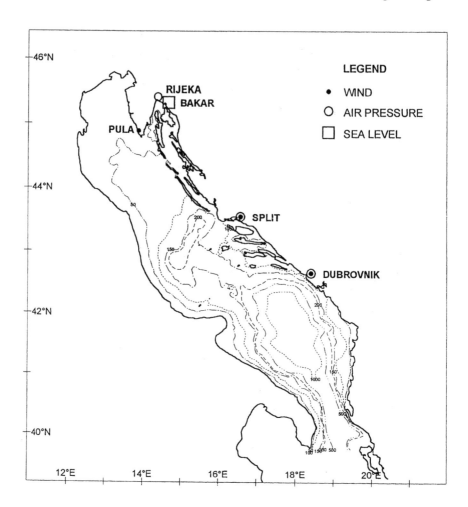

**FIGURE 1.1**   The Adriatic Sea position and topography. Also shown are locations of the meteorological and tide-gauge stations.

a narrow pipe. Although there exist a number of more modern instruments for sea level monitoring, e.g., bottom pressure recorders or satellite altimeters, the classical instruments are still indispensable if one is interested, as we are here, in a broad range of processes and therefore needs long time series. The function of stilling well is to damp short-period oscillations—in the case of Bakar, those having periods smaller than about 1 min. In order to determine hourly values, one must filter out oscillations with periods smaller than 2 h. At Bakar, these are mostly related to standing waves—seiches—of the Bakar Bay, having a maximum period of about 20 min [3].

After extracting hourly sea levels from a tide-gauge record, one is confronted with the variability extending over a wide frequency range and a number of physical processes that control the variability. This is illustrated by a spectrum computed from 16-year time series collected at Bakar (Fig. 1.2). There are some lines in the spectrum, related to tides. There are also some broad maxima, the most notable being the one at the near-diurnal period. Finally, there is an increase of energy toward the lowest frequencies. The plan for this chapter is to isolate, by using various filtering procedures, different phenomena that control the Bakar sea level variability, and to explore them starting from the smallest periods and progressing toward the largest. After the analysis, a synthesis will be attempted in the concluding section with the aim of explaining the 96 cm elevation observed on February 1, 1986. While the processes considered are to some extent influenced by the Adriatic and Mediterranean environments in which they

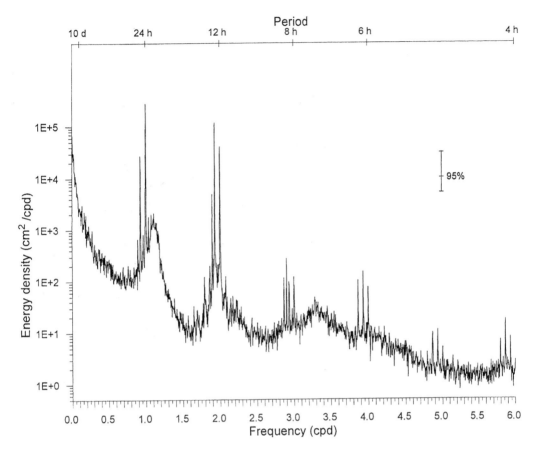

**FIGURE 1.2**   Spectrum computed from the hourly sea level heights recorded at Bakar between 1983 and 1999. The spectrum was determined by using the Parzen window, with 63 degrees of freedom.

develop, they become of increasingly general character with an increase in period and, it is hoped, of interest to a broad circle of readers.

## Tides

As is obvious from Fig. 1.2, tidal lines in the Bakar sea level spectrum are numerous. Yet, the greatest amplitudes are related to the three near-diurnal periods (25.82, 24.07, and 23.93 h) and to the four near-semidiurnal periods (12.66, 12.42, 12.00, and 11.97 h). A usual approach is to fit the sum of the so-called harmonic terms, characterized by the above-mentioned periods, to the sea level time series, and to determine corresponding amplitudes and phases—a procedure known as harmonic analysis. Harmonic synthesis then enables tidal signal to be isolated from the other contributions to the original time series. The tides thus obtained for the two-month interval encompassing the episode of February 1, 1986 are shown in Fig. 1.3. They are of a mixed type, semidiurnal at the new and full moon, diurnal at the first- or last-quarter moon. On February 1, 1986 the tides culminated at 01 h EMT and contributed 20 cm to the sea level maximum.

Empirical and theoretical investigations of tides of the world oceans and seas are summarized in several excellent review articles [4, 5] and books [6, 7] published on the subject. As for the Adriatic, Galilei [8] noticed that its tides were large in comparison with the tides occurring elsewhere in the Mediterranean (although they are much smaller than the tides observed in some other basins). Harmonic analysis, performed for various Adriatic ports by Kesslitz [9], as well as the first maps depicting both the corange

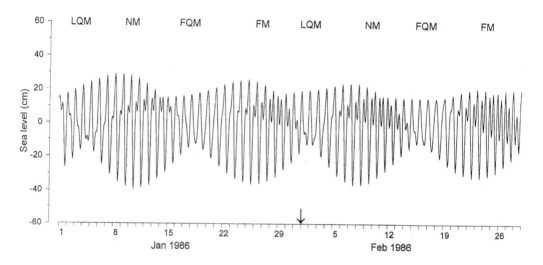

**FIGURE 1.3**   Tides recorded at Bakar in January and February 1986. Indicated are phases of the moon (NM—new moon, FQM—first-quarter moon, FM—full moon, LQM—last-quarter moon). In this, as well as in all subsequent figures, the arrow points to 01 h EMT on February 1, 1986.

and cotidal lines in the Adriatic, constructed by Polli [10], corroborated the early observations. Airy [11] was the first to interpret the relatively large tides of the Adriatic in terms of the resonant excitation of its normal modes by the open Mediterranean tides. The interpretation was substantiated by the numerical models of the Adriatic, inaugurated by Sterneck [12], in which periodical forcing was imposed at the open boundary, in the Otranto Strait, and the tides were computed for the basin interior. It has received further support from the fine-resolution, two-dimensional modeling of the whole Mediterranean, which showed that the Adriatic takes tidal energy from the eastern basin through the Otranto Strait and that this energy is dissipated by bottom friction [13].

## Storm Surges and Seiches

Periods close to, but somewhat greater than tidal, characterize processes related to the synoptic-scale atmospheric disturbances. According to meteorological analyses [14], the upper limit of the periods may be placed at 10 d. Consequently, sea level time series, from which the tidal signal had been subtracted, has been subjected to a high-pass digital filter having a cutoff period of 10 d. The resulting sea levels are depicted in Fig. 1.4, along with the air pressure and wind data simultaneously collected along the east Adriatic coast. Maximum sea level height, amounting to 57 cm, occurred on February 1, 1986 at 01 h EMT. It was related to the low air pressure and strong southerly wind blowing over the Adriatic, and they, in turn, were due to a cyclone that approached the Adriatic on January 30, 1986 [15]. Two days later a front swept over the sea [15] bringing about a sudden decrease of the wind speed (Fig. 1.4). This resulted in the lowering of sea level and generation of persistent free oscillations having close-to-diurnal periods. The initial sea level rise, controlled by the meteorological agents, is termed storm surge, while the subsequent free oscillations represent Adriatic-wide seiches.

   Some in-depth overviews of the storm surge [16–18] and seiche [19] research are available in the literature. The first comparison of the Adriatic sea level with the air pressure and wind was completed by Bučić [20]. Later on, Kesslitz clearly distinguished between the storm surges and seiches [21, 22]. Early spectral analyses of the sea level time series enabled 21.7 and 10.8 h to be pinpointed as periods of the first- and second-mode Adriatic seiches [23]. Maxima may be noticed at the same periods in Fig. 1.2, along with another maximum at about 7 h. In a recent study [24] decay of the seiches has been studied and attributed partly to the bottom-friction control in the Adriatic and partly to the energy loss through

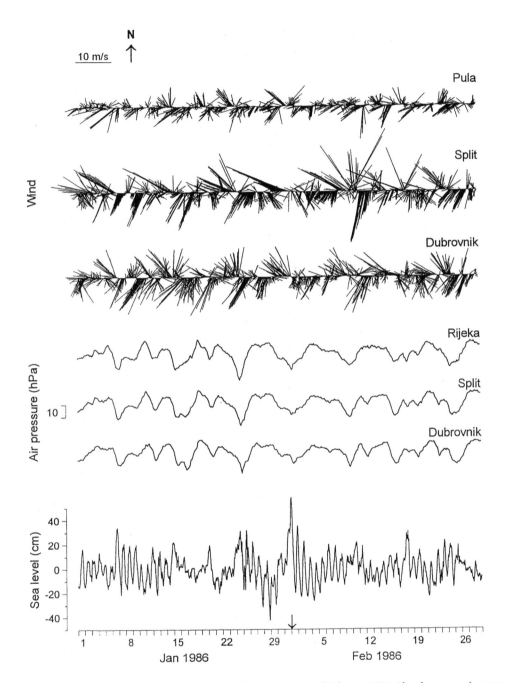

**FIGURE 1.4** Storm surges and seiches recorded at Bakar in January and February 1986. Also shown are air pressure and wind data simultaneously collected at various stations distributed along the east Adriatic coast.

Otranto Strait. Numerical modeling of the Adriatic seiches was pioneered by Sterneck [12] and it took more than 50 years to extend it to the storm surges [25]. In both instances, nodal line was assumed at the model open boundary, in the Otranto Strait. This assumption could be relaxed when the seiche [26] and storm surge [27] modeling was extended to the whole Mediterranean. The former model reproduced the observed seiche periods with unprecedented accuracy. The latter showed that the difference between computed and observed sea level heights may reach 15 cm.

## Response of the Sea to Planetary-Scale Atmospheric Forcing

The next frequency band of interest extends from about 0.01 to 0.1 cpd. Accordingly, 30-day moving averages had been subtracted from the original Bakar sea levels, and the time series thus obtained has been low-pass filtered with a cutoff period placed at 10 d. Figure 1.5 shows the result, along with the air pressure and wind data collected at the nearby meteorological stations and processed in the same way. On February 1,1986 at 01 h EMT, sea level was at its local maximum, equaling about 7 cm, obviously due to the low air pressure and southerly winds prevailing above the Adriatic. The latter could be related to the geopotential heights of the 500 hPa surface [15], which show that at the time, a trough of the planetary atmospheric wave was positioned over the western Mediterranean and that its front side was influencing the Adriatic. Figure 1.5 also demonstrates that sea level variability is greater in winter than in summer, that it is opposed in phase to the air pressure fluctuations, and that it surpasses the variability that would be expected from the simple inverted barometer rule (according to which an air pressure change of 1 hPa corresponds to a sea level change of 1 cm). Cross-spectral analysis (results not shown) reveals that these findings may be mostly interpreted in terms of the coherent influence of the air pressure and wind on the sea.

Analyses of subsynoptic sea level variability, which take into account both the meteorological parameters and planetary atmospheric waves controlling them, are rare. In the Celtic Sea a change in the gradient of the sea level spectrum above and below 0.1 cpd has been noticed and attributed to the dominance of eastward propagating weather patterns above the delimiting frequency as compared with the symmetric propagation of weather patterns below it [28]. As for the Adriatic and Mediterranean Seas, in a series of papers of increasing spatial coverage [29–31] the subsynoptic sea level variability was related to the air pressure fluctuations, which in turn were ascribed to the planetary atmospheric waves traveling above the sea. When the analyses were extended to the longer time series [32], the departure from the inverted barometer response has become obvious at low frequencies. A recent two-input (air pressure, wind) single-output (sea level) cross-spectral analysis has shown that the wind may be responsible for the apparent inverted barometer overshoot, but also that due to the errors inherent in the wind data, the parameters describing response of the sea to both the air pressure and wind are determined with a bias [33]. A numerical model that would enable this interpretation to be checked is still lacking.

## Seasonal and Year-to-Year Variability

By shifting further along the frequency axis one enters the realm of seasonal changes and their interannual variability. These have been analyzed on the basis of monthly mean sea levels from which a 12-month moving average had been subtracted. Figure 1.6 shows the resulting time series superimposed on the average seasonal course and corresponding standard deviations computed for almost 50 years of continuous measurements at Bakar. Sea levels are generally low in late winter or early spring, high in late autumn. Also visible are considerable departures of monthly mean values from the long-term averages, i.e., anomalies. There is no regularity in their occurrence: sometimes the anomalies of different sign follow each other, on other occasions similar anomalies may persist for years. In January and February 1986 the sea level anomaly was positive, being related to the low air pressure anomaly, which at first had extended all over the Europe and then shifted to the Mediterranean area [34]. In the beginning of February 1986 these processes contributed 12 cm to the Bakar sea level height.

An extensive analysis of the annual sea level cycle throughout the world, based on the classical measurements, was published by Pattullo and co-authors [35]. Recently, it has been supplemented by results derived from satellite altimetry [36]. Theoretical studies have indicated that seasonal sea level changes may depend on the air pressure and wind forcing [37] and on both the isostatic and nonisostatic buoyancy-flux effects [38]. Year-to-year variability received considerable attention in the equatorial regions, particularly by researchers of the El Niño Southern Oscillation (ENSO) phenomenon [39, 40]. Detailed investigations of interannual sea level changes in the Adriatic started recently [41, 42]. While the changes have been successfully related to the corresponding atmospheric fluctuations, the origin of the latter still represents an open question. Meteorological analyses suggest that the atmosphere above

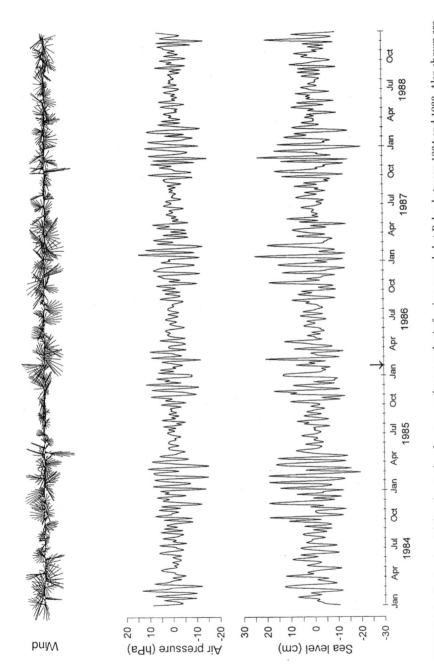

**FIGURE 1.5** Sea level variability due to the planetary-scale atmospheric forcing, recorded at Bakar between 1984 and 1988. Also shown are corresponding air pressures registered at Rijeka and winds measured at Pula.

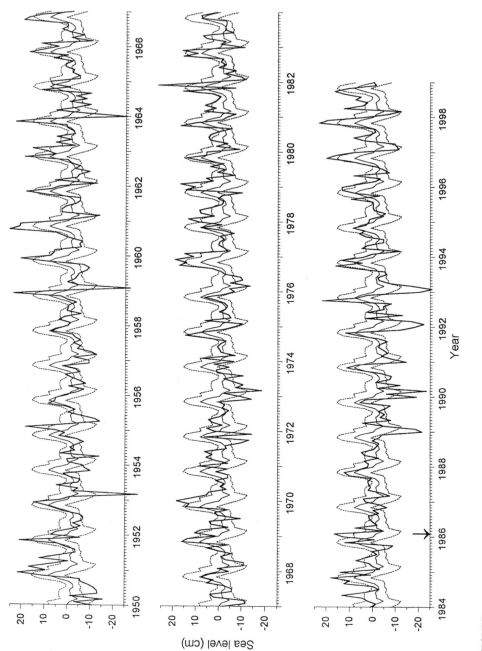

**FIGURE 1.6** Average seasonal course of sea level (thin solid line), corresponding standard deviations (dashed lines), and year-to-year sea level variations (thick solid line) recorded at Bakar between 1950 and 1998.

Europe may be sensitive to the sea-surface temperature variability, both in the northwest Atlantic [43] and equatorial Pacific [44]. It appears that warmer-than-usual surface waters in these areas tend to be associated with anomalously low air pressures over Europe (and vice versa). Yet, air pressure anomalies computed from the observed sea-surface temperature anomalies underestimate observed values, implying that the atmosphere-sea coupling at the frequencies considered here is still not well understood.

## Interdecadal Variability

In order to investigate interdecadal variability, long-term sea level trend, to be analyzed in the next section, had been removed from the Bakar monthly means, and the time series has been subjected to averaging over a 12-month sliding window. Figure 1.7 shows that some year-to-year variability has leaked through the low-pass filter applied. While computation of annual mean values is routine in sea level analysis, response characteristics of such a filter are far from ideal. Figure 1.7 also reveals a well-pronounced oscillation having close to a 20 year period and dominating all the other signals in the frequency band considered. By fitting a 20-year cosine function to the sea levels of Fig. 1.7, one obtains the amplitude of about 2 cm and maxima in the years 1961 and 1981. In the beginning of 1986, however, this oscillation contributed only 1 cm to the total sea level height observed at Bakar.

The 20-year oscillation does not represent the nodal tide; neither the amplitude nor the phase agree with the theoretical values for the equilibrium nodal tide [45]. On the other hand, the oscillation may be related to bidecadal signal, which has been observed globally and which manifested itself in the Mediterranean area in the low air pressure [46–48] that occurred in the early 1960s and 1980s simultaneously with the high air [46, 49, 50] and sea-surface [47, 48, 51] temperatures; moreover, low salinity was recorded in the eastern Mediterranean in the beginning of 1980s [52]. All the factors conspired to raise the Mediterranean and Adriatic sea level in the years mentioned. Consequently, the bidecadal cycle visible in Fig. 1.7 may be attributed partly to the air pressure (and possibly wind) forcing and partly to the steric influence and related nonisostatic water-flux effect, which in turn represent regional manifestations of the global bidecadal signal. First observation of the bidecadal oscillation in the Adriatic was reported by Polli [53], whereas its connection with the global processes has been considered only recently [54]. Understanding interdecadal variability appears to be crucial to differentiating natural

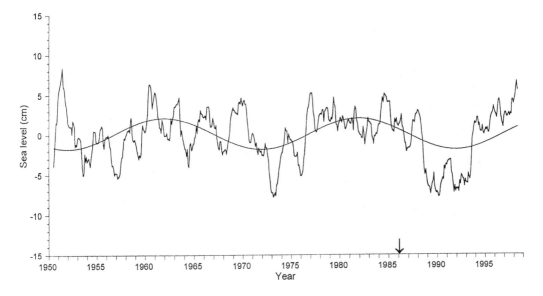

**FIGURE 1.7** Sea level heights measured at Bakar between 1950 and 1998, detrended and smoothed by a 12-month moving average (thick line). Also drawn is a 20-year cosine function fitted to the data (thin line).

climate changes from those due to anthropogenic forcing, and therefore the variability is investigated worldwide [55].

## Sea Level Trend

The last process to be considered is sea level change that extends over many decades. The basic data are annual mean sea levels collected at Bakar between 1930 and 1938 and from the year 1950 onward. Although the measurements were interrupted during the World War II, the data referred to the same datum level before and after the war. The linear trend computed from the data equals 0.7 mm/year (Fig. 1.8), i.e., is much smaller than the values typical for the world oceans and seas. The reason becomes obvious if one fits a second-degree polynomial to the annual mean values, thus achieving a marginally better approximation than by a simple linear fit (Fig. 1.8). It appears that the Bakar sea level rise slowed down over the years. Its contribution to the total sea level height was only about 1 cm at the beginning of 1986. While interpreting these findings one must bear in mind that the tide-gauge measurements are performed relatively, and therefore may be influenced by both the sea and land movements.

The literature on the sea level rise is vast and has been reviewed in a number of articles [56–59] and several books [6, 60]. Linear trends were extracted for the first time from the Adriatic tide-gauge data extending over extended intervals by Polli [61], and nonlinearities began to be considered by Mosetti [62]. Presently, there is a consensus that global sea level has risen by between 10 and 25 cm over the past century, mostly due to thermal expansion of sea water and melting of low-latitude glaciers. Comparison of this finding with the results of Fig. 1.8 indicates that, besides the global phenomenon, some regional and local processes manifest themselves at Bakar as well. According to a recent interpretation [54], deceleration observed at Bakar may be ascribed to at least two regional processes (multidecadal thermal variations recorded in the north Atlantic area and sea-surface lowering due to man-induced reduction of freshwater input to the Mediterranean) and two local effects (long-term atmospheric forcing and upward acceleration of the crustal movements in the vicinity of Bakar). The contribution of all these processes to the relative sea level change is small if compared with the $50 \pm 30$ cm global rise that is expected over the next century due to anthropogenic increase of the atmospheric concentration of greenhouse gases and related warming of the earth [59].

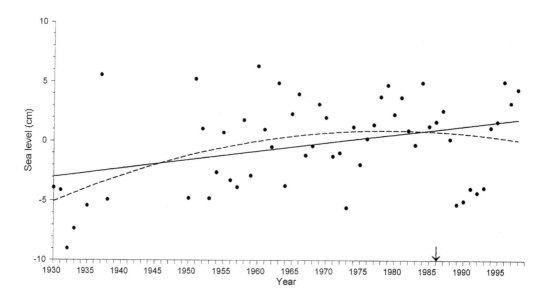

**FIGURE 1.8**  Linear (solid line) and nonlinear (dashed line) fit to the annual mean sea levels (dots) recorded at Bakar between 1930 and 1938 and from the year 1950 onward.

## Conclusion

Results of analysis reported in the previous sections may be used to synthesize the Bakar sea level height for 01 h EMT on February 1, 1986. The contributions of tides (20 cm), storm surges and seiches (57 cm), planetary-scale disturbances (7 cm), seasonal and interannual variability (12 cm), bidecadal signal (1 cm), and long-term trend (1 cm) sum up to 98 cm. This is slightly greater than the observed 96 cm, due to the rounding errors and the fact that the filters applied did not always complement each other. Yet, the exercise served the purpose—namely, to illustrate the broad range of sea level related processes in an illuminating example. It is obvious from the results of analysis that the maximum sea level height could be higher, e.g., if the storm-surge high water (Fig. 1.4) occurred at the time of spring, rather than neap tides (Fig. 1.3) or if it coincided with a higher planetary-scale maximum (Fig. 1.5). It should be borne in mind, however, that not all combinations are possible, as different processes may depend one on the other. Thus, for example, it is well known that the surface cyclonic disturbances in the atmosphere usually deepen when finding themselves to the east of an upper-level trough. This would tend to limit sea level maxima due to the combined synoptic- and planetary-scale atmospheric forcing.

The basic assumption of the present analysis is that the various contributions to the sea level height are linearly superimposed. At Bakar, there are some indications of nonlinear processes at work. In particular, high-frequency tidal lines, visible in Fig. 1.2, may be attributed to nonlinear shallow water effects. However, they appear to be sufficiently weak to warrant the procedure followed in this chapter, thus making the Adriatic ideally suited for an introductory overview of sea level variability. For some other stations, located in shallower seas and subject to stronger tidal forcing, decomposition of the sea level time series should allow not only for the distinct processes similar to those considered here, but for the effects of their nonlinear interaction as well.

The results of sea level research, briefly summarized herein, have been used to develop empirical or theoretical forecasting schemes for some of the processes of practical importance. Thus, tides are successfully forecasted since the end of the 19th century, and storm surges since the 1950s. Excellent predictions of interannual variations, particularly those related to the ENSO phenomenon, marked an important advance in geophysical fluid dynamics over the last decade. Long-term sea level rise also received considerable attention recently but, as already pointed out, the projections are characterized by a wide range of uncertainty. If the mean sea level rises, response of the sea to the planetary-scale atmospheric forcing would become of paramount interest to forecasters, as it would tend to support long-lasting flooding events. The analyses and attempts to predict interdecadal variability are likewise related to the possible sea level rise, as they could help to distinguish between natural climate changes and anthropogenic effects.

## Acknowledgment

Sea level data were taken from the data bank of the Andrija Mohorovičić Geophysical Institute, Faculty of Science, University of Zagreb. Air pressure and wind time series were provided by the State Hydrometeorological Institute of the Republic of Croatia. I am indebted to Ms. Miroslava Pasarić for help in processing the data.

## References

1. Tomasin, A. and Frassetto, R., Cyclogenesis and forecast of dramatic water elevations in Venice, in *Marine Forecasting*, Nihoul, J. C. J., Ed., Elsevier, Amsterdam, 1979, 427.
2. Karmelić, J., The Bakar tide gauge (in Croatian), *Jugolinija*, 31, 23, 1992.
3. Goldberg, J. und Kempni, K., Ueber die Schwingungen der Bucht von Bakar und das allgemeine Problem der Seiches von Buchten, *Bulletin International de l'Académie Yougoslave des Sciences et des Beaux-Arts, Classe des Sciences Mathématiques et Naturelles*, 31, 74, 1937.
4. Cartwright, D. E., Oceanic tides, *Reports on Progress in Physics*, 40, 665, 1977.
5. Hendershott, M. C., Long waves and ocean tides, in *Evolution of Physical Oceanography*, Warren, B. A. and Wunsch, C., Eds., MIT Press, Cambridge, MA, 1981, chap. 10.

6. Pugh, D. T., *Tides, Surges and Mean Sea-Level,* John Wiley & Sons, New York, 1987, 59.
7. Cartwright, D. E., *Tides—A Scientific History,* Cambridge University Press, Cambridge, 1999, 1.
8. Galilei, G., *Dialogue Concerning the Two Chief World Systems—Ptolemaic and Copernican* (trans. by S. Drake), University of California Press, Berkley, 1967, 416.
9. Kesslitz, W., Die Gezeitenerscheinungen in der Adria, I. Teil, Die Beobachtungsergebnisse der Flutstationen, *Akademie der Wissenschaften in Wien, Denkschriften Matematisch-Naturwissenschaftliche Klasse,* 96, 175, 1919.
10. Polli, S., La propagazione delle maree nell'Adriatico, *IX Convegno della Associazione Geofisica Italiana,* Associazione Geofisica Italiana, Roma, 1959, 1.
11. Airy, G. B., Tides and waves, *Encyclopaedia Metropolitana,* 5, 241, 1845.
12. Sterneck, R., Die Gezeitenerscheinungen in der Adria, II. Teil, Die theoretische Erklaerung der Beobachtungs-Tatsachen, *Akademie der Wissenschaften in Wien, Denkschriften Matematisch-Naturwissenschaftliche Klasse,* 96, 277, 1919.
13. Tsimplis, M. N., Proctor, R., and Flather, R. A., A two-dimensional tidal model for the Mediterranean Sea, *Journal of Geophysical Research,* 100, 16, 223, 1995.
14. Palmén, E. and Newton, C. W., *Atmospheric Circulation Systems,* Academic Press, New York, 1969, 140.
15. *Europaeischer Wetterbericht fuer das Jahr 1986,* Deutscher Wetterdienst, Offenbach am Main, 1986.
16. Heaps, N. S., Storm surges, *Oceanography and Marine Biology—An Annual Review,* 5, 11, 1967.
17. Heaps, N. S., Storm surges, 1967–1982, *Geophysical Journal of the Royal Astronomical Society,* 74, 331, 1983.
18. Murty, T. S., Storm surges—meteorological ocean tides, *Canadian Bulletins of Fisheries and Aquatic Sciences,* 212, 1, 1984.
19. Wilson, B. W., Seiches, *Advances in Hydroscience,* 8, 1, 1972.
20. Bučić, G., Hoehe des Meeresspiegels und des Luftdruckes, *Uebersichten der Witterung in Oesterreich und einigen Auswaertigen Stationen im Jahre 1860,* November, 47, 1861.
21. Kesslitz, W., Das Gezeitenphaenomen im Hafen von Pola, *Mitteilungen aus dem Gebiete des Seewesens,* 38, 445, 1910.
22. Kesslitz, W., Die Sturmflut am 15. und 16. November 1910 in Pola, *Mitteilungen aus dem Gebiete des Seewesens,* 39, 157, 1911.
23. Sguazzero, P., Giommoni, A., and Goldmann, A., An Empirical Model for the Prediction of the Sea Level in Venice, Technical Report 25, IBM, Venice, 1972.
24. Cerovečki, I., Orlić, M., and Hendershott, M. C., Adriatic seiche decay and energy loss to the Mediterranean, *Deep-Sea Research I,* 44, 2007, 1997.
25. Accerboni, E., Castelli, F., and e Mosetti, F., Sull'uso di modelli matematici idrodinamici per lo studio dell'acqua alta a Venezia, *Bollettino di Geofisica Teorica ed Applicata,* 13, 18, 1971.
26. Schwab, D. J. and Rao, D. B., Barotropic oscillations of the Mediterranean and Adriatic Seas, *Tellus,* 35A, 417, 1983.
27. de Vries, H., Breton, M., de Mulder, T., Krestenitis, Y., Ozer, J., Proctor, R., Ruddick, K., Salomon, J. C., and Voorrips, A., A comparison of 2D storm surge models applied to three shallow European seas, *Environmental Software,* 10, 23, 1995.
28. Pugh, D. T. and Thompson, K. R., The subtidal behaviour of the Celtic Sea, I. Sea level and bottom pressures, *Continental Shelf Research,* 5, 293, 1986.
29. Penzar, B., Orlić, M., and Penzar, I., Sea-level changes in the Adriatic as a consequence of some wave occurrences in the atmosphere, *Thalassia Jugoslavica,* 16, 51, 1980.
30. Orlić, M., On the frictionless influence of planetary atmospheric waves on the Adriatic sea level, *Journal of Physical Oceanography,* 13, 1301, 1983.
31. Lascaratos, A. and Gačić, M., Low-frequency sea level variability in the northeastern Mediterranean, *Journal of Physical Oceanography,* 20, 522, 1990.
32. Pasarić, M. and Orlić, M., Response of the Adriatic sea level to the planetary-scale atmospheric forcing, in *Sea Level Changes—Determination and Effects,* Woodworth, P. L., Pugh, D. T., DeRonde J. G., Warrick, R. G., and Hannah, J., Eds., American Geophysical Union, Washington, 1992, chap. 3.

33. Pasarić, M., Pasarić, Z., and Orlić, M., Response of the Adriatic sea level to the air pressure and wind forcing at low frequencies (0.01–0.1 cpd), *Journal of Geophysical Research,* 105, 114, 23, 2000.

34. *Europaeischer Grosswetterlagen fuer das Jahr 1986,* Deutscher Wetterdienst, Offenbach am Main, 1986.

35. Pattullo, J., Munk, W., Revelle, R., and Strong, E., The seasonal oscillation in sea level, *Journal of Marine Research,* 14, 88, 1955.

36. Knudsen, P., Global low harmonic degree models of the seasonal variability and residual ocean tides from TOPEX/POSEIDON altimeter data, *Journal of Geophysical Research,* 99, 24, 643, 1994.

37. Gill, A. E. and Niiler, P. P., The theory of the seasonal variability in the ocean, *Deep-Sea Research,* 20, 141, 1973.

38. Orlić, M., A simple model of buoyancy-driven seasonal variability in the oceans, *Bollettino di Oceanologia Teorica ed Applicata,* 11, 93, 1993.

39. Philander, S. G., *El Niño, La Niña, and the Southern Oscillation,* Academic Press, San Diego, CA, 1990, 1.

40. Allan, R., Lindesay, J., and Parker, D., *El Niño Southern Oscillation and Climatic Variability,* CSIRO Publishing, Collingwood, 1996, 3.

41. Crisciani, F., Ferraro, S., and Raicich, F., Evidence of recent climatic anomalies at Trieste (Italy), *Climatic Change,* 28, 365, 1994.

42. Orlić, M. and Pasarić, M., Adriatic sea level and global climatic changes (in Croatian, abstract in English), *Pomorski Zbornik,* 32, 481, 1994.

43. Palmer, T. N. and Sun, Z., A modelling and observational study of the relationship between sea surface temperature in the north-west Atlantic and the atmospheric general circulation, *Quarterly Journal of the Royal Meteorological Society,* 111, 947, 1985.

44. Palmer, T. N., Response of the UK Meteorological Office General Circulation Model to sea-surface temperature anomalies in the tropical Pacific Ocean, in *Coupled Ocean-Atmosphere Models,* Nihoul, J. C. J., Ed., Elsevier, Amsterdam, 1985, 83.

45. Rossiter, J. R., An analysis of annual sea level variations in European waters, *Geophysical Journal of the Royal Astronomical Society,* 12, 259, 1967.

46. Mann, M. E. and Park, J., Joint spatiotemporal modes of surface temperature and sea level pressure variability in the Northern Hemisphere during the last century, *Journal of Climate,* 9, 2137, 1996.

47. White, W. B. and Cayan, D. R., Quasi-periodicity and global symmetries in interdecadal upper ocean temperature variability, *Journal of Geophysical Research,* 103, 21, 335, 1998.

48. Venegas, S. A., Mysak, L. A., and Straub, D. N., An interdecadal climate cycle in the South Atlantic and its links to other ocean basins, *Journal of Geophysical Research,* 103, 24723, 1998.

49. Mann, M. E. and Park, J., Global-scale modes of surface temperature variability on interannual to century timescales, *Journal of Geophysical Research,* 99, 25819, 1994.

50. Ovchinnikov, I., Winter hydrological conditions in the Adriatic Sea and water exchange through the Otranto Strait: results of Soviet researches during 1959–1990, in *Oceanography of the Adriatic Sea,* Cushman-Roisin, B., Ed., Abdus Salam International Centre for Theoretical Physics, Trieste, Italy, 1998, 1.

51. White, W. B., Lean, J., Cayan, D. R., and Dettinger, M. D., Response of global upper ocean temperature to changing solar irradiance, *Journal of Geophysical Research,* 102, 3255, 1997.

52. Lascaratos, A., Interannual variations of sea level and their relation to other oceanographic parameters, *Bollettino di Oceanologia Teorica ed Applicata,* 7, 317, 1989.

53. Polli, S., Analisi periodale delle serie dei livelli marini di Trieste e Venezia, *Rivista di Geofisica Pura ed Applicata,* 10, 29, 1947.

54. Orlić, M. and Pasarić, M., Sea-level changes and crustal movements recorded along the east Adriatic coast, Il Nuoro Cimento, in press.

55. O'Brien, J. J., Oceanic Interdecadal Climate Variability, IOC Technical Series 40, UNESCO, Paris, 1992.

56. Barnett, T. P., Low-frequency changes in sea level and their possible causes, in *The Sea, Vol. 9B,* Le Méhauté, B. and Hanes, D. M., Eds., John Wiley & Sons, New York, 1990, chap. 24.

57. Woodworth, P. L., A review of recent sea-level research, *Oceanography and Marine Biology—An Annual Review*, 31, 87, 1993.

58. Warrick, R. A., Climate and sea level change—a synthesis, in *Climate and Sea Level Change—Observations, Projections and Implications*, Warrick, R. A., Barrow, E. M., and Wigley, T. M. L., Eds., Cambridge University Press, Cambridge, 1993, chap. 1.

59. Warrick, R. A., Le Provost, C., Meier, M. F., Oerlemans, J., and Woodworth, P. L., Changes in sea level, in *Climate Change 1995*, Houghton, J. T., Meira Filho, L. G., Callander, B. A., Harris, N., Kattenberg, A., and Maskell, K., Eds., Cambridge University Press, Cambridge, 1996, chap. 7.

60. Emery, K. O. and Aubrey, D. G., *Sea Levels, Land Levels, and Tide Gauges*, Springer-Verlag, New York, 1991, 1.

61. Polli, S., Livelli medi, capisaldi di livellazione e ampiezze della marea nel Porto di Trieste, *Memorie, R. Comitato Talassografico Italiano*, 253, 1, 1938.

62. Mosetti, F., Sulla tendenza secolare del livello medio marino a Trieste, *Atti dell'Istituto Veneto di Scienze, Lettere ed Arti, Classe di Scienze Matematiche e Naturali*, 119, 425, 1961.

## 1.2   Guidance and Control Systems for Marine Vehicles

*Zoran Vukić and Bruno Borović*

### Introduction

The guidance and control systems for marine vehicles will be described in as simple as possible terms in order for the reader to be introduced to this very interesting area of research, which is advancing very quickly. We are witnessing rapid development and use of control systems onboard marine vehicles for many reasons. Modern control systems ensure improved system performances, reliability, operational security, economical savings due to less maintenance costs, fuel consumption, and other operational costs. They make sailing, life of the crew, or realization of the mission much easier under different weather conditions, minimizing the possibility of human mistakes or failure of the task. Development of modern control theory and especially technology will in the future allow more sophisticated control systems to be designed and implemented onboard marine vehicles for the sake of improving their capabilities in primary tasks for which they are designed.

### Positions, Forces, and States of a Marine Vehicle

To be able to design guidance control systems for the marine vehicle, a mathematical description of all vehicle motions is needed. Marine vehicle dynamics are traditionally determined from first principles using Newton's laws of motion applied to a rigid body moving in fluid(s).

The main difficulty is to determine the hydrodynamic forces acting on the hull. These are usually obtained from tests with scale models in hydrodynamics basins. Internal (inherent) vibrations of the marine vehicle hull are neglected. Two coordinate systems are normally used: one fixed to the moving body of a vehicle (body-fixed), and another fixed in space (inertial or Earth-fixed) coordinate system (see Fig. 1.9). The moving (body-fixed) coordinate system has the origin placed in the hull center of symmetry (ITTC maneuvering standard[1]), with the x-axis placed along the main symmetry axis of the body. The rigid body can provide six degrees of freedom (6DOF): three translational and three rotational motions. The body-fixed coordinate system is useful for describing "local" ("body-related") motions of the marine vehicle, while the inertial coordinate system is useful for "global" motions of the marine vehicle in space. Table 1.1 (ITTC) represents the notation of forces, moments, and motions (movements) for all 6DOF. Lateral motions: surge, sway, and heave with corresponding velocities and angular movements: roll, pitch, and yaw with corresponding angular velocities are described from the body-fixed reference frame. A passenger onboard can feel these motions. By integrating the angular velocities, the

---

[1]International Towing Tank Conference.

**TABLE 1.1**  Notation of Motions, Forces, and Moments Related to Particular DOF (ITTC Maneuvering Standard)

| Degree of Freedom | Motion Name | Forces/Torque | Velocities Vector $\boldsymbol{\nu}$ | Position/Orientation $\boldsymbol{\eta}$ |
|---|---|---|---|---|
| 1. | *Surge* | X | u | x |
| 2. | *Sway* | Y | v | y |
| 3. | *Heave* | Z | w | z |
| 4. | *Roll* | K | p | $\varphi$ |
| 5. | *Pitch* | M | q | $\theta$ |
| 6. | *Yaw* | N | r | $\psi$ |

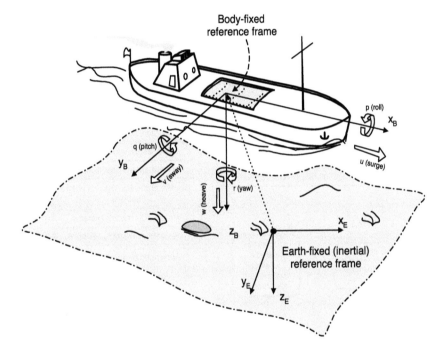

**FIGURE 1.9**  Definition of positions, forces, and states for a ship, with reference to inertial (Earth-fixed) and body-fixed coordinate systems.

vehicle orientation angles can be obtained, while by integrating lateral velocities, relative position from the origin of the body-fixed coordinate system can be found. However, the marine vehicle position in space is, in practice, more important. The Earth-fixed (inertial) reference frame (see Figure 1.9) enables description of the vehicle's motions relative to Earth. Contrary to the body-fixed coordinate system, the inertial coordinate system can be oriented in space according to control system requirements. The marine vehicle moves if some external forces and torque are acting on its hull.

These forces and torque come from applying forces and moments produced by actuators or prime movers (propellers, propulsors, rudders, etc.) or they could be the result of environmental forces such as waves, winds, sea currents, passing ship disturbance, etc. The desired forces and moments that are commanded by the autopilot (or by the helmsman) are called the control forces or moments and form the control vector, while the others, produced by the environment, are called external (environmental) disturbances or simply disturbances. Actuators such as: the rudder, hydroplane, propellers, jet propulsors, etc., usually create a control vector. The control vector forces depend on propeller (thruster) and rudder type. However, torque also depends on the position of force generators along the hull.

## Classification of a Marine Vessel's Control Systems

Control systems for marine vehicles are related to the type of the vehicle. Ships will have different control systems in use. Crude carriers, tankers, drilling ships, gas carriers, or many other types have some systems in common and some systems that differ because of specific features of each. For instance, systems related to the type of cargo will have different control systems for cargo handling and cargo monitoring, while other systems will be very similar such as navigation, guidance, or power generation systems. The same can be said for submarines, underwater vehicles, and other marine vehicles. If it is necessary to somehow present the marine vehicle as a system consisting of many subsystems, then it seems natural to focus on some "common denominator" for all of them. Figure 1.10 shows several groups of different marine vehicle subsystems, which are common to a majority of them. Five main groups can be identified here:

1. *Systems related to movement of the vehicle.* The main purpose of these systems is to help in guiding the marine vehicle from one place to another with improved accuracy, less energy consumption, or some other desired performance specifications. The following systems form this group: guidance systems in a narrow sense formed of course-keeping or track-keeping, closed-loop control systems with an autopilot as a controller; navigation systems for defining the position of the vehicle; speed control systems for keeping the desired speed of the marine vehicle; roll-stabilization systems

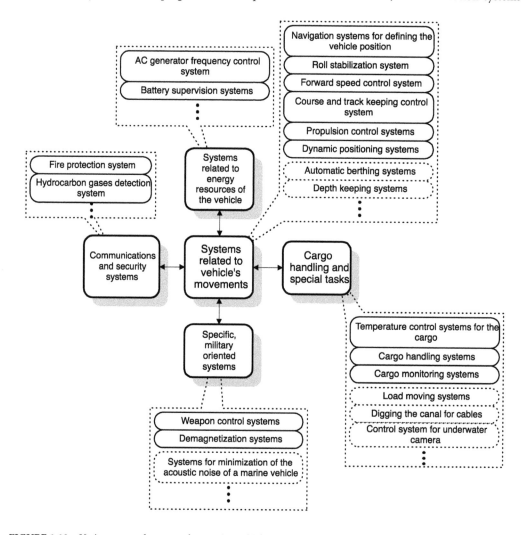

**FIGURE 1.10**   Various control systems for marine vehicles.

whose main purpose is to minimize the roll of the vehicle and make the journey more comfortable or task more manageable; propulsion (diesel engine, gas turbine, DC or AC motor, or other type) system, which serves for generating necessary forces or moments for moving the marine vehicle. Other systems include: dynamic positioning systems (for special purpose ships or platforms) whose role is to keep the vehicle at a desired position and orientation with minimal movement around it; and automatic berthing systems and automatic depth-keeping systems (for submarines or underwater vehicles). In a broader sense, the guidance system also includes the navigation system, because the track-keeping is not possible without the exact knowledge of the marine vehicle's own position.

2. *Systems related to cargo handling and special tasks.* The main purpose of these systems is to help in handling the cargo or some specific load, so that the cargo is transferred from the port or from the ship (or inside the ship in case of ammunition for instance), as fast as possible or with high accuracy in case of special loads. The cargo should not jeopardize the security of the ship so these systems should consider the proper trim and ballast while maintaining the good condition of the cargo. This group includes the following systems: cargo handling systems that load the vessel and consider trim and ballast; temperature control systems for the cargo; cargo monitoring systems that monitor and sound an alarm if something threatens the cargo; special control systems dedicated to specific tasks such as moving the load (or sample from the bottom of the sea for instance), digging the canal for underwater cables, moving an underwater camera, etc. Conditionally we can also include that part of the alarm system responsible for the cargo and trimming control system of a submarine.

3. *Systems related to energy resources of the vehicle.* Every marine vehicle must have enough energy resources to achieve its goal. The majority of marine vehicles have their own energy source, and only some specific marine vehicles, such as remotely operated underwater vehicles or tractors, have the energy source in some other place (usually on the command platform). So, for the majority of marine vehicles the energy source is autonomous and cannot rely on some other source in case of failure, as is the case with land power systems. Here, the power plant system (AC generators, DC batteries, etc.) is related to all consumers onboard and must be properly dimensioned and controlled.

4. *Systems related to communication and security.* Communication systems have the role of ensuring open communication channels within the marine vehicle as well as with the world. Here, we can also include the alarm and security systems, such as fire protection systems, because their primary role is to react to everything that may endanger the vehicle, and to pass that information to those responsible for that particular part of the system. Systems that monitor the forces acting on the bow during sailing in rough weather (on ships such as large oil tankers or ultralarge crude carriers), also belong to this group of systems. When very large ships are sailing in rough weather, the forces acting on the hull and the bow are so severe (slamming) that the officer on duty does not have the ability to assess the severity of the damage on the bow, so he/she needs such a system to help him/her in decision-making during the voyage.

5. *Specific military-oriented systems.* Weapon control systems on warships, submarines, minehunters, and other navy marine vehicles belong to this category including systems for demagnetization of the warship and systems for minimizing the acoustic noise of the warship or submarine. These systems are specific to navy marine vehicles.

Generally, every control system is designed to meet desired quality performance specifications and economic benefits for the owner. Control systems are usually implemented as separate subsystems. Ships have at least one guidance control system implemented. For instance, cargo ships will usually be supplied with course-keeping and track-keeping systems. Some types of special ships, such as supply vessels or drilling ships, will have additional control systems such as a dynamic positioning system. Implemented control systems depend on the type of marine vehicle. For instance, depth-keeping control systems will not be used for ships, but for submarines or unmanned underwater vehicles they are indispensable.

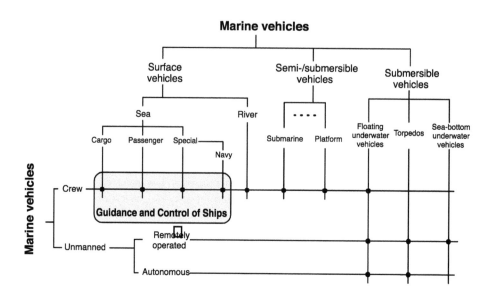

**FIGURE 1.11**   Main types of marine vehicles.

## Types of Marine Vehicles

Many different types of marine vehicles are available. They could be classified according to the type of vessel, type of power used for moving, speed, medium in which they primarily sail, type of control systems applied, etc. Three main groups of marine vehicles categorized according to the type of control systems implemented on them and their primary sailing conditions are:

- Surface vessels—ships (cargo, passenger, and special)
- Semisubmersible/submersible vehicles—submarines, platforms, etc.
- Submersible vehicles—unmanned underwater vehicles (autonomous and remotely operated)

The main types of marine vehicles are given in Fig. 1.11.

   Ships are the most widely used marine vehicles from ancient times. This paper will focus on main guidance and control systems of sea surface vehicles (shaded in Fig. 1.11). There are three main types of ships: cargo, passenger, and special. Many special type ships can be found in operation today, of which navy ships are quite numerous. The main characteristic of a ship's guidance control system is that it is almost always designed for only three degrees of freedom (DOF): surge, sway, and yaw. The reason for this is the fact that the ships are surface marine vehicles moving mainly in 3DOF along the border of two fluids (air and water). Also, in many cases the coupling effects, which usually exist for instance between roll and sway-yaw dynamics or between yaw and pitch and roll motions, are usually neglected. This is justified for some types of ships such as tankers, but not for others such as high-speed container ships. However, in some special cases such as in roll stabilization and for special types of ships (high-speed craft), the control of other degrees of freedom (mainly roll and pitch) have to be taken into consideration. Forces and moments needed for ship guidance are usually generated by rudder, propeller, and active thrusters (the thruster is usually a propeller, but it could be of a different type such as a jet engine).

   Offshore platforms can be classified as submersible vehicles and also as special types of ships because they operate on the sea surface, but during their exploitation they are fixed on the particular position. Offshore platforms are usually supplied with dynamic positioning systems. Their thrusters are similar to those used for underwater vehicles.

   Unmanned underwater vehicles include of all types of unmanned vehicles, which are used above or on the seabed. Underwater vehicles could be remotely operated (ROV) or autonomous (AUV). Remotely operated underwater vehicles can be connected to the command platform (mother-ship) with an umbilical cable that supplies the energy and information signals, or with a tether cable that supplies only the information signals while the energy source is autonomous. On the other hand, autonomous underwater vehicles operate without power supply or guidance (information) connection to a mother-ship. As a consequence of a variety of applications, underwater vehicles are supplied with a variety of different control systems. Underwater vehicles almost always have thrusters for generating moments and forces needed for proper guidance and control usually in all six degrees of freedom.

Submarines are almost always warships. However lately, submarines for tourist purposes have become popular. The development of submarines dates from the end of the 19th century. The basic requirement for a high-quality submarine guidance system is the proper trimming of the submarine. A submarine must be trimmed in roll and pitch. This requirement must be fulfilled in order to allow other control systems to work correctly. Other control systems usually implemented on submarines are: course-keeping, depth-keeping, and track-keeping systems. Forces and moments needed for submarine guidance are usually generated by rudders, propellers, hydroplanes, and many tanks allocated along the submarine hull.

### Classification of Control Systems Used in Guidance of the Marine Vehicle

Classification of control systems used in marine vehicles can also be done according to various criteria. The purpose of the control system seems most appropriate here. The consequence of a variety of specific requirements on control systems, primarily due to their role and use, results in a variety of possible control systems. The following control systems classification, based on most often used guidance systems, can be suggested:

- Course-keeping control systems
- Track-keeping (path or trajectory following) control systems
- Roll stabilization systems
- Speed control systems
- Dynamic positioning systems

All these control system can form a subset of a hypothetical control system responsible for all six degrees of freedom of a marine vehicle. Depending on specific requirements for the marine vessel's guidance, some of them will be implemented. For underwater vehicles, trim and depth control systems are also important.

## Mathematical Model of Marine Vehicle Dynamics

A mathematical model describes in mathematical terms the dynamics of the marine vehicle. The mathematical model can be determined from first principles using Newton's laws of motion with the main difficulty being the determination of hydrodynamic forces acting on the hull. They are usually obtained from tests with scale models. The mathematical model is, in general, very complex and could be divided into several submodels such as: the model of vehicle dynamics, the model of vehicle kinematics, models of closed-loop control system elements (actuators and sensors), and models of disturbances acting upon the vehicle. Each model is related to another model, and the complexity stems from the fact that in reality the model is time variant and nonlinear. The dynamic characteristics of the ship (marine vehicle) change due to vessel speed, draft, trim, heel, underkeel clearance, or other operating conditions. The mathematical model of a marine vehicle is indispensable because most of control design methods rely on the knowledge of the mathematical model. Unconventional control design such as fuzzy control design methods do not rely on the mathematical models.

Mathematical models are usually made before the ship (or other marine vehicle) is built. The results obtained from scale model tests are rescaled and normalized in order to obtain mathematical models that are relevant for the full-scale marine vehicle. This is done because naval architects (ship designers) also need insight into the ship's dynamics in order to analyze how his/her design is related to the project specifications concerning the nominal speed, resistance, maneuvering characteristics, and other relevant parameters for design of the ship (marine vehicle). When the ship (marine vehicle) is built, test voyages

(Zig-Zag, turning circle, etc.) are performed in order to check all specifications of the design and sometimes to improve the mathematical model of the ship's dynamics.

For the purpose of control system design, four basic mathematical models are needed:

- vehicle dynamics model,
- vehicle kinematics model,
- control vector (force and moments) generation model,
- model of disturbances.

Wave, wind, current, passing ship disturbance, and bottom and bank effects (i.e., environmental distur-bance models) are quite useful for control system design.

### Equations of Motion of a Marine Vehicle in Body-Fixed Frame

It is common to describe vehicle dynamics in a body-fixed reference frame. The mathematical model is given with the following differential matrix equation [1]:

$$\mathbf{M}\dot{v} + \mathbf{C}(v)v + \mathbf{D}(v)v + \mathbf{g}(\eta) = \tau_{\text{ext}} \tag{1.1}$$

where:

$\mathbf{M}$—inertial matrix – $\dim(\mathbf{M}) = 6 \times 6$,
$\mathbf{C}(v)$—Coriolis and centripetal matrix – $\dim(\mathbf{C}) = 6 \times 6$
$\mathbf{D}(v)$—hydrodynamic damping matrix – $\dim(\mathbf{D}) = 6 \times 6$
$\mathbf{g}(\eta)$—vector of gravitational forces and torque (restoring forces) – $\dim(\mathbf{g}) = 6 \times 1$
$\tau_{\text{ext}}$—vector of excitation forces (control and disturbance vector) – $\dim(\tau_{\text{ext}}) = 6 \times 1$
$v = [u\ v\ w\ p\ q\ r]^T$—vector of linear and angular velocities in body-fixed coordinate system
$\eta = [x\ y\ z\ \phi\ \theta\ \psi]^T$—vector of positions and orientations in Earth-fixed coordinate system

Differential matrix Eq. (1.1) represents six matrix equations of first order. Matrices $\mathbf{M}$, $\mathbf{C}(v)$, $\mathbf{D}(v)$, and $\mathbf{g}(\eta)$ depend on vehicle parameters and vectors $v$ and $\eta$. Vector of excitation forces consists of control and disturbance vector $\tau_{\text{ext}} = \tau_c + \tau_d$. Matrix $\mathbf{M}$ is given by $\mathbf{M} = \mathbf{M}_{\text{RB}} + \mathbf{M}_{\text{A}}$, where $\mathbf{M}_{\text{RB}}$ represents the rigid body inertia matrix and $\mathbf{M}_{\text{A}}$ represents the inertia of so-called added masses. According to separation of the matrix $\mathbf{M}$, the Coriolis and centripetal matrix $\mathbf{C}$ could also be separated as $\mathbf{C}(v) = \mathbf{C}_{\text{RB}}(\mathbf{v}) + \mathbf{C}_{\text{A}}(\mathbf{v})$. Matrix $\mathbf{D}$ consists of matrices describing all kinds of damping effects. Damping effects are forces generated by vessel motion through fluid (water) and they usually act opposite of the vehicle motion. The effects of hydrodynamic forces on vessel motion are very complicated and are the topic of experiments in hydrodynamic basins [2, 3]. The vector of restoring forces $\mathbf{g}(\eta)$ represents forces and torque depending on the ship's orientation in space, center of gravity, and center of buoyancy. It is related to ship stability. For instance, when someone inclines a boat in roll DOF and then lets go, the restoring forces will bring the boat to equilibrium (roll angle is zero) through the damped oscillatory motions.

### Equations of Motion of a Marine Vehicle in Earth-Fixed Frame

To complete the mathematical model of marine vehicle motions, the connection between the vector of linear and angular velocities in a body-fixed coordinate system $v$ and the vector of positions and orien-tations in an Earth-fixed coordinate system $\eta$ have to be described. This connection between velocities in body-fixed and Earth-fixed frames could be described by Jacobian matrix:

$$\dot{\eta} = \mathbf{J}(\eta_2)v \tag{1.2}$$

where $v = [u\ v\ w\ p\ q\ r]^T$ is the velocity vector in body-fixed frame, $\dot{\eta} = [\dot{\eta}_1\ \dot{\eta}_2]^T$ is the velocity vector in inertial frame, $\eta_1 = [x\ y\ z]^T$ is position vector, and $\eta_2 = [\theta\ \varphi\ \psi]^T$ is orientation vector. From Equations (1.1) and (1.2) follows the Earth-fixed representation of motion of the marine vehicle [1, 3]:

$$\mathbf{M}_{\eta}(\eta)\ddot{\eta} + \mathbf{C}_{\eta}(v, \eta)\dot{\eta} + \mathbf{D}_{\eta}(v, \eta)\dot{\eta} + \mathbf{g}_{\eta}(\eta) = \tau_{\eta} \tag{1.3}$$

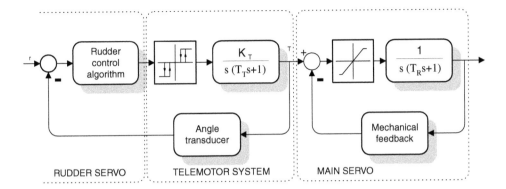

**FIGURE 1.12** Block diagram of the steering servosystem.

A large number of hydrodynamic derivatives are necessary for 6DOF nonlinear equations of motion. The number of hydrodynamic derivatives can be reduced if we use body symmetry.

## Mathematical Model of Control Vector Generation

Mathematical modeling of control vector generation can be divided into thruster/propeller forces modeling and control surfaces/rudder forces modeling. It is usually very complex and beyond the scope of this text, but superficial insight into input-output and dynamic relationships inside these devices will be given.

Two main groups of propellers are in common use on marine vehicles (ships): fixed blade propellers and controllable pitch propellers. Propeller thrust $T = f(n, V_a, \theta_p)$ and torque on shaft $Q = f(n, V_a, \theta_p)$ are generally a function of propeller rotational speed ($n$), advance speed of the vehicle ($V_a$), and propeller blade pitch ($\theta_p$). A fixed blade propeller can change its thrust by changing the shaft (propeller) rotational speed, while keeping the pitch of the blade constant. The controllable pitch propeller operates with constant shaft rotational speed (for instance, powered by an AC electrical drive) and its thrust can be controlled by change of propeller pitch. Advance speed also depends on the shape of ship's hull.

The rudder produces forces and torque when water flows around it. The produced force is a complicated function of water flow speed $U$, the rudder surface (area) $A_R$, rudder deflection $\delta$, water density $\rho$, and rudder force coefficient $C_R$. The rudder force can be described by:

$$F_R = C_R \rho A_R U^2 = \frac{a \sin\delta}{b + c \sin\delta} \rho A_R U^2 \qquad (1.4)$$

Coefficients $a$, $b$, and $c$ can be found in [4]. $F_R \cos\delta$ and $F_R \sin\delta$ give rudder forces in surge and sway directions, respectively. Hence, if $x_R$ is the distance between the rudder shaft and the center of gravity of a marine vehicle (ship), the torque around the vertical axis will be given by $F_R x_R \sin\delta$. The hydroplane (diving planes) forces and torque can be derived in a similar way.

For control system design it is necessary to know the dynamics of the steering machine. Usually, the block diagram of the steering machine as given in Fig. 1.12 is used for that purpose [1].

The simplified block diagram of the steering machine used very often in simulations of yawing motions of the marine vehicle is given in Fig. 1.13. The rudder limiter is determined by the mechanical constraints of the steering machine. Usually, the maximum rudder deflection is $\pm 35[^0]$, while the maximum rudder speed is between *2.5* and *7*[$^0$/s]. The minimum value is determined by the classification societies[2] demanding that the rudder move from *35*[$^0$] port to *35*[$^0$] starboard within *30*[s].

Mainly due to the nonlinearities of the steering machine, the limit cycle behavior of the course-keeping control system often emerges [5, 6].

---

[2]Lloyd, DetNorske Veritas, etc.

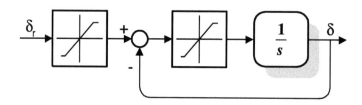

**FIGURE 1.13**    Simplified block diagram of the steering servosystem.

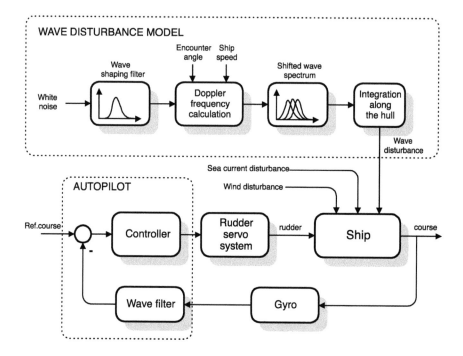

**FIGURE 1.14**    Block diagram of a ship course control system and disturbances.

## Disturbances Caused by Waves, Wind, and Sea Currents

Mathematical modeling of interactions between the vehicle and its environment is also very important if we want to have better knowledge of the behavior of the marine vehicle in its surroundings. The interactions among environmental forces, torque, and the ship are very complex. The complexity involved here can be represented by Fig. 1.14.

From this figure it is evident that the influence of the wave motions on a ship can be analytically computed using strip theory.[3] First, the description of waves should be simplified by disregarding the different directions of propagation, assuming that only unidirectional waves are acting on the ship, and assuming that we are dealing with fully developed seas. Second, the wave frequency has to be changed because the ship speed and angle of attack change the wave frequencies to encounter frequencies, which are felt by the ship. Then the underwater part of ship's hull has to be divided into several segments (strips) and the influence of the waves on each segment computed separately, followed by integration over all hull segments [7]. The wind and current influence are also direction-dependent as well as speed and hull shape dependent.

_____

[3]Strip theory is not the only one to use for this purpose.

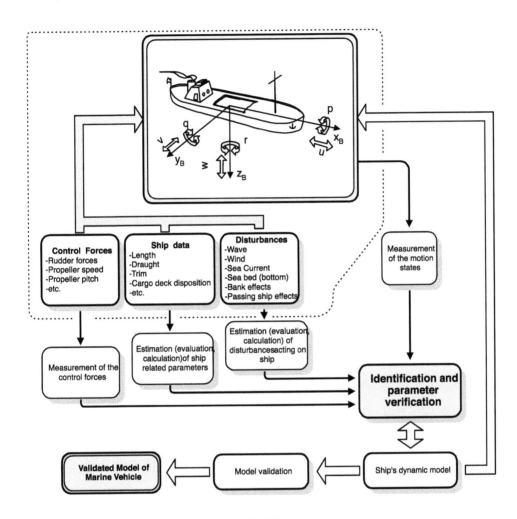

**FIGURE 1.15**    Functional diagram of the ship's model validation process.

## Model Validation

Modern control system design relies heavily upon simulation. In simulations, controller parameters are tuned and dynamics of the process analyzed in various circumstances. Simulation requires good mathematical description of all relevant components and, because of that, model validation is necessary. Figure 1.15 graphically represents the process of model validation where, through comparison of data from the mathematical model with data obtained in experiments, constant improvements of mathematical models are obtained. Finally, the validated model can be used in simulations dealing with the controller design.

In Figure 1.15 shaded arrow lines represent excitations felt by a ship (marine vehicle), white arrow lines represent data acquisition and data processing paths, while black arrow lines represent measured data lines. A validated model of a marine vehicle is used in simulations of the closed-loop control system—Fig. 1.16. It should be stressed that the control structure for guidance of a marine vehicle will follow the same philosophy as that shown in Figs. 1.15 and 1.16.

## Course- and Track-Keeping Control Systems

A main difficulty in the ship's control problem is wave filtering. Ship motions caused by waves could be described as the sum of two motions. Oscillatory motions around a desired value caused by so-called 1st order wave forces and wave drift forces called 2nd order wave disturbances. The aim is to counteract

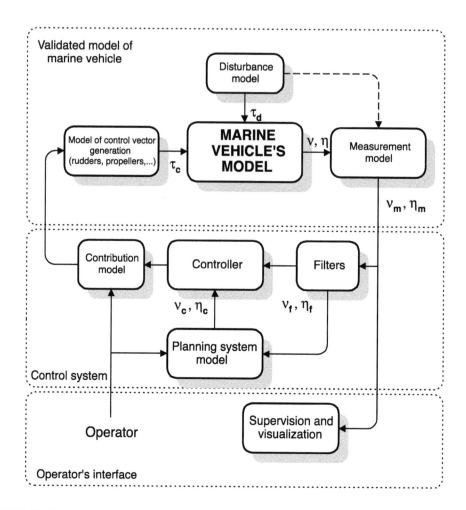

**FIGURE 1.16**  Simulation structure of marine vehicle closed-loop control system.

only the drift forces in order to avoid high-frequency rudder (actuator) motions. The motions caused by 1st order wave forces could be filtered by a few techniques—dead zone before actuator, low and bandstop (notch) filters, and by the principle of separation of motions. The last approach is the most popular, but it requires a model of disturbances.

For the purposes of control system design, the simplified models of disturbances are usually good enough. Only the consequences of environmental disturbances on a marine vehicle are of interest here. The main consequences of disturbance activity are change of vehicle position, orientation, and velocity. Disturbances are additive and multiplicative to the dynamic equations of motion, therefore, simplified models, under the assumption of validity of the superposition principle, are usually obtained by super-positioning the model of the marine vehicle dynamics and the model of disturbances (Fig. 1.17). For instance, constant speed sea current can be modeled as a superposition of sea current speed and ship's speed. The simplified modeling of waves and wind can be realized in a similar way. The effect of disturbances can be modeled as exponentially correlated disturbances with the use of first- or second-order shaping filters.

### Course-Keeping Control System

The main aim of the course-keeping control system is to maintain the reference course of the vehicle ($\psi_{ref} = const.$). The course-keeping control system prevails among control systems used in guidance of marine vehicles.

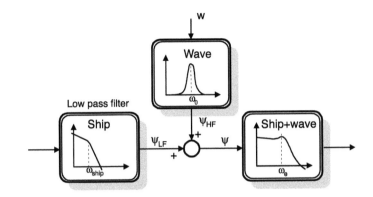

**FIGURE 1.17**  Disturbance included in simulation model assuming validity of superposition principle.

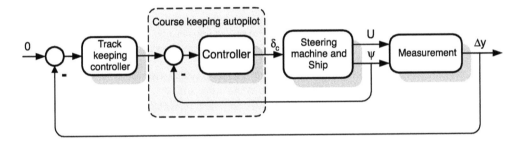

**FIGURE 1.18**  Simplified block diagram of a marine vehicle track-keeping control system.

The conventional block diagram of the course-keeping, closed-loop control system is given in Fig. 1.14. It consists of an autopilot (controller + wave filter), rudder servosystem (actuator), ship as a control object, measuring subsystem, and disturbances caused by waves, wind, sea currents, etc.

The vehicle's course is usually measured by gyrocompass. Sometimes, in low cost implementations, the magnetic compass is used instead of a gyrocompass. More complex course-keeping systems use other state variables to improve the course-keeping capabilities of the marine vehicle. Here, the yaw rate (usually measured by rate gyro) and surge speed are most often used. Surge speed is important because the hydrodynamic parameters of the vehicle depend on it. The necessity for course-keeping control systems has become important because of safety and economic reasons. The first autopilots implemented on ships were PID-based autopilots [8]. Later, new types were introduced, including gain scheduling adaptive autopilots, where autopilot parameters change with the speed of the vehicle. The first model reference and self-tuning adaptive autopilots appeared in the mid-1970s and 1980s, respectively [9, 10]. During the 1990s, the research was focused on fuzzy and neuro course-keeping control systems [11, 12, 13]. However, PID-type autopilots are still very popular and represent a majority of autopilots in use.

## Track-Keeping Control Systems

In addition to course-keeping, conventional autopilots can also follow a predefined reference path (track-keeping operation) (see Fig 1.18). In this case the additional control loop in the control system with the position feedback must be active. The system is usually designed in such a way that the ship can move forward with constant speed $U$, while the sway position $y$ is controlled. Hence, the ship can be made to track a predefined reference path, which can again be generated by some route management system. The desired route is most easily specified by way points [14]. If weather data are available, the optimal route can be generated such that the ship's wind and water resistance are minimized and, consequently, fuel is saved. Many track controllers are based on low-accuracy positioning systems such as *Decca*, *Omega*, and *Loran C*.

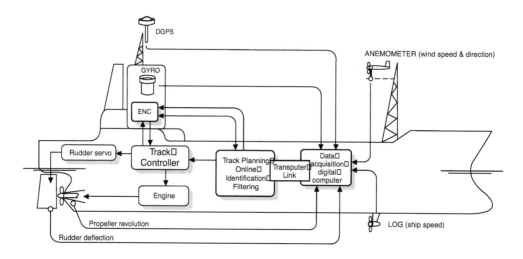

**FIGURE 1.19**  Ship's track-keeping control system with measuring instruments.

These systems are usually combined with a low-gain PI controller in cascade with the autopilot. The output from the autopilot will then represent the desired course angle.

Unfortunately, such systems result in tracking errors up to *300 m*, which are only satisfactory in open seas. Recently, more sophisticated high-precision track controllers have been designed. These systems are based on optimal control theory utilizing *Navstar GPS* (Global Positioning System). *Navstar GPS* consists of *21* satellites in six orbital planes, with three or six satellites in each plane, together with three active spares. By measuring the distance to the satellite, the global position $(x, y)$ of the vessel can be computed by application of the Kalman filter algorithm. Tracking errors can then be reduced tenfold.

It is a well-known fact that a control system should process measured data collected from various instruments. In Fig. 1.19 the track-keeping control system for a ship is presented with the main instruments from which the control system collects data (ship position—DGPS, ship speed—LOG, ship course—GYRO, wind speed and direction—ANEMOMETER, propeller rotational speed and rudder deflection). Sometimes rate gyro is also used to improve capabilities of a control system. The electronic navigation chart (ENC), collects all relevant data about ship movements during the journey. A mathematical model of every component in Fig. 1.19 must be known in order to design a good control system for track keeping.

## Roll Stabilization Systems

Roll stabilization is used on passenger ships to improve comfort, while on merchant ships it is used to prevent cargo damage and to increase the efficiency of the crew. Large roll motions can cause people to become seasick, tired, and prone to mistakes. For naval ships, certain operations such as landing a helicopter, rescue operations, or the effectiveness of the crew during combat are severely hampered by excessive roll. Therefore, roll reduction is very important, and improving roll stabilization is an interesting research area.

Roll reduction systems can be divided into two main groups: systems using active stabilizers and systems using passive stabilizers. Active stabilizers implement the roll stabilization control system with actuators such as rudders, fin stabilizers, etc. The main goal of an active roll stabilizing control system is to reduce the roll of the ship without compromising quality of course or track keeping.

On the other hand, passive roll reduction systems are implemented without control system actuators. They use fixed fins along the hull or bilge keels. They increase the hull resistance and are not as effective at low speed. They also change ship dynamics.

## Active Stabilization Systems

Three main systems are customarily used for roll reduction:

- Anti-rolling tanks
- Fin stabilizers
- Rudder roll stabilization

Several anti-rolling tank types are used on ships. The most common anti-rolling tanks in use are free surface tanks, U-tube tanks, and diversified tanks. Damping of roll motion is also possible with them at low speeds. Their disadvantages include a reduction in metacenter height due to free water surface effects and the large amount of space required. The earliest versions were installed in about 1874.

Fin stabilizers provide considerable roll damping if the speed of the ship is not too low. They increase the hull resistance (except for some systems that are retractable) and their installation is expensive, including at least two new hydraulic systems. The fins are not effective at low speed and they cause drag and underwater noise. The patent for fin stabilizers was granted to John I. Thornycroft in 1889.

Rudder roll stabilization systems (RRS) use relatively fast rudders ($\dot{\delta} = 5 \div 20[^0/s]$) to stabilize the roll. RRS is relatively inexpensive compared to fin stabilizers, has satisfactory (50 to 70%) roll reduction, and causes no drag. The disadvantage is that the RRS will not be effective if the ship speed is low. Early references discussing the possible use of RRS were reported in the early 1970s [15, 16].

## Forward Speed Control Systems

Forward speed controls systems consist of all control systems and devices related to thrust devices and machinery in the marine vehicle propulsion system. These systems include shaft speed and/or propeller pitch control systems, ship speed control systems, and thrust control systems. Properly designed and optimized speed control systems enable significant fuel savings. Design of the forward speed control system requires knowledge of propeller models, engine models, and ship dynamics models. The first part of the forward speed control system is the prime mover control system. A number of different engines are used as marine vehicle prime movers. Most modern ships have standard diesel engines as the prime mover. Diesel engines for ships are mainly of medium (150 to 500 rpm) and slow rotational speeds (25 to 125 rpm) [1, 17, 18]. For smaller and faster ships, high rotational speed engines and jet propulsors are common. Submarines have mainly combined electrical-diesel drives. The dynamics of the prime mover and its control system are tightly coupled to the speed dynamics of the ship. In the majority of applications, the dynamics of the diesel engine can be approximated with the first order transfer function [19], first order transfer function with time delay [1, 17], and the second order transfer function [20]. More about ship propeller and prime mover mathematical models can be found in [1, 17, 20].

The dynamics of the diesel engine can be written as [1, 18]:

$$I_m \dot{n} = Q_m - Q - Q_f \tag{1.5}$$

where $n$ is the shaft speed $[rad/s]$, $I_m$ is the inertia of the rotating parts of engine including the inertia and added inertia of the propeller ($kgm^2$), $Q$ is the propeller torque, $Q_m$ is the torque produced by diesel engine, and $Q_f$ is the friction torque (torque are in $Nm$). Friction torque usually can be neglected for the purpose of control system synthesis. Generally, the ship's prime mover control system can be divided into propeller pitch control (controllable pitch propeller) and shaft speed control (fixed blade propeller) systems. Depending on the type of engine, one of them would be implemented. Fixed blade propellers are usually used when it is possible to control the shaft speed (usually diesel engine drives). On the other hand, controllable pitch propellers are used when shaft speed is constant or almost constant. The input to the control system in both cases is thrust demand, which is transformed to commanded pitch or shaft speed, respectively. When designing the prime mover control system, the engine constraints have to be

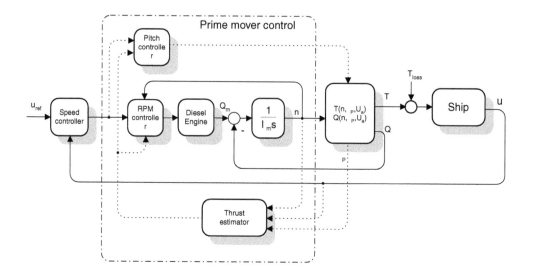

**FIGURE 1.20** Simplified block diagram of the speed control system.

considered. In an overall speed control context, the prime mover controller can be treated as an ideal device with constraints taken into consideration. Detailed controller dynamics can be neglected.

After the prime mover control system is designed, the ship's speed control system should be designed (Fig. 1.20). The speed controller compares commanded and feedback speed and gives a commanded thrust demand control signal. Commanded thrust demand is the reference signal for the prime mover control system. A block diagram of the speed-propulsion control system with (hypothetical) propeller pitch and shaft speed control is shown in Fig. 1.20

It is desirable to know the propeller and ship models in order to estimate thrust and ship's speed. Thrust feedback is implemented in order to improve performances of the speed control system and it can be used instead of shaft speed and pitch feedback. Propeller models are usually complex, so simplified models are used to describe the dynamic relationship between shaft speed, pitch, advance speed, thrust, and engine torque. Thrust feedback can be implemented if robust estimation of the thrust is available.

## Dynamic Positioning Systems

Dynamic positioning systems (DPS) are relatively new systems. They have been commercially available for drilling vessels, platforms, supply vessels, and mine hunters since the 1960s. Det Norske Veritas defines a dynamically positioned vessel as "a vessel which maintains its position (fixed location or predetermined track) exclusively by means of active thrusters." DPS systems are also used for underwater vehicles. Most dynamic positioning systems use thrusters to maintain the vessel's position and heading while heave, pitch, and roll are not controlled by the DPS. DP systems consists of: sensors (motion and environmental), actuators, and a controller (computer) with DP algorithms. As motion sensors, the following systems are used: stationary satellites that can send the position of the vessel continuously (GPS, DGPS, GLONASS), coastal radar, or radio navigation systems (Loran, Decca). Hydrophones can also be used in shallow waters. A gyroscopic compass sends the yaw information. The environmental sensors are anemometers for measuring the wind speed and direction, electromagnetic sensors to measure the sea current speed and direction, and floating radar, which continuously measures the wave speed and height. Actuators (thrusters) are usually propellers. DP ships have from three to six propellers. Some of them are azimuth propellers. Offshore platforms have eight to twelve propellers.

Two main problems are encountered in DP system design. The first is filtering of HF ship's motions related to position and heading measurements caused by first-order wave disturbances. The second is optimization of a contribution for the propulsion configuration. The aim of DPS is to counteract the

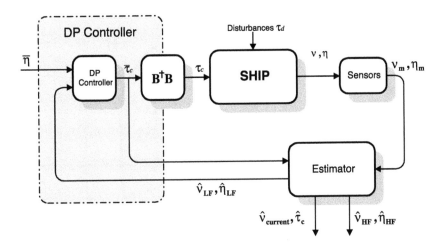

**FIGURE 1.21**   Simplified block diagram of DP control system with extraction of LF signal.

drift of the vehicle from the desired position. These motions are caused by current, wind, and second-order wave forces. Unfortunately, first-order wave forces cause oscillatory ship motions around the commanded position. Thrusters try to counteract the first-order wave disturbances, which are relatively large. This activity, called thruster modulation, will increase wear and tear on thrusters and consequently damage thrusters over time. Position can be maintained almost as well if thrusters counteract only the drift forces. The problem can be resolved by using low pass or notch filters for motions caused by first-order wave disturbances [21]. Unfortunately, this method introduces additional phase lag into the control loop.

Modern control techniques can solve these problems. Today, the most popular is application of stochastic optimal control theory [22, 23]. The LQG controller consists of the optimal state regulator and Kalman filter, which is used to estimate low frequency ship motions. Only LF state estimation is used in the controller for dynamic positioning. The simplified block diagram of DPS with LF extraction by the estimator is shown in Fig. 1.21. As a result of this approach, first-order wave disturbances are filtered out and thrusters are no longer forced in modulation. The major drawback of this approach is that kinematics equations of motions must be linearized about a set of predefined constant yaw angles to cover the heading envelope. New, promising research is going on in this area using nonlinear passive control and observer design [24, 25].

The second problem is optimization of a contribution for the propulsion configuration (see block $\mathbf{B^+B}$ in Fig. 1.21). The dynamic positioning system tries to maintain the vessel's position by using thrusters that must be situated in proper places along the hull to be able to produce forces and moments in directions of desired degrees of freedom. For instance, if we want to maintain the vessel's position $(x, y)$ and yaw angle $(\psi)$, we have to be able to produce forces in surge (along the x-axis), sway (along the y-axis), and moment (around the z-axis) to move the vessel according to desired specifications. In Fig. 1.22 the contribution model (described with matrix $\mathbf{B^+}$) of the commanded control vector $\bar{\tau}_c$ to the propeller configuration described by matrix $\mathbf{B}$ (matrix $\mathbf{B}$ consists of thruster's parameters and parameters of thrusters' position on the ship's hull) is shown.

The output of the dynamic positioning controller is the reference control vector of forces and torque $\bar{\tau}_c = [XYN]^T$ (see Fig. 1.21). The reference control vector is distributed to each thruster through matrix $\mathbf{B^+}$. Thruster dynamics usually can be neglected in comparison with dynamics of controlled motions in DOFs. The resulting vector of commanded thruster's forces $F_i (i = 1, 2, \ldots 4$ in Fig. 1.22) is, in fact, the output from the dynamic positioning controller. The matrix $\mathbf{B^+}$ is part of the DP controller and, if the number of thrusters is equal to the number of controlled DOF (three for ships), matrix $\mathbf{B^+}$ has to be an inverse of matrix $\mathbf{B}$, so ideally their multiplication should be the identity matrix. If the number of thrusters

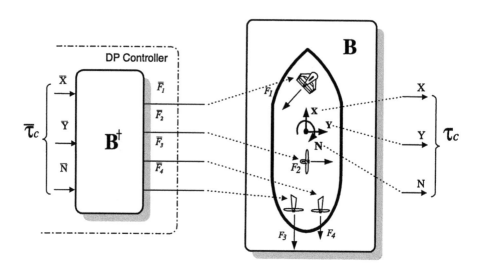

**FIGURE 1.22** Optimization problem of thruster allocation in DP.

is greater than the number of controlled DOFs (as in Figure 1.22), then there are an infinite number of solutions, and we need to find one solution that satisfies specified criteria. We are then optimizing the contribution matrix $\mathbf{B}^+$. The optimization can be done in a number of ways. The most popular is the energy consumption performance index. The optimization is done online, while the DP control system is working.

It is desirable (if possible) to include the feed-forward control loops in the DP control system. Usually it is wind feed-forward control, but current and wave feed-forward control could also be found in practical applications. Feed-forward loops give additional improvements of DP control system performance. They can save a lot of energy by making the positioning corrections before the environmental forces really hit the hull (platform), i.e., before the marine vehicle acquires momentum.

## Integrated Ship Control Systems

Modern ships have many control systems for various processes. The majority are computer controlled, and historically three phases of development can be identified: centralized computer control systems in the late 1970s, decentralized computer control systems in the 1980s, and integrated control systems in the early 1990s. Each phase brought advancements and today complex control system is unthinkable without digital computers. However, due to the specific nature of shipbuilding, each vendor usually supplied its equipment with a proprietary interface, and communication between different vendors' equipment became an obstacle for integration of equipment in a consistent and reliable way. Another challenge is that each vendor's user interface tends to be based on different principles, inviting possible confusion in critical situations.

The main goal of an integrated ship control system (ISCS) is to enable smooth integration, communication, and operation of various control systems. This task is not easy because of the complexities involved. Fig. 1.23 shows the open automation architecture where communication is needed among various layers as well as among various systems of the same level. Here, "open" means that hardware and software modules for specific tasks (such as control modules, process input/output modules, communication modules, etc.) can be easily defined by the user. Each functional module is defined by Device Description Language (DDL), and at least one microprocessor or microcontroller is included in each module. The communication standards, such as field bus and the role that MiTS forum is trying to achieve by defining the communication mechanism between control computers on the ship, will contribute to ISCS.

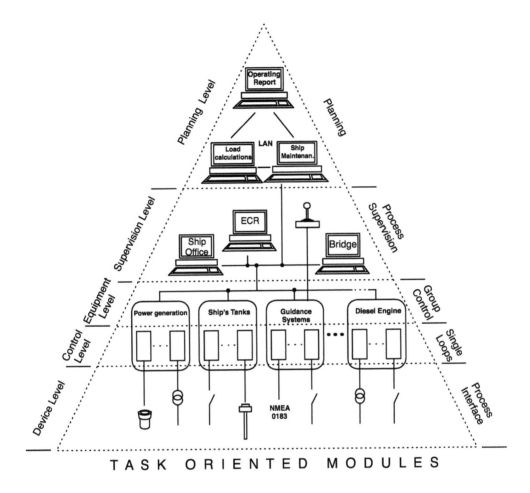

**FIGURE 1.23** Integrated ship control system.

Modular structure is based upon hierarchical functional modules. Each functional module is built from basic elements. Communication between modules of the same level is direct and horizontal, while communication with modules on higher levels is vertical and global and is controlled by the modules at the higher levels. ISCS should have the following functions:

- vendor-independent communication mechanism for control modules,
- gateways between equipment, supervision, and planning levels,
- intelligent sensors in the device level,
- new capabilities of the control level (like fault tolerance control, reconfigurable control, intelligent control, etc.),
- distributed databases where application programs from any level can retrieve information from the control computers,
- consistent man-machine interface, which should enable manufacturers to design operator interface with common look and feel.

The user should not be distressed by internal functions of the components that form the module (operating systems, software for intelligent input/output subsystems, communication interface, communication protocols, etc.) and instead should concentrate on the main task of the module and improve its functioning by implementing better algorithms. The synergistic effect, which this new approach allows, will have a great impact on the automation of ships in the future.

## Defining Terms

**Added mass:**  The total hydrodynamic force, per unit acceleration, exerted on a marine vehicle in phase with and proportional to the acceleration.

**Control of marine vehicle:**  The activity enabling the marine vehicle to fulfill the goal for which it is designed. Guidance is only one of many activities needed for this purpose. For instance, there exist many ships' subsystems that must be controlled, such as: power system, propulsion system, guidance system, cargo system, etc. All these have to act reliably if the ship is to perform the task of transferring the cargo from one port to another.

**Course-keeping control system:**  The process of automatic maintenance of the desired course of a marine vehicle by use of a closed-loop control system.

**Guidance of marine vehicle:**  An activity of giving the set of requirements for generating specific forces and torque with the aim of moving the marine vehicle in some predetermined way. Hence, when the wheelman is driving a marine vehicle it is simply called guidance of that vehicle. Moving the vessel from one point to another and/or trying to hold the direction of sailing and/or holding the predetermined depth are all referred to as guidance of a marine vehicle. Guidance is always related to the movement of the vehicle. If guidance is automatic, it is called the guidance control system.

**Kalman filter:**  Optimal linear estimator. The best estimator for stochastic Gaussian signal.

**LQG controller:**  Linear Quadratic Gaussian controller.

**Marine vehicle:**  Every vehicle whose movements and/or use are related to the sea, lake, or river. Marine vehicles are ships, submarines, autonomous and nonautonomous underwater vehicles, underwater and floating robots, underwater tractors, semisubmersible drilling vessels, etc. Offshore drilling platforms are also marine vehicles.

**MiTS:**  Maritime IT Standard. An initiative by a group of manufacturers and supporters of standardization for ships for the benefit of integrated ship control systems.

**Slamming:**  A shock-like blow caused by a severe impact of the water surface and the side or bottom of a hull.

**Track-keeping control system:**  The process of automatic maintenance of a minimal distance from the desired path (trajectory) that a marine vehicle has to follow. This is also realized through a closed-loop control system.

## References

1. Fossen, T. I., *Guidance and Control of Ocean Vehicles*, John Wiley & Sons, New York, 1994.
2. Faltinsen, O. M., *Sea Loads on Ships and Offshore Structures*, University Press, Oxford, 1990.
3. Abkowitz, M. A., Lectures on Ship Hydrodynamics—Steering and Manoeuvrability, Hydro-og Aerodynamisk Laboratorium, Report no. Hy-5, 1964.
4. Bhattacharaya, R., *Dynamics of Marine Vehicles*, John Wiley & Sons, New York, 1978.
5. Reid, R., Youhanaie, M., Blanke, M., and Nørtoft-Thømsen, J. C., Energy losses due to steering gear installations on merchant ships, in *Proc. Ships Cost and Energy Symposium*, New York, 1984.
6. Vukić, Z., Kuljača, Lj., and Milinović, D., Predictive gain scheduling autopilot for ships, in *Proc. 8th Mediterranean Electrotechnical Conference—Melecon '96*, Bari, Italy, 1996, 1133.
7. Gerritsma, J., *Behavior of a Ship in a Sea-Way*, Nederlands Scheepsstudiecentrum, TNO, Rep. no. 84s, 1966.
8. Minorsky, N., Directional stability of automatically steered bodies. *J. Amer. Soc. of Naval Engineers*, 34, 2, 280, 1992.
9. Van Amerongen, J., Adaptive Steering of Ships—A Model Reference Approach to Improved Maneuvering and Economical Course Keeping, Ph.D. thesis, Delft University of Technology, Delft, 1982.
10. Åström, K. J. and Wittenmark, B., *Adaptive Control*, Addison Wesley, New York, 1989.
11. Sutton, R. and Roberts, G. N., Approaches to fuzzy autopilot design optimization, in *Proc. of 4th IFAC Conference on Maneuvering and Control of Marine Craft—MCMC '97*, Brijuni, Croatia, Z. Vukić and Roberts, G. N., Eds., Pergamon, New York, 1997, 77.

12. Vukić, Z., Omerdić, E., and Kuljača, Lj., Fuzzy autopilot for ships experiencing shallow water effect in manoeuvering, in *Proc. of 4th IFAC Conference on Maneuvering and Control of Marine Craft—MCMC '97*, Brijuni, Croatia, Z. Vukić and Roberts, G. N., Eds. Pergamon, New York, 1997, 99.

13. Hearn, G. E., Zhang, Y., and Sen, P., Alternative designs of neural network based autopilots: A comparative study, in *Proc. of 4th IFAC Conference on Maneuvering and Control of Marine Craft—MCMC '97*, Brijuni, Croatia, Z. Vukić and Roberts, G. N., Eds. Pergamon, New York, 1997, 83.

14. Holzhüter, T. and Schultze, R., Operating experience with a high precision track controller for commercial ships, in *Proc. of 3rd IFAC Workshop on Control Applications in Marine Systems*, Trondheim, Norway, 1995, 278.

15. Källström, C. G., Control of yaw and roll by a rudder/fin stabilization system, in *Proc. 6th Ship Control System Symposium*, Ottawa, Canada, 1981.

16. Blanke, M. and Christensen, A. C., *Rudder-Roll Damping Autopilot Robustness to Sway-Yaw-Roll Couplings*, Dept. of Control Engineering, Aalborg University, Rep. no. P93-4026, 1993.

17. Blanke, M., Ship Propulsion Losses Related to Automated Steering and Prime Mover Control, Ph.D. thesis, The Technical University of Denmark, Lyngby, 1981.

18. Blanke, M., Optimal speed control for cruising, in *Proc. 3rd International Conference Maneuvering and Control of Marine Craft—MCMC '94*, Roberts, G. N. and Pourzanjani, M. M. A., Eds., Southampton, U.K., 1994, 125.

19. Horigome, M., Hara, M., Hotta, T., and Ohtsu, K., Computer control of main diesel engine speed for merchant ships, in *Proc. of ISME Kobe90*, Vol.2, Kobe, Japan, 1990.

20. Ohtsu, K. and Ishizuka, M., Statistical identification and optimal control of marine engine, in *Proc. of 2nd IFAC Workshop on Control Applications in Marine Systems*, Genova, Italy, 1992, 25.

21. Sælid, S., Jenssen, N. A., and Balchen, J. G., Design and analysis of a dynamic positioning system based on Kalman filtering and optimal control, *IEEE Trans. Automatic Control*, AC-28(3), 331, 1983.

22. Balchen, J. G., Jenssen, N. A., and Sælid, S., Dynamic positioning using Kalman filtering and optimal control theory, *IFAC/IFIP Symposium on Automation in Offshore Oil Field Operation*, Amsterdam, Holland, 1976, 183.

23. Grimble, M. J. and Johnson, M. A., *Optimal Control and Stochastic Estimation. Theory and Applications*, John Wiley & Sons, New York, 1989.

24. Fossen, T. I. and Strand, J. P., Nonlinear Ship Control, Tutorial session at IFAC Conference on Control Applications in Marine Systems—CAMS '98, Fukuoka, Japan, 1998.

25. Fossen, T. I., Nonlinear Passive Control and Observer Design for Ships, Tutorial workshop TT1 at European Control Conference—ECC '99, Karlsruhe, Germany, 1999.

# 1.3   Sea Ambient Noise—An Example from the Middle Adriatic

*Dario Matika*

In this work the noise sources and properties of the noise spectrum of the Adriatic Sea are analyzed. Sea noise sources and noise spectrum characteristics are defined, and the results of the Adriatic Sea noise research performed at a "central Adriatic" location, 200 m from the coast down to sea depth of 40 m, in 10 to 50 kHz frequency range are given. The result of this work is the representative noise spectrum of the Adriatic Sea for the given location, which is compared with results obtained for other seas by other researchers. Three characteristic spectral parts of the Adriatic Sea noise spectrum are described. This work represents a contribution to the modeling of acoustic properties at particular locations in the Adriatic Sea.

"The World of Silence" is a traditional, yet incorrect description of the sound situation existing below the sea surface. In fact, this medium abounds in sounds of various intensities which, under certain conditions, can be compared to those existing in the air.

The sounds in sea water in most cases originate as a consequence of various natural phenomena, such as activities of sea organisms, or as a result of human activities, where producing the sound is not their sole purpose. Therefore, these sounds often consist of time-variant components that vary in a random and mutually independent fashion. Sea noise is often referred to as ambient noise, emphasizing that it is a "by-product" of certain activities, and not the result of a direct source, so the noise has neither the explicit directionality nor does its intensity change considerably with a change in the listener's position.

For hydroacoustic systems, sound is the carrier of information—whether the response (echo) is to an induced source, or the sound produced by moving vessels, intentionally or unintentionally. Such sounds are called signals. Sea noise disturbs the reception of useful signals. Despite the fact that the spectrum of useful signal differs from the spectrum of sea noise, the sea noise represents an energetic mask of the signal, which is one of the causes of the limited range of hydroacoustic systems.

Analyzing many published sea noise research results, it can be concluded that great importance is given to this problem. These research results are applicable to various sea environments, because a number of derived laws and regularities are not strictly dependent on the location of the experiment. In general, it can be said that proper and effective use of modern hydroacoustic systems requires the extensive knowledge of hydroacoustic parameters of the sea. It is therefore of special interest to explore the relationships among various parameters, and their behavior with respect to time and space. The change of parameters that characterizes certain hydroacoustic environments are influenced by large numbers of variables that are often of random nature, so it is not always possible to describe the laws of change of sea hydroacoustic parameters.

The basic parameters that define the hydroacoustic environment are: vertical profile of the speed of sound, sound propagation, underwater noise, reverberation, reflection from the sea surface and bottom, and absorption of the sound in water. The most influential parameter, which defines the effectiveness of hydroacoustic system, is underwater noise. Three types of underwater noise can be defined: ship noise, self-induced noise, and sea noise. *Ship noise* is generated by ships, boats, and submarines as a consequence of thrust engines and movement through the water. This sound can be detected from a large distance and represents a useful signal for passive sonars. It is often referred to as the target noise. *Self-induced noise* is generated by a hydroacoustic transducer and can be the result of water flow around the transcuder, trembling of the connecting cables or transducer, or waves splashing. It is also generated by the carrier ship, whether it is moving or not. Self-induced noise represents an unwanted effect that reduces the efficiency of the sonar. *Sea noise* is a property of the environment where the sonar is located. If all the other sources of noise are reduced or eliminated, this noise represents the lowest limiting factor for signal detection. For this reason, great importance is given to sea noise.

First results on sea noise research were published in 1944, when Knudsen et al. [1] published an extensive study on sea noise research that took place during the Second World War. In their report the sea properties in frequency range 100 to 25 kHz, near the coast and harbors were given. Their major contribution was the diagram of spectral density of sea noise pressure with respect to sea state as the parameter.

Later investigations, of which Wenz's research in 1962 [2] is the most significant and extensive, have confirmed Knudsen's results to some extent, but also broadened the knowledge of some other sources and properties of sea noise. Most of the research was conducted in a deep sea (ocean) environment and, because of that, some differences can be observed with respect to Knudsen's results. Sea noise is today, and will be in the future, an important research topic, as the development and use of new hydroacoustic systems in various locations—from very shallow to very deep seas, from harbors and small seas to oceans, under surfaces covered with ice, etc.—require further investigation. Currently, the main efforts are focused on research of sources and properties of the sea noise such as turbulence, depth dependency, directionality, spatial-temporal correlation, influence of distant traffic, etc.

Often, when speaking of sea noise, the terms *deep sea noise* and *shallow sea noise* are encountered. It is usually understood that the shallow sea is the sea in coastal areas, and the deep sea is in regions distant from the coast. In hydroacoustics, we primarily speak of acoustically deep and acoustically shallow seas, because the conditions for sound propagation differ in these two environments. The border between the deep and shallow sea depends on the ratio of the sea depth and wave length of specific signals, therefore it depends on product $kd$, where $k$ is the wave number and $d$ is the sea depth. In practice, the sea with $kd \leq 10$ is considered a shallow sea. It is obvious that in higher frequency ranges, we have acoustically deep seas in coastal environments and, vice versa, in low frequency ranges deep sea is treated as acoustically shallow sea.

Sea noise measurement is an especially complicated problem. The measurement system is, as a rule, the result of numerous requirements and limitations. It is usually required that the measurement is conducted in a wide frequency range from below 1 up to 100 kHz and higher. In this frequency range the noise spectrum has the dynamics of more than 120 dB. A very low intensity signal corresponds to the minimal level of noise. Therefore, the proper choice of measurement hydrophone and preamplifier is crucial for design of the measurement system. The hydrophone is required to be of high reception sensitivity, and the preamplifier must have a low level of self-induced noise. The special properties of hydrophones and preamplifiers are designed for measurement of minimum sea noise level.

Special requirements, electrical and mechanical, are imposed on a signal underwater cable, which conducts the signal from the hydrophone with preamplifier to a signal processing unit. Correct sea noise measurement also requires eliminating or reducing self-induced noise to the lowest possible level, such as trembling of hydrophone cable and the hydrophone itself, broom, water flow around the hydrophone, sea organism population growth on the hydrophone, self-induced noise from the measuring vessel or other vessels near the measurement location, etc. It is also of great importance that all the parameters that describe the conditions in which the measurement is conducted are collected and updated constantly and objectively.

## Sea Noise Sources

The total sea noise is a result of simultaneous activity of numerous sources of various types. The level of noise and the shape of the spectrum are determined by sources influenced by differrent circumstances. Noise sources can be permanent or periodical, and dependent on time and position. A number of sources and mechanisms determine the shape and level of sea noise spectrum in a specific moment and place.

In the deep sea, noise sources are primarily dependent on the hydrometeorological conditions (wind, waves, rain, currents, etc.) and the level of noise can be assessed fairly accurately. In the shallow sea—especially close to the coast and harbors—other very variable sources may exist due to sea traffic and industrial activity, etc. They change from time to time and depending on location.

In general, sea noise sources can be divided into:

- hydrodynamic sources caused by the action of wind or waves, turbulent flows in the sea, the atmosphere, rain, etc.
- biological sources originating from various sea species, fish, shellfish, crabs, etc.
- technical sources, as a consequence of human activities both on sea and coast,
- seismic sources, a consequence of volcanic and tectonic activities,
- thermal sources, caused by chaotic motion of molecules in the sea,
- noise below ice surface, originating from ice crust cracking or iceberg crash.

In many seas (Adriatic, Mediterranean, etc.) the ice-covered sea surfaces are not encountered and will not be further considered in this text. The up-to-date research of this kind of environment has shown that the level and shape of the noise spectrum differs considerably from those encountered in seas without surface ice. The level and shape of the spectrum depend on the amount of ice, wind speed, and air temperature [15].

### Hydrodynamic Sources

Hydrodynamic sources contribute to the total level and shape of the sea noise spectrum in frequencies ranging from below 1 to more than 100 kHz. There exist a number of hydrodynamic processes in the sea that can be considered as sound sources. Hydrodynamic noise is present at every point in the sea under all hydrometeorological conditions. Other sources produce noise that is dependent on location and time. Therefore, biological sources are related to certain geographic regions, while technical sources are concentrated in coastal areas and are significantly time varying, similar to seismic sources.

Characteristics of hydrodynamic noise are bound to hydrometeorological parameters of specific aquatorium. The wind speed, sea state, direction and speed of currents, and precipitation amount are the main parameters determining hydrodynamical noise sources.

### Absolute Sea Level Fluctuations

Absolute sea level fluctuations cause relatively large fluctuations of hydrostatic pressure level. These fluctuations originate from ebb and flow and surface waves. Pressure variations caused by ebb and flow are very slow and it can be said that they have discrete components in the spectrum equal to one or two changes per day. Somewhat higher, but still below 1 Hz, are the fluctuations of the hydrostatic pressure caused by surface waves. In general, we can say that with an increase of wind force, the amplitude of waves increases—but the wave length also increases so the maximum of the frequency spectrum moves toward lower frequencies. The pressure level caused by surface waves decreases rapidly with increases in sea depth. In shallow seas the effect of damping is partially reduced. Because of that, it is possible that in the shallow sea the fluctuations of pressure caused by surface waves influence the total sea noise spectrum in the frequency range up to 0.3 Hz.

When two waves with the same wave length propagate in opposite directions, the so-called second-order pressure fluctuations appear. The resulting pressure has half the wave length and the amplitude, which is equal to the product of amplitudes of each wave. These types of waves are encountered in open seas and in the coastal areas as a consequence of reflection from the shore. Pressure fluctuations of second order do not decrease with increases in sea depth. Their contribution to the total sea noise spectrum is in the frequency range below 10 kHz.

### Turbulence

Turbulence is an irregular random flow of particles of the media. Turbulence in the sea changes with respect to time and location. It is the consequence of the movement of layers of water streaming over each other at different speeds. Turbulence is expected at the bottom of the sea, especially in coastal areas, channels and harbors, river deltas, at the sea surface due to surface movement, and in the area between the surface and the bottom as a result of vertical and horizontal currents.

The noise originating from a turbulent flow around the hydrophone at a measurement point is more appropriate to consider as self-induced noise than as sea noise, because turbulence causes trembling and clinking of the hydrophone. The major energy content of turbulent flows is in the frequency range below 1 Hz. However, large and slow turbulences break into smaller and smaller ones, so part of the energy transfers up to higher frequencies. The level of turbulent noise generated in this way, which can be theoretically evaluated, rapidly decreases with distance, so this type of noise does not contribute significantly to the total sea noise level. The most significant acoustic effect of turbulence is caused by fluctuations of pressure inside the turbulent area. The frequency range of turbulent noise is from 1 to 100 Hz.

### Surface Waves

Waves are one of the most important noise sources in surface layers. Wind acting on the sea surface generates waves. At the air-sea border, the time varying interaction is established, during which the sea is receiving energy from the wind field, through the turbulent border layer.

Surface waves have a dominant influence on the shape and level of the sea noise spectrum in a frequency range from 10 to 25 kHz in shallow seas, and from 500 to 25 kHz in deep seas. The sea noise measurements obtained during the Second World War, analyzed by the end of the war by Knudsen et al. [1], showed that the level of the sea noise spectrum is dependent on wind speed and sea state.

Although it is well known that the wavy sea surface is the major source, or noise generator in a given frequency range, our knowledge of mechanisms generating this type of noise is not yet satisfactory. The most acceptable explanation is wave crest dispersion. However, this is not the only mechanism of noise generation, as it is found that the level of sea noise spectrum increases with wave height and the speed of wind, even when the wave height is smaller, i.e., when crests do not appear.

## Bubbles

The surface layer contains a number of air bubbles. The development, oscillations, splitting, and joining of air bubbles generates sound. Bubbles can oscillate freely, or can be forced by the influence of wind and waves. The natural frequency of oscillations is [3]:

$$f_0 = \frac{1}{2\pi r}\sqrt{\frac{3\,\gamma p_0}{\rho}} \qquad\qquad (1.6)$$

where:

$f_0$ natural oscillation frequency [Hz],
$\gamma$ ratio of specific heat of gasses in the bubble,
$p_0$ static pressure [Pa],
$\rho$ sea density [kg/m³],
$r$ bubble radius [m].

For air bubbles in the surface layers of the sea, this relation [1] can be simplified to:

$$f_0 = \frac{3,26}{r} \qquad\qquad (1.7)$$

Natural frequency $f_0$ is inversely proportional to the bubble size. Bubble size is, naturally, limited. The frequency of the maximum of the noise spectrum originating from bubble oscillations is determined by the mean expected value of bubble size. The maximum lies in the range from 300 to 1000 Hz, which corresponds to the mean expected value of bubble radius from 1.1 to 0.33 cm. By moving toward lower frequencies, the spectrum decreases rapidly by about $-8$ to $-12$ dB/oct, which is explained by the steep decrease in the amount of bubbles with larger dimensions. By shifting toward higher frequencies, the spectrum decreases by $-6$ dB/oct. This decrease is primarily the consequence of decreased emitted sound energy as well as possible decrease of number of bubbles.

## Water Droplets

Dropping of water droplet "curtains" onto the sea surface is a significant mechanism of underwater noise generation. Droplets can emerge from wave crest scattering, or from waves breaking on the shore, or as a result of strong wind, when a layer of water "dust" is formed above the surface level. Similar effects are caused by rain.

Research of the noise originating from water droplets falling on the sea surface was published by Franz [4]. He found two principal mechanisms of creating this type of noise. When water droplets hit the sea surface, a short impulse is generated. Passing of droplets through the water often creates air vesicles that vibrate and thus create noise. The impulse of the hit is proportional to the droplet's kinetic energy and has a wide spectrum with a low maximum peak at frequency $f = v/r$ ($v$ is the speed of droplet at a moment of hit m/s, and $r$ is the radius of a droplet m). The noise originating in air vesicle vibration has the spectrum with more distinct maximum and is fairly independent of the droplet's speed and size. The typical spectrum of the noise generated by these two mechanisms is between 100 and 500 Hz and the frequency range is between 10 and 10 kHz. The total noise spectrum is found by superposition.

## Influence of Rain

In contrast to water droplets from wave crest scattering whose speed is several m/s, raindrops have significantly higher speeds, thus producing a noise whose spectrum has a mild maximum peak at higher frequencies, in a range between 1 and 10 kHz. The noise level increases with an increase of precipitation. Differences in noise level are possible due to the size of raindrops and their speed at the moment of impact with the surface. In situations with large amounts of precipitation, the level of noise generated by rain significantly exceeds the level of noise generated by wind or waves.

TABLE 1.2    Values of Constants $D$ and $E$

| Frequency Range [Hz] | D | E |
|---|---|---|
| 300–600 | 75,6 | 13,9 |
| 600–1200 | 74,0 | 14,7 |
| 1200–2400 | 74,5 | 15,8 |
| 2400–4800 | 74,1 | 16,3 |
| 4800–9600 | 71,9 | 16,1 |

If the precipitation amount is known, noise level $L_s$ can be approximately determined using the following relation [3]:

$$Ls = D + 1{,}4E\log R\,[dBr//\mu Pa/\sqrt{Hz}]\ ^4 \qquad (1.8)$$

where $D$ and $E$ are constants that depend on frequency, and $R$ is the precipitation amount in mm/h. Constants $D$ and $E$ are given in Table 1.2.

### Cavitation

It is known that the surface layer of the sea is saturated with dissipated air and that it contains a multitude of microscopically small vesicles. Under the influence of surface waves, turbulent flows are developed in surface levels, which in turn cause fluctuations of static pressure. With a decrease in local static pressure, these microscopic vesicles increase in size. Their size increases up to a certain critical value, when they disintegrate producing a short but intensive impulse. Due to the impulse nature of vesicle disintegration, large amounts of such disintegration with temporally random distribution produces noise with a spectrum in the frequency range from 100 to 20 kHz. The spectrum of cavitation noise in the surface layer has a maximum at a frequency determined by the time of decay of the largest bubbles ($f_m$). For $f < f_m$ the spectrum increases by 6 to 12 dB/oct and for $f > f_m$ it decreases by $-6$ dB/oct.

## Biological Sources

It is an established fact that many sea animal species are sources of sound. The sounds of biological origin are more intensive in warmer seas, especially in tropical and subtropical seas, at shallower depths. Individual sounds are of short duration or impulse character, like snapping, scraping, clicking, etc., and are very diverse. In some areas the sounds of large numbers of these sources merge into a monotonous noise that is easily distinguished from other noise sources. Underwater sounds are produced by various species of crustaceans (crabs, shellfish, etc.), fish, and sea mammals. More detailed identification is difficult because of the problems of recording the pictures and sounds of certain species in their natural environment. Among the crustaceans, the loudest are the small crabs from the Alpheidae family [1, 6]. They reside in tropical and subtropical regions, on a rocky or sandy sea bottom containing sediments that originate from shellfish, which provide them with a safe shelter. They can be found in a shallow sea down to maximum depths of 50 m and can grow up to 25 cm in length. They produce characteristic crackling sounds with strong snapping of their claws. A single snap of claws produces a sound of about 146 dB re 1 $\mu$Pa at a distance of 1 m [6]. When there is a large concentration of these crabs at a specific location (there can be 200 crabs per square meter in some areas), they produce an intensive and continuous sound similar to the crackling of a forest fire. The frequency spectrum of this type of noise is dominant in a range from 500 to 20 kHz. The known fact is that these crabs are not very migratory so the level of noise they produce is fairly constant throughout the year. The only variations in level occur daily, immediately after sunset and before sunrise, when the level of noise increases by 3 to 5 dB.

Among the fish, some species produce a discontinuous series of dabbing or throbbing sounds, similar to the sound produced by a woodpecker. The intensity depends on a concentration of fish in the area.

---

[4]Measuring unit accepted in hydroacoustics.

The maximum of the noise spectrum caused by fish lies between 200 and 1000 Hz. The level of noise produced by a single fish can be up to 140 dB re 1 $\mu$Pa, measured at a distance of 1 m [5, 6].

When speaking of sea mammals, sounds are produced for purposes of communication and orientation by echo, by whales and dolphins. They also produce sounds by breathing. The level of these sounds reaches up to 160 dB re 1 $\mu$Pa at a distance of 1 m [5, 6].

## Technical Sources

Technical sources include the sources of noise resulting from human activities. The possible sources of this type are ships passing the measuring location, distant sea traffic, industrial activity at the coast and on the sea bottom, explosions, and other human activities. The main characteristic of this type of noise is that it is variable with respect to space and time, and also with respect to level and shape of the spectrum.

### Near Sea Traffic
One or more ships passing near the measuring location generate the noise registered by a measurement hydrophone. The noise level depends on the type of ship, its speed and distance from the hydrophone, and varies with the speed of a passing ship relative to a measurement point. The spectrum is characterized by discrete components in a range below 200 Hz, generated by drive engines (engine noise) and propeller (rotational noise), as well as broadband cavitation noise, which decreases toward higher frequencies by $-6$ dB/oct.

### Distant Sea Traffic
The influence of distant sea traffic on total noise level depends on number, type, and spatial distribution of ships, as well as sound propagation conditions in the specific location. The spectrum of noise originating from ships decreases at higher frequencies by $-6$ dB/oct.

In frequency range below 500 Hz the spectrum takes different shapes, but in most cases there is a blunt maximum in a region of about 100 Hz. The sound propagation conditions in sea medium have an influence on this form of frequency spectrum when measured from larger distances in such a way that higher frequency components are usually damped. For the majority of ships, the sound source is placed at 3 to 10 m below the surface, so the source and its reflected image (because of reflection from the surface) create a dipole, causing a spectrum below 50 Hz to have slope of $+6$ dB/oct.

The spectrum of distant sea traffic noise has a maximum at about 50 to 100 Hz, and can be dominant in the total sea noise spectrum in the range from 20 to 500 Hz. This type of noise has horizontal directionality and is independent of sea state and wind speed. Under certain circumstances, distant sea traffic noise can be detected at distances greater than 100 km.

### Industrial Noise
Industrial activity on the coast and at the sea bottom, such as the activity of various machinery, riveting in shipyards, sea bottom drilling, and explosions also generates sea noise. For each specific location the characteristics of this type of noise are different, but it can be said that it is most intensive near the coast. The influence of industrial noise on the total sea noise level is highest in the low frequency range below 300 Hz.

In harbors and areas around harbors, the noise components originating from auxiliary engines of docked or anchored ships, and from docking maneuvers and cargo loading and unloading are dominant. In this case the noise also has the highest intensity in a range below 300 Hz.

## Seismic Sources

As a consequence of tectonic or volcanic activities or during artificially invoked explosions on coast or sea, seismic waves, which can emit a substantial amount of energy to the sea, are developed. One form of such activity are the microseismic processes that exhibit periodic behavior with frequencies of approximately 1/7 Hz and amplitude of order $10^{-6}$m. Equivalent sound pressure equals 120 dB re 1$\mu$Pa, which is similar to the level of sea noise measured at frequencies below 1 Hz. Microseismic disturbances are hence the major noise sources at low frequencies. Earthquakes and volcanic eruptions contribute to the total sea noise level in the form of transient processes that are temporally and spatially limited. In some cases seismic sources can influence the level of sea noise in a frequency range up to 100 Hz.

**Thermal Sources**

If all the other sources of sea noise were "turned off," there still remains the energy of the chaotic movement of water molecules that impact the active surface of a hydrophone. Tabain [9] has shown that the thermal noise is a limiting factor on the high frequency dynamics of a hydrophone. The equivalent level of sound pressure is the level that would be detected by an undirected hydrophone, with efficiency coefficient of 100%. For such a hydrophone, the value of spectral density of thermal noise $Ls$ can be determined by using the following relation [3]:

$$Ls = 120 + 20\log\left(\frac{2f}{c}\sqrt{\pi kT\rho c}\right) \quad [dBr//\mu Pa/\sqrt{Hz}] \tag{1.9}$$

where:

$f$  frequency [Hz],
$c$  speed of sound in sea [m/s],
$k$  Boltzmann constant ($k = 1.38\ 10^{-23}$ [J/K]),
$T$  absolute temperature [K],
$\rho$  sea density [kg/m³].

In practice, for sea temperatures from 0 to 30 °C, the following relation is used [3]:

$$Ls = -15 + 20\log f \quad [dBr//\mu Pa/\sqrt{Hz}] \tag{1.10}$$

where $f$ is the frequency in kHz.

Equation (1.10) shows that the thermal noise increases by 6 dB/oct and it determines a minimum of sea noise level at frequencies over 20 kHz.

## Properties of Sea Noise

A multitude of sources and a large number of parameters such as measurement location, time of experiment, and hydrometeorological conditions influence the forming of the sea noise field. Therefore, they determine the properties of sea noise.

In Fig. 1.24 some of the noise sources are shown with respect to location of their origin. Basic parameters that influence the character and properties of sea noise are:

- **Measurement location**—sea depth, distance from the coast, acoustic properties of the sea bottom, biological characteristics of location, distance from traffic routes, speed of sea currents, sound propagating conditions from close and distant locations around the measurement point.
- **Time of measurement**—time of day or night, season of the year.
- **Hydrometeorological parameters**—speed, direction and persistence of the wind, surface condition and sea state, existence of turbulence in sea and air and its energetic spectrum, vertical profile of the speed of sound, degree of saturation of the sea with dispersed air, amount and distribution of air bubbles, amount and speed of precipitation.
- **Conditions and methods of measurement**—depth of hydrophone location, speed of water flow around the hydrophone, the type and stability of hydrophone and cable installation, influence of sea surface or bottom on hydrophone direction when it is situated near the surface or bottom.

Sea noise research tries to establish the relationships among sea noise characteristics and the mentioned parameters, as well as provide a way for reliable prediction of noise level and its spectrum form at specific locations. Sea noise level and its spectrum form are usually expressed with respect to the following parameters: wave height, speed of wind, and sea state.

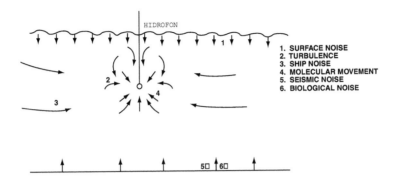

**FIGURE 1.24** Sea noise sources.

## Sea Noise Spectrum

Most of the information about the character and properties of sea noise can be obtained from the description of the noise signal in the frequency domain. This has been the most frequently used method for processing sea noise signals. By frequency analysis of a signal we obtain a spectrum that can be expressed in two ways, as a level of sound pressure in specific frequency ranges or as a spectral density of power. Frequency analysis of a signal is conducted by splitting a frequency range of interest into a series of narrow frequency bands, and then the amplitude of a signal is measured in each of these bands. The results are usually shown in a frequency diagram where the measured amplitude of a signal is shown at the central frequency of each band. Regardless of frequency range width, the sea noise can be calculated for 1 Hz wide bands, and in this case it is referred to as the spectral density of sea noise.

The frequency range for which extensive measurements were conducted in all of the world seas reaches from below 1 up to 100 kHz. The well-known research by Wenz and Knudsen are considered as the most extensive sea noise research, which resulted in the sea noise spectrum shown in Fig. 1.25. Based on the influence of the different noise sources, generally, four frequency ranges in the sea noise spectrum can be identified (they sometimes overlap due to differences in individual sources):

1. *Low frequency range* (1 to 100 Hz): the spectral level decreases with a slope of $-10$ dB/oct. In the shallow sea the spectral level is mostly influenced by wind speed, i.e., sea state. The most probable noise-generating mechanisms are turbulence-pressure fluctuations.
2. *Medium frequency range* (10 to 500 Hz): the spectrum is characterized by the maximum between 20 and 200 Hz and an abrupt decrease of the spectral level after maximum. The most probable noise source is distant and close sea traffic.
3. *High frequency range* (100 to 20 kHz): the spectrum depends on the sea state, i.e., wind speed. The maximum exists between 100 and 1 kHz, and afterwards the spectrum decreases with a slope of $-6$ dB/oct. The spectral level for the deep sea is 5 dB lower compared to the shallow sea at the same wind speed. The most probable noise-generating mechanisms are bubbles and spindrift. In the shallow sea, in tropical and subtropical seas, the influence of the biological sources is significant, and the spectrum can have a maximum in the range between 5 and 10 kHz.
4. *Very high frequency range* (above 30 kHz): the spectrum is determined by thermal noise, and has a negative slope of 6 dB/oct.

## Results of Adriatic Sea Noise Measurement

The Adriatic Sea, as can be seen from Fig 1.26, is a closed sea surrounded by Italy, Slovenia, Croatia, Yugoslavia (Montenegro), and Albania. There exists a single entrance to the Adriatic Sea and the sea currents enter along the Albanian coast, pass along the Yugoslavian (Montenegrian) and Croatian coasts, and exit along the Italian peninsula. The Croatian coast line is very well indented, with more than 1000 islands and very rich flora and fauna. In Table 1.3 the list of organisms that are potential noise sources is given. A number of winds are encountered in the Adriatic. The wind rosette is shown in Fig. 1.27.

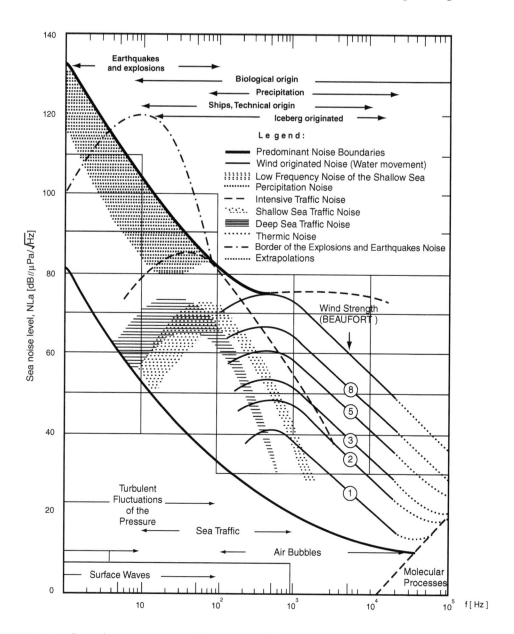

**FIGURE 1.25**   Sea noise spectrum according to Wenz [2] and Knudsen et al. [1].

   In Table 1.4. the sea state scale for the Adriatic Sea and the relation to the ocean scale is shown. The data presented shows that the Adriatic Sea has a number of different sea state modes, so for the optimization of performance of hydroacoustic detection and localization systems, it is of great interest to explore the sources and properties of noise in the Adriatic Sea. Based on results of measuring noise in the Adriatic Sea [7, 8], the noise spectra for the central Adriatic location were obtained, as shown in Fig. 1.28. As can be seen in Fig. 1.28, the measurement was conducted in both low- and high-frequency ranges. Considering the dependence on wave height, two tendencies can be observed:

   a. in the low-frequency range, the noise level increases with increases in sea state (worse hydrometeorological conditions).
   b. in the high-frequency range, the sea noise level decreases with increases in sea state, i.e., it is higher in calm than in rough seas.

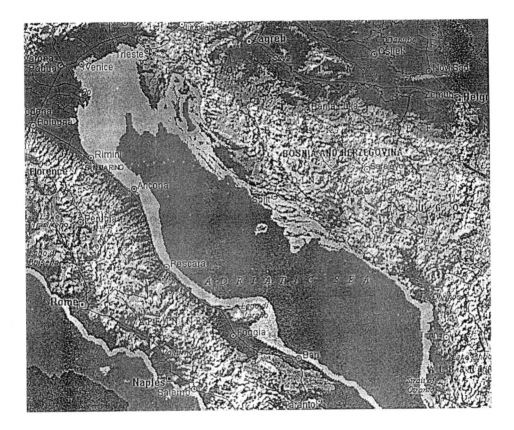

**FIGURE 1.26**   Adriatic Sea.

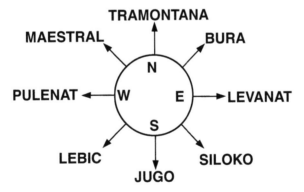

**FIGURE 1.27**   Wind rosette of the Adriatic Sea.

In Table 1.5 the parameters of normally developed waves, based on which the sea state according to Tabain [9] can be obtained, are shown.

The results of experiment in three low-frequency ranges (200 to 1 kHz, 100 to 1 kHz, and 100 to 3 kHz) are shown in Fig. 1.29. The results of experiments in the high-frequency range (22.5 to 23.5 kHz) are presented in Fig. 1.30. The typical spectrum of sea noise with respect to sea state is shown in Fig. 1.31. The same tendencies as in Fig. 1.28 can be observed, i.e., at lower frequencies the level of noise increases with an increase of the sea state, and vice-versa at higher frequencies, the noise level decreases with sea state increase.

In the following figures the noise spectra are compared to results of other authors.

**TABLE 1.3**    Organisms in Adriatic Sea—Potential Sea Noise Sources

| Mekušci<br>Croatian Name | Mollusca<br>Latin Name |
| --- | --- |
| Školjke | Bivalvia |
| Kamenica | Ostrea edulis Linnaeus |
|  | Spondylus gaederopus Linnaeus |
|  | Lima hians Gmelin |
| Jakopska kapica | Pecten jacobaeus Linnaeus |
|  | Chlamys opercularis Lamarck |
|  | Chlamus varius Lamarck |
| Dagnja | Mytilus galloprovincialis Lamarck |
|  | Modiolus barbatus Linnaeus |
| Kunjka | Arca noae Linnaeus |
|  | Pectunculus glycimeris Forbes & Huxley |
| Kamotočac | Pholas dactylus Linnaeus |
|  | Saxicava arctica Deshayes |
| Glavonošci | Cephalopoda |
| Sipa | Sepia officinalis Linnaeus |
| Liganj | Loligo vulgaris (Lamarck) |
| Lignjun | Ommatostrephes sagittatus (Lamarck) |
| RAKOVI | CRUSTACEA |
| Vitičtari | Cirripedia |
|  | Balanus amphitrite comunis Darwin |
|  | Balanus eburneus Gould |
|  | Balanus perforratus Bruguiere |
|  | Chthamalus stellatus (Poli) |
|  | Chthamalus depressus (Poli) |
|  | Chelonibia testudinaria (Linnaeus) |
| Desetonošci | Decapoda |
|  | Synalpheus gambarelloides (Nardo) |
|  | Alpheus dentipes Guerin |
|  | Alpheus machrocheles (Hailstone) |
|  | Alpheus glaber (Olivi) |
|  | Typton spongicola Costa |
| Kuka | Scyllarides latus (Latreille) |
| Zezavac | Scyllarus arctus (Linnaeus) |
| Jastog | Palinurus elephas (Fabricius) |
| Hlap | Homarus gammarus (Linnaeus) |
| Rak samac | Paguristes oculatus (Fabricius) |
|  | Pagurus cuanensis Bell |
|  | Pagurus prideauxi Leach |
| Rakovica velika, račnjak | Maja squinado Herbst |
| Račnjak bradavičasti | Maja verrucosa H. Milne-Edwards |
| Runjavac | Pilumnus spinifer H. Milne-Edwards |
|  | Stomatopoda |
| Vabić | Squilla mantis Fabricius |
| BODLJIKAŠI | ECHINODERMATA |
| Morske zvijezde | Asteroidea |
| Narančasta zvijezda | Astropected arantiacus (Linnaeus) |
| Ježinci | Echinoidea |
| Ježinac crni | Arbacia lixula (Linnaeus) |
| Ježinac kamenjar | Paracentrotus lividus (Lamarck) |
| Ježinac pjegavi | Sphaerechinus granularis (Lamarck) |
|  | Echinocardium cordatum (Pennant) |
| RIBE | PISCES |

**TABLE 1.3**    Organisms in Adriatic Sea—Potential Sea Noise Sources (continued)

| Mekušci<br>Croatian Name | Mollusca<br>Latin Name |
| --- | --- |
| Hrskavičnjače | Chondrichthyes |
| Žutulja | Trygon pastinaca Cuvier |
| Koštunjače | Osteichthyes |
| Srdela | Sardina pilchardus Walbaum |
| Srdela golema | Sardinella aurita Cuvier et Valenciennes |
| Papalina | Sprattus sprattus (Linnaeus) |
| Kostorog | Engraulis encrasicholus Linnaeus |
| Zeleniš batelj | Atherina boyeri Risso |
| Zeleniš šiljan | Atherina hepsetus Linnaeus |
| Skočac zlatac | Mugil auratus Risso |
| Skočac glavaš | Mugil cephalus Cuvier |
| Skočac putnik, cipal | Mugil chelo Cuvier et Valenciennes |
| Glavoč blatar | Gobius jozo Linnaeus |
| Pauk bijelac | Trachinus draco Linnaeus |
| Bežmek | Uranoscopus scaber Linnaeus |
| Šnjur | Trachurus linnaei Malmgren |
| Špar kolorep | Sargus annularis Geoffroy |
| Siljac | Charax puntazzo Cuvier et Vallenciennes |
| Oštrulja | Maena chrysellis Cuvier et Vallenciennes |
| Gira oblica | Maena amaris (Linnaeus) |
| Modrak | Maena maena (Linnaeus) |
| Crnelj | Chromis chromis Linnaeus |
| Drozd | Labrus turdus (Linnaerus) |
| Martinka | Crenilabrus ocellatus Cuvier et Valenciennes |
| Inac | Crenilabrus cinereus V. Crs. |
| Cučin | Trigla aspera Cuvier et Valenciennes |
| Lastavica prasica | Trigla obscura L.Bl. |
| Vrana | Corvina nigra Cuvier |
| Grb šarac | Umbrina cirrhosa Linnaeus |
| Brancin | Morone labrrax Linnaeus |
| Pirka | Serranellus scriba (Linnaeus) |
| Kirnja golema | Serranus gigas (Brünn.) |
| Kirnja glavulja | Polyprion americanum (Schm.) |
| Orada | Chrysophris surata Linnaeus |
| Trlja batoglavka | Mullus barbatus Linnaeus |
| Trlja kamenjarka | Mullus surmuletus Linnaeus |
| Skuša | Scomber scombrus Linnaeus |
| Plavica | Scomber japonicus Houtt. |
| GMIZAVCI | REPTILIA |
| Kornjače | Chelonia |
| Želva glavata | Caretta caretta Linnaeus |
| PTICE | AVES |
| Gnjurci | |
| Gnjurci ćubasti | Podiceps cristatus (Linnaeus) |
| Kormoran | Phalacrocorax carbo (Linnaeus) |
| SISAVCI | MAMMALIA |
| Kitovi | Cetacea |
| Pliskavica prava, deflin | Delphinus delphis Linnaeus |
| Pliskavica dobra | Tursiops tursio Faber |

**TABLE 1.4**   Sea State Scale for the Adriatic Sea [16]

| Weather Condition (Bf) | Mean Wind Speed (kn) | Characteristic Wave Heights (m) | | | Significant Periods Range(s) | Peak Spectrum Period(s) | Mean T | λ |
|---|---|---|---|---|---|---|---|---|
| 0 | — | — | — | — | — | — | — | — |
| 1 | 2 | 0,03 | 0,05 | 0,06 | 0,3–1,2 | 0,9 | 0,7 | 1 |
| 2 | 5 | 0,1 | 0,2 | 0,26 | 0,8–2,5 | 0,8 | 1,4 | 3 |
| 3 | 9 | 0,3 | 0,5 | 0,6 | 1,4–3,8 | 2,9 | 2,2 | 6 |
| 4 | 13 | 0,5 | 0,8 | 1,0 | 1,8–5,0 | 3,9 | 2,9 | 8 |
| 5 | 19 | 0,8 | 1,3 | 1,7 | 2,4–6,6 | 5,2 | 3,8 | 12 |
| 6 | 24 | 1,1 | 1,9 | 2,4 | 2,8–7,8 | 6,2 | 4,4 | 16 |
| 7 | 30 | 1,6 | 2,6 | 3,3 | 3,2–9,2 | 7,3 | 5,2 | 21 |
| 8 | 37 | 2,1 | 3,5 | 4,5 | 3,5–10,7 | 8,6 | 5,9 | 28 |
| 9 | 44 | 2,8 | 4,6 | 5,9 | 3,9–12,3 | 9,8 | 6,7 | 36 |
| 10 | 52 | 3,5 | 5,9 | 7,6 | 4,3–13,8 | 10,9 | 7,4 | 45 |
| 11 | 60 | 4,4 | 7,3 | 9,3 | 4,7–15,4 | 11,7 | 8,1 | 55 |
| 12 | 68 | 5,3 | 8,8 | 11,2 | 5,0–17,0 | 12,4 | 8,8 | 65 |

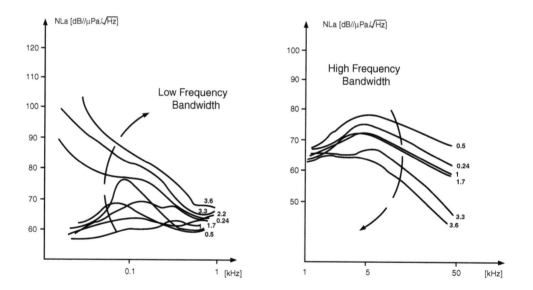

**FIGURE 1.28**   Adriatic sea noise spectra in relation to wave height $H_{1/3}$ [7, 8].

## Comparison with Results of Other Authors

Sea noise properties, as well as form and level of measured spectra at specific locations, need to be compared with results of other researches who explored the sea noise in similar conditions. The same sources (turbulence, waves, biological sources) are active in other locations, so similar results should be expected. A comparison with research in other locations can be useful for prediction of sea noise levels in a specific location knowing the conditions that are present there.

In Fig. 1.32a comparative diagram of the Adriatic Sea noise spectra and the spectra given by Wenz is shown, in Fig. 1.33a diagram comparing the spectra measured by Wenz and Knudsen, and in Fig. 1.34 comparative diagrams involving spectra measured by Painter [11], Knudsen [14], Widener [12] and Piggott [13].

**TABLE 1.5** Meteorological Conditions According to Tabain [9]

| Degree of Adriatic and/or Bf Degree 1 | Average Wind Speed 2 | | Sea Surface Appearance 3 | Relationship with SMO (WMO) Scale for Oceans | | |
|---|---|---|---|---|---|---|
| | | | | Degree of Sea State 4 | $H_{1/3}$ (m) 5 | Sea State 6 |
| 0 | m/s<br>knots<br>km/h | 0–0,2<br>0–2,0<br>0–2,0 | Surface is calm and smooth as a mirror. Smoke raises vertically. | 0 | — | Glossy |
| 1 | m/s<br>knots<br>km/h | 0,3–1,5<br>1,0–3,0<br>1,0–5,0 | Ripples like fish scales without crests. Wind direction can be detected by smoke movement, but not by wind gauge. | 1 | 11,2% 0,0–0,1 | Rippled |
| 2 | m/s<br>knots<br>km/h | 1,6–3,3<br>4,0–6,0<br>6,0–11,0 | Very small waves, short, but more intensive. Crests have a glossy appearance but do not break. Wind can be sensed on the face, and detected by wind gauge. | 2 | 0,1–0,5 | Wavelets |
| 3<br>4 | m/s<br>knots<br>km/h<br>m/s<br>knots<br>km/h | 3,4–5,4<br>7,0–10,0<br>12,0–19,0<br>5,5–7,9<br>11,0–16,0<br>20,0–28,0 | Very small waves, crest begins to scatter, crest spume has a glassy appearance, sometimes white (white horses). Light flags start to wave. Small waves becoming longer, more frequent waves with white spume (many white horses) on crests. | 3 | 31,7% 0,5–1,25 | Slight |
| 5<br>6 | m/s<br>knots<br>km/h<br>m/s<br>knots<br>km/h | 8,0–10,7<br>17,0–21.0<br>29,0–38,0<br>10,8–13,8<br>22,0–27,0<br>39,0–49,0 | Moderate waves of elongated shape, many waves with white spume on the crests. Sea spray can appear. Larger waves begin to develop, white spume on all crests, sea spray can appear. | 4 | 40,2% 1,25–2,5 | Moderate |
| 7<br>8 | m/s<br>knots<br>km/h<br>m/s<br>knots<br>km/h | 13,9–17,1<br>28,0–33,0<br>50,0–61,0<br>17,2–20,7<br>34,0–40,0<br>62,0–74,0 | Accumulation of waves, white foam from breaking waves. The foam is carried by wind in the form of elongated streaks. Moderately high waves with longer length, crest edges break into sea smoke, spume is elongating into distinctive streaks along the wind direction. | 5 | 12,8% 2,5–4,0 | Rough |
| 9<br>10 | m/s<br>knots<br>km/h<br>m/s<br>knots<br>km/h | 20,8–24,4<br>41,0–47,0<br>75,0–88,0<br>24,5–28,4<br>48,0–55,5<br>89,0–102.0 | High waves with dense streaks along the wind direction, wave crests are rolling, turning over and breaking, sea spray can reduce visibility. Very high waves with long crests that overcast waves like a plume, foam in large patches elongates in broad white streaks along the wind direction, whole surface has a whitish appearance, breaking of waves becomes very strong. Reduced visibility. | 6 | 3,0% 4,0–6,1 | Very rough |

*(Continued)*

**TABLE 1.5**   Meteorological Conditions According to Tabain [9] (continued)

| Degree of Adriatic and/or Bf Degree 1 | Average Wind Speed 2 | | Sea Surface Appearance 3 | Relationship with SMO (WMO) Scale for Oceans | | | |
|---|---|---|---|---|---|---|---|
| | | | | Degree of Sea State 4 | $H_{1/3}$ (m) 5 | Sea State 6 | |
| 11 | m/s<br>knots<br>km/h | 28,5–32,6<br>56,0–63,0<br>103–117 | Extremely large waves (smaller and medium ships can be temporarily lost from sight), sea is completely covered with long white patches in direction of wind. Crest edges are blown into froth, reduced visibility. | 7 | 0,9% | 6,1–9,1 | High |
| 12 | m/s<br>knots<br>km/h | 32,7–36,9<br>64,0–71,0<br>118–133 | Air filled with foam and spray. The sea is completely white visibility is greatly reduced. | | | | |

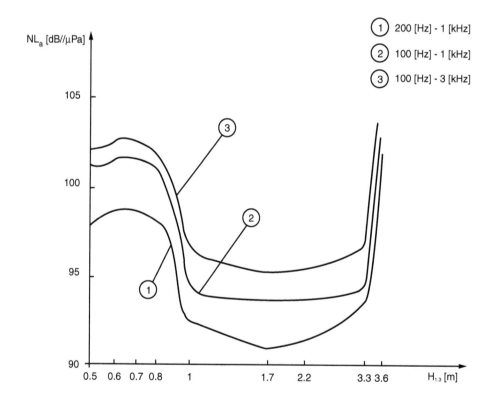

**FIGURE 1.29**   Sea noise level in low-frequency range in relation to wave height $H_{1/3}$ [10].

## Conclusion

Based on the presented material, it can be concluded that there exist three characteristic components of the Adriatic Sea noise spectrum.

- *Low frequency* (10 to 150 Hz): the main noise source are turbulence fluctuations caused by surface waves. The form of the spectrum is changeable, depending on the sea depth at the place of the measurement. The spectral level increases with the increase of the sea state.
- *Medium frequency* (150 to 1 kHz): the noise is exclusively determined by the sources in the surface layer of the sea. The main sources are surface waves and wind. The mechanism of the noise

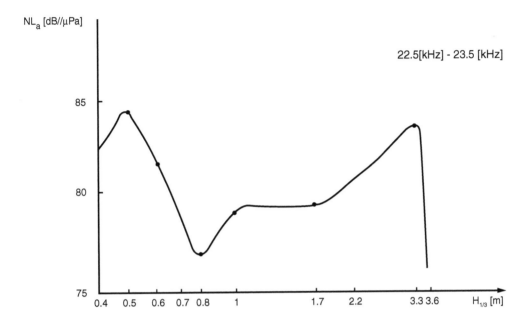

**FIGURE 1.30**   Sea noise level in high-frequency range in relation to wave height $H_{1/3}$ [10].

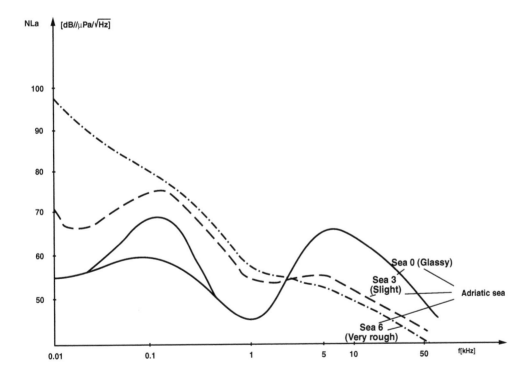

**FIGURE 1.31**   Typical sea noise spectrum [7, 8].

generation by the surface waves has not been fully established yet, however, the form of the spectrum and its level point to those such as oscillations of air bubbles in the surface layer under the influence of the waves, i.e., turbulent pressure fluctuations, breaking of crested waves, and cavitation. A strong dependence of the spectral level on the sea state is evident.

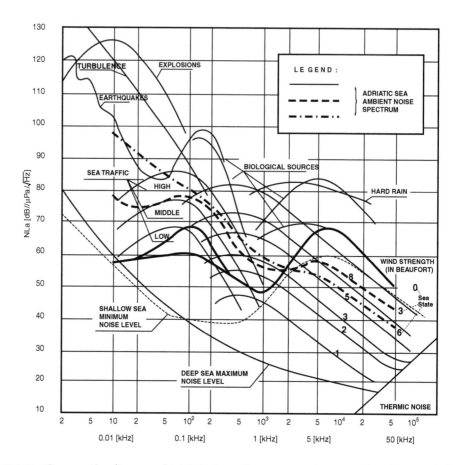

**FIGURE 1.32**  Comparative diagram of Adriatic Sea noise spectra and spectra given by Wenz [2].

- *High frequency* (above 1 kHz): characterized by a bell-shaped maximum at about 5 kHz. the characteristic sound, the form and level of the spectra, its dependence on the distance from the shore, and the change of the noise level during the day undoubtedly point to the biological source. The spectral level decreases with the increase of the sea state.

Hydroacoustic signals that carry useful information about the presence of vessels in the sea are, as a rule, interlaced with disturbances (sea noise), which are masking the useful signal and therefore falsely "hint" at the existence of vessels in a specific location. One cannot detect a vessel in a whole useful signal and disturbance frequency range because signal processing methods that can differentiate useful signals from natural and artificial disturbances (called sea noise) do not exist.

When choosing the optimal detection frequency range, one must take into account the characteristics of the spectrum, the most important being:

- Is the sea noise level minimal in this frequency range?
- Is the level of spectrum of relatively constant slope, i.e., change?
- Does the sea noise level significantly depend on a sea depth or distance from the coast?
- Can the level of sea noise be determined from atmospheric (rain and wind) and biological conditions in the environment?

The significance of this research is that it contributes to the optimization of the frequency range for detection of vessels in the Adriatic Sea.

The results obtained in this research can be applied in research of other seas having characteristics similar to the Adriatic Sea.

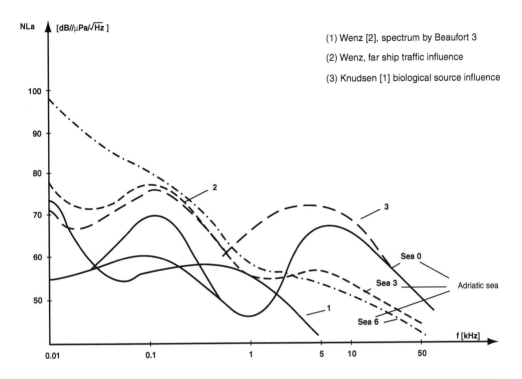

**FIGURE 1.33** Comparative diagram sea noise spectra and spectra measured by Wenz [2] and Knudsen et al. [1].

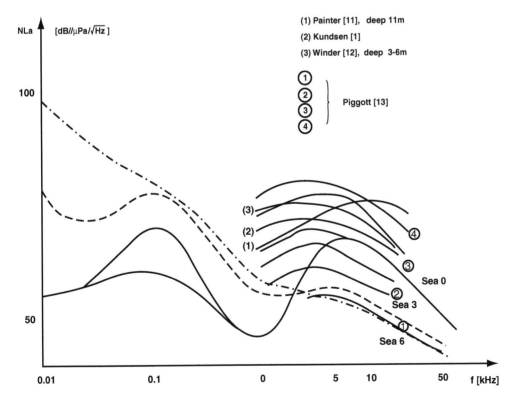

**FIGURE 1.34** Comparative diagram sea noise spectra and spectra measured by Painter [11], Knudsen et al. [14], Widener [12], and Piggott [13].

## References

1. Knudsen, V. O., Alford, R. S., and Emling, J. W., *Survey of Underwater Sound-Ambient Noise* (*Report No. 3*), Office of Scientific Research and Development, National Defence Research Committee, Division 6, Section 6.1, 1944.
2. Wenz, G. M., Acoustic ambient noise in the ocean: spectra and sources, *J. Acoust. Soc. Amer.*, 34, 1936, 1962.
3. Saje, Z., A contribution to the hydroacoustic sea noise research, Masters thesis, University of Zagreb, FER, 1982, (in Croatian).
4. Franz, G. J., Splashes and sources of sound in liquids, *J. Acoust. Soc. Amer.*, 31, 1080, 1959.
5. Brehovskih, L. M., *Acoustics of Ocean,* Izdatelstvo 'Nauka', Moscow, 1974, (in Russian).
6. Tavogla, W. N., *Marine Bioacoustics,* Pergamon Press, New York, 1964.
7. Matika, D., A Contribution to the research of the adriatic sea noise spectrum, *Brodogradnja* 44, 121–125, 1996.
8. Matika, D. and Ožbolt, H., Adriatic sea noise spectrum measurement and polynomial approximation, *Oceans '98, Conf. Proc.,* Nice, France, 1998, 1364–1367.
9. Tabain, T., *Wave Characteristics,* Brodarski Institut-Zagreb, 1977.
10. Matika, D., A contribution to identification of vessels in acoustic sense, Doctoral Thesis, University of Maribor, 1996, (in Croatian).
11. Painter, D. W., Ambient noise in a coastal lagoon, *J. Acoust. Soc. Amer.,* 35,1458, 1962.
12. Widener, M. W., Ambient-noise levels in selected shallow water off Miami, Florida, *J. Acoust. Soc. Amer.,* 42, 904, 1967.
13. Piggott, C. L., Ambient sea noise at low frequencies in shallow water, *J. Acoust. Soc. Amer.,* 36, 2152, 1965.
14. Knudsen, V. O., Alford, R. S., and Emling J. W., *Survey of Underwater Sound-Ambient Noise* (*Report No. 3*), Office of Scientific Research and Development, National Defence Research Commitee, Division 6, Section 6.2, 1944.
15. Milne, A. R., Shallow water under-ice acoustics in Burrow Strait, *J. Acoust. Soc. Amer.,* 32,1007, 1960.
16. *Naval encyclopaedia,* Adriatic Sea, JLZ, Zagreb, 1976, (in Croatian).
17. Urick, R. J., *Principles of Underwater Sound for Engineers,* New York, McGraw-Hill, 1967.
18. McCarthy, E., Acoustic characterization of submerged aquatic vegetation, SACLANTCEN Proc. Series CP-45, *High frequency acoustics in shallow water,* Lerici, Italy, 363–369, 1997.
19. Brekhovskikh, L. and Lysanov, Y., *Fundamentals of Ocean Acoustics,* Springer-Verlag, New York, 1982.
20. Jensen, F. B., Kuperman, W. A., Porter, M. B., and Schmidt, H., *Computational Ocean Acoustics,* AIP Press, Woodbury, New York, 1994.
21. Munk, W., Worcester, P., and Wunsch, C., *Ocean Acoustic Tomography,* Cambridge Univ. Press, Cambridge, U.K., 1995.

# 1.4   Basic Shipboard Instrumentation and Fixed Automatic Stations for Monitoring in the Baltic Sea

*Siegfried Krueger*

The Institut fuer Ostseeforschung (IOW) is the German Centre for Baltic Sea Research. IOW carries out basic research in the semi-enclosed brackish water ecosystem of the Baltic Sea, where natural processes and the effects of human activities can be studied particularly well. The most important anthropogenic influences are fisheries, input of nutrients and harmful substances, as well as the building of bridges and dams in the transition area between the Baltic and the North Sea. In order to understand the functioning of the system and to identify anthropogenic effects, basic research is required comprising field campaigns, modeling, and laboratory experiments. By analyzing the sedimentary records, the natural variability and the effects of human activities in the recent history of the Baltic Sea can be detected. The interaction of physical, chemical, biological, and geological processes in the system of the Baltic requires an

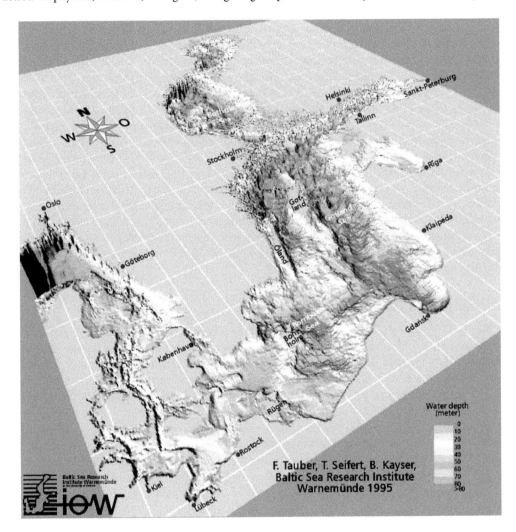

**FIGURE 1.35** The Baltic Sea.

**The Baltic Sea**

| | |
|---|---|
| Area: | 412,560 km² |
| In comparison— | |
| Area of Germany: | 356,957 km² |
| Volume: | 21,631 km³ |
| South-north-<br>expansion: (54°–66° N) | ca. 1300 km |
| West-east-<br>expansion: (10°–30° E) | ca. 1000 km |
| Max. width: | ca. 300 km |
| Mean depth: | 52 m |
| Max. depth: | 460 m<br>(Landsort deep) |

interdisciplinary approach. Close interdisciplinary cooperation is an important feature of the IOW research program. One of the main goals is the development of ecosystem models of the Baltic Sea, which can only be achieved through a sound understanding of the fundamental processes that control the system. Closely connected with this basic research is an interdisciplinary monitoring program aimed at routine observations of key parameters, which provides the basic knowledge for a sustained use of the marine environment by the riparian countries. Simultaneously with the basic Baltic Ecosystem Research of the IOW, this program is carried out by contract and on behalf of the Federal Maritime and Hydrographical Agency in Hamburg (BSH). Using the scientific environment of a research institute, IOW is responsible for the German contribution to the international coordinated Baltic Monitoring Programme (BMP) of HELCOM. HELCOM is the governing body of the Helsinki Convention, the Baltic Marine Environment Protection Commission (since 1974). In 1992, a new convention was signed by all the countries bordering the Baltic Sea and by the European Economic Community. The international cooperation for the protection of the marine environment of the Baltic Sea area has been recognized since the 1970s as a valuable example of cooperation between countries to protect the environment in the region, thus confirming the belief that the deterioration of the Baltic Sea can be arrested and the state of the marine environment improved. The 1992 convention was ratified by the European Community, Germany, Latvia, and Sweden in 1994, by Estonia and Finland in 1995, by Denmark in 1996, and by Lithuania in 1997 (new actions see [1]). In the framework of HELCOM, the IOW carries out an intensive, internationally coordinated ship-borne monitoring program. In addition to that, the IOW is responsible for the establishment and operation of four automatic stations of the Marine Monitoring Network (MARNET) of the BSH in front of the German Baltic Sea coast.

## Basic Shipboard Measurement Systems

For the monitoring work, basic equipment is needed on board research vessels. IOW operates two research vessels. For basic research as well as for monitoring purposes, two basic measuring systems are permanently installed on board these vessels: a fixed data collection, storage, and distribution system for weather, navigation, and surface water parameters and an integrated CTD-probe system with rosette sampler. Figure 1.36 shows a block diagram of the IOW shipboard data collection system with an integrated weather station of the German Weather Service (DWD). Figure 1.37 shows the special IOW-SBE911 plus-System (Seabird Electronics, U.S.A.), a modification with a compact SBE-Carousel (1 × Ø1m) up to 16 FreeFlow water samplers of 5 l (IOW-HYDROBIOS, Germany). The SBE911 has a ducted and pumped C/T-measurement sensor assembly avoiding salinity spiking, which is nearly perfect in waters with high gradients like in the Baltic Sea [2, 3]. The FreeFlow samplers have very good water exchange while the probe is lowered, resulting in a good match of the water samples and the *in situ* measured data. At CTD-profiling, all parameters are measured 24 times per second. Later all raw data are transferred to the IOW database. For operational use, 1 or 0.5 m bins (mean values) are calculated regularly. On sea daily comparison measurements with reversing thermometers, water sampling for lab analysis of salinity (AUTOSAL) and oxygen (WINKLER), as well as on-deck pressure sensor registrations in comparison with the actual air pressure ensure the security and long-term reliability of the basic oceanographic measurements. Exchangeable, pre-calibrated spare temperature and conductivity sensor modules are always on board in case of a fault. Regular lab checks and calibrations for the CTD sensors (via the system provider or in a high precision calibration lab) are necessary every year and on special request (before and after longer cruises) [4]. IOW has its own calibration lab in the instrumentation department, which works in close cooperation with the national bureau of standards (PTB) and guarantees the WOCE standard in the lab for the three basic parameters of temperature, salinity (conductivity), and pressure. A new CTD technology was introduced in IOW in 97/98 with the so-called towed fish technology. A towed vehicle (GMI, Denmark) in the form of an airplane wing, with a highly sophisticated inner control and a built-in SBE-CTD, is towed 5 to 6 knots behind the ship, undulating and measuring up and down.

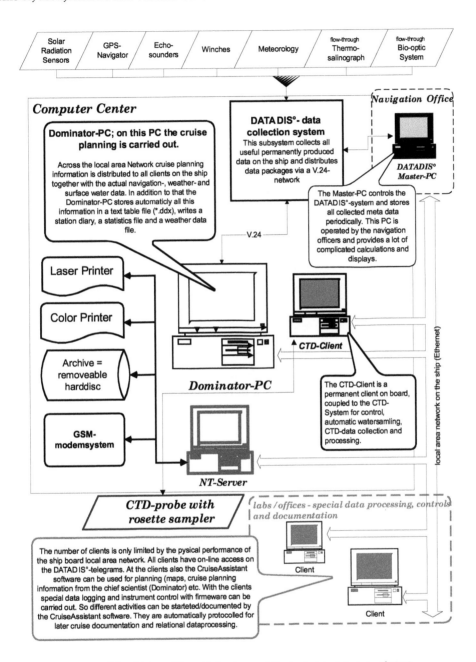

**FIGURE 1.36** Structure of the shipboard Data Collection and CruiseAssistant System of IOW.

## The IOW Automatic Station Network in the Southern Baltic Sea As a Part of the Marine Monitoring Network (MARNET) of the BSH

The Federal Maritime and Hydrographical Agency in Hamburg (BSH) operates an Automatic Marine Monitoring Network (MARNET) of presently 4 stations in the North Sea and 4 stations in the Baltic Sea. Plans are to add one more station in the North Sea and two more in the Baltic. Hourly messages with meteorological and hydrographical data are sent from each station via satellite (METEOSAT) to the database in Hamburg. The Institute of Baltic Sea Research in Warnemuende (IOW), by contract with the BSH, is in charge of the installation and maintenance of the four most easterly stations in the Baltic Sea.

**FIGURE 1.37**    CTD-probe SBE 911 plus-IOW Carousel 16 × 3 or 51 (photo by K.P. Wlost).

**FIGURE 1.38**    ScanFish MKII-IOW with SBE 911 and Fluorometer (photo by S. Krueger).

The IOW-stations are of urgent importance for the permanent observation of the water exchange between the Baltic Sea and the North Sea, as well as the exchange between different shallow basins and the central Baltic. Two new stations have been in operation for several years now (Mast DARSS SILL 1993, Discus Buoy ODER-BANK 1996) [5, 6]. Two more shall be established after 2000 (ARKONA SEA, MECKLENBURG BIGHT). The so-called "Darss Sill" is an important barrier limiting the water exchange between the North Sea and the Baltic Sea. Near the bottom, inflowing salt-rich North Sea water is normally

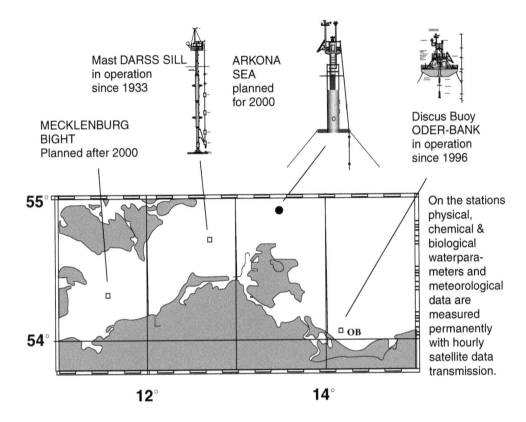

**FIGURE 1.39** MARNET—IOW Stations in the Southern Baltic Sea.

impeded by the shallow Darss Sill in its flow into the deeper basins. Only at specific meteorological conditions can the salt water overflow the Darss Sill.

The monitoring and observation of these processes are the main tasks of the automatic station Darss Sill. The actual design is based on the so-called IOW-Articulated Mast (Fig. 1.40). The basic idea of this design is a cigar-like mast of different air-filled aluminium tube modules. This "cigar" is fixed on the bottom in a flexible way by a heavy bottom weight and a universal joint. In the water, the buoyancy force erects the mast, so that the top comes out of the water and can carry a platform. The buoyancy force carries the mast like a reversed pendulum in the water. For a water depth on the Darss Sill of 21 m, a 25-m long "cigar" was designed. The diameter of the biggest tube module is 75 cm. Because the Baltic Sea has practically no tide the platform is always about 4 m above the surface. The platform can carry more than 500 kg of equipment. The mean inclination is about ±3°, in very heavy weather seldom more than ±10°. The "roll period" is about 15 seconds. Higher wave frequencies have practically no energy (Baltic Sea mean: 10 sec).

For the Mast station, the IOW-instrumentation department developed a new instrumentation concept. The instrumentation is supplied by environmental energy only (wind, solar). The station records 30 meteorological, hydrological, and housekeeping parameters every 10 minutes. In four main levels, the water temperature, the conductivity, oxygen, and current are measured. In some extra levels the system measures additional temperatures. The complete instrumentation is modular and expandable. All instruments have their own backup battery, which is normally "overwritten" by the main power supply. They are equipped with their own intelligence, their own time base and with long-term storage facilities. All instruments and communication units are connected via RS232 lines with the platform-host-computer (multitasking handheld). Every hour a complete data set of mean values is transmitted via a METEOSAT satellite to the database in Hamburg. An additional bidirectional online connection via GSM-radiotelephone

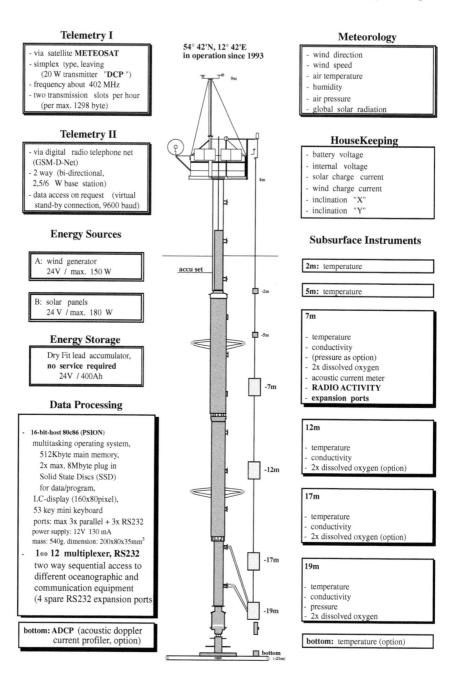

**Telemetry I**

- via satellite **METEOSAT**
- simplex type, leaving
  (20 W transmitter "**DCP** ")
- frequency about 402 MHz
- two transmission slots per hour
  (per max. 1298 byte)

**Telemetry II**

- via digital radio telephone net
  (GSM-D-Net)
- 2 way (bi-directional,
  2,5/6 W base station)
- data access on request (virtual
  stand-by connection, 9600 baud)

**Energy Sources**

A: wind generator
24V / max. 150 W

B: solar panels
24 V / max. 180 W

**Energy Storage**

Dry Fit lead accumulator,
**no service required**
24V / 400Ah

**Data Processing**

- **16-bit-host 80c86 (PSION)**
  multitasking operating system,
  512Kbyte main memory,
  2x max. 8Mbyte plug in
  Solid State Discs (SSD)
  for data/program,
  LC-display (160x80pixel),
  53 key mini keyboard
  ports: max 3x parallel + 3x RS232
  power supply: 12V 130 mA
  mass: 540g. dimension: 200x80x35mm³
- **1⇔12 multiplexer, RS232**
  two way sequential access to
  different oceanographic and
  communication equipment
  (4 spare RS232 expansion ports)

**bottom: ADCP** (acoustic doppler
current profiler, option)

54° 42'N, 12° 42'E
in operation since 1993

9m

4m

accu set

-2m

-5m

-7m

-12m

-17m

-19m

bottom
(-21m)

**Meteorology**

- wind direction
- wind speed
- air temperature
- humidity
- air pressure
- global solar radiation

**HouseKeeping**

- battery voltage
- internal voltage
- solar charge current
- wind charge current
- inclination "X"
- inclination "Y"

**Subsurface Instruments**

**2m:** temperature

**5m:** temperature

**7m**
- temperature
- conductivity
- (pressure as option)
- 2x dissolved oxygen
- acoustic current meter
- **RADIO ACTIVITY**
- **expansion ports**

**12m**
- temperature
- conductivity
- 2x dissolved oxygen (option)

**17m**
- temperature
- conductivity
- 2x dissolved oxygen (option)

**19m**
- temperature
- conductivity
- pressure
- 2x dissolved oxygen

**bottom:** temperature (option)

**FIGURE 1.40**  Baltic Station—Darss Sill.

is available from IOW for service, extra data transmissions, and event handling. The main power supply system is a balanced combination of a wind generator, 4 solar panels, and a 400 Ah dry-fit lead battery package. This system can supply about 1.5 A at 24 V, permanently. The actual instrumentation needs about 0.5 A. With fully charged batteries, all systems can operate a maximum of 50 days without any wind or solar energy input. An extraordinary advantage of the IOW-Mast is the ability to fill some tube sections with water, to put down the complete carrier on the seabed in the case of sea ice, as was necessary from February to April 1996. In 2000 the articulated mast, after permanent operation of its main mechanical parts for more than 10 years, will be exchanged for a new one.

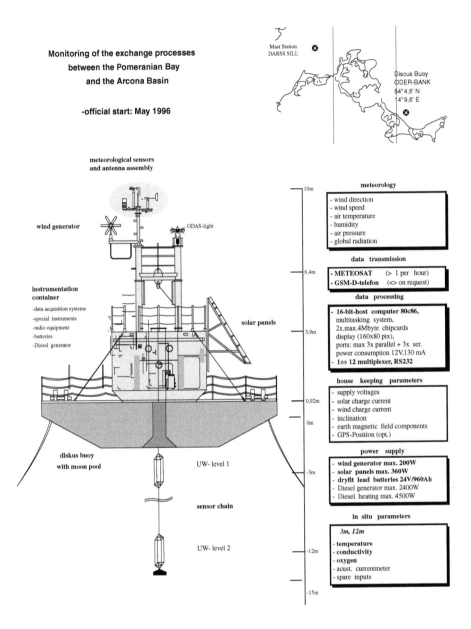

**FIGURE 1.41**    Baltic Station—ODER-BANK.

The second IOW-station near the "ODER-BANK" (Pomeranian Bay) started in May 1996 (Fig. 1.41). The station helps to monitor the complicated water exchange processes between the Pomeranian Bay and the Arkona Sea. The platform of this station is a heavy discus buoy (ø10 m) with an instrumentation container (4 m × ø4 m), developed by the BSH. This concept was chosen as a more robust design for a shallower region with ice problems every year. The container gives more room for special instrumentation in the future. The instrumentation concept is similar to the system on the Mast "Darss Sill". Twenty-five parameters are measured now, the underwater parameters in two main levels. The similar environmental energy supply has twice the capacity for extensions. An extra diesel generator and heating are available. The sampling concept for the underwater sensors is a little bit different according to the effects of the faster moving buoy on the measurements (one burst measurement for 10 minutes every hour, min. 4 Hz sampling frequency). The station "Oder-Bank" played an important role in the heavy ODER-flood in Northeast Germany and Poland in 1997.

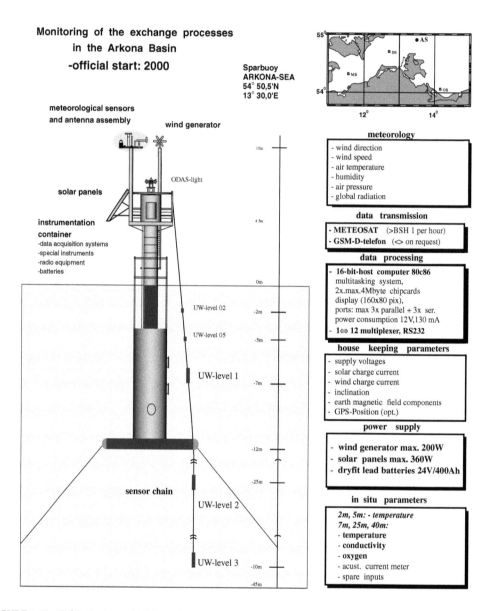

Monitoring of the exchange processes
in the Arkona Basin
-official start: 2000

Sparbuoy
ARKONA-SEA
54° 50,5'N
13° 30,0'E

meteorological sensors
and antenna assembly
wind generator

ODAS-light

solar panels

instrumentation
container
-data acquisition systems
-special instruments
-radio equipment
-batteries

UW-level 02
UW-level 05
UW-level 1
sensor chain
UW-level 2
UW-level 3

**meteorology**

- wind direction
- wind speed
- air temperature
- humidity
- air pressure
- global radiation

**data transmission**

- METEOSAT (>BSH 1 per hour)
- GSM-D-telefon (<> on request)

**data processing**

- 16-bit-host computer 80c86
  multitasking system,
  2x.max.4Mbyte chipcards
  display (160x80 pix),
  ports: max 3x parallel + 3x ser.
  power consumption 12V,130 mA
- 1 ⇔ 12 multiplexer, RS232

**house keeping parameters**

- supply voltages
- solar charge current
- wind charge current
- inclination
- earth magnetic field components
- GPS-Position (opt.)

**power supply**

- **wind generator max. 200W**
- **solar panels max. 360W**
- **dryfit lead batteries 24V/400Ah**

**in situ parameters**

*2m, 5m: - temperature*
*7m, 25m, 40m:*
- **temperature**
- **conductivity**
- **oxygen**
- acust. current meter
- spare inputs

**FIGURE 1.42** Baltic Station—ARKONA SEA.

The third IOW-station (ARKONA SEA, Fig. 1.42) is under construction now. For this more central position with a water depth of 45 m, a new type of an instrument carrier was developed by IOW, BSH, and the Federal Waterbilding Agency (BAW). This carrier is a mixture of a mast and a buoy combining the advantages of both. In the shape of a slim, long spar buoy with heavy ballast in the foot, it forms a semi-diving carrier. The spar buoy design was optimized by the results of model tests in a wave channel of the Technical University in Berlin. The characteristic frequencies of diving and rolling were designed at the lower end of the Baltic Sea wave spectrum. The buoy, in nature 16.5 m high with a mass of 32 t, will be anchored with two ropes. Nearly three quarters will be under water, damping high waves to minimum diving and rolling motions. The naked buoy is now equipped in the harbor of Rostock. The IOW-MARNET group started the design and construction of the measurement and energy supply system. It will be similar to the systems of the "Darss Sill" and "Oder-Bank." Twenty-eight parameters will be measured here initially with the underwater parameters in three main levels. The sampling concept of the underwater sensors will be the same as on the "Oder-Bank" according to the effects of a moving buoy on the measurements (one burst measurement for 10 minutes every hour, min. 4 Hz sampling frequency). The deployment of the complete ARKONA-SEA-Station is planned for 2000.

Here, for the first time as a new technology in the German Marine Monitoring Network, the regular operation an AUV, a Shallow Water Profiling Instrument Carrier (SWAPIC) with CTD-sensors will be introduced [7]. The SWAPIC Profiler is made of glass and has a hydraulic system to change its buoyancy. It can carry commercial instruments as payload, in this case a pumped, high accuracy, expandable CTD with high sampling rate (4 or 8 Hz). It works independently of the mother station using a simple guiding underwater mooring, which is easy to recover and to deploy for maintenance. The system is parked on the bottom. For preprogrammed profiling it is switching on the pumped CTD and moves up and down. The data are then processed to one meter bins with statistics and are written into the acoustic underwater modem. On request they are transmitted to the main station. With this new technology a new quality of underwater measurements on fixed automatic stations for monitoring purposes is expected, reducing maintenance costs at the same time. The MARNET-Stations will play an especially important role in the contribution of Germany to the Global Ocean Observing System.

## References

1. The Baltic Sea Joint Comprehensive Environmental Action Programme—Recommendations for Updating and Strengthening, *Baltic Sea Environment Proceedings No: 72*, Helsinki, 1998.
2. Pedersen, A. M. and Gregg, M. C., Development of a small *in-situ* conductivity instrument, *IEEE Journal of Oceanic Engineering*, 4, 3, 69–75, 1979.
3. Gregg, M. C. and Hess, W. C., Dynamic response calibration of sea-bird temperature and conductivity probes, *JAOT*, 2, 3, 304–313, 1985.
4. Larson, N. G., Calibration of oceanographic CTD instruments: methods and traceability, *Journal of Advanced Marine Technology Conference*, 7, 3–13, 1993.
5. Krueger, S., The Mast Station of the IOW on the "Darss Sill," Instrumentation Group (Physical Department), Institut für Ostseeforschung Warnemuende, *Proceedings of the 18th Conference of the Baltic Oceanographers*-18 CBO 23-27/11/92, St. Petersburg, Russia.
6. Krueger, S., Roeder, W., and Wlost, K.-P., The IOW–Baltic Stations DARSS SILL & ODER-BANK – part of the German Marine Monitoring Network – MARNET *Proceedings – Baltic Marine Science Conference*, October 21–26, 1996, Rønne, Denmark.
7. Krueger, S., Roeder, W., and Wlost, K.-P., Physical Oceanography, Instrumentation Dept., Baltic Sea Research Institute Rostock-Warnemuende (IOW); Koch, M., Kaemmerer, H. – 4H Jena Engineering GmbH, Knutz, T. – Institute for Applied Physics Kiel (IAP), Germany; Autonomous Instrumentation Carrier (APIC) with acoustic transmission for shallow water profiling – Oceanology International 98 – The Global Ocean – 10–13 March 1998, Brighton, U.K., *Conference Proceedings Vol. 2*, 149–158.

# 2

# Modeling Considerations

Hisaaki Maeda
*University of Tokyo*

Ferial El-Hawary
*BH Engineering Systems, Ltd.*

Matiur Rahman
*Dalhousie University*

## 2.1   Marine Hydrodynamics and Dynamics of a Floating Structure

*Hisaaki Maeda*

The behavior of a structure in natural environmental conditions is important to consider in designing the structure with enough safety and function. This section focuses on the structure of a mobile offshore unit and its related marine facilities such as semisubmersible rigs, TLPs, pontoons, FPSO vessels with riser pipes and mooring lines (position-keeping facilities), and so on. Natural environmental conditions consist mainly of wind, waves, and current.

The frequency range for the natural environmental conditions varies approximately from 0(steady) and 0.01 to 10 Hz.

Fluid–structure interaction and structure responses due to natural environmental conditions are the topics of this section. Among fluid–structure interactions, wave excitation force, radiation force, and drag force are the most important, while lift force and friction force are not always as important. These forces act as external loads on a structure and lead to various kinds of responses that are referred to as motion (displacement, deflection), acceleration, internal load (bending moment, shear force), pressure distribution, mooring line force, riser tension, etc.

In the following section, marine hydrodynamics are described including external hydrodynamic loads on a structure and dynamics of a structure under natural environmental conditions.

## Marine Hydrodynamics

Marine hydrodynamics covers hydrodynamics on an offshore or coastal structure itself and on related marine facilities such as mooring lines, riser pipes, propellers, rudders, and so on. Among them, we focus only on hydrodynamics on a floating structure and slender line structures such as mooring lines and riser pipes.

Generally speaking, linear phenomena are dominant for dynamics of a floating structure except in cases of catastrophic phenomena such as capsizing. Linear hydrodynamic force is important as it relates to an ideal fluid. Of course, nonlinear hydrodynamic forces such as drag force or viscous damping force are important in some cases, but we can treat them independently. The free surface makes this ideal fluid phenomenon more interesting. The general ocean waves are unsteady, irregular, and directional. We need to know not only the wave excitation force, but also the reaction hydrodynamic force due to oscillation of a structure (so-called radiation force, such as added mass or wave-making damping force).

Drag force is important for a bluff body, riser string, or mooring line in wind and current. Lift force is important for a vessel and a deck of a semisubmersible in wind, and for a vessel in current. Drag force is also important for a line structure such as riser pipes in waves, which behave like oscillating flow. Drag force is caused mainly by separated flow in viscous fluid. Drag force is a function of the Reynolds number and surface roughness of a body, and the Keulegan-Carpenter number, especially in oscillating flow. A bilge keel of a general vessel produces mainly drag force, which acts as a damping force to reduce the roll motion of a vessel.

Hydrodynamics of an ideal fluid is described in detail by Newman [12], Mei [11], and Faltinsen [8]. Viscous force such as drag force is treated in detail by Hoerner [9] and Sarpkaya and Isaacson [14]. General hydrodynamics for offshore structures is explained by Chakrabarti [4], and Clauss, Lehman, and Oestergaard [5].

### Estimation of Hydrodynamic Forces

Three options are available to us for estimating hydrodynamic forces acting on offshore structures at a preliminary design stage. These are the model test, the full scale or prototype model test, and the numerical test. The state-of-the-art for these three options are described in the proceedings of the ITTC (International Towing Tank Conference), which is held every three years (in 1999, 1996, 1993, etc.). The home page address of the ITTC is as follows; http://www.kriso.re.kr/ITTC.

Since all of these three options include some uncertainties, the integration of these three is important to extract a reliable estimation of hydrodynamical data. The relationship of these three options is shown in Fig. 2.1. The numerical test is called CFD (computational fluid dynamics), and the model test in a experimental tank is called EFD (experimental fluid dynamics).

Today, powerful PCs with large memories enable us to make more reliable and practical numerical estimations. If we formulate a reliable mathematical model that corresponds to a real physical phenomenon, it is not difficult to solve the mathematical model and obtain final results in both frequency and time domains. However, the verification and validation is still important to make the results more reliable. The validation requires the corresponding experiment in model scale or in full scale.

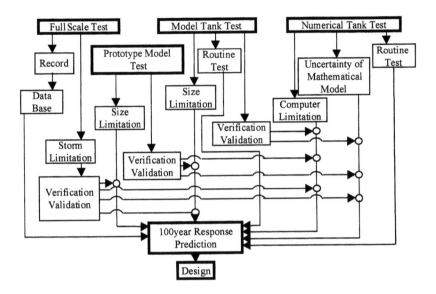

FIGURE 2.1   Integration of Model Test, Numerical Tank Test, and Prototype Model or Full Scale Test.

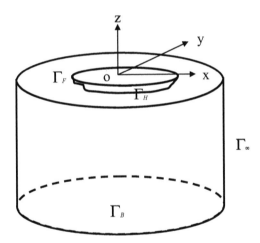

FIGURE 2.2   Coordinate system.

Generally speaking, flow field in space and time is expressed by several physical parameters such as velocity of fluid particles, pressure, density, and temperature, which are functions of space and time. These physical parameters are derived from mass conservation law, momentum conservation law, energy conservation law, and equations of state.

As to dynamics of offshore structures at ocean, the density and temperature can be assumed as constant in space and time. Therefore, what we need to know is the velocity field and pressure field in the fluid domain. The fundamental equations can be derived from mass conservation law and momentum conservation law, which are represented by equations of motion of fluid.

The coordinate system shown in Fig. 2.2 is the right-hand, Cartesian system in which the x-y plane is horizontal at the calm water surface and the z-axis is vertically upward.

If the fluid is inviscid, the velocity potential can be introduced, which derives the velocity field, and the velocity potential can be obtained from the Laplace equation with full boundary conditions, which represents the mass conservation law. The corresponding pressure field can be derived from the generalized Bernoulli equation, which corresponds to the momentum conservation law and is a function of velocity potential and pressure. Boundary conditions are given on a structure hull surface $\Gamma_H$, free surface $\Gamma_F$, infinitely far field $\Gamma_\infty$, and sea bottom $\Gamma_B$. The initial condition is also given.

(Laplace equation)

$$\frac{\partial^2 \phi}{\partial x^2} + \frac{\partial^2 \phi}{\partial y^2} + \frac{\partial^2 \phi}{\partial z^2} = \nabla^2 \phi = 0, \tag{2.1}$$

(Bernoulli equation)

$$\frac{\partial \phi}{\partial t} + \frac{1}{2}\nabla\phi \cdot \nabla\phi + g\zeta + \frac{p}{\rho} = 0 \tag{2.2}$$

If the fluid is viscid, the fundamental equations are the Navier-Stokes equation and the equation of continuity that is the mass conservation law. The boundary or initial conditions are the same as those of the inviscid fluid. Once the velocity vector **u** and $p$ at time $t = t$ are obtained, the velocity vector **u** at time $t = t + \Delta t$ is derived from the Navier-Stokes equation. Once the velocity vector **u** at time $t = t + \Delta t$ is obtained, the pressure $p$ can be derived from the Navier-Stokes equation, considering the equation of continuity at $t = t$, while the equation of continuity is not usually satisfied at $t = t + \Delta t$.

(Navier-Stokes equation)

$$\frac{\partial \mathbf{u}}{\partial t} + \mathbf{u} \cdot \nabla\mathbf{u} = -\frac{1}{\rho}\nabla p - \mathbf{g} + \nu\nabla^2\mathbf{u} \tag{2.3}$$

(Equation of continuity)

$$\nabla \cdot \mathbf{u} = 0 \tag{2.4}$$

Once the velocity field and pressure field in the entire fluid region are obtained, the hydrodynamic forces on a structure can be derived easily from this information. The hydrodynamic force $\mathbf{F}_p$ due to pressure is obtained from the pressure integral on a hull of a structure, which is expressed as follows:

$$\mathbf{F}_p = \iint_{\Gamma_H} p \cdot \mathbf{n} \, dS \tag{2.5}$$

where "$p$" is pressure in the fluid region, $\Gamma_H$ is a hull surface, "**n**" is normal cosine, and $dS$ is the surface element of a hull. If we consider the generalized normal vector "**n**," then the force "$\mathbf{F}_p$" includes hydrodynamic moment as the generalized force. In case of friction force "$\mathbf{F}_f$," "$p$" is replaced by friction force "$\mathbf{f}_p$" per area, and "**n**" is replaced by the tangential vector "s" on the hull.

### Numerical Schemes to Solve Fundamental Equation

There are four general numerical schemes that exist to solve the fundamental equation such as the Laplace equation or the Navier-Stokes equation. These schemes are the Eigenfunction Expansion Method (EEM), Boundary Integral Method (BIM), Finite Element Method (FEM), and Finite Difference Method (FDM). Each of these four methods consists of many variations as shown in Fig. 2.3. Generally speaking, EEM and BIM are applied mainly to the inviscid flow problem, while FDM is used for the viscous fluid problem. FEM is sometimes utilized for both viscous and inviscid problems.

The EEM is mainly based on the series expansion of sinusoidal functions (Fourier series), orthogonal functions, or eigen functions. The problem is to determine the coefficient of the series of the complete set of orthogonal functions, which forms the simultaneous linear equations, considering the boundary or initial conditions.

The BIM is based on the integral equation along the boundaries with the so-called Green Functions. The strength of the Green Function on the boundary is determined by satisfying the boundary conditions

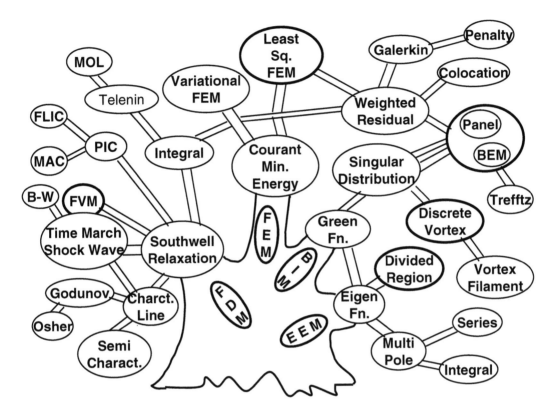

**FIGURE 2.3** Numerical calculation tools for marine hydrodynamics.

and initial condition. The integral equation is discretized and transformed into simultaneous linear equations. The strength of the Green Function only on the boundaries gives all information on the fluid velocities and pressures in the entire fluid region.

In the FEM and FDM, all the velocities or potentials in full fluid region are unknown and they are directly determined through the discretized simultaneous linear equation derived from the Navier-Stokes or Laplace equation. The number of unknowns is usually larger than those of the BIM and EFE.

As a practical method, several hybrid methods of the above-mentioned four algorithms are applied in the case of a structure with a complicated configuration or nonlinear free surface conditions. In the inner region, BIM with fundamental singularity, or EEM or FEM are used, while in the outer region, BIM with Green Function or EEM are adopted.

**Practical Prediction of Hydrodynamic Forces**

From a practical point of view, three practical schemes are introduced in the following: the "Strip method," "Drag force, Lift force," and "Morrison equation."

*Strip Method*
It is easier to calculate 2D (two-dimensional) hydrodynamic force than 3D. If a structure is slender along the **x**-axis, then 3D hydrodynamic force $\mathbf{F}_{3D}$ can be approximated by the superposition of 2D sectional forces $\mathbf{F}_{2D}$ shown in the following equation.

$$\mathbf{F}_{3D} = \int_L \mathbf{F}_{2D}(x)dx \tag{2.6}$$

where

$$\mathbf{F}_{2D}(x) = \int_C p_{2D}(x) \cdot \mathbf{n}_{2D}(x)ds(x) \tag{2.7}$$

and $p_{2D}(x)$ is the sectional pressure at the longitudinal position $x$, $\mathbf{n}_{2D}(x)$ is the sectional normal vector, and $ds(x)$ is the line element on the contour at position $x$.

### Drag Force and Lift Force

If a bluff body is located in a steady current or wind, the fluid flow is separated just behind the body and turns to wake with separated vorticity. The drag force and lift force occur on the body. The drag force $\mathbf{F}_D$ is in-line force and the lift force $\mathbf{F}_L$ is transverse force with regard to the fluid flow direction, which are expressed respectively as follows;

$$\left.\begin{array}{r} F_D \\ F_L \end{array}\right\} = \frac{1}{2}\rho \left\{\begin{array}{c} C_D \\ C_L \end{array}\right\} S_A \cdot V^2 \tag{2.8}$$

where $\rho$ is the fluid density, the drag coefficient $C_D$ and lift coefficient $C_L$ are the function of configuration of a structure, $S_A$ is the projected area of a structure, and $V$ is the relative velocity of a fluid. Strictly speaking, the coefficients $C_D$ and $C_L$ are dependent on the Reynolds number, however, in full-scale case, the Reynolds number is relatively high in turbulence range and the $C_D$, $C_L$ are almost functions of the configuration of a structure. The practical value for $C_D$ is as follows; $C_D = 1.17$ for a 2D circular cylinder, 2.05 for a 2D rectangular cylinder, 0.5 for a 3D sphere, and 1.05 for a 3D cube. For more details on $C_D$, $C_L$, see Hoerner [9].

### Morrison Equation

If line structures such as riser pipes or semisubmersible columns, the representative length of which is relatively smaller than the wave length of incoming waves are located in waves, the wave exciting force $F_W$ on the structure is expressed by the so-called Morrison equation. If the axis of the line structure is parallel to the vertical $z$-axis, the horizontal wave force on the line structure $dF_W/dz$ is formulated as follows;

$$dF_W/dz = C_M \rho \frac{\pi}{4} D^2 \dot{u} + C_D \rho \frac{D}{2} u|u| \tag{2.9}$$

where $C_M$ is the inertia coefficient, $C_D$ is the drag coefficient, $D$ is the diameter of the section of the line structure, and "u" is the horizontal velocity of a wave particle. The hydrodynamic coefficients $C_D$ and $C_L$

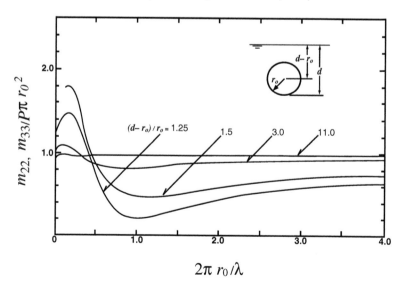

**FIGURE 2.4**    Added mass of a 2D submerged circular cylinder.

are the function of the Reynolds number Re $= uD/v$ and the Keulegan-Carpenter number $K_C = uT/D$, where $v$ is the dynamic viscous coefficient and $T$ is the period of incident waves. The practical value for $C_D$ is the same as that for drag force, while $C_M = 2.0$ for a 2D circular cylinder, 2.19 for a 2D rectangular cylinder, 1.5 for a 3D sphere, 1.67 for a 3D cube. The details are available in Sarpkaya and Isaacson [14].

## Some Examples of Hydrodynamic Forces on a Structure

Hydrodynamic radiation forces on a 2D submerged circular cylinder are shown in Figs. 2.4 and 2.5 which are added mass "m" and wave amplitude ratio $\bar{A}$ of radiated wave, respectively. The results of sway mode "$m_{22}$, $\bar{A}_2$" are equivalent to those of heave mode "$m_{33}$, $\bar{A}_3$" in the case of a submerged circular cylinder.

3D cylindrical columns with a submerged sphere are shown in Fig. 2.6, which are called waveless configurations. The added mass $m_3$ and wave damping coefficient $N_3$ in heave mode are shown in Fig. 2.7, and the sway mode added mass $m_2$ and wave excitation $e_2$ are shown in Fig. 2.8.

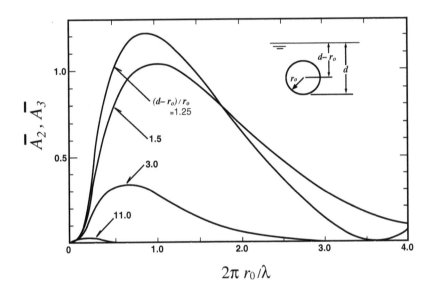

**FIGURE 2.5**   Radiation wave amplitude of a 2D submerged circular cylinder.

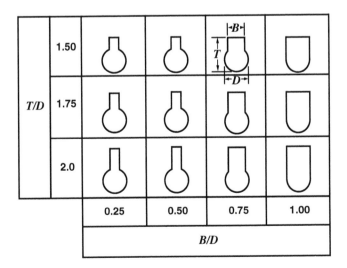

**FIGURE 2.6**   Series of 3D waveless forms [19].

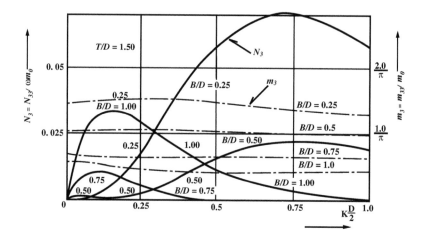

**FIGURE 2.7**   Added mass and wave making damping for 3D waveless form (heave mode) [19].

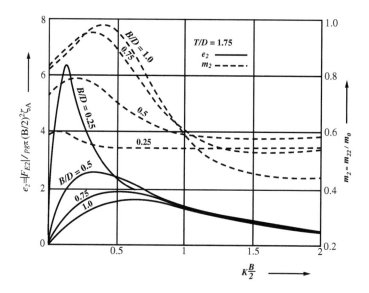

**FIGURE 2.8**   Added mass and wave exciting force (sway mode) [19].

Wave excitation sway force on an 8-column, 2-lower hull semisubmersible is shown in Fig. 2.9, and the corresponding heave added mass in Fig. 2.10. The ratio of L/B for the semisubmersible is 1.2 and the scale ratio is 1/50. The dashed line in Fig. 2.9 is based on the experimental radiation forces.

From the standpoint of the preliminary design stage, calculated results based on linear theory are sufficient, even though there are some discrepancies between the calculated one and the corresponding experimental one. In order to make the results more reliable, nonlinear and viscous effect should be introduced more deeply.

Wave drifting force coefficient $F_D$ on a 2D floating body is shown in Fig. 2.11. The higher the wave frequency, the larger the wave drift force. The maximum value for the 2D wave drifting force coefficient is 1.0, while the 3D wave drifting force coefficient is easily over 1.0, depending on the definition of the coefficient. The denominator is defined as the structure width. The wave front width, which affects the wave drift force, is over the structure width.

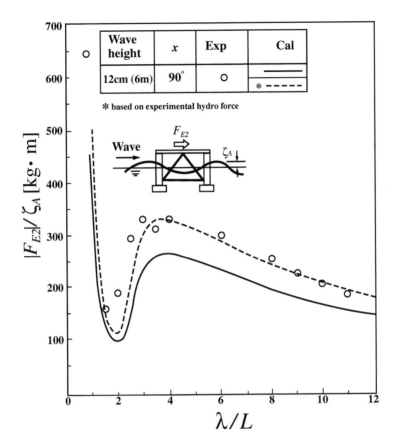

**FIGURE 2.9** Sway mode wave exciting force on an 8-column, 2-lower hull semisubmersible [19].

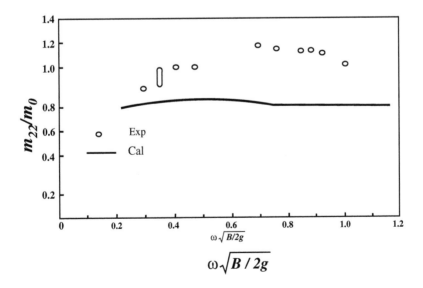

**FIGURE 2.10** Sway mode added mass on an 8-column, 2-lower hull semisubmersible [19].

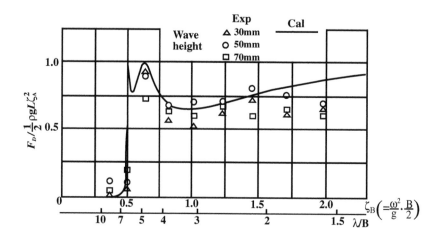

**FIGURE 2.11**   Steady wave drifting force on a 2D floating structure [19].

## Dynamics of a Floating Structure

The general description of dynamics of a floating structure is shown in detail in Hooft [10] and Faltinsen [8]. The dynamics of a line structure such as a mooring line or riser pipe is described in detail by Berteaux [1]. Hydroelastic behavior is treated by Bishop and Price [2]. The irregular responses of a floating structure in ocean waves are described in detail by Price and Bishop [13].

### Equation of Motion

We first assume rigid body motions in waves, wind, and current without forward velocity of the structure. The coordinate system can be assumed as a body fixed system, which corresponds to a space fixed system to express the linear equivalent dynamics except capsizing. We consider a moored floating structure for which the linear hydrodynamics and linear motions are dominant. Nonlinear hydrodynamics are introduced as viscous damping or second-order wave excitation to the equation of motions, if necessary.

The general equation of motions for a rigid body is expressed as follows;

$$\sum_{j-1}^{6}[(M_{ij} + m_{ij}(\omega))\ddot{x}_j(t) + N_{ij}(\omega)\dot{x}_j(t) + D_{ij}|\dot{x}_j|\dot{x}_j(t) + (A_{ij} + G_{ij})x_j(t)]$$
$$= F_{w0} + F_{w1}(\omega;t) + F_{w2}(\omega;t), \quad i = 1,2,...6, \tag{2.10}$$

where, $x_j(t)$ is the *jth* mode displacement of rigid body in the time domain, $\dot{x} = dx/dt$, $\ddot{x} = dt^2/dx^2$, $M_{ij}$ is the generalized inertia term of a rigid body, $M_{ij}(\omega)$ is the *ith* mode generalized added mass (radiation potential force) due to the *jth* mode motion on a body, $N_{ij}(\omega)$ is the generalized wave damping force (radiation potential force), $D_{ij}$ is the generalized viscous damping coefficient, $A_{ij}$ is the generalized restoring force coefficient due to hydrostatic force, $G_{ij}$ is the generalized restoring force due to the mooring system, $F_{w0}$ is the steady external force due to steady current or steady wind velocity, $F_{w1}(\omega)$ is the generalized linear wave excitation, and $F_{w2}(\omega; t)$ is the generalized second-order wave excitation. The subscript $i,j = 1,2,...5$, 6 stands for surge, sway, heave, roll, pitch, and yaw, respectively. The restoring force coefficient $A_{ij}$, $G_{ij}$ may be nonlinear with regard to a large amplitude of displacement $x_j(t)$ as follows:

$$A_{ij} = A_{0ij} + A_{1ij} \cdot |x_{ij}(t)| + A_{2ij} \cdot x_{ij}^2(t) \tag{2.11}$$

$$G_{ij} = G_{0ij} + G_{1ij} \cdot |x_{ij}(t)| + G_{2ij} \cdot x_{ij}^2(t) \tag{2.12}$$

which occur in rolling motion for $A_{ij}$ and slack mooring system for $G_{ij}$. If the dolphin-fender mooring system is adopted, the restoring force coefficient $G_{ij}$ behaves like hysteresis.

Strictly speaking, the general equation of motion, Eq. (2.1), is held only in regular waves, steady current, and steady wind, which correspond to the frequency domain equation, because the generalized added mass $m_{ij}(\omega)$ and generalized wave damping coefficient $N_{ij}(\omega)$ are functions of frequency $\omega$, the 2nd order wave excitation $F_{w2}(\omega; t)$ is in so-called difference frequency (low frequency) in irregular waves, which is far from wave frequency, and the motion in a regular wave is linear dominant for a general large-scale floating structure. That is to say, x(t) must be as follows:

$$x(t) = \bar{x} \cdot \exp(i\omega t) \tag{2.13}$$

What shall we do in case of irregular waves in time domain? If Eq. (2.1) is only linear, then the superposition principle holds and the response can be decomposed to the superposition of the responses of corresponding regular waves. If Eq. (2.1) can be expressed in an equivalent linear form, the superposition principle works well, too. However, in case of the general nonlinear Eq. (2.1), $m_{ij}(\omega)$, $N_{ij}(\omega)$ are linear radiation potential forces in the frequency domain, the corresponding time domain terms can be easily derived through the Fourier transformation as follows;

$$K_{ij}(\tau) = \frac{2}{\pi} \int_0^\infty N_{ij}(\omega) \cos \omega \tau \, d\omega \tag{2.14}$$

$$m_{ij}(\omega) = m_{ij}(\infty) + \frac{1}{\omega} \int_0^\infty K_{ij}(\tau) \sin \omega \tau \, d\tau \tag{2.15}$$

where $K_{ij}(\tau)$ is the memory function or Kernel function, which gives the radiation force through the convolution integral, and $m_{ij}(\infty)$ is the added mass at infinite frequency. Then the exact equation of motion in time domain is expressed as follows;

$$\sum_{j-1}^{6} \left[ (M_{ij} + m_{ij}(\infty)) \ddot{x}_j(t) + \int_0^t K_{ij}(t - \tau) \dot{x}_j(\tau) d\tau + D_{ij} |\dot{x}_j| \dot{x}_j(t) + (A_{ij} + G_{ij}) x_j(t) \right]$$
$$= F_{w0} + e_1(t) + e_2(t), \qquad i = 1, 2, \dots \dots, 6 \tag{2.16}$$

$$e_1(t) = \int_0^t Z_1(t - \tau) \cdot \zeta(\tau) d\tau \tag{2.17}$$

$$Z_1(t) = \frac{1}{\pi} \int_0^\infty E_1(\omega) \cdot \exp(i\omega t) d\omega \tag{2.18}$$

$$e_2(t) = \int_0^t \int_0^t Z_2(t - \tau_1, t - \tau_2) \cdot \varsigma(\tau_1) \varsigma(\tau_2) d\tau_1 d\tau_2 \tag{2.19}$$

$$Z_2(t_1, t_2) = \frac{1}{2\pi^2} \int_0^\infty \int_0^\infty E_2(\omega_1, \omega_2) \cdot \exp(i\omega_1 t_1 + i\omega_2 t_2) d\omega_1 d\omega_2 \tag{2.20}$$

where $e_1(t)$, $e_2(t)$ are the wave excitation of first and second order in time domain respectively, which are derived from the impulse response function of first and second order $Z_1(t)$ and $Z_2(t_1, t_2)$, respectively. The first and second order impulse functions are transformed through the Fourier transformation of response transfer functions of first and second order $E_1(\omega)$, $E_2(\omega_1, \omega_2)$, respectively, where $E_1(\omega)$ is the linear transfer function and $E_2(\omega_1, \omega_2)$ is the quadratic transfer function of unit incident wave of

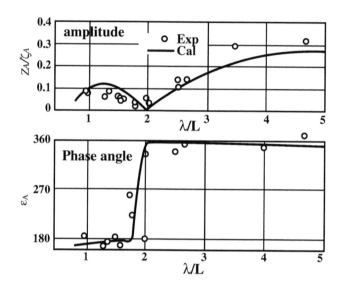

**FIGURE 2.12**   Heave of 4-column footing semisubmersible [19].

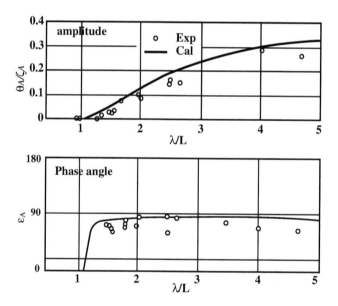

**FIGURE 2.13**   Pitch of 4-column footing semisubmersible [19].

frequency $\omega$, and two component incident waves with the frequency $\omega_1$ and $\omega_2$, respectively. $\zeta(t)$ is the incident wave surface elevation in time domain.

With regard to a four-column footing semisubmersible, the heave response with phase shift is shown in Fig. 2.12, the pitching in Fig. 2.13 and the surging in Fig. 2.14. In these results the calculated results are in good agreement with the corresponding experimental ones, though the interaction effects of multiple columns are not considered.

### Some Relationships among Hydrodynamic Forces

Interesting relationships exist among hydrodynamic forces based on potential flow, such as the reciprocal relation, the Haskind-Newman relation, and the Kramers-Kroenig's relation. The following is the

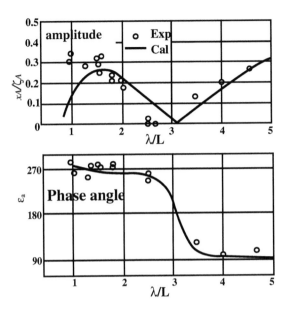

**FIGURE 2.14**   Surge of 4-column footing semisubmersible [19].

reciprocal relation among the radiation potential forces,

$$m_{ij} = m_{ji}, \qquad i,j = 1,2,3,....,5,6 \tag{2.21}$$

$$N_{ij} = N_{ji} \qquad i,j = 1,2,3,....,5,6 \tag{2.22}$$

If the Kochin function $H_j(\beta)$ is expressed in the following form,

$$H_j(\beta) = \begin{cases} \dfrac{1}{D}\displaystyle\int_{\Gamma_H}\left\{\dfrac{\partial\phi_j}{\partial n} - \phi\dfrac{\partial}{\partial n}\right\}\phi_0(\beta + \pi)ds \\[3mm] \dfrac{1}{D}\displaystyle\int\int_{\Gamma_H}\left\{\dfrac{\partial\phi_j}{\partial n} - \phi_j\dfrac{\partial}{\partial n}\right\}\phi_0(\beta + \pi)ds \end{cases} \tag{2.23}$$

then the Haskind relation is the relation between the $j$th mode of wave excitation $E_j$ and the $j$th mode of Kocin function $H_j$ as follows:

$$E_j = \rho g a D \cdot H_j(\beta + \pi) \tag{2.24}$$

where the Kochin function $H_j$ corresponds to the $j$th mode of amplitude of the radiation wave at infinite far field, "$a$" is the incident wave amplitude, "$\beta$" is the incident angle of incident waves, "$\rho$" is density of water, "$g$" is the gravitational acceleration, and "$D$" is the shallow water parameter in the following relation.

$$D = (\tanh Kh + Kh\,\mathrm{sech}^2 Kh) \tag{2.25}$$

where "$h$" is water depth and "$K$" is the wave number. $\phi_j$, $\phi_0$ are the radiation velocity potential and incident wave potential with unit velocity amplitude and unit wave amplitude, respectively.

The Ocean Engineering Handbook

The Haskind-Newman relation is the relation between the $j$th mode of wave excitation and the $j$th mode of wave making damping coefficient $N_{jj}$ as follows;

$$|E_j|^2 = \frac{\rho(ga)^2}{\omega}N_{jj}. \qquad (2.26)$$

The Kramers-Kroenig's relation is the relation between the in-phase and out-of-phase component of the radiation force as follows;

$$m_{kj}(\omega) - m_{kj}(\infty) = \frac{2}{\pi}\int_0^\infty \frac{N_{kj}(\omega')}{\omega'^2 - \omega^2}d\omega' \qquad (2.27)$$

$$N_{kj}(\omega) = -\frac{2}{\pi}\int_0^\infty \frac{\{m_{kj}(\omega') - m_{kj}(\infty)\}\omega'^2}{\omega'^2 - \omega^2}d\omega' \qquad (2.28)$$

The generalized added mass satisfies the following equation.

$$\int_0^\infty \{m_{kj}(\omega) - m_{kj}(\infty)\}d\omega = 0 \qquad (2.29)$$

The above-mentioned relationship of Eqs. (2.21) to (2.29) can be held not only in a rigid floating body, but also in a flexible floating body, as described in the next section.

Once the Bessho's reverse time potential is introduced, the diffraction potential is related to the radiation potential, considering the corresponding radiation Kochin function.

$$\phi_j^* = \begin{cases} \phi_j + i\left\{H_j^*\left(\frac{\pi}{2}\right)\cdot\phi_s\left(-\frac{\pi}{2}\right) + H_j^*\left(-\frac{\pi}{2}\right)\cdot\phi_s\left(\frac{\pi}{2}\right)\right\} \\ \phi_j + \frac{iK}{2\pi}\int_{-\pi}^\pi \phi_s(\beta+\pi)\cdot H_j^*(\beta)d\beta \end{cases} \qquad (2.30a,b)$$

The same kind of relation of the diffraction potential and diffraction Kochin function is also derived as follows;

$$\phi_s^*(\beta) = \begin{cases} \phi_s(\beta+\pi) + i\left\{H_d^*\left(\beta,\frac{\pi}{2}\right)\cdot\phi_s\left(-\frac{\pi}{2}\right) + H_d^*\left(\beta,\frac{\pi}{2}\right)\cdot\phi_s\left(\frac{\pi}{2}\right)\right\}, \\ \phi_s(\beta+\pi) + \frac{iK}{2\pi}\int_{-\pi}^\pi \phi_s(\alpha+\pi)\cdot H_\alpha^*(\beta,\alpha)d\alpha, \end{cases} \qquad (2.31a,b)$$

where "*" denotes the complex conjugate; the Bessho's reverse time potential $\phi^*$ is defined as the complex conjugate of the regular time potential $\phi$; $\phi_s = \phi_0 + \phi_d$ is the scattering potential, which is the summation of incident potential $\phi_0$ and the diffraction potential $\phi_d$; $H_j$ and $H_d$ are the radiation Kochin function of $j$th mode and the diffraction Kochin function, respectively; the argument $\alpha$ or $\beta$ is the direction of incident wave or radiated wave; $K = 2\pi/\lambda = \omega^2/g$ is the wave number; and "$i$" stands for the imaginary unit.

## Dynamics of a Very Large Floating Structure

Representative length of a general offshore oil floating structure is around 100 m or so. The structure can be considered as a rigid body for which internal loads due to waves are estimated through the pressure distribution on the rigid body. However, a floating airport, more than 1000 m length, is not defined as a rigid body, but as a flexible body requiring hydroelastic behavior. Compared with Eq. (2.10), which represents displacements only around a center of gravity, the equation of dynamics of a flexible floating body should be expressed as an equilibrium of external and internal forces on each point of a structure. An average, very large floating structure can be considered as a flat plate with a very small draft, since the horizontal dimension of the structure is far larger than the vertical displacement. Out-of-plane deformation is important, while in-plane deformation is negligible.

The equation of deformation of a very large floating structure is expressed as follows;

$$m'(x)\ddot{\zeta}(x, t) + \rho g b(x)\zeta(x, t) + EI(x)\frac{d^4\zeta(x, t)}{dx^4} = F_w(x, t) + F_R(x, t) \tag{2.32}$$

where a beam case is assumed, and $m'(x)$ is distributed mass at $x$ point, $\zeta$ is deformation or displacement at $x$ point, $b$ is breadth or water plane area at point $x$, and $EI(x)$ is rigidity of the structure at point $x$. On the left-hand side of Eq. (2.30), the first term is an inertia term, the second term is a hydrostatic force term, and the third term is bending reaction force of the flexible structure. On the right-hand side of Eq. (2.30), $F_w$ and $F_R$ are wave excitation and hydrodynamic radiation force at point $x$, respectively.

Two practical methods for solving Eq. (2.32) are modal analysis and the influence function method. The former method expresses the deformation as the superposition of the Eigen function of the dry modes of the flexible vibration equation of a beam. In the latter method, the deformation of a beam is represented by the superposition of unit displacement of each point on a beam. The following is the Eigen function method. The deformation of point $x$ is expressed by the superposition of Eigen functions with principal coordinates as a kind of Fourier coefficient as follows:

$$\zeta(x, t) = \sum_{r=1}^{\infty} q_r(t)\eta_r(t) \tag{2.33}$$

The final formulation of the equation of flexible vibration is expressed as follows:

$$\sum_{r=1}^{\infty}[(a_{rs}\delta_{rs} + A_{rs})\ddot{q}_r(t) + N_{rs}\dot{q}_r(t) + (b_{rs}\delta_{rs} + C_{rs})q_r(t)] = E_s(t), \quad s = 1,2,3,\ldots\ldots,\infty \tag{2.34}$$

where,

$$a_{rs} = \int_L M'(x)\eta_r(x)\eta_s(x)dx \tag{2.35}$$

$$\delta_{rs} = \begin{cases} 1, & s = r \\ 0, & s \neq r \end{cases} \tag{2.36}$$

$$A_{rs} = \int_L m'_{rs}(x)\eta_r(x)\eta_s(x)dx \tag{2.37}$$

$$N_{rs} = \int_L N'_{rs}(x)\eta_r(x)\eta_s(x)dx \tag{2.38}$$

$$b_{rs} = \int_L \rho g b(x)\eta_r(x)\eta_s(x)dx \tag{2.39}$$

$$C_{rs} = \int_L EI(x)\frac{d^4\eta_r(x, t)}{dx^4}\eta_s(x)dx \tag{2.40}$$

$$E_s(t) = \int_L F_W(x, t)\eta_s(x)dx \tag{2.41}$$

The mode shape $\eta_r(x)$ is derived from the equation of flexible vibration beam in air as an Eigen function. In the following, the more general case for a flat plate is described. The equation for flat plate vibration in air is written as follows:

$$D_x\frac{\partial^4 \eta_r}{\partial x^4} + 2H\frac{\partial^4 \eta_r}{\partial x^2 \partial y^2} + D_y\frac{\partial^4 \eta_r}{\partial y^4} + m'\frac{\partial^2 \eta_r}{\partial t^2} = 0 \tag{2.42}$$

where $D_x$, $D_y$ are bending rigidity along the $x$ and $y$ axes, respectively,

$$H = D_I + 2D_{xy} \tag{2.43}$$

$D_I$ is coupled bending rigidity
$D_{xy}$ is torsional rigidity.

In the one-dimensional free-free beam case, the mode shape can be obtained analytically as follows:

$$\eta_r(x) = \frac{1}{2}\left[\begin{array}{l}\cosh \beta_r\left(\frac{L}{2}+x\right)+\cos \beta_r\left(\frac{L}{2}+x\right)\\[2mm] -\frac{\cosh \beta_r L - \cos \beta_r L}{\sinh \beta_r L - \sin \beta_r L}\left\{\sinh \beta_r\left(\frac{L}{2}+x\right)+\sin \beta_r\left(\frac{L}{2}+x\right)\right\}\end{array}\right] \tag{2.44}$$

where

$$1 - \cos \beta_r L \cdot \cosh \beta_r L = 0 \tag{2.45}$$

In case of a flat plate, mode shape $\eta_r(x,y)$ can be expressed by the product of $\eta_r(x)$ and $\eta_r(y)$.

The calculated results of a very large floating pontoon are shown in the following. As to the model dimensions, the length to width ratio L/B is 4.0, while the draft is almost zero. The vertical deflection at the weather side end is shown in Fig. 2.15 and the bending moment at the midship point is shown in Fig. 2.16. The EI-A denotes hard elasticity, while the EI-C corresponds to very soft elasticity. When a structure is more flexible, the deflection at the weather side end increases and the bending moment

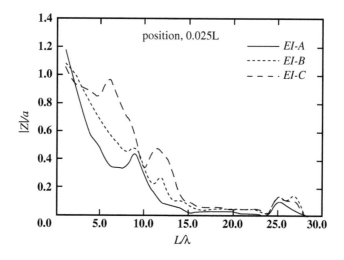

**FIGURE 2.15** Vertical deflection at weather side end of VLFS (very large floating structure).

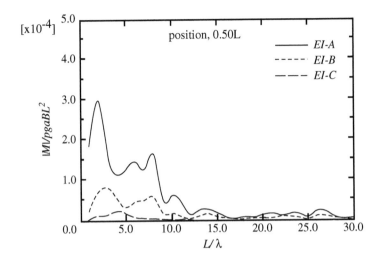

**FIGURE 2.16** Bending moment at midpoint of VLFS.

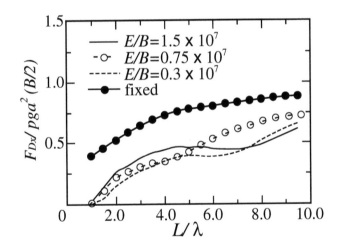

**FIGURE 2.17** Steady drift force on VLFS.

decreases. As the wave drift coefficient $F_{DX}$ decreases, the structure rigidity becomes more flexible as shown in Fig. 2.17.

### Dynamics of Underwater Vehicles and FPSO (A Single Point Moored Vessel)

Two coordinate systems are used to analyze dynamics of underwater vehicles or FPSO including the space-fixed coordinate system and the body-fixed coordinate system. The location of the vessel is expressed in the space-fixed coordinate system, while the motion dynamics of the vessel in current and waves is formulated in the so-called maneuvering equation of a ship. The general maneuvering equation in 3D is written as follows;

$$\begin{Bmatrix} \dot{\mathbf{V}} \\ \dot{\omega} \end{Bmatrix} = [[M] + [\mu]]^{-1} \begin{Bmatrix} \mathbf{F} - m\omega \times \mathbf{V} \\ \Gamma - m\omega \times \mathbf{h} \end{Bmatrix} \tag{2.46}$$

where $\mathbf{V} = \{u, v, w\}$ and $\omega = \{p, q, r\}$ are velocity of a body and angular velocity on the body-fixed coordinate system, respectively. $\dot{\mathbf{V}} = \partial V/\partial t$ and so on.

$$[M] = \begin{bmatrix} m & 0 & 0 & \\ 0 & m & 0 & [0] \\ 0 & 0 & m & \\ & [0] & & [I] \end{bmatrix} \tag{2.47}$$

$$[I] = \begin{bmatrix} I_{xx} & 0 & -I_{xz} \\ 0 & I_{yy} & 0 \\ -I_{zx} & 0 & I_{zz} \end{bmatrix} \tag{2.48}$$

where "$m$" is mass of a body, "$I_{ik}$" is general moment of inertia, "$\mathbf{F}$," "$\Gamma$" are external force and moment, respectively, "$\mathbf{h}$" is the angular momentum vector, and $[\mu]$ is generalized added mass. "$\mathbf{F}$" and "$\Gamma$" consist of wave excitation, wave making damping, viscous force, tension due to mooring line, hydrostatic force, and so on.

Once Eq. (2.46) is solved, the vector "$\mathbf{V}$," "$\omega$" is obtained and then the location of the body in space can be derived through the transformation of the body-fixed coordinate and the space-fixed coordinate using the Eulerian angle as follows:

$$\begin{Bmatrix} \dot{\mathbf{x}} \\ \dot{\theta} \end{Bmatrix} = \begin{bmatrix} [T] & [0] \\ [0] & [S] \end{bmatrix} \begin{Bmatrix} \mathbf{V} \\ \omega \end{Bmatrix} \tag{2.49}$$

$$[T] = \begin{bmatrix} c_5 c_6 & s_4 s_5 s_6 - c_4 s_6 & c_4 s_5 s_6 + s_4 s_6 \\ c_5 c_6 & s_4 s_5 s_6 + c_4 s_6 & c_4 s_5 c_6 - s_4 s_6 \\ -s_5 & s_4 c_5 & s_4 c_5 \end{bmatrix} \tag{2.50}$$

$$c_i = \cos x_i, \qquad s_i = \sin x_i \tag{2.51}$$

$$[S] = \begin{bmatrix} 1 & s_4 \tan x_5 & c_4 \tan x_5 \\ 0 & c_4 & -s_4 \\ 0 & s_4/c_5 & c_4/c_5 \end{bmatrix} \tag{2.52}$$

The above equations hold in 3D and are applicable to the motion of underwater vehicles, while in the horizontal plane motion, only "$u, v, r$," which are x-direction velocity, y-direction velocity, and angular velocity around the z-axis, are the key parameters.

## Conclusion

In this chapter, marine hydrodynamics were described covering external hydrodynamic loads on a structure and dynamics of a structure under natural environmental conditions.

The dynamics of a rather large-scale floating structure such as a ship or offshore oil structure are governed mainly by linear terms, while nonlinear terms such as viscous effect or higher order terms play as auxiliary terms on the motion of a main floating body. This does not mean that the nonlinear terms are not important, since viscous damping is important for the prediction of low frequency motions of a moored floating body, while VIV (vortex induced vibration) is important for riser pipe or mooring line dynamics, and so on.

It is not difficult to solve an equation of motion itself such as Eqs. (2.10),(2.16),(2.34), and (2.46), using some numerical algorithms even in the time domain.

In this chapter, the prediction methods for estimating hydrodynamic forces on an offshore structure or riser pipes and dynamics of those structures in wave, current and wind were introduced. Those prediction tools are useful at a preliminary design stage of offshore structures or very large floating structures, while reliable and practical prediction methods to estimate drag/lift forces on a slender structure with regard to the ratio of structure length and wave length are still based on experimental data, even though CFD has achieved the big progress.

We have to remember that even precise numerical calculation tools contain some uncertainties that should be verified and validated by other corresponding numerical results or the corresponding model tests or prototype model tests. Even model tests contain uncertainties, and we should therefore consider the integration of the numerical calculation, model tests and prototype model or full-scale tests.

## References

1. Berteaux, H. O., *Buoy Engineering*, John Wiley & Sons, New York, 1976.
2. Bishop, R. E. D. and Price, W. G., *Hydroelasticity of Ships*, Cambridge University Press, Cambridge, 1979.
3. Clayton, B. R. and Bishop, R. E. D., *Mechanics of Marine Vehicles*, Gulf Publishing, 1982.
4. Chakrabarti, S. K., *Hydrodynamics of Offshore Structures*, Computational Mechanics Publications, Springer-Verlag, Berlin, 1987.
5. Clauss, G., Lehmann, E., and Oestergaard, C., *Offshore Structures, Vol. 1, Conceptual Design and Hydromechanics*, Springer-Verlag, Berlin, 1992.
6. Comstock, J. P., *Principle of Naval Architecture*, SNAME (American Society of Naval Architects and Marine Engineers), 1967.
7. Ertekin, R. C. and Kim, J. W., *Proceedings of the Third International Workshop on Very Large Floating Structures* (VLFS '99), Vols. 1 and 2, University of Hawaii at Manoa, 1999.
8. Faltinsen, O. M., *Sea Loads on Ships and Offshore Structures*, Ocean Technology Series, Cambridge University Press, Cambridge, 1990.
9. Hoerner, S. F., *Fluid Dynamic Drag*, S. F. Hoerner, New Jersey, 1958.
10. Hooft, J. P., *Advanced Dynamics of Marine Structures*, John Wiley & Sons, New York, 1982.
11. Mei, C. C., *The Applied Dynamics of Ocean Surface Waves*, Advanced Series on Ocean Engineering, Vol. 1, World Scientific, 1989.
12. Newman, J. N., *Marine Hydrodynamics*, MIT Press, Cambridge, MA, 1977.
13. Price, W. G. and Bishop, R. E. D., Probabilistic Theory of Ship Dynamics, Chapman and Hall, London, 1974.
14. Sarpkaya, T. and Isaacson, M., *Mechanics of Wave Forces on Offshore Structures*, Van Nostrand Reinhold, 1981.
15. USAF : USAF Stability and Control DATCOM, Flight Control Division, Air Force Flight Dynamics Laboratory, Wright-Patterson Air Force Base, Dayton, Ohio, 1975.
16. Gresho, P. M., Some current CFD issues relevant to the incompressible Navier-Stokes equations, *Comput. Methods Appl. Mech. Eng.*, 87, 201–252, 1991.
17. Fletcher, C. A. J., *Computational Techniques for Fluid Dynamics*, Springer-Verlag, Berlin, 1988.
18. Harlow, F. H. and Welch, J. E., Numerical calculation of time-dependent viscous incompressible flow with free surface, *Phys. Fluids*, 8,12, 2182, 1965.
19. Motora, S., Kayama, T., Fujino, M., and Maeda, H., *Dynamics of Ships and offshore Structures* (in Japanese) Seizando Publisher, Tokyo, 1997.

## 2.2   Mathematical Modeling of Ocean Waves

*Matiur Rahman*

This report contains the investigation of mathematical modeling of ocean waves. In the first part of this report, we try to reexamine the wave-wave interactions of four progressive waves traveling with four different wave numbers and frequencies. It has been demonstrated by many previous researchers including

Webb [19], Phillips [16] and Hasselmann [3] that this set of four waves, called a quadruplet, could exchange energy if they interact nonlinearly such that the resonant conditions are satisfied (this will be explained later). In this report, however, we shall be concerned with the approximate solutions of the nonlinear transfer action functions that satisfy a set of first-order ordinary differential equations using the JONSWAP spectrum as initial conditions. We will demonstrate three simple methods to compute the nonlinear transfer action functions. The first method is due to Picard, the second method due to Bernoulli (these two are analytical methods), and the third is a numerical integration method using a fourth-order Runge-Kutta scheme. We compare these results in graphical forms that show excellent agreement.

The second part of this report is devoted mainly to the determination of the analytical solutions for the Eulerian currents present in ocean circulation. A variety of solutions that satisfy the required boundary and initial conditions are obtained. A Laplace transform method in conjunction with the convolution concept is used as a solution technique. Some of the solutions are graphically illustrated, and the physical meaning is described.

## Nonlinear Wave Interactions

This report discusses the mathematical methods of ocean waves. Masson and Leblond [14] and Longuet-Higgins [13] suggested that ocean wave spectra within the MIZ obey the energy balance equation for wave spectra

$$\left(\frac{\partial}{\partial t} + \mathbf{C}_g \cdot \nabla\right) E(f, \theta) = (S_{in} + S_{ds})(1 - f_i) + S_{nl} + S_{ice} \tag{2.53}$$

where the two-dimensional wave spectrum $E(f, \theta)$ is a function of frequency $f$, (cycles/sec, Hertz), and direction $\theta$, time $t$ and position $\mathbf{x}$, and where $\mathbf{C}_g$ is the group velocity; $\nabla$ is a gradient operator. Ocean surface waves are central to the atmosphere-ocean coupling dynamics at the air-sea interface. The dominant physical process that determines ocean surface waves are the source functions given on the right-hand side of Eq. (2.53). Here $S_{in}$ is the input of energy due to wind, $S_{ds}$ is the energy dissipation due to white capping, $S_{nl}$ is the nonlinear transfer between spectral components due to wave-wave interactions, $S_{ice}$ is the change in energy due to wave interactions with ice floes, and $f_i$ is the fraction of area of the ocean covered by ice. If $f_i = 0$, it implies that the ocean is free from ice-cover and so $S_{ice} = 0$; and when $f_i = 1$, the ocean is completely covered by ice and consequently $(S_{in} + S_{ds}) \times 0 = 0$, and $S_{nl}$ and $S_{ice}$ will contribute to the energy balance (see Perrie and Hu [15]).

Equation (2.53) can be written as

$$\frac{dE(f, \theta)}{dt} = (S_{in} + S_{ds})(1 - f_i) + S_{nl} + S_{ice} \tag{2.54}$$

where $\frac{d}{dt}$ denotes total differentiation when traveling with the group velocity. The accepted formulation for wind input $S_{in}$, as parameterized in the WAM model of Hasselmann et al. [5] and Komen et al. [10], suggests that $S_{in}$ should be represented as

$$S_{in}(f, \theta) = \beta \omega E(f, \theta) \tag{2.55}$$

where $\beta$ is a nondimensional function of sea state maturity and $\omega = 2\pi f$ represents the angular (radian) frequency, which is related to the wave number $k$ through the deep water dispersion relation $\omega^2 = gk$ for deep ocean and $\omega^2 = gk \tanh kh$ for finite depth ocean.

Hasselmann et. al. [5] (WAM) defined the parameter $\beta$ as (empirical formula)

$$\beta = \max\left\{0, 0.25\epsilon\left(28\frac{U_*}{C_p}\cos(\Delta\theta) - 1\right)\right\} \tag{2.56}$$

where $\Delta\theta$ = the difference between the wind and wave directions, and $\epsilon = \frac{\rho_a}{\rho_w}$ is the ratio of the densities of air and water. Many authors including Komen et al. [10] have modified the formulation of $\beta$ to simulate the coupling feedback between waves and wind.

Komen et al. [9], as parameterized in the WAM model of Hasselmann et al. [5] and Komen et al. [10] suggested the formulation of the wave dissipation $S_{ds}$ in the following form

$$S_{ds} = -C_{ds}\left(\frac{\hat{\alpha}}{\hat{\alpha}_{PM}}\right)^2 \left(\frac{\omega}{\overline{\omega}}\right)^2 \overline{\omega} E(f, \theta) \tag{2.57}$$

where $\hat{\alpha} = m_0 \overline{\omega}^4/g^2$, and $m_0$ is the zeroth moment of the variance spectrum, $\overline{\omega}$ is the mean radian frequency

$$\overline{\omega} = \frac{\int\int E(\omega, \theta)d\omega\, d\theta}{E_{\text{total}}}, \tag{2.58}$$

in which $E_{\text{total}}$ is the total spectral energy. Tuning is achieved by a filtering parameter, $C_{ds}$, and $\hat{\alpha}/\hat{\alpha}_{PM}$ is an overall measure of steepness in the wave field.

The empirical formulation for the change in energy due to wave interaction with ice floes, $S_{ice}$, has been described by Isaacson [7], Masson and LeBlond [14]. With the MIZ, the ice term $S_{ice}$ is expressed in terms of a transformation tensor $T_{fl}^{ij}$:

$$S(fl, \theta_i)_{ice} = E(f_e, \theta_j)T_{fl}^{ij} \tag{2.59}$$

where space $\mathbf{x}$ and time $t$ coordinates are important, and summation is over all $j$ angle bands of discretization. The transformation tensor $T_{fl}^{ij}$ is expressed as

$$T_{fl}^{ij} = A^2[\beta|D(\theta_{ij})|^2 \Delta\theta + \delta(\theta_{ij})(1 + |\alpha_c D(0)|^2) + \delta(\pi - \theta_{ij})|\alpha_c D(\pi)|^2] \tag{2.60}$$

where $\delta$ is the Dirac delta function and $\Delta\theta$ is the angular increment in $\theta$: and $\theta_{ij} = |\theta_i - \theta_j|$. Other parameters arising in this formulation have been described by Masson and LeBlond [14], and will not be repeated here.

The formulation of the nonlinear transfer between spectral components due to wave-wave interactions, $S_{nl}$, is rather complicated. It was demonstrated by Hasselmann et al. [5] that the energy spectrum $E(f, \theta)$ is actually proportional to the action spectrum $N(f, \theta)$ such that $E(f, \theta) = \omega N(f, \theta)$, where the proportionality constant is the radian frequency $\omega$. Therefore, Eq. (2.54) can be written in two ways:

For energy:

$$\frac{dE}{dt} = (S_{nl} + S_{ds})_e(1 - f_i) + (S_{nl})_e + (S_{ice})_e \tag{2.61}$$

For action:

$$\frac{dN}{dt} = (S_{nl} + S_{ds})(1 - f_i) + S_{nl} + S_{ice} \tag{2.62}$$

Equation (2.62) is most basic because $S_{nl}$ is expressed in terms of action. We shall proceed with the evaluation of the nonlinear wave-wave interaction $S_{nl}$ with the use of the following approximate nonlinear simultaneous differential equation as described in the preceding paragraph. Hasselmann et al. [5] constructed a nonlinear interaction operator by considering only a small number of neighboring and finite distance interactions. It was found that, in fact, the exact nonlinear transfer could be well simulated by just one mirror-image pair of intermediate range interaction configurations. In each configuration, two wave numbers were taken as identical $\mathbf{k}_1 = \mathbf{k}_2 = \mathbf{k}$. The wave numbers $\mathbf{k}_3$ and $\mathbf{k}_4$ are of different magnitude and lie at an angle to the wave number $\mathbf{k}$, as required by the resonance conditions. The second configuration is obtained from the first by reflecting the wave numbers $\mathbf{k}_3$ and $\mathbf{k}_4$ with respect to the $\mathbf{k}$-axis. (see [10], p. 226). The scale and direction of the reference wave number are allowed to vary continuously in wave number space. For configurations

$$\left.\begin{array}{rcl} \omega_1 = \omega_2 & & = \omega \\ \omega_3 = \omega(1 + \lambda) & = & \omega_+ \\ \omega_4 = \omega(1 - \lambda) & = & \omega_- \end{array}\right\} \tag{2.63}$$

where $\lambda = 0.25$, a constant parameter, satisfactory agreement with exact computation was found. From the resonance conditions, the angles $\theta_3$, $\theta_4$ of the wave numbers $\mathbf{k}_3$ and $(\mathbf{k}_4)$ and $\mathbf{k}_4$ $(\mathbf{k}_-)$ relative to $\mathbf{k}$ are found to be $\theta_3 = 11.5°$ and $\theta_4 = -33.6°$.

The discrete interaction approximation has its most simple form for deep ocean for the rate of change in time of the action density in wave number space. The balanced equations can be written as (see Hasselmann et al. [5])

$$\frac{d}{dt}\left\{\begin{array}{c} N \\ N_+ \\ N_- \end{array}\right\} = \left\{\begin{array}{c} -2 \\ 1 \\ 1 \end{array}\right\} Cg^{-8}f^{19}[N^2(N_+ + N_-) - 2NN_+ N_-]\Delta\mathbf{k} \tag{2.64}$$

where $\frac{dN}{dt}$, $\frac{dN_+}{dt}$, and $\frac{dN_-}{dt}$ are the rate of change in action at wave numbers $\mathbf{k}$, $\mathbf{k}_+$, and $\mathbf{k}_-$, respectively due to the discrete interactions within the infinitesimal interaction phase-space element $\Delta\mathbf{k}$ and $C$ is a numerical constant. The new source function $S_{nl}$ can be obtained by summing Eq. (2.64) over all wave numbers, directions, and interaction configurations. Equation (2.64) is only valid for deep water oceans. Numerical computations by Hasselmann and Hasselmann [4] of the full Boltzmann integral for water of an arbitrary depth have shown that there is an approximate relationship between the transfer rate for deep water and water of finite depth. For a frequency direction spectrum, the transfer for a finite depth ocean is identical to the transfer for infinite depth, except for a scaling factor $R$:

$$S_{nl} \text{ (finite depth)} = R(\bar{k}h)S_{nl} \text{ (infinite depth)} \tag{2.65}$$

where $\bar{k}$ is the mean of the wave numbers. This scaling relation holds in the range $\bar{k}h > 1$, where the scaling factor can be expressed as

$$R(x) = 1 + \frac{5.5}{x}\left(1 - \frac{5x}{6}\right)\exp\left(-\frac{5x}{4}\right) \tag{2.66}$$

with $x = \frac{3}{4}\bar{k}h$. The WAM model uses this approximation.

In this report, we shall try to reexamine this nonlinear transfer function from a different perspective. In the first part of this report, we will demonstrate three easily available methods to compute the nonlinear transfer action function. The first method is due to Picard, the second due to Bernoulli, and the third is a numerical integration using the fourth-order Runge-Kutta scheme. We shall compare these results with numerical computations. The second part of this report will be devoted mainly to determining the analytic solutions for Eulerian currents present in ocean circulation. A variety of solutions with different boundaries and initial conditions will be examined and the physical inference will be derived.

## Nonlinear Wave-Wave Interaction

This section will be devoted to the solution techniques of the nonlinear wave-wave interactions. We have already described the mathematical model valid for deep water oceans in Eq. (2.64) and for finite depth oceans in Eq. (2.65). This problem is of paramount importance to the oceanographer and many previous researchers in this area have tried to obtain simple and formidable solutions, but until now, no one has been successful in achieving this goal. We wish to demonstrate some advancements from the scientific and analytical viewpoints in this report. We will discuss three elegant methods:

- Picard's method of successive approximations (analytical)
- Bernoulli's method of integrating factors (analytical)
- Fourth-order Runge-Kutta numerical scheme

The solutions obtained by these three methods will be compared with available field data and a conclusion will be drawn with respect to their efficiency. For convenience, Eqs. (2.64) are explicitly rewritten with their initial conditions as follows:

$$\frac{dN}{dt} = \alpha_1[N^2(N_+ + N_-) - 2NN_+N_-] \tag{2.67}$$

$$\frac{dN_+}{dt} = \alpha_2[N^2(N_+ + N_-) - 2NN_+N_-] \tag{2.68}$$

$$\frac{dN_-}{dt} = \alpha_3[N^2(N_+ + N_-) - 2NN_+N_-] \tag{2.69}$$

where

$$\left.\begin{array}{l} \alpha_1 = -2Cg^{-8}f^{19}\Delta k \\[2mm] \alpha_2 = Cg^{-8}f^{19}\Delta k_+ \\[2mm] \alpha_3 = Cg^{-8}f^{19}\Delta k_- \end{array}\right\} \tag{2.70}$$

The initial conditions are

$$\text{at} \quad t = 0: \quad \left.\begin{array}{l} N(0, f) = N_0(f) \\[2mm] N_+(0, f) = N_{+0}(f_+) \\[2mm] N_-(0, f) = N_{-0}(f_-) \end{array}\right\} \tag{2.71}$$

The specific algebraic forms of these initial values will be stated later.

## Picard's Method of Successive Approximations

It can be easily observed that Eqs. (2.67), (2.68), and (2.69) are related to each other as follows:

$$\frac{dN_+}{dt} = \left(\frac{\alpha_2}{\alpha_1}\right)\frac{dN}{dt} \tag{2.72}$$

$$\frac{dN_-}{dt} = \left(\frac{\alpha_3}{\alpha_2}\right)\frac{dN}{dt} \tag{2.73}$$

Thus, if we can determine the solution for $N(t, f)$, then the solutions for $N_+(t, f)$ and $N_-(t, f)$ can be easily determined by interaction from Eqs. (2.72) and (2.73). However, we shall integrate Eqs. (2.67), (2.68), and (2.69) using Picard's successive approximation method.

### First Approximation

Since the equation is highly nonlinear, it is not an easy matter to integrate it at one time. In the first approximation we shall replace the source terms on the right-hand side of the equations by their initial values, which will be integrated at once. For instance, Eq. (2.67) can be written as

$$\frac{dN}{dt} = \alpha_1[N_0^2(N_{+0} + N_{-0}) - 2N_0N_{+0}N_{-0}] \tag{2.74}$$

The right-hand side of Eq. (2.74) is a constant and can be integrated immediately, with respect to time from $t = 0$ to $t = t$:

$$N(t, f) = N_0(f) + \alpha_{10}t \tag{2.75}$$

where

$$\alpha_{10} = \alpha_1[N_0^2(N_{+0} + N_{-0}) - 2N_0N_{+0}N_{-0}] \tag{2.76}$$

Similarly, the solutions for Eqs. (2.68) and (2.69) yield:

$$N_+(t, f) = N_{+0}(f_+) + \alpha_{20}t \tag{2.77}$$

$$N_-(t, f) = N_{-0}(f_-) + \alpha_{30}t \tag{2.78}$$

where

$$\alpha_{20} = \alpha_2[N_0^2(N_{+0} + N_{-0}) - 2N_0N_{+0}N_{-0}] \tag{2.79}$$

$$\alpha_{30} = \alpha_3[N_0^2(N_{+0} + N_{-0}) - 2N_0N_{+0}N_{-0}] \tag{2.80}$$

Equations (2.75), (2.77), and (2.78) are the first approximate solutions.

### Second Approximation

To determine the second approximate solutions we shall update the source function by the first approximate solutions and then integrate. To do this we need to calculate the expression $\{N^2(N_+ + N_-) -$

$2NN_+N_-$}, and this yields:

$$
\begin{aligned}
N^2(N_+ + N_-) - 2NN_+N_- &= [N_0^2(N_{+0} + N_{-0}) - 2N_0N_{+0}N_{-0}] \\
&\quad + [N_0^2(\alpha_{20} + \alpha_{30}) + 2\alpha_{10}N_0(N_{+0} + N_{-0}) \\
&\quad - 2\{N_0(\alpha_{30}N_{+0} + \alpha_{20}N_{-0}) + \alpha_{10}N_{+0}N_{-0}\}]t \\
&\quad + [2\alpha_{10}(\alpha_{20} + \alpha_{30})N_0 + \alpha_{10}^2(N_{+0} + N_{-0}) \\
&\quad - 2\{\alpha_{20}\alpha_{30}N_0 + \alpha_{10}(\alpha_{30}N_{+0} + \alpha_{20}N_{-0})\}]t^2 \\
&\quad + [\alpha_{10}^2(\alpha_{20} + \alpha_{30}) - 2\alpha_{10}\alpha_{20}\alpha_{30}]t^3
\end{aligned}
\tag{2.81}
$$

Thus, the differential Eq. (2.67) can be written as

$$
\begin{aligned}
\frac{dN}{dt} &= \alpha_1[N^2(N_+ + N_-) - 2NN_+N_-] \\
&= a_0 + a_1t + a_2t^2 + a_3t^3
\end{aligned}
\tag{2.82}
$$

where

$$
\begin{aligned}
a_0 &= \alpha_1[N_0^2(N_{+0} + N_{-0}) - 2N_0N_{+0}N_{-0}] = \alpha_{10} \\
a_1 &= \alpha_1[(\alpha_{20} + \alpha_{30})N_0^2 + 2\alpha_{10}N_0(N_{+0} + N_{-0}) \\
&\quad - 2\{N_0(\alpha_{30}N_{+0} + \alpha_{20}N_{-0}) + \alpha_{10}N_{+0}N_{-0}\}] \\
a_2 &= \alpha_1[2\alpha_{10}(\alpha_{20} + \alpha_{30})N_0^2 + \alpha_{10}^2(N_{+0} + N_{-0}) \\
&\quad - 2\{\alpha_{20}\alpha_{30}N_0 + \alpha_{10}(\alpha_{30}N_{+0} + \alpha_{20}N_{-0})\}] \\
a_3 &= \alpha_1[\alpha_{10}^2(\alpha_{20} + \alpha_{30}) - 2\alpha_{10}\alpha_{20}\alpha_{30}]
\end{aligned}
\tag{2.83}
$$

Integrating Eq. (2.82) with respect to $t$ from $t = 0$ to $t = t$, we obtain

$$
N(t, f) = N_0(f) + a_0t + a_1\frac{t^2}{2} + a_2\frac{t^3}{3} + a_3\frac{t^4}{4}
\tag{2.84}
$$

It is worth noting that at $t = 0$, $N(0, f) \approx N_0(f)$ and

$$
\frac{dN}{dt}\bigg|_{t=0} = a_0 = \alpha_1[N_0^2(N_{+0} + N_{-0}) - 2N_0N_{+0}N_{-0}] = \alpha_{10}
\tag{2.85}
$$

The integrals for $N_+(t, f)$ and $N_-(t, f)$ are simply

$$
N_+(t, f) = N_{+0}(f_+) + \left(\frac{\alpha_2}{\alpha_1}\right)\left[a_0 + a_1\frac{t^2}{2} + a_2\frac{t^3}{3} + a_3\frac{t^4}{4}\right]
\tag{2.86}
$$

$$
N_-(t, f) = N_{-0}(f_-) + \left(\frac{\alpha_3}{\alpha_1}\right)\left[a_0 + a_1\frac{t^2}{2} + a_2\frac{t^3}{3} + a_3\frac{t^4}{4}\right]
\tag{2.87}
$$

Equations (2.84), (2.85), and (2.86) are the second approximate solutions. These are nonlinear.

**Third Approximation**

In this approximation, we shall update the differential Eq. (2.67) and then Eqs. (2.68) and (2.69) by the new expression $N$, $N_+$, and $N_-$ obtained in Eqs. (2.84), (2.85), and (2.86). The differential Eq. (2.67) becomes

$$
\begin{aligned}
\frac{dN}{dt} &= \alpha_1[N^2(N_+ + N_-) - 2NN_+N_-] \\[6pt]
&= \alpha_1\Bigg[\{N_0^2(N_{+0} + N_{-0}) - 2N_0N_{+0}N_{-0}\} + \left\{\frac{\alpha_2\alpha_3 N_0^2}{\alpha_1} + 2N_0(N_{+0} + N_{-0})\right. \\[6pt]
&\quad\left. - 2\left(N_{+0}N_{-0} + \left(\frac{N_0}{\alpha_1}\right)(\alpha_3 N_{+0} + \alpha_2 N_{-0})\right)\right\}A \\[6pt]
&\quad + \left\{\frac{2\alpha_2\alpha_3 N_0}{\alpha_1} + (N_{+0} + N_{-0}) - 2\left(\frac{N_0\alpha_2\alpha_3}{\alpha_1^2} + \frac{N_{+0}\alpha_3 + N_{-1}\alpha_2}{\alpha_1}\right)\right\}A^2 \\[6pt]
&\quad + \left\{\frac{\alpha_2\alpha_3}{\alpha_1} - \frac{2\alpha_2\alpha_3}{\alpha_1^2}\right\}A^3\Bigg]
\end{aligned}
$$

Or this can be simply written as

$$
\frac{dN}{dt} = \alpha_1[\beta_0 + \beta_1 A + \beta_2 A^2 + \beta_3 A^3] \tag{2.88}
$$

where

$$
\left.
\begin{aligned}
\beta_0 &= N_0^2(N_{+0} + N_{-0}) - 2N_0N_{+0}N_{-0} \\[6pt]
\beta_1 &= 2N_0^2(N_{+0} + N_{-0}) + \frac{\alpha_2\alpha_3 N_2^0}{\alpha_1} \\[6pt]
&\quad - 2\left\{N_{+0}N_{-0} + \left(\frac{N_0}{\alpha_1}\right)(\alpha_3 N_{+0} + \alpha_2 N_{-0})\right\} \\[6pt]
\beta_2 &= (N_{+0} + N_{-0}) + \frac{2N_0\alpha_2\alpha_3}{\alpha_1} \\[6pt]
&\quad - 2\left\{\frac{N_0\alpha_2\alpha_3}{\alpha_1^2} + \frac{\alpha_3 N_{+0} + \alpha_2 N_{-0}}{\alpha_1}\right\} \\[6pt]
\beta_3 &= \frac{\alpha_2\alpha_3}{\alpha_1} - \frac{2\alpha_2\alpha_3}{\alpha_1^2} \quad \text{with} \\[6pt]
A &= a_0 t + a_1\frac{t^2}{2} + a_2\frac{t^3}{3} + a_3\frac{t^4}{4}
\end{aligned}
\right\} \tag{2.89}
$$

Integrating Eq. (2.88) with respect to time $t$, from $t = 0$ to $t = t$, we obtain

$$
N(t, f) = N_0(f) + \alpha_1\left[\beta_0 t + \beta_1\int_0^t A\,dt + \beta_2\int_0^t A^2\,dt + \beta_3\int_0^t A^3\,dt\right] \tag{2.90}
$$

We show the calculations as follows:

$$\int_0^t A\,dt = \int_0^t \left( a_0 t + a_1 \frac{t^2}{2} + a_2 \frac{t^3}{3} + a_3 \frac{t^4}{4} \right) dt$$

$$= a_0 \frac{t^2}{2} + a_1 \frac{t^3}{6} + a_2 \frac{t^4}{12} + a_3 \frac{t^5}{20}$$

$$\int_0^t A^2\,dt = \int_0^t \left( a_0 t + a_1 \frac{t^2}{2} + a_2 \frac{t^3}{3} + a_3 \frac{t^4}{4} \right)^2 dt$$

$$= \frac{a_0^2 t^3}{3} + \frac{a_0 a_1 t^4}{4} + \left( \frac{a_1^2}{20} + \frac{2a_0 a_2}{15} \right) t^5$$

$$+ \left( \frac{a_0 a_3}{12} + \frac{a_1 a_2}{18} \right) t^6 + \left( \frac{a_1 a_3}{28} + \frac{a_2^2}{63} \right) t^7$$

$$+ \frac{a_2 a_3 t^8}{48} + \frac{a_3^2 t^9}{144}$$

$$\int_0^t A^3\,dt = \int_0^t \left( a_0 t + a_1 \frac{t^2}{2} + a_2 \frac{t^3}{3} + a_3 \frac{t^4}{4} \right)^3 dt$$

$$= a_0^3 \frac{t^4}{4} + \frac{3a_0^2 a_1 t^5}{10} + \left( \frac{a_0 a_1^2}{8} + \frac{a_0^2 a_2}{6} \right) t^6$$

$$+ \left( \frac{a_1^3}{56} + \frac{3a_0^2 a_3}{28} + \frac{a_0 a_1 a_2}{7} \right) t^7$$

$$+ \left( \frac{3a_0 a_1 a_3}{32} + \frac{a_1^2 a_2}{32} + \frac{a_0 a_2^2}{24} \right) t^8$$

(2.91)

$$+ \left( \frac{a_0 a_2 a_3}{18} + \frac{a_1 a_2^2}{54} + \frac{a_1^2 a_3}{48} \right) t^9$$

$$+ \left( \frac{3a_0 a_3^2}{160} + \frac{a_1 a_2 a_3}{40} + \frac{a_2^3}{270} \right) t^{10}$$

$$+ \left( \frac{3a_1 a_3^2}{352} + \frac{a_2^2 a_3}{132} \right) t^{11} + \frac{a_2 a_3^2}{192} t^{12}$$

$$+ \frac{a_3^3}{832} t^{13}$$

Equation (2.90) is the third approximate solution and is highly nonlinear in $t$, a polynomial of degree thirteen. The parameters are defined above and they are functions of $N_0$, $N_{+0}$, and $N_{-0}$, i.e., they are functions of frequency $f$. Next we shall demonstrate another approximate analytic method due to Bernoulli.

## Bernoulli's Method of Integrating Factors

It can easily be observed from Eqs. (2.67), (2.68), and (2.69) that if we assume the functions $N_+$ and $N_-$ are constant using the initial values in Eq. (2.71), then these equations submit to Bernoulli's form of nonlinear ordinary differential equations. We know that analytic closed-form solutions exist for these types of ordinary differential equations. Therefore, let us consider

$$\left. \begin{array}{r} 2\alpha_1 N_+ N_- = P(t) \\ \text{and} \quad \alpha_1(N_+ + N_-) = Q(t) \end{array} \right\} \tag{2.92}$$

As the first approximation, we use the initial conditions $N_+$ and $N_-$, which will yield

$$\left. \begin{array}{r} 2\alpha_1 N_{+0} N_{-0} = P_0 = \text{constant} \\ \text{and} \quad \alpha_1(N_{+0} + N_{-0}) = Q_0 = \text{constant} \end{array} \right\} \tag{2.93}$$

So differential Eq. (2.67) can be rewritten in Bernoulli's form as

$$\frac{dN}{dt} + P_0 N = Q_0 N^2 \tag{2.94}$$

with the initial condition

$$t = 0 \qquad N(0, f) = N_0(f) \tag{2.95}$$

Equation (2.94) can be rearranged as

$$\frac{1}{N^2} \frac{dN}{dt} + \frac{P_0}{N} = Q_0 \tag{2.96}$$

by setting

$$y(t) = 1/N \tag{2.97}$$

Equation (2.96) can be written in a simple form

$$\frac{dy}{dt} - P_0 y = -Q_0 \tag{2.98}$$

which is a linear first-order ordinary differential equation and can be integrated at once by the integrating factor method (or by any elementary method) to yield

$$y(t) = \frac{Q_0}{P_0} + ce^{P_0 t} \tag{2.99}$$

where $c$ is an integration constant. Using the initial condition of Eq. (2.95), we obtain the solution as

$$y = \frac{Q_0}{P_0}(1 - e^{P_0 t}) + y_0 e^{P_0 t} \tag{2.100}$$

where $y_0 = 1/N_0$. The solution in terms of $N$ and $N_0$ becomes

$$
\left.
\begin{aligned}
N(t, f) &= \frac{P_0 N_0}{Q_0 N_0 + (P_0 - Q_0 N_0)e^{P_0 t}} \\[2mm]
\frac{dN}{dt} &= -\frac{N_0 P_0^2 (P_0 - N_0 Q_0)e^{P_0 t}}{[Q_0 N_0 + (P_0 - Q_0 N_0)e^{P_0 t}]^2}
\end{aligned}
\right\}
\tag{2.101}
$$

Similarly, solutions for $N_+$ and $N_-$ can be easily written as

$$
N_+(t, f) = N_{+0}(f_+) + \left(\frac{\alpha_2}{\alpha_1}\right)\left(\frac{(N_0 P_0 - Q_0 N_0^2)(1 - e^{P_0 t})}{Q_0 N_0 + (P_0 - Q_0 N_0)e^{P_0 t}}\right)
\tag{2.102}
$$

$$
N_-(t, f) = N_{-0}(f_-) + \left(\frac{\alpha_3}{\alpha_1}\right)\left(\frac{(N_0 P_0 - Q_0 N_0^2)(1 - e^{P_0 t})}{Q_0 N_0 + (P_0 - Q_0 N_0)e^{P_0 t}}\right)
\tag{2.103}
$$

The parameters $P_0$, $Q_0$, $\alpha_1$, $\alpha_2$, and $\alpha_3$ are already defined and they are functions of frequency, $f$. It can be easily shown that

$$
\left.\frac{dN}{dt}\right|_{t=0} = \alpha_1[N_0^2(N_{+0} + N_{-0}) - 2N_0 N_{+0} N_{-0}]
\tag{2.104}
$$

and
$$
\left.\frac{dN}{dt}\right|_{t \to \infty} = 0
\tag{2.105}
$$

The physical implication of this result may be that when $t \to \infty$,

$$
N(\infty, f) \to \left(\frac{2N_{+0} N_{-0}}{(N_{+0} + N_{-0})}\right)
$$

which is a constant and remains the same for large values of time confirming that the time gradient is zero and the curve is asymptotically parallel to the $t$-axis. This is the consequence of the first approximation of Bernoulli's method. We shall try to see this important result graphically.

Note: It can be concluded that the polynomial solution is equivalent to the integrating factor method. However, by the polynomial method, the values $N$ and $\frac{dN}{dt}$ cannot be determined at infinite time, whereas the integrating factor method easily provides these values at $t \to \infty$.

The next approximate solution can be obtained by solving the differential equation

$$
\frac{dy}{dt} - P(t)y = -Q(t)
\tag{2.106}
$$

where $y(t) = 1/N(t)$, and $P(t)$ and $Q(t)$ are functions of $t$ but are not constant. By the use of the integrating factor method, the solution can be written at once in integral form as

$$
y(t) = y_0 e^{\int_0^t P(t)dt} - e^{\int_0^t P(t)dt}\left\{\int_0^t Q(t)e^{-\int_0^t P(t)dt}dt\right\}
\tag{2.107}
$$

where

$$
\left.\begin{array}{l}
P(t) = 2\alpha_1 N_+ N_- \\
\text{and} \quad Q(t) = \alpha_1(N_+ N_-)
\end{array}\right\} \tag{2.108}
$$

The explicit form of $P(t)$ and $Q(t)$ can be written as

$$
P(t) = 2\alpha_1\left(\left(N_{+0} - \frac{\alpha_2 N_0}{\alpha_1}\right)\left(N_{-0} - \frac{\alpha_3 N_0}{\alpha_1}\right)\right.
$$

$$
+ \left\{\left(\frac{\alpha_3 N_{+0} + \alpha_2 N_{-0}}{\alpha_1}\right) - \frac{2\alpha_2\alpha_3 N_0}{\alpha_2^2}\right\}
$$

$$
\times \left\{\frac{P_0 N_0}{Q_0 N_0 + (P_0 - Q_0 N_0)e^{P_0 t}}\right\}
$$

$$
\left.+ \left(\frac{\alpha_2\alpha_3}{\alpha_1^2}\right)\left\{\frac{P_0 N_0}{Q_0 N_0 + (P_0 - Q_0 N_0)e^{P_0 t}}\right\}^2\right) \tag{2.109}
$$

and

$$
Q(t) = \alpha_1\left(\left\{(N_{+0} + N_{-0}) - N_0\left(\frac{\alpha_2 + \alpha_3}{\alpha_1}\right)\right\}\right.
$$

$$
\left.+ \left(\frac{\alpha_2 + \alpha_3}{\alpha_1}\right)\left\{\frac{P_0 N_0}{Q_0 N_0 + (P_0 - Q_0 N_0)e^{P_0 t}}\right\}\right) \tag{2.110}
$$

Exact solutions may not be possible from Eq. (2.107) by using these expressions for $P(t)$ and $Q(t)$. As a result, we do not want to pursue this matter beyond this point. We next demonstrate the fourth-order Runge-Kutta method.

## Numerical Solution Using the Fourth-Order Runge-Kutta Method

This section is devoted to determining the solutions of Eqs. (2.67), (2.68), and (2.69) with their initial conditions [Eq. (2.71)]. We shall very briefly state the numerical scheme used to solve this initial value problem. For a given set of initial conditions, we will try to solve these highly nonlinear first-order ordinary differential equations. The scheme is as follows:

Rewriting Eqs. (2.67), (2.68), and (2.69) in functional form, we have

$$
\frac{dN}{dt} = f(t, N, N_+, N_-) \tag{2.111}
$$

$$
\frac{dN_+}{dt} = g(t, N, N_+, N_-) \tag{2.112}
$$

$$
\frac{dN_-}{dt} = h(t, N, N_+, N_-) \tag{2.113}
$$

where

$$f = \alpha_1[N^2(N_+ + N_-) - 2NN_+N_-] \tag{2.114}$$

$$g = \alpha_2[N^2(N_+ + N_-) - 2NN_+N_-] \tag{2.115}$$

$$h = \alpha_3[N^2(N_+ + N_-) - 2NN_+N_-] \tag{2.116}$$

The fourth-order Runge-Kutta integration scheme implies the solution of the $(j + 1)$th time step as

$$N^{j+1} = N^j + \frac{1}{6}(k_1 + 2k_2 + 2k_3 + k_4) \tag{2.117}$$

$$N_+^{j+1} = N_+^j + \frac{1}{6}(l_1 + 2l_2 + 2l_3 + l_4) \tag{2.118}$$

$$N_-^{j+1} = N_-^j + \frac{1}{6}(m_1 + 2m_2 + 2m_3 + m_4) \tag{2.119}$$

where

$$\left. \begin{aligned} k_1 &= (\Delta t)f(t, N, N_+, N_-) \\ l_1 &= (\Delta t)g(t, N, N_+, N_-) \\ m_1 &= (\Delta t)h(t, N, N_+, N_-) \end{aligned} \right\} \tag{2.120}$$

$$\left. \begin{aligned} k_2 &= (\Delta t)f\left(t + \frac{\Delta t}{2}, N + \frac{k_1}{2}, N_+ + \frac{l_1}{2}, N_- + \frac{m_1}{2}\right) \\ l_2 &= (\Delta t)g\left(t + \frac{\Delta t}{2}, N + \frac{k_1}{2}, N_+ + \frac{l_1}{2}, N_- + \frac{m_1}{2}\right) \\ m_2 &= (\Delta t)h\left(t + \frac{\Delta t}{2}, N + \frac{k_1}{2}, N_+ + \frac{l_1}{2}, N_- + \frac{m_1}{2}\right) \end{aligned} \right\} \tag{2.121}$$

$$\left. \begin{aligned} k_3 &= (\Delta t)f\left(t + \frac{\Delta t}{2}, N + \frac{k_2}{2}, N_+ + \frac{l_2}{2}, N_- + \frac{m_2}{2}\right) \\ l_3 &= (\Delta t)g\left(t + \frac{\Delta t}{2}, N + \frac{k_2}{2}, N_+ + \frac{l_2}{2}, N_- + \frac{m_2}{2}\right) \\ m_3 &= (\Delta t)h\left(t + \frac{\Delta t}{2}, N + \frac{k_2}{2}, N_+ + \frac{l_2}{2}, N_- + \frac{m_2}{2}\right) \end{aligned} \right\} \tag{2.122}$$

and

$$\left. \begin{aligned} k_4 &= (\Delta t)f(t + \Delta t, N + k_3, N_+ + l_3, N_- + m_3) \\ l_4 &= (\Delta t)g(t + \Delta t, N + k_3, N_+ + l_3, N_- + m_3) \\ m_4 &= (\Delta t)h(t + \Delta t, N + k_3, N_+ + l_3, N_- + m_3) \end{aligned} \right\} \tag{2.123}$$

The Ocean Engineering Handbook

Equations (2.117) to (2.119) specify the action transfer in the air-sea momentum exchange. Once the values of $N$, $N_+$, and $N_-$ at the $(j + 1)$th time step have been determined, the time derivative $\left(\frac{dN}{dt}\right)$ can be obtained from the equation

$$\left(\frac{dN}{dt}\right)^{j+1} = \alpha_1[N^2(N_+ + N_-) - 2NN_+N_-]^{j+1}. \tag{2.124}$$

We shall carry out this numerical integration for the range from $t = 0$ to $t = 36{,}000$ seconds. This result will be compared with the analytic solutions obtained in the previous two sections.

## Results and Discussion

We shall present our results from these three methods in graphical and tabular form. We will also compare these results with the results of previous researchers along with the available field data. The initial conditions used in these calculations are as follows:

At $t = 0$, we use the JONSWAP (Joint North Sea Wave Project) Spectrum as the initial condition. The expression of this spectrum is given below for ready reference:

$$N(f) = \alpha g^2 \frac{f^{-5}}{(2\pi)^4} \exp\left\{-\frac{5}{4}\left(\frac{f}{f_p}\right)^{-4}\right\} \gamma^{\exp\left[-\frac{(f-f_p)^2}{2\tau^2 f_p^2}\right]} \tag{2.125}$$

where $\alpha = 0.01$, $\gamma = 3.3$, $\tau = 0.08$, and $f_p = 0.3$. Here, $f_p$ is called the peak frequency of the JONSWAP spectrum.

Similarly, the initial conditions for $N_+$ and $N_-$ are the following:

$$N_+(f_+) = N((1 + \lambda)f) \tag{2.126}$$

$$N_-(f_-) = N((1 - \gamma)f) \tag{2.127}$$

The corresponding spreading of the directional spectrum $N(f, \theta)$ was found to depend primarily on $\frac{f}{f_p}$ (see Hasselmann et al. [5])

$$N(f, \theta) = \frac{\beta}{2}N(f)\operatorname{sech}^2\beta(\theta - \bar{\theta}(f)) \tag{2.128}$$

where $\bar{\theta}$ is the mean wave direction and

$$\beta = \begin{cases} 2.61\left(\frac{f}{f_p}\right)^{1.3} & \text{for} \quad 0.56 < \frac{f}{f_p} < 0.95 \\ 2.28\left(\frac{f}{f_p}\right)^{-1.3} & \text{for} \quad 0.95 < \frac{f}{f_p} < 1.6 \\ 1.24 & \text{otherwise} \end{cases} \tag{2.129}$$

The parameters involved in this problem for a one-dimensional deep ocean case are given below:

$$\alpha_1 = -2Cg^{-8}f^{19}\Delta k$$
$$\alpha_2 = Cg^{-8}f^{19}\Delta k_+$$
$$\alpha_3 = Cg^{-8}f^{19}\Delta k_-$$
$$C = 3 \times 10^7$$
$$g = 9.8 \text{ m/s}^2$$

in which

$$\Delta k = \frac{8\pi^2 f \Delta f}{g}$$

$$\Delta k_+ = \frac{8\pi^2 f(1+\lambda)^2 \Delta f}{g}$$

$$\Delta k_- = \frac{8\pi^2 f(1+\lambda)^2 \Delta f}{g}$$

The computation then continues for the frequency range $f = 0.0$ Hz to $f = 2$ Hz with a step size of $\Delta f = 0.001$. We will present the results in tabular and graphical form. To determine the net total action transfer, $N_{total}$, we use the following formula:

$$N_{total} = \int_{\theta=0}^{2\pi} \int_{f=0}^{\infty} N(f, \theta) df \, d\theta \tag{2.130}$$

But in a practical situation, the limit of the infinite integral takes on finite values

$$N_{total} = \int_{\theta=0}^{2\pi} \int_{f=0}^{2} N(f, \theta) df \, d\theta \tag{2.131}$$

The upper limit of the f-integral is assumed to be 2 Hz, which seems to be a realistic cut-off frequency instead of infinity, with the understanding that the contribution to the integral from 2 to infinity is insignificant. The $N(f)$ at a certain time for different frequencies is highly nonlinear. Thus, to plot $N_{total}$ in the time scale, we first obtain the values at different time scales and then graph these values against the time. From the analytical solutions for $N(f)$ we can find the values for $N_{total}$ using either Gaussian quadrature or the $\frac{3}{8}$ Simpson's rule of integration. This idea will be explored later in the computation. Directional spreading is, for the time being, beyond the scope of this writing. This analysis is valid only for deep water oceans. For finite depths we extend this result using Eq. (2.65). The graphical results of the action transfer function $N(f)$ are plotted in Figs. 2.18, 2.19, and 2.20 using these three methods. Excellent agreements have been noted from these figures.

## Wave-Induced Surface Currents

In this section, we shall first formulate the Eulerian current equations in two dimensions. Given wind fields $U_{10}$ at 10 m reference height, the wave balance Eq. (2.54) gives the two-dimensional spectral wave energy $E(f, \theta)$. The associated Stokes drift is the mean velocity following a fluid particle, and therefore, by definition, is a Lagrangian property. Let us consider that $\mathbf{U}_s$ is the Lagrangian velocity of a particle at initial position $(\mathbf{x}, c, t = 0)$, and the Stokes drift at time $t$ is simply given by

$$\mathbf{U}_s(\mathbf{x}, t) = 4\pi \iint f\mathbf{K}e^{2kc}E(f, \theta) df \, d\theta \tag{2.132}$$

following Huang [6] and Jenkins [8]. The vertical Lagrangian coordinate $c$ corresponds to the usual vertical Eulerian coordinate $z$ at the initial time $t = 0$. The quasi-Eulerian current $\mathbf{U}_E$ satisfies the following partial differential equation as described by Jenkins [8]:

$$\frac{\partial \mathbf{U}_E}{\partial t} + \mathbf{f} \times \mathbf{U}_E = \frac{\partial}{\partial c}\left(\nu \frac{\partial \mathbf{U}_E}{\partial c}\right) - \mathbf{f} \times \mathbf{U}_s$$

$$- 2\pi \int df \int f\mathbf{K}S_{ds} 2kNe^{2kc} d\theta \tag{2.133}$$

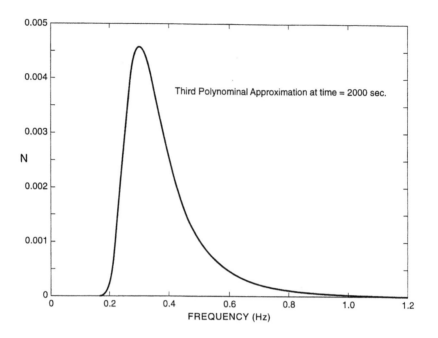

**FIGURE 2.18**   Analytical solution of the action transfer function $N(f)$ vs. the frequency $f$ using Picard's method (third approximation).

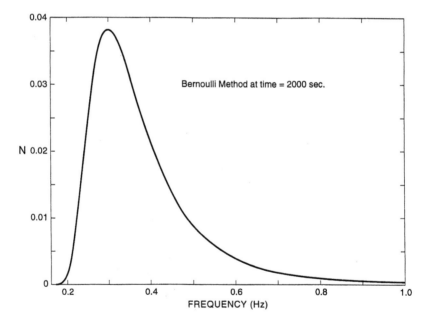

**FIGURE 2.19**   Analytical solution of the action transfer function $N(f)$ vs. the frequency $f$ using Bernoulli's method (first approximation).

where ice floes are not assumed to be present. The quasi-Eulerian current $\mathbf{U}_E$ can be thought of as being the Eulerian mean current $\mathbf{U}_e$ with reference to a Lagrangian coordinate system and so $\mathbf{U}_E = \mathbf{U}_L - \mathbf{U}_s = \vec{U}_e$, where $\mathbf{U}_L$ is the Lagrangian mean-current. Other variables of Eq. (2.133) are $\nu$, the eddy viscosity, and $\mathbf{f}$, the Coriolis acceleration $|\mathbf{f}| = 2\Omega \sin\phi$, where $\Omega$ is the earth's angular velocity and $\phi$ is the latitude.

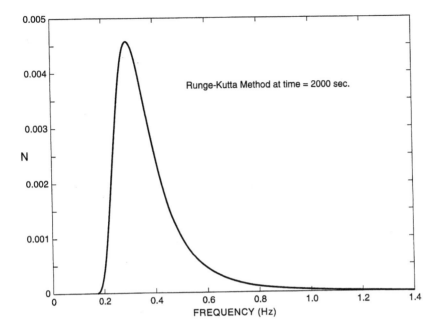

**FIGURE 2.20** Numerical solution of the action transfer function $N(f)$ vs. the frequency $f$ using the fourth order Runge-Kutta Method.

The vector **K** is defined as $\mathbf{K} = (\cos(\frac{\pi}{2} - \theta), \sin(\frac{\pi}{2} - \theta))$, which is related to wave number $k$ by

$$\mathbf{k} = k\mathbf{K} = k[\sin\theta, \cos\theta]$$

The integral expression of Eq. (2.133) represents the generation of $\mathbf{U}_E$ from the waves through the wave dispersion $S_{ds}$. The coefficient $N$ represents momentum transfer from waves to current, and it can be assumed that $N = 1$ (see Jenkins [8]). Finally, the partial derivative term $\frac{\partial}{\partial c}\left(\nu\frac{\partial \vec{U}_E}{\partial c}\right)$ represents the vertical transport of momentum by viscous shear stress. The boundary condition at the sea surface is

$$\nu\frac{\partial \mathbf{U}_E}{\partial c}\bigg|_{c=0} = \frac{\vec{\tau}}{\rho_\omega} - 2\pi\int df\int f\mathbf{K}S_{in}\,d\theta \qquad (2.134)$$

where

$$\tau = \sqrt{\tau_x^2 + \tau_y^2} = |\vec{\tau}|$$
$$= \rho_a U_*^2 = \rho_a C_D |U_{10}^2|$$

is the wind stress on the water surface, $\rho_a$ is the air density, $\rho_\omega$ is the water density, and $C_D$ is the air-water drag coefficient. The wind stress is assumed to be in the direction of the wind. The integral on the right-hand side of Eq. (2.134) represents the momentum transfer from wind into the waves.

The drag coefficient $C_D$, can be parameterized as

$$C_D = \begin{cases} C_s + \left(1.85 - \dfrac{2.24c_p}{(U_{10}\cos(\theta - \theta_{10}))}\right) \times 10^{-3}, & \text{where } \left(\dfrac{c_p}{x_{10}\cos\theta}\right) < 0.82 \\ C_s, & \text{otherwise} \end{cases} \qquad (2.135)$$

where $C_s$ is the open ocean long-fetch drag coefficient, $\theta_{10}$ is the wind direction, $c_p = \frac{g}{2\pi f_p}$ is the phase-speed, and $f_p$ is the spectral peak frequency. For numerical computations, the following information with regard to some other parameters are important. The eddy viscosity $\nu$ is defined as

$$\nu = -0.4 u_*^{\omega} \tag{2.136}$$

between 1 cm and 12 m depth where $u_*^{w}$ is the water friction velocity. The water friction velocity may be expressed as

$$u_*^{\omega} = \sqrt{\frac{T_a}{T_\omega}} U_* \tag{2.137}$$

where $U_*$ is the air friction velocity. The bottom boundary condition is the no-slip condition and is given by

$$U_E = 0 \quad \text{at} \quad c = \begin{cases} -\infty \\ -h \end{cases} \tag{2.138}$$

The Coriolis parameter $|\mathbf{f}|$ was set to $1.07 \times 10^{-4}\ s^{-1}$, corresponding to 50° N latitude. We shall now discuss the solution behavior for Eq. (2.133) under different simplified conditions. Before we proceed with the types of solutions, we will rewrite governing Eq. (2.133) and the boundary conditions, Eqs. (2.134) and (2.138) in terms of respective components as follows. We define

$$
\begin{aligned}
\mathbf{f} &= (0,\ 0,\ \lambda) \\
\mathbf{U}_E &= (u_e,\ v_e,\ 0) \\
\mathbf{U}_s &= (u_s,\ v_s,\ 0) \\
\mathbf{S}_d &= (S_{dx},\ S_{dy},\ 0) \\
\mathbf{S}_\omega &= (S_{\omega x},\ S_{\omega y},\ 0) \\
\vec{\tau} &= (\tau_x,\ \tau_y,\ 0) \\
\mathbf{K} &= \left( \cos\left(\frac{\pi}{2} - \theta\right),\ \sin\left(\frac{\pi}{2} - \theta\right),\ 0 \right) \\
&= (\sin\theta,\ \cos\theta,\ 0)
\end{aligned}
\tag{2.139}
$$

where

$$
\begin{aligned}
\mathbf{S}_d &= 2\pi \int df \int f \mathbf{K} S_{ds} 2kN e^{2kc} d\theta \\
\mathbf{S}_\omega &= 2\pi \int df \int f \mathbf{K} S_{in} d\theta
\end{aligned}
\tag{2.140}
$$

and

$$\mathbf{U}_s = 4\pi \in \int f \mathbf{K} e^{2kc} E(f,\ \theta) d\theta\, df$$

We calculate the cross product as follows:

$$\mathbf{f} \times \mathbf{U}_E = \begin{vmatrix} \mathbf{i} & \mathbf{j} & \mathbf{k} \\ 0 & 0 & \lambda \\ u_e & v_e & 0 \end{vmatrix} = \mathbf{i}(-\lambda v_e) + \mathbf{j}(\lambda u_e) + K(0)$$

$$\mathbf{f} \times \mathbf{U}_s = \begin{vmatrix} \mathbf{i} & \mathbf{j} & \mathbf{k} \\ 0 & 0 & \lambda \\ u_s & v_s & 0 \end{vmatrix} = \mathbf{i}(-\lambda v_s) + \mathbf{j}(\lambda u_s) + K(0)$$

With this information, we can explicitly write the governing equations with their boundary conditions as (considering usual vertical coordinate $z$)

$$\frac{\partial u_e}{\partial t} - \lambda v_e = \frac{\partial}{\partial z}\left(v \frac{\partial u_e}{\partial z}\right) + \lambda v_s - S d_x$$

$$\frac{\partial v_e}{\partial t} + \lambda u_e = \frac{\partial}{\partial z}\left(v \frac{\partial u_e}{\partial z}\right) - \lambda u_s - S d_y$$

(2.141)

The boundary conditions can be written as:
**Surface boundary conditions:**

$$\text{at } z = 0: \quad \left.\begin{array}{l} v \dfrac{\partial u_e}{\partial z} = \dfrac{\tau_x}{\rho_\omega} - S_{x\omega} \\[2mm] v \dfrac{\partial v_e}{\partial z} = \dfrac{\tau_x}{\rho_\omega} - S_{y\omega} \end{array}\right\}$$

(2.142)

**Bottom boundary conditions:**

$$\text{at } z = \left\{\begin{array}{ll} -\infty \text{ (infinite depth)} \\ -h \text{ (finite depth)} \end{array}\right. ; \quad \left.\begin{array}{l} u_e = 0 \\ v_e = 0 \end{array}\right\}$$

(2.143)

The initial condition is assumed to be

$$t = 0: \quad \left.\begin{array}{l} u_e = 0 \\ v_e = 0 \end{array}\right\}$$

(2.144)

Throughout this investigation, we will assume that $u_e$ and $v_e$ are functions of the vertical coordinate $z$ and the time $t$ only; and they do not depend upon the horizontal coordinates $x$ and $y$. We shall start our investigation with a very simple problem of Eulerian current in ocean circulation.

## Unsteady Eulerian Current in One Dimension

### Case I: Deep Ocean

We shall consider first the deep ocean Eulerian current in one dimension. In this case the Coriolis force does not play any part. We assume that the wave dispersion term is negligible. We also assume that the wave input term is negligible and the eddy viscosity $v$ is constant. Equation (2.141) with its boundary

The Ocean Engineering Handbook

conditions, Eqs. (2.142) and (2.143), and initial condition, Eq. (2.144), are as follows:

$$\frac{\partial u_e}{\partial t} = \nu \frac{\partial^2 u_e}{\partial z^2} \tag{2.145}$$

$$z = 0: \qquad \frac{\partial u_e}{\partial z} = \frac{\tau_x}{\nu \rho_\omega} \tag{2.146}$$

$$z = -\infty: \qquad u_e = 0 \tag{2.147}$$

$$\text{and } t = 0: \qquad u_e = 0 \tag{2.148}$$

**Solution:**

The Laplace transform method will be suitable for this problem. Define the Laplace transform of $u_e(z, t)$ as $\mathcal{L}\{u_e(z, t)\} = \int_0^t e^{-st} u_e(z, t)dt$ such that $\mathcal{L}\{\frac{\partial u_e}{\partial t}\} = s\mathcal{L}\{u_e\}$. Equations (2.145) through (2.147) can be transformed as follows:

$$\frac{d^2}{dz^2}\mathcal{L}\{u_e\} - \left(\frac{s}{\nu}\right)\mathcal{L}\{u_e\} = 0 \tag{2.149}$$

$$z = 0: \qquad \frac{d}{dz}\mathcal{L}\{u_e\} = \frac{\tau_x}{\nu \rho_\omega}\left(\frac{1}{s}\right) \tag{2.150}$$

$$z = -\infty: \qquad \mathcal{L}(u_e) = 0 \tag{2.151}$$

The solution to the ordinary differential Eq. (2.149), is simply

$$\mathcal{L}\{u_e\} = Ae^{\sqrt{\frac{s}{\nu}}z} + Be^{-\sqrt{\frac{s}{\nu}}z} \tag{2.152}$$

where $A$ and $B$ are arbitrary constants. The boundary condition, Eq. (2.151), implies that $B$ must be zero and we are left with the solution

$$\mathcal{L}\{u_e\} = Ae^{\sqrt{\frac{s}{\nu}}z} \tag{2.153}$$

We differentiate this equation with respect to $z$ and use the boundary condition in Eq. (2.150) to obtain

$$A = \frac{\tau_x}{\rho_\omega s \sqrt{s\nu}} \tag{2.154}$$

Therefore the solution, Eq. (2.153) becomes

$$\mathcal{L}\{u_e\} = \left(\frac{\tau_x}{\rho_\omega \sqrt{\nu}}\right)\left\{\frac{e^{\sqrt{\frac{s}{\nu}}z}}{s\sqrt{s}}\right\} \tag{2.155}$$

The Laplace inverse formula is given by

$$\mathcal{L}^{-1}\left\{\frac{e^{-a\sqrt{s}}}{s\sqrt{s}}\right\} = 2\sqrt{\frac{t}{\pi}}e^{-\frac{a^2}{4t}} - a\ \mathrm{erfc}\left(\frac{a}{2\sqrt{t}}\right)$$

where

$$\mathrm{erf}(x) = \text{error function of } x$$

$$= \frac{2}{\sqrt{\pi}}\int_0^x e^{-\eta^2}d\eta$$

and $\mathrm{erfc}(x) = $ error complementary function of $x$

$$= \frac{2}{\sqrt{\pi}}\int_x^\infty e^{-\eta^2}d\eta$$

such that $\mathrm{erf}(x) + \mathrm{erfc}(x) = 1$. Thus, using the inversion formula, we obtain the solution to Eq. (2.155) as

$$u_e(z, t) = \left(\frac{\tau_x}{\rho\omega\sqrt{\nu}}\right)\left[2\sqrt{\frac{t}{\pi}}e^{-\frac{z^2}{4\nu t}} + \frac{z}{\sqrt{\nu}}\mathrm{erfc}\left(-\frac{z}{2\sqrt{\nu t}}\right)\right] \qquad (2.156)$$

At the sea surface the Eulerian current takes on the deep water case

$$u_e(0, t) = \frac{2\tau_x}{\rho_\omega}\left(\sqrt{\frac{t}{\pi\nu}}\right). \qquad (2.157)$$

The total momentum transfer per unit area between air and sea can be calculated from the following integral

$$\int_{-\infty}^0 \rho_\omega u_e dz = \left(\frac{\tau_x\rho_\omega}{\rho_\omega\sqrt{\nu}}\right)\left\{2\sqrt{\frac{t}{\pi}}\int_{-\infty}^0 e^{-\frac{z^2}{4\nu t}}dz + \frac{1}{\nu}\int_{-\infty}^0 z\ \mathrm{erfc}\left(-\frac{z}{2\sqrt{\nu t}}\right)dz\right\}$$

$$= \left(\frac{2\tau_x}{\sqrt{\nu\pi}}\right)t\sqrt{\nu\pi} + \left(\frac{\tau_x}{\nu}\right)\int_{-\infty}^0 z\ \mathrm{erfc}\left(-\frac{z}{2\sqrt{\nu t}}\right) \qquad (2.158)$$

$$= 2\tau_x t + \left(\frac{\tau_x}{\nu}\right)\int_{-\infty}^0 z\ \mathrm{erfc}\left(-\frac{z}{2\sqrt{\nu t}}\right)dz$$

The momentum transfer increases linearly with respect to time, and it becomes unbounded for large amounts of time. We will see whether it is a realistic solution. This is a heat conduction type of solution.

## Case II: Finite Depth Ocean

In this case the bottom boundary condition at $z = -h$ where $h$ is the depth of the ocean is given by Eq. (2.147), and the Laplace transform solution, Eq. (2.152), can be assumed in the following manner:

$$\mathcal{L}\{u_e\} = A\cosh\left(\sqrt{\frac{s}{\nu}}z\right) + B\sinh\left(\sqrt{\frac{s}{\nu}}z\right) \qquad (2.159)$$

where $A$ and $B$ are once again arbitrary constants. Differentiating this equation with respect to $z$ yields

$$\frac{d}{dz}\mathcal{L}\{u_e\} = \sqrt{\frac{s}{\nu}}\left[A\sinh\left(\sqrt{\frac{s}{\nu}}z\right) + B\cosh\left(\sqrt{\frac{s}{\nu}}z\right)\right] \tag{2.160}$$

Using the sea-surface and sea-bottom conditions, we obtain

$$B = \frac{\tau_x}{\nu\rho_\omega}\sqrt{\frac{\nu}{s}} = \frac{\tau_x}{\rho_\omega\sqrt{s\nu}} \tag{2.161}$$

and

$$A\cosh\left(\sqrt{\frac{s}{\nu}}h\right) = B\sinh\left(\sqrt{\frac{s}{\nu}}h\right)$$

or

$$A = \frac{\tau_x}{\rho_\omega\sqrt{s\nu}}\left(\frac{\sinh(\sqrt{\frac{s}{\nu}}h)}{\cosh(\sqrt{\frac{s}{\nu}}h)}\right) \tag{2.162}$$

Equation (2.159) then becomes

$$\mathcal{L}\{u_e\} = \left(\frac{\tau_x}{\rho_\omega\sqrt{\nu}}\right)\frac{\sinh(\sqrt{\frac{s}{\nu}}(z+h))}{\sqrt{\nu}\cosh(\sqrt{\frac{s}{\nu}}h)} \tag{2.163}$$

We know the Laplace inverse formula

$$\mathcal{L}^{-1}\left(\frac{\sinh x\sqrt{s}}{\sqrt{s}\cosh a\sqrt{s}}\right) = \frac{2}{a}\sum_{n=1}^{\infty}(-1)^{n-1}e^{-(2n-1)^2\pi^2 t/4a^2}\sin\left(\frac{(2n-1)\pi x}{2a}\right)$$

and applying this inverse in our problem, we have

$$u_e(z,t) = \left(\frac{\tau_x}{\rho_\omega\sqrt{\nu}}\right)\left[\frac{2\sqrt{\nu}}{h}\sum_{n=1}^{\infty}(-1)^{n-1}e^{-\frac{(2n-1)^2\pi^2\nu t}{4h^2}}\sin\left(\frac{(2n-1)\pi(z+h)}{2h}\right)\right] \tag{2.164}$$

This solution is valid for the finite depth ocean. At the ocean surface $z = 0$,

$$u_e(0,t) = \left(\frac{2\tau_x}{\rho_\omega h}\right)\sum_{n=1}^{\infty}(-1)^{n-1}e^{-\frac{(2n-1)^2\pi^2\nu t}{4h^2}}\sin\left(\frac{(2n-1)\pi}{2}\right) \tag{2.165}$$

The total momentum transfer per unit area is given by

$$\int_{-h}^{0} \rho_\omega u_e(z, t)\, dz = \left(\frac{2\tau_x}{h}\right) \sum_{n=1}^{\infty} e^{-\frac{(2n-1)^2 \pi^2 \nu t}{4h^2}} \int_{-h}^{0} \sin\frac{(2n-1)\pi(z+h)}{2h}\, dz$$

$$= \left(\frac{2\tau_x}{h}\right) \sum_{n=1}^{\infty} (-1)^{n-1} \left(\frac{2h}{(2n-1)\pi}\right) e^{-\frac{(2n-1)2\pi^2 \nu t}{4h^2}} \left(1 - \cos\frac{(2n-1)\pi}{2}\right)$$

$$= \left(\frac{4\tau_x}{\pi}\right) \sum_{n=1}^{\infty} (-1)^{n-1} \left(\frac{1}{(2n-1)}\right) e^{-\frac{(2n-1)^2 \pi^2 \nu t}{4h^2}} \tag{2.166}$$

The momentum transfer in Eq. (2.166) implies that it has a finite value when $t$ is very large.

## Case III: Shallow Water Ocean

In this case the Laplace transform solution, Eq. (2.163), should be modified by taking into consideration the fact that $h \to 0$ or $(z + h) \to 0$. So the transform solution becomes, (to the order $O(h^3)$):

$$\mathcal{L}\{u_e\} = \left(\frac{\tau_x}{\rho_\omega\sqrt{\nu}}\right) \frac{\sqrt{\frac{s}{\nu}}(z+h)}{\sqrt{s}\left(1 + \frac{s}{2\nu}h^2\right)}$$

$$= \frac{2\tau_x}{\rho_\omega \nu} \frac{\nu(z+h)}{h^2\left(s + \frac{2\nu}{h^2}\right)} \tag{2.167}$$

$$\mathcal{L}\{u_e\} = \frac{2\tau_x}{\rho_\omega h^2} \frac{z+h}{\left(s + \frac{2\nu}{h^2}\right)}$$

The inverse of this transform is simply

$$u_e(z, t) = \frac{2\tau_x(z+h)}{\rho_\omega h^2} e^{-\frac{2\nu t}{h^2}} \tag{2.168}$$

at the sea surface $z = 0$

$$u_e(0, t) = \frac{2\tau_x}{\rho_\omega h} e^{-\frac{2\nu t}{h^2}}$$

The total momentum transfer per unit area is given by the value of the following integral

$$\int_{-h}^{0} \rho_\omega u_e(z, t)\, dz = \int_{-h}^{0} 2\tau_x \frac{(z+h)}{h^2} e^{-\frac{2\nu t}{h^2}}\, dz$$

$$= \frac{\tau_x e^{-\frac{2\nu t}{h^2}}}{h^2} [z+h]^2 \Big|_{-h}^{0} = \tau_x e^{-\frac{2\nu t}{h^2}}. \tag{2.169}$$

The momentum transfer in this case is a finite quantity and when $h = 0$, there is no momentum transfer between air and sea because there is no water to transfer momentum between these two media.

Note: It is worth mentioning here that the shallow water solution does not satisfy the governing equation, the surface boundary condition, or the initial condition. It only satisfies the bottom boundary condition. To obtain a satisfactory result, an alternate formulation must be adopted as demonstrated by Rahman [18].

## Steady Two-Dimensional Eulerian Currents in Ocean Circulation

### Method of Real Variables

In this section we investigate the steady Eulerian currents when Stokes drift, the transfer function due to wind and the transfer function due to wave dispersion are negligible. We assume that the eddy viscosity is constant. It is interesting to note that the Coriolis acceleration will be dominant in this case. The governing equations with their boundary conditions can be rewritten as

$$
\left.
\begin{aligned}
-\lambda v_e &= \nu \frac{d^2 u_e}{dz^2} \\
+\lambda u_e &= \nu \frac{d^2 v_e}{dz^2}
\end{aligned}
\right\}
\tag{2.170}
$$

The surface boundary conditions are:

$$
z = 0: \quad
\left.
\begin{aligned}
\nu \frac{du_e}{dz} &= \frac{\tau_x}{\rho_\omega} \\
\nu \frac{dv_e}{dz} &= \frac{\tau_y}{\rho_\omega}
\end{aligned}
\right\}
\tag{2.171}
$$

The bottom boundary conditions are :

$$
z =
\left\{
\begin{array}{lll}
-\infty & \text{(infinite depth),} & u_e = 0 \\
-h & \text{(finite depth),} & v_e = 0
\end{array}
\right\}
\tag{2.172}
$$

Eliminating $v_e$ from coupled Eq. (2.170), we obtain a single fourth-order ordinary differential equation in $u_e$ as shown below

$$
\frac{d^4 u_e}{dz^4} + \left(\frac{\lambda}{\nu}\right)^2 u_e = 0
\tag{2.173}
$$

This is our first real variable approach. We show the complex variable approach later.

### Case I: Deep Ocean Currents

In this case the auxiliary equation of Eq. (2.173) can be written as

$$
m^4 + \left(\frac{\lambda}{\nu}\right)^2 = 0
\tag{2.174}
$$

The four roots of this equation are given by

$$m = \beta(1 \pm i), \; -\beta(1 \pm i) \tag{2.175}$$

$$\text{where } \beta^2 = \left(\frac{\lambda}{2\nu}\right) \tag{2.176}$$

The solution for $u_e$ is given by

$$\begin{aligned} u_e(z) &= e^{\beta z}(\alpha_1 \cos \beta z + \alpha_2 \sin \beta z) \\ &+ e^{-\beta z}(\alpha_3 \cos \beta z + \alpha_4 \sin \beta a) \end{aligned} \tag{2.177}$$

The boundary condition at $z \to -\infty$ inplies that $\alpha_3 = 0$ and $\alpha_4 = 0$. Therefore, Eqs. (2.170) yield

$$u_e(z) = e^{\beta z}(\alpha_1 \cos \beta z + \alpha_2 \sin \beta z) \tag{2.178}$$

The dependent variable $v_e$ satisfies the same differential Eq. (2.173) and the solution for $v_e$ satisfing the boundary condition at $z \to -\infty$ can be obtained from Eq. (2.170) as

$$\begin{aligned} v_e &= -\left(\frac{\nu}{\lambda}\right)\frac{d^2 u_e}{dz^2} \\ &= e^{\beta z}(\alpha_1 \sin \beta z - \alpha_2 \cos \beta z) \end{aligned} \tag{2.179}$$

The arbitrary constants $\alpha_1$ and $\alpha_2$ can be determined by using the surface boundary conditions at $z = 0$. The following two equations are obtained:

$$\begin{aligned} \alpha_1 + \alpha_2 &= \frac{\tau_x}{\rho_\omega \beta \nu} \\ \alpha_1 - \alpha_2 &= \frac{\tau_y}{\rho_\omega \nu \beta} \end{aligned} \tag{2.180}$$

Solving for $\alpha_1$ and $\alpha_2$, gives

$$\alpha_1 = \frac{\tau_x + \tau_y}{2\rho_\omega \nu \beta}, \qquad \alpha_2 = \frac{\tau_x - \tau_y}{2\rho_\omega \nu \beta} \tag{2.181}$$

The trajectories of the Eulerian currents in deep ocean are found to be a circular spiral

$$u_e^2 + v_e^2 = r^2 \tag{2.182}$$

where $r$ is the radius and is given by

$$r = \sqrt{\alpha_1^2 + \alpha_2^2} \, e^{\beta z} \tag{2.183}$$

It can be easily seen that at the bottom of the ocean, $r = 0$ and the contour is a point circle. As we move upward along the vertical $z$-axis, the trajectory becomes a pronounced circular orbit with different radii at different $z$-points, and at the surface $z = 0$, the radius of the circle takes its maximum value and is given by $r = \sqrt{\alpha_1^2 + \alpha_2^2}$. This type of phenomenon was discovered by Ekman [1] through experimentation.

Equations (2.178) and (2.179) can be rewritten in a concise form:

$$u_e(z) = \sqrt{\alpha_1^2 + \alpha_2^2}\, e^{\beta z} \cos(\beta z - \theta)$$
$$v_e(z) = \sqrt{\alpha_1^2 + \alpha_2^2}\, e^{\beta z} \sin(\beta z - \theta) \tag{2.184}$$

where $\theta = \tan^{-1}\left(\frac{\alpha_2}{\alpha_1}\right)$ is the phase angle with the amplitude of a sinusoidal wave, i.e., $\sqrt{\alpha_1^2 + \alpha_2^2}\, e^{\beta z}$. It is very important to note that in the case of a deep ocean, the wave number plays a very important role. This parameter, $\beta = \sqrt{\frac{\lambda}{2\nu}}$ predicts how deep the ocean should be. If we set the Coriolis parameter $\lambda = 1.07 \times 10^{-4}\ s^{-1}$, corresponding to 50° N latitude and eddy viscosity $\nu = 10^6\ m^2/s$, then $\beta$ will be of order $\beta = O(10^{-5}/m)$. We know for deep water oceans, $\beta h > \pi$ and so $h > \pi \beta^{-1} = O(10^5 \pi m)$. So with this set of parameters, the depth of the ocean will be on the order of $(10^5 \pi)$ meters. This is very important information for infinite ocean depth (see Lamb [11]).

The horizontal momentum transfers per unit area of the surface are given below.

**x-momentum transfer:**

$$\int_{-\infty}^{0} \rho_\omega u_e(z)\,dz = \rho_\omega \int_{-\infty}^{0} e^{\beta z}(\alpha_1 \cos \beta z + \alpha_2 \sin \beta z)\,dz$$

$$= \alpha_1 \rho_\omega \left[ \frac{e^{\beta z}(\beta \cos \beta z - \beta \sin \beta z)}{2\beta^2} \right]_{-\infty}^{0}$$

$$+ \alpha_2 \rho_\omega \left[ \frac{e^{\beta z}(\beta \sin \beta z - \beta \cos \beta z)}{2\beta^2} \right]_{-\infty}^{0} \tag{2.185}$$

$$= \frac{\alpha_1 \rho_\omega}{2\beta} - \frac{\alpha_2 \rho_\omega}{2\beta} = \frac{\rho_\omega(\alpha_1 - \alpha_2)}{2\beta}$$

**y-momentum transfer:**

$$\int_{-\infty}^{0} \rho_\omega v_e\,dz = \rho_\omega \int_{-\infty}^{0} e^{\beta z}(\alpha_1 \sin \beta z - \alpha_2 \cos \beta z)\,dz$$

$$= \alpha_1 \rho_\omega \left[ \frac{e^{\beta z}(\beta \sin \beta z - \beta \cos \beta z)}{2\beta^2} \right]_{-\infty}^{0}$$

$$\times \alpha_2 \rho_\omega \left[ \frac{e^{\beta z}(\beta \cos \beta z + \beta \sin \beta z)}{2\beta^2} \right]_{-\infty}^{0} \tag{2.186}$$

$$= -\frac{\alpha_1 \rho_\omega}{2\beta} - \frac{\alpha_2 \rho_\omega}{2\beta} = -\frac{\rho_\omega(\alpha_1 - \alpha_2)}{2\beta}$$

### Case II: Finite Depth Ocean Currents

In this case the solution for $u_e(z)$, Eq. (2.177) can be constructed as

$$u_e(z) = (a_1 \cosh \beta z + a_2 \sinh \beta z)(b_1 \cos \beta z + b_2 \sin \beta z) \tag{2.187}$$

where $a_1$, $a_2$, $b_1$, and $b_2$ are arbitrary constants. It can be easily verified that the velocity component $v_e(z)$ can be obtained from Eq. (2.187) using the relation $v_e = \left(\frac{\nu}{\lambda}\right)\frac{d^2 u_e}{dz^2}$, and the expression for $v_e(z)$ is

$$v_e(z) = (a_1 \sinh \beta z + a_2 \cosh \beta z)(b_1 \sin \beta z - b_2 \cos \beta z) \tag{2.188}$$

Using the surface boundary conditions at $z = 0$, we obtain

$$\left.\begin{array}{rcl} a_1 b_2 + a_2 b_1 &=& \dfrac{\tau_x}{\rho_\omega \beta \nu} \\[2ex] -a_1 b_2 + a_2 b_1 &=& \dfrac{\tau_y}{\rho_\omega \beta \nu} \end{array}\right\} \tag{2.189}$$

Thus, the solutions are

$$\left.\begin{array}{rcl} a_1 b_2 &=& \dfrac{\tau_x - \tau_y}{2\rho_\omega \beta \nu} \\[2ex] a_2 b_1 &=& \dfrac{\tau_x + \tau_y}{2\rho_\omega \beta \nu} \end{array}\right\} \tag{2.190}$$

Using the boundary conditions at $z = -h$, we obtain

$$(a_1 \cosh \beta h - a_2 \sinh \beta h)(b_1 \cos \beta h - b_2 \sin \beta h) = 0$$
$$(-a_1 \sinh \beta h + a_2 \cosh \beta h)(-b_1 \sin \beta h - b_2 \cos \beta h) = 0$$

and these two equations can be arranged as

$$a_1 b_1 \cosh \beta h \cos \beta h + a_2 b_2 \sinh \beta h \sin \beta h = \gamma_1$$
$$a_1 b_1 \sinh \beta h \sin \beta h - a_2 b_2 \cosh \beta h \cos \beta h = \gamma_2 \tag{2.191}$$

where $\gamma_1$ and $\gamma_2$ are known and are given by

$$\gamma_1 = a_1 b_2 \cosh \beta h \sin \beta h + a_2 b_1 \sinh \beta h \cos \beta h$$
$$\gamma_2 = -a_1 b_2 \sinh \beta h \cos \beta h + a_2 b_1 \cosh \beta h \sin \beta h \tag{2.192}$$

Solving for $a_1 b_1$ and $a_2 b_2$ by using Cramer's rule, we obtain

$$a_1 b_1 = \dfrac{a_1 b_2 \cosh \beta h \sin \beta h + a_2 b_1 \cosh \beta h \sinh \beta h}{\cosh^2 \beta h - \sin^2 \beta h}$$
$$a_2 b_2 = \dfrac{a_1 b_2 \cosh \beta h \sinh \beta h - a_2 b_1 \cos \beta h \sin \beta h}{\cosh^2 \beta h - \sin^2 \beta h} \tag{2.193}$$

So, in principle all the unknown constants are determined. However, there is one problem: how to obtain the values of individual constants $a_1$, $a_2$, $b_1$, and $b_2$ explicitly. We obtain these values using a little trick.

$$\text{Let}\quad \left.\begin{array}{rcl} a_1 b_1 &=& k_{11} \\ a_1 b_2 &=& k_{12} \\ a_2 b_1 &=& k_{21} \\ a_2 b_2 &=& k_{22} \end{array}\right\} \tag{2.194}$$

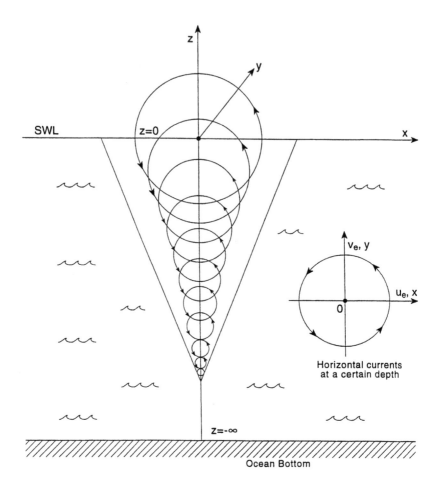

**FIGURE 2.21**   Eulerian currents in circular orbit for a deep ocean.

where $k_{11}$, $k_{12}$, $k_{21}$, and $k_{22}$ are known functional values given by Eqs. (2.190) and (2.193). There are four equations with four unknowns that can be solved. After a little thought, we find that a unique solution exists. Let us choose $a_1 = b_1$ such that $a_1^2 = k_{11}$ and therefore $a_1 = \sqrt{k_{11}}$, so $b_1 = \sqrt{k_{11}}$. Thus, from the second of Eqs. (2.194), $b_2 = \frac{k_{12}}{\sqrt{k_{11}}}$, and from the third, $a_2 = \frac{k_{21}}{\sqrt{k_{11}}}$. Hence, all the values are determined. It is interesting to note that from the fourth part of Eqs. (2.194),

$$\frac{k_{12}k_{21}}{k_{11}} = k_{22} \qquad \text{or} \qquad k_{12}k_{21} = k_{11}k_{22}$$

which is amazingly true from Eq. (2.194). Summarizing the situation, we have

$$\left. \begin{array}{ll} a_1 = \sqrt{k_{11}} & b_1 = \sqrt{k_{11}} \\[2ex] a_2 = \dfrac{k_{21}}{\sqrt{k_{11}}} & b_2 = \dfrac{k_{12}}{\sqrt{k_{11}}} \end{array} \right\} \qquad (2.195)$$

The solutions for $u_e(z)$ and $v_e(z)$ are given by Eqs. (2.187) and (2.188). It is interesting to see the steady Eulerian current trajectories or the orbital path in two-dimensional ocean circulation in a finite depth ocean if there is an influence of Coriolis acceleration $\lambda$ with the eddy viscosity $\nu$. It can easily be verified that the orbital path for a finite depth ocean is an elliptical spiral,

$$\frac{u_e^2}{\left(A\sqrt{b_1^2 + b_1^2}\right)^2} + \frac{v_e^2}{\left(B\sqrt{b_1^2 + b_2^2}\right)^2} = 1 \tag{2.196}$$

where

$$\begin{aligned} A &= a_1 \cosh \beta z + a_2 \sinh \beta z \\ B &= a_1 \sinh \beta z + a_2 \cosh \beta z \end{aligned} \tag{2.197}$$

Here, $A\sqrt{b_1^2 + b_2^2}$ and $B\sqrt{b_1^2 + b_2^2}$ are respectively called the semimajor and semiminor axes of the ellipse, $a_1, a_2, b_1,$ and $b_2$ are constants, but $A$ and $B$ depend upon the vertical coordinate $z$. It can easily be verified that $A$ and $B$ are not equal at any finite depth and so the contour is an ellipse. At the sea surface, $z = 0$, $A = a_1$ and $B = a_2$ where $a_1$ is different from $a_2$ and so the contour of the Eulerian current at the sea surface remains an ellipse.

It can be observed that the trajectories can also be in the form of a hyperbola. The contour for $u_e$ and $v_e$ is obtained as

$$\frac{u_e^2}{\left(A_1\sqrt{a_1^2 - a_2^2}\right)^2} - \frac{v_e^2}{\left(B_1\sqrt{a_1^2 - a_2^2}\right)^2} = 1 \tag{2.198}$$

where

$$\begin{aligned} A_1 &= b_1 \cos \beta z + b_2 \sin \beta z \\ B_1 &= -b_1 \sin \beta z + b_2 \cos \beta z \end{aligned} \tag{2.199}$$

Equation (2.198) can be plotted and the trajectories are shown in Fig. 2.22 at a certain depth.

From the mathematical analysis we have discovered that two-dimensional Eulerian currents produce two types of trajectories in the finite depth ocean. It is interesting to see whether these mathematical conjectures can be proven to be true in real world situations.

### Case III: Shallow Water Ocean
In this case $h \to 0$ or $(z + h) \to 0$, and therefore Eqs. (2.187) and (2.188) are modified, and can be written as (up to $O(z)$ or $O(h)$)

$$\begin{aligned} u_e(z) &= (a_1 + a_2(\beta z))(b_1 + b_2(\beta z)) \\ v_e(z) &= (a_1(\beta z) + a_2)(b_1(\beta z) - b_2) \end{aligned} \tag{2.200}$$

and

$$\left. \begin{aligned} a_1 b_1 &= \beta h (a_1 b_2 + a_2 b_1) \\ a_2 b_2 &= \beta h (a_1 b_2 - a_2 b_1) \\ \text{and} \quad a_1 b_2 &= \frac{\tau_x - \tau_y}{2 \rho_\omega \beta \nu} \\ a_2 b_1 &= \frac{\tau_x + \tau_y}{2 \rho_\omega \beta \nu} \end{aligned} \right\} \tag{2.201}$$

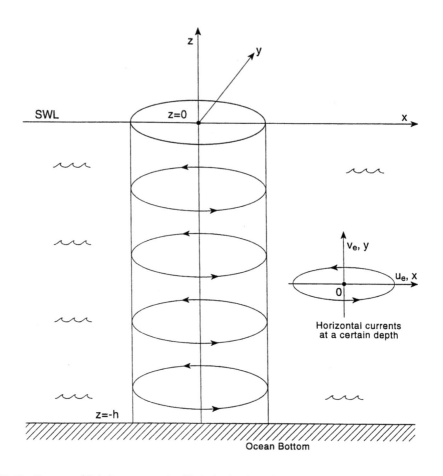

**FIGURE 2.22**  Contour of Eulerian currents in elliptical orbit for a finite depth ocean.

Thus,

$$\left.\begin{aligned}
a_1 b_1 &= \frac{\tau_x h}{\rho_\omega \nu} \\[4pt]
a_2 b_2 &= -\frac{\tau_x h}{\rho_\omega \nu} \\[4pt]
a_1 b_2 &= \frac{\tau_x - \tau_y}{2\rho_\omega \nu} \\[4pt]
a_2 b_1 &= \frac{\tau_x + \tau_y}{2\rho_\omega \beta \nu}
\end{aligned}\right\} \tag{2.202}$$

Also, Eq. (2.200) can be rewritten as (up to $O(z)$):

$$\begin{aligned}
u_e(z) &= a_1 b_1 + (a_1 b_2 + a_2 b_1)(\beta z) \\[4pt]
&= \frac{\tau_x h}{\rho_\omega \nu} + \frac{\tau_x z}{\rho_\omega \nu} \\[4pt]
&= \frac{\tau_x}{\rho_\omega \nu}(z + h)
\end{aligned} \tag{2.203}$$

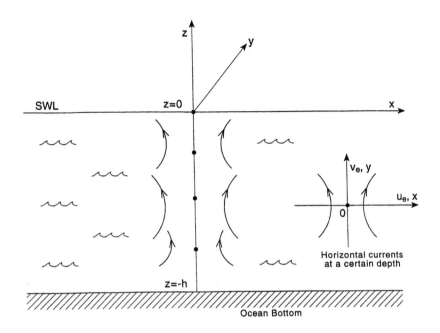

**FIGURE 2.23**   Contours of Eulerian currents in hyperbolic orbit for a finite depth ocean.

$$v_e(z) = -a_2 b_2 + (a_2 b_1 - a_1 b_2)(\beta z)$$

$$= \frac{\tau_y h}{\rho_\omega \nu} + \frac{\tau_y z}{\rho_\omega \nu} \tag{2.204}$$

$$= \frac{\tau_y}{\rho_\omega \nu}(z + h)$$

Hence, for a shallow water ocean the Eulerian currents are the linear profiles with respect to the vertical coordinate $z$. This solution satisfies only the sea bed condition.

**Method of Complex Variables**

In this subsection we will demonstrate a method of complex variables to obtain the solution of the boundary value problem given by Eqs. (2.170) to (2.172). These equations may be combined into a single equation with their boundary conditions if we let

$$\left.\begin{array}{l} u_e + i v_e = w_e \\ \tau_x + i \tau_y = T \end{array}\right\} \tag{2.205}$$

where $w_e$ is a complex function of $z$ and $T$ is a complex quantity. The set of Eqs. (2.170) to (2.172) can be rewritten as

$$i\lambda w_e = \nu \frac{d^2 w_e}{dz^2} \tag{2.206}$$

with surface boundary condition:

$$z = 0: \quad \frac{dw_e}{dz} = \frac{T}{\nu \rho_\omega} \tag{2.207}$$

bottom boundary condition:

$$z = \begin{cases} -\infty \\ -h \end{cases} : \quad w_e = 0 \qquad (2.208)$$

We investigate the solutions for three cases as we did in the previous section by the method of real variables.

### Case I: Deep Ocean

We can write the solution at once in this case, which yields

$$w_e(z) = A e^{\sqrt{i\frac{\lambda}{\nu}z}} + B e^{-\sqrt{i\frac{\lambda}{\nu}z}} \qquad (2.209)$$

The boundary condition at the bottom $(z = -\infty)$ suggests that $B = 0$, and so

$$w_e(z) = A e^{\sqrt{i\frac{\lambda}{\nu}z}}$$

and differentiating with respect to $z$ gives

$$\frac{dw_e}{dz} = \sqrt{i\frac{\lambda}{\nu}}\, A e^{\sqrt{i\frac{\lambda}{\nu}z}}$$

Using the surface boundary condition at $z = 0$, we obtain

$$A = \frac{T}{\rho_\omega \nu} \frac{\sqrt{\nu}}{\sqrt{i\lambda}} = \frac{T}{\rho_\omega \sqrt{i\lambda\nu}} \qquad (2.210)$$

Therefore, the solution is

$$w_e(z) = \frac{T}{\rho_\omega \sqrt{i\lambda\nu}}\, e^{\sqrt{i\frac{\lambda}{\nu}z}} \qquad (2.211)$$

We know that

$$i^{\frac{1}{2}} = \frac{1}{\sqrt{2}}(1 + i)$$

and

$$i^{-\frac{1}{2}} = \frac{1}{\sqrt{2}}(1 - i)$$

and therefore, Eq. (2.211) can be written as

$$w_e(z) = \frac{T(1 - i)}{\rho_\omega \sqrt{2\lambda\nu}}\, e^{(1 + i)\sqrt{\frac{\lambda}{\nu}}z} \qquad (2.212)$$

Let us define $\frac{\lambda}{2\nu} = \beta^2$ such that $\sqrt{2\lambda\nu} = 2\beta\nu$. Equation (2.212) yields

$$w_e(z) = \frac{T(1-i)}{2\rho_\omega\beta\nu}e^{(1+i)\beta z} \tag{2.213}$$

Now equating the real and imaginary parts,

$$
\left.
\begin{aligned}
u_e(z) &= \frac{\sqrt{2(\tau_x^2 + \tau_y^2)}}{2\rho_\omega\beta\nu}e^{\beta z}\cos(\beta z - \theta) \\[2mm]
v_e(z) &= \frac{\sqrt{2(\tau_x^2 + \tau_y^2)}}{2\rho_\omega\beta\nu}e^{\beta z}\sin(\beta z - \theta)
\end{aligned}
\right\} \tag{2.214}
$$

where $\theta = \tan^{-1}(\frac{\tau_x - \tau_y}{\tau_x + \tau_y})$ is the phase angle. These results are identical to Eq. (2.184), and they are obtained without much calculation. The motion is practically confined to a surface stratum whose depth is of order $\beta^{-1}$. The direction of the flow at the surface deviates $\theta = \tan^{-1}(\frac{\tau_x - \tau_y}{\tau_x + \tau_y})$ (if $\tau_y = 0$, $\theta = 45$ degrees) to the right (in the northern hemisphere) from that of the Coriolis force, i.e., the earth's rotational force. The total momentum per unit area of the surface, on the other hand, is

$$
\begin{aligned}
\int_{-\infty}^{0}\rho_\omega(u_e + iv_e)dz &= \int_{-\infty}^{0}\rho_\omega w_e dz \\[2mm]
&= \int_{-\infty}^{0}\frac{T(1-i)}{2\beta\nu}e^{(1+i)\beta z}dz \\[2mm]
&= \frac{T(1-i)}{2\beta^2\nu(1+i)} \\[2mm]
&= -\frac{2Ti}{4\nu\beta^2} = \frac{-Ti}{2\beta^2\nu} = -\frac{Ti}{\lambda}
\end{aligned} \tag{2.215}
$$

Thus the total momentum is

$$
\begin{aligned}
\int_{-\infty}^{0}\rho_\omega w_e(z)dz &= \frac{\tau_y - i\tau_x}{\lambda} \\[2mm]
&= \frac{\sqrt{\tau_x^2 + \tau_y^2}}{\lambda}e^{-i\tan^{-1}\left(\frac{\tau_x}{\tau_y}\right)}
\end{aligned} \tag{2.216}
$$

The direction of this momentum at the surface stratum is at $\tan^{-1}\left(\frac{\tau_x}{\tau_y}\right)$ degrees (at a right angle if $\tau_y = 0$) to that of the Coriolis force. In this investigation we have shown that the steady-state currents known as Eulerian drift can be produced when the Earth's rotation is taken into account.

Here again, it should be noted that the results have practical value only if we replace $\nu$ by a coefficient of turbulence. With the ordinary value of $\nu$ for water, $\beta^{-1}$ would only be of order (see Lamb [11])

$$O\left(\sqrt{\frac{2\nu}{\lambda}}\right) = O\left(\sqrt{\frac{2 \times 1.084 \times 10^{-5}}{1.07 \times 10^{-4}}}\right)$$

$$= O\left(\sqrt{\frac{0.2168}{1.07}}\right) = O(0.45)\text{ ft} = O(14\text{ cm})$$

where $\nu = 1.084 \times 10^{-5}$ ft²/sec and $\lambda = 1/07 \times 10^{-4}$/sec.

### Case II: Finite Depth Ocean

In this case the general Eq. (2.209) has been modified and we rewrite as follows:

$$w_e(z) = A \cosh\left(\sqrt{\frac{i\lambda}{\nu}}z\right) + B \sinh\left(\sqrt{\frac{i\lambda}{\nu}}z\right) \tag{2.217}$$

If we introduce $\beta^2 = \frac{\lambda}{2\nu}$, then this equation can be written as

$$w_e(z) = A \cosh[(1+i)\beta z] + B \sinh[(1+i)\beta z] \tag{2.218}$$

The boundary conditions are:

$$\left.\begin{array}{l} z = 0: \quad \dfrac{dw_e}{dz} = \dfrac{T}{\rho_\omega \nu} \\[2mm] z = -h: \quad w_e = 0 \end{array}\right\} \tag{2.219}$$

Using these two conditions to determine $A$ and $B$, we obtain the final solution as

$$w_e(z) = \frac{T(1-i)}{2\rho_\omega \beta \nu} \frac{\sinh[(1+i)\beta(z+h)]}{\cosh[(1+i)\beta h]} \tag{2.220}$$

Equating the real and imaginary parts we obtain the Eulerian currents in finite depth

$$u_e(z) = \frac{P(z)\cosh\beta h\cos\beta h + Q(z)\sinh\beta h\sin\beta h}{2\rho_\omega \beta \nu(\cosh^2\beta h - \sin^2\beta h)} \tag{2.221}$$

$$v_e(z) = \frac{Q(z)\cosh\beta h\cos\beta h - P(z)\sinh\beta h\sin\beta h}{2\rho_\omega \beta \nu(\cosh^2\beta h - \sin^2\beta h)} \tag{2.222}$$

where

$$\begin{aligned} P(z) = {} & (\tau_x + \tau_y)\sinh[\beta(z+h)]\cos[\beta(z+h)] \\ & + (\tau_x - \tau_y)\cosh[\beta(z+h)]\sin[\beta(z+h)] \end{aligned} \tag{2.223}$$

$$\begin{aligned} Q(z) = {} & (\tau_x + \tau_y)\cosh[\beta(z+h)]\sin[\beta(z+h)] \\ & -(\tau_x - \tau_y)\sinh[\beta(z+h)]\cos[\beta(z+h)] \end{aligned} \tag{2.224}$$

These solutions are identical to those obtained by the method of real variables.

### Case III: Shallow Water Ocean

In this case $\beta h \to 0$ or $\beta(z+h) \to 0$, and using these approximations

$$\begin{aligned} P(z) &\approx (\tau_x + \tau_y)\beta(z+h) + (\tau_x - \tau_y)\beta(z+h) \\ &= 2\tau_x\beta(z+h) \\ Q(z) &\approx (\tau_x + \tau_y)\beta(z+h) - (\tau_x - \tau_y)\beta(z+h) \\ &= 2\tau_y\beta(z+h). \end{aligned}$$

Thus, the Eulerian currents are:

$$u_e(z) = \frac{2\tau_x \beta(z+h) + 2\tau_y \beta(z+h)[\beta h]^2}{2\rho_\omega \beta \nu}$$

$$v_e(z) = \frac{2\tau_y \beta(z+h) - 2\tau_x \beta(z+h)[\beta h]^2}{2\rho_\omega \beta \nu}$$

and up to $O(\beta h)$:

$$\left.\begin{array}{l} u_e(z) \approx \dfrac{\tau_x}{\rho_\omega \nu}(z+h) \text{ and} \\[2ex] v_e(z) \approx \dfrac{\tau_y}{\rho_\omega \nu}(z+h) \end{array}\right\} \tag{2.225}$$

## Time Dependent Eulerian Currents Under the Influence of the Earth's Rotation

In this section we investigate the Eulerian currents in ocean circulation for the unsteady state situation. We will assume that Coriolis acceleration has an important influence on these currents. We neglect the effects of Stokes' drift, wave dispersion, and the momentum transfer from wind to waves. Let us consider that

$$u_e + iv_e = w_e$$
$$\tau_x + i\tau_y = T$$

as defined previously in Eq. (2.205). Equations (2.141) to (2.144) can be rewritten as

$$\frac{\partial w_e}{\partial t} + i\lambda w_e = \nu \frac{\partial^2 w_e}{\partial z^2} \tag{2.226}$$

Boundary conditions:

$$z = 0: \quad \frac{\partial w_e}{\partial t} = \frac{T}{\rho_\omega \nu} \tag{2.227}$$

$$z = \left\{ \begin{array}{l} -\infty \\ -h \end{array} \right. \quad w_e = 0$$

Initial conditions:

$$t = 0 \quad w_e = 0 \tag{2.228}$$

The complex Eulerian current $w_e$ is a function of the vertical coordinate $z$, and time $t$ and the system is linear. Therefore, the Laplace transform method is suitable. We shall study the solutions for three cases as before starting with a deep water ocean.

**Case I: Deep Ocean**

We define the Laplace transform of $w_e(z, t)$ as

$$\mathcal{L}\{w_e\} = \int_0^\infty e^{-st} w_e(z, t) dt \quad \text{such that}$$

$$\mathcal{L}\left\{\frac{\partial w_e}{\partial t}\right\} = s\mathcal{L}\{w_e\} - w_e(z, 0)$$

In this problem the initial value of $w_e$ is given to be zero. Thus, the Laplace transform of Eqs. (2.226) to (2.228) yields:

$$\frac{d^2}{dz^2}\mathcal{L}\{w_e\} - \left(\frac{s + i\lambda}{\nu}\right)\mathcal{L}\{w_e\} = 0 \tag{2.229}$$

$$z = 0: \quad \frac{d}{dz}\mathcal{L}\{w_e\} = \frac{T}{\rho_\omega \nu}\frac{1}{s} \tag{2.230}$$

$$z = -\infty: \quad \mathcal{L}\{w_e\} = 0 \tag{2.231}$$

The solution of ordinary differential Eq. (2.229) is

$$\mathcal{L}\{w_e\} = Ae^{\sqrt{\frac{s + i\lambda}{\nu}}z} + Be^{-\sqrt{\frac{s + i\lambda}{\nu}}z} \tag{2.232}$$

The boundary condition at $z = -\infty$ implies that for a bounded solution we must have $B = 0$. So, Eq. (2.232) reduces to

$$\mathcal{L}\{w_e\} = Ae^{\sqrt{\frac{s + i\lambda}{\nu}}z} \tag{2.233}$$

Using the surface boundary condition at $z = 0$, we obtain the transformed solution

$$\mathcal{L}\{w_e\} = \frac{T}{\rho_\omega\sqrt{\nu}}\left(\frac{e^{\sqrt{\frac{s + i\lambda}{\nu}}z}}{s\sqrt{s + i\lambda}}\right) \tag{2.234}$$

The inverse of Eq. (2.234) yields

$$w_e(z, t) = \frac{T}{\rho_\omega\sqrt{\nu}}\mathcal{L}^{-1}\left\{\frac{e^{\sqrt{\frac{s + i\lambda}{\nu}}z}}{s\sqrt{s + i\lambda}}\right\}$$

$$= \left(\frac{Te^{-i\lambda t}}{\rho_\omega\sqrt{\nu}}\right)\mathcal{L}^{-1}\left\{\frac{e^{\sqrt{\frac{s}{\nu}}z}}{(s - i\lambda)\sqrt{s}}\right\} \tag{2.235}$$

We know from the Laplace inverse table [17] that

$$\mathcal{L}^{-1}\left\{\frac{e^{-k\sqrt{s}}}{s\sqrt{s}}\right\} = 2\sqrt{\frac{t}{\pi}}e^{-\frac{k^2}{4t}} - k\,\text{erfc}\left(\frac{k}{2\sqrt{t}}\right)$$

and also

$$\mathcal{L}\left\{\frac{e^{-k\sqrt{s}}}{\sqrt{s}}\right\} = \frac{1}{\sqrt{\pi t}}e^{-\frac{k^2}{4t}}$$

To get the inverse of Eq. (2.235) requires the use of the convolution integral such that

$$w_e(z, t) = \left(\frac{Te^{-i\lambda t}}{\rho_\omega\sqrt{\nu}}\right)\int_0^t e^{i\lambda(t-\tau)}\left\{\frac{1}{\sqrt{\pi\tau}}e^{-\frac{z^2}{4\nu\tau}}\right\}d\tau$$

$$= \frac{T}{\rho_\omega\sqrt{\pi\nu}}\int_0^t e^{-i\lambda\tau}\frac{1}{\tau}e^{-\frac{z^2}{4\nu\tau}}d\tau \tag{2.236}$$

By equating the real and imaginary parts, we obtain

$$u_e(z, t) = \left(\frac{\sqrt{\tau_x^2 + \tau_y^2}}{\rho_\omega\sqrt{\pi\nu}}\right)\int_0^t \cos(\lambda\tau - \theta)\frac{e^{-\frac{z^2}{4\nu\tau}}}{\sqrt{\tau}}d\tau \tag{2.237}$$

$$v_e(z, t) = \left(\frac{\sqrt{\tau_x^2 + \tau_y^2}}{\rho_\omega\sqrt{\pi\nu}}\right)\int_0^t \sin(\lambda\tau - \theta)\frac{e^{-\frac{z^2}{4\nu\tau}}}{\sqrt{\tau}}d\tau \tag{2.238}$$

where $\theta = \tan^{-1}\left(\frac{\tau_y}{\tau_x}\right)$. If $\tau_y = 0$, then $\theta = 0$.

## Case II: Finite Depth Ocean

In this case the exponential Eq. (2.232) is modified to yield

$$\mathcal{L}\{w_e\} = A\cosh\left(\frac{\sqrt{s+i\lambda}}{\nu}z\right) + B\sinh\left(\frac{\sqrt{s+i\lambda}}{\nu}z\right) \tag{2.239}$$

Differentiating Eq. (2.239) with respect to $z$ and using the surface boundary condition at $z = 0$, we obtain

$$B = \frac{T}{\rho_\omega\sqrt{\nu s}\sqrt{s+i\lambda}} \tag{2.240}$$

and at $z = -h$, $\mathcal{L}\{w_e\} = 0$ gives

$$A = B\frac{\sinh\left(\frac{\sqrt{s+i\lambda}}{\nu}h\right)}{\cosh\left(\frac{\sqrt{s+i\lambda}}{\nu}h\right)} \tag{2.241}$$

Therefore, the solution can be put as

$$\mathcal{L}\{w_e\} = \left(\frac{T}{\rho_\omega\sqrt{\nu}}\right)\frac{\sinh\left(\frac{\sqrt{s+i\lambda}}{\nu}(z+h)\right)}{s\sqrt{s+i\lambda}\cosh\left(\frac{\sqrt{s+i\lambda}}{\nu}h\right)} \tag{2.242}$$

The inverse of this transform is simply

$$
w_e(z, t) = \left(\frac{Te^{-i\lambda t}}{\rho_w\sqrt{\nu}}\right)\mathcal{L}^{-1}\left\{\frac{\sinh\left(\sqrt{\frac{s}{\nu}}(z + h)\right)}{(s - i\lambda)\sqrt{s}\cosh\left(\sqrt{\frac{s}{\nu}}h\right)}\right\} \tag{2.243}
$$

We know from the Laplace inverse table that

$$
\mathcal{L}^{-1}\left\{\frac{\sinh x\sqrt{s}}{\sqrt{s}\cosh a\sqrt{s}}\right\} = \frac{2}{a}\sum_{n=1}^{\infty}(-1)^{n-1}e^{-(2n-1)^2\pi^2 t/4a^2}\sin\left(\frac{(2n-1)\pi t}{2a}\right)
$$

and using this formula in conjunction with the convolution integral,

$$
w_e(z, t) = \frac{T}{\rho_w h}\int_0^t e^{-i\lambda\tau}\left\{\sum_{n=1}^{\infty}(-1)^{n-1}e^{-\frac{(2n-1)^2\pi^2\nu\tau}{4h^2}}\sin\left(\frac{(2n-1)\pi(z+h)}{2h}\right)\right\}d\tau \tag{2.244}
$$

Equating real and imaginary parts:

$$
u_e(z, t) = \left(\frac{2}{\rho_w h}\right)\int_0^t (\tau_x\cos\lambda\tau + \tau_y\sin\lambda\tau)
$$

$$
\times \sum_{n=1}^{\infty}(-1)^{n-1}e^{-\frac{(2n-1)^2\pi^2\nu\tau}{4h^2}}\sin\left(\frac{(2n-1)\pi(z+h)}{2h}\right)d\tau \tag{2.245}
$$

$$
v_e(z, t) = \left(\frac{2}{\rho_w h}\right)\int_0^t (\tau_y\cos\lambda\tau - \tau_x\sin\lambda\tau)
$$

$$
\times \sum_{n=1}^{\infty}(-1)^{n-1}e^{-\frac{(2n-1)^2\pi^2\nu\tau}{4h^2}}\sin\left(\frac{(2n-1)\pi\nu(z+h)}{2h}\right)d\tau \tag{2.246}
$$

Equations (2.245) and (2.246) and can be compactly written as

$$
u_e(z, t) = \left(\frac{2\sqrt{\tau_x^2 + \tau_y^2}}{\rho_w h}\right)\int_0^t \cos(\lambda\tau - \theta)f(z, \tau)d\tau \tag{2.247}
$$

$$
v_e(z, t) = \left(\frac{-2\sqrt{\tau_x^2 + \tau_y^2}}{\rho_w h}\right)\int_0^t \sin(\lambda\tau - \theta)f(z, \tau)d\tau \tag{2.248}
$$

where

$$
\theta = \tan^{-1}\left(\frac{\tau_y}{\tau_x}\right)
$$

$$
f(z, t) = \sum_{n=1}^{\infty}(-1)^{n-1}e^{-\frac{(2n-1)^2\pi^2\nu\tau}{4h^2}}\sin\left(\frac{(2n-1)\pi(z+h)}{2h}\right) \tag{2.249}
$$

These are the solutions for the unsteady Eulerian currents produced under the influence of the Earth's rotation.

## Case III: Shallow Water Ocean

In this case Eq. (2.243) must be modified by considering that $h \to 0$ and $(z + h) \to 0$, and keeping the solution up to $O(h)$, we obtain

$$
w_e(z, t) = \left( \frac{Te^{-i\lambda t}}{\rho_\omega \sqrt{\nu}} \right) \mathcal{L}^{-1} \left\{ \frac{\sqrt{\frac{s}{\nu}}(z + h)}{(s - i\lambda)\sqrt{s}} \right\}
$$

$$
= \frac{T(z + h)}{\rho_\omega \nu}
$$

and therefore,

$$
\left. \begin{array}{l}
u_e(z, t) = \dfrac{\tau_x}{\rho_\omega \nu}(z + h) \\[4mm]
v_e(z, t) = \dfrac{\tau_y}{\rho_\omega \nu}(z + h)
\end{array} \right\} \tag{2.250}
$$

These solutions can be improved if we consider the following approximation (up to $O(h^2)$):

$$
w_e(z, t) = \frac{Te^{-i\lambda t}}{\rho_\omega \sqrt{\nu}} \mathcal{L}^{-1} \left\{ \frac{\sqrt{\frac{s}{\nu}}(z + h)}{(s - i\lambda)\sqrt{s}\left(1 + \frac{sh^2}{2\nu}\right)} \right\}
$$

$$
= \frac{2Te^{-i\lambda t}}{\rho_\omega h^2} \mathcal{L}^{-1} \left\{ \frac{(z + h)}{(s - i\lambda)\left(s + \frac{2\nu}{h^2}\right)} \right\}
$$

$$
= \frac{2T(z + h)}{\rho_\omega(2\nu + i\lambda h^2)} \left\{ 1 - e^{-\left(i\lambda + \frac{2\nu}{h^2}\right)t} \right\}
$$

Equating the real and imaginary parts:

$$
u_e(z, t) = \frac{2(z + h)}{\rho_\omega(4\nu^2 - i\lambda h^4)} \left\{ (2\nu\tau_x + \lambda\tau_y h^2)\left( 1 - \cos\lambda t\, e^{-\frac{2\nu t}{h^2}} \right) \right.
$$

$$
\left. + (\lambda\tau_x h^2 - 2\nu\tau_y)\sin\lambda t\, e^{-\frac{2\nu t}{h^2}} \right\} \tag{2.251}
$$

$$
v_e(z, t) = \frac{2(z + h)}{\rho_\omega(4\nu^2 - \lambda h^4)} \left\{ (2\nu\tau_x + \lambda\tau_y h^2)\sin\lambda t\, e^{-\frac{2\nu t}{h^2}} \right.
$$

$$
\left. - (\lambda\tau_x h^2 - 2\nu\tau_y)\left( 1 - \cos\lambda t\, e^{-\frac{2\nu t}{h^2}} \right) \right\} \tag{2.252}
$$

These shallow water solutions appear to be much improved. However, the reader should be warned that they only satisfy the bottom boundary condition. Even the governing equations are not satisfied.

## Effects of the Earth's Rotation, Stokes Drift, Energy Due to Wind, and Energy Dissipation Due to Breaking of Waves on Unsteady Eulerian Currents

In this section we will study the full set of Eqs. (2.141) to (2.144). In doing so we shall assume that Stokes drift, energy dissipation due to waves, and energy due to wind stress are all constant parameters. It is interesting to note that, with these assumptions, analytical solutions exist. We define the following quantities to produce only one dependent variable:

$$\text{Let} \quad \left.\begin{array}{c} u_e + iv_e = w_e \\ u_s + iv_s = w_s \\ S_{dx} + iS_{dy} = S_d \\ S_{wx} + iS_{wy} = S_w \\ \tau_x + i\tau_y = T \end{array}\right\} \tag{2.253}$$

The complex Eulerian current $w_e$ is a function of $z$ and $t$. Equations (2.141) to (2.144) can be rewritten as shown below.

The governing equation:

$$\frac{\partial w_e}{\partial t} + i\lambda w_e = \nu\frac{\partial^2 w_e}{\partial z^2} - i\lambda w_s - S_d \tag{2.254}$$

The surface boundary condition:

$$z = 0: \quad \frac{\partial w_e}{\partial z} = \frac{T - \rho_\omega S_w}{\rho_\omega \nu} \tag{2.255}$$

The bottom boundary condition:

$$z = \begin{cases} -\infty: \\ -h: \end{cases} \quad w_e = 0 \tag{2.256}$$

The initial condition:

$$t = 0: \quad w_e = 0 \tag{2.257}$$

We use the Laplace transform method defined by

$$\mathcal{L}\{w_e\} = \int_0^\infty e^{-st} w_e(z, t)\,dt$$

and

$$\mathcal{L}\left\{\frac{\partial w_e}{\partial t}\right\} = s\mathcal{L}\{w_e\} - w_e(z, 0)$$

Equations (2.254) to (2.257) can be transformed as

$$\frac{d^2}{dz^2}\mathcal{L}\{w_e\} - \left(\frac{s + i\lambda}{\nu}\right)\mathcal{L}\{w_e\} = \frac{i\lambda w_s + s_d}{\nu s} \tag{2.258}$$

$$z = 0: \quad \frac{d}{dz}\mathcal{L}\{w_e\} = \frac{T - \rho_\omega S_w}{\rho_\omega \nu s} \tag{2.259}$$

$$z = \begin{cases} -\infty: & \mathcal{L}\{w_e\} = 0 \\ -h: \end{cases} \tag{2.260}$$

As before, we will investigate three cases.

## Case I: Deep Ocean

The solution to Eq. (2.258) is given by

$$\mathcal{L}\{w_e\} = Ae^{\sqrt{\frac{s + i\lambda}{\nu}}z} + Be^{-\sqrt{\frac{s + i\lambda}{\nu}}z} - \frac{i\lambda w_s + s_d}{s(s + i\lambda)} \tag{2.261}$$

As we are interested in the convergent solution, then $B$ must be zero to satisfy the boundary condition at $z = -\infty$. Hence Eq. (2.261) reduces to

$$\mathcal{L}\{w_e\} = Ae^{\sqrt{\frac{s + i\lambda}{\nu}}z} - \frac{i\lambda w_s + s_d}{s(s + i\lambda)} \tag{2.262}$$

Differentiating with respect to $z$ and using Eq. (2.259), we get

$$A = \frac{T - \rho_\omega S_w}{\rho_\omega \sqrt{\nu s}\sqrt{s + i\lambda}}$$

and therefore, Eq. (2.262) can be written as

$$\mathcal{L}\{w_e\} = \frac{(T - \rho_\omega S_w)e^{\sqrt{\frac{s + i\lambda}{\nu}}z}}{\rho_\omega \sqrt{\nu s}\sqrt{s + i\lambda}} - \frac{i\lambda w_s + s_d}{s(s + i\lambda)} \tag{2.263}$$

The inverse transform is obtained by using the Laplace transform table in conjunction with the convolution integral as

$$w_e(z, t) = \left(\frac{T - \rho_\omega S_w}{\rho_\omega \sqrt{\nu}}\right)e^{-i\lambda t}\mathcal{L}^{-1}\left\{\frac{e^{\sqrt{\frac{s}{\nu}}z}}{(s - i\lambda)\sqrt{s}}\right\} - \mathcal{L}^{-1}\left\{\frac{i\lambda w_s + S_d}{s(s + i\lambda)}\right\} \tag{2.264}$$

$$= \left(\frac{T - \rho_\omega S_w}{\rho_\omega \sqrt{\nu}}\right)e^{-i\lambda t}\int_0^t e^{-i\lambda(t - \tau)}\frac{1}{\sqrt{\pi\tau}}e^{-\frac{z^2}{4\nu\tau}}d\tau - \left(\frac{i\lambda w_s + S_d}{i\lambda}\right)[1 - e^{-i\lambda t}] \tag{2.265}$$

Equating real and imaginary parts

$$u_e(z, t) = \left(\frac{2\gamma}{\rho_\omega \sqrt{\pi\nu}}\right)\int_0^{\sqrt{t}} \cos(\lambda\beta^2 - \theta)e^{-\frac{z^2}{4\nu\beta^2}}d\beta$$

$$- \frac{1}{\lambda}[(1 - \cos\lambda t)(\lambda u_s + S_{dy}) - (\lambda v_s - S_{dx})\sin\lambda t] \tag{2.266}$$

$$v_e(z, t) = \left(\frac{2\gamma}{\rho_\omega \sqrt{\pi \nu}}\right)\!\int_0^{\sqrt{t}} \sin(\lambda\beta^2 - \theta)e^{-\frac{z^2}{4\nu\beta^2}}d\beta$$

$$-\frac{1}{\lambda}[(1 - \cos\lambda t)(\lambda\nu_s - S_{dx}) - (\lambda u_s + S_{dy})\sin\lambda t] \qquad (2.267)$$

where

$$\theta = \tan^{-1}\left(\frac{T_y - \rho_\omega S_{wy}}{T_x - \rho_\omega S_{wx}}\right) \qquad (2.268)$$

$$\gamma = \sqrt{(T_x - \rho_\omega S_{wx})^2 + (T_y - \rho_\omega S_{wy})^2} \qquad (2.269)$$

These are the Eulerian currents in deep water ocean circulation.

## Case II: Finite Depth Ocean

In this case we consider Eq. (2.261) in the following form:

$$\mathcal{L}\{w_e\} = A\cosh\left(\sqrt{\frac{s + i\lambda}{\nu}}z\right) + B\sinh\left(\sqrt{\frac{s + i\lambda}{\nu}}z\right) - \frac{i\lambda w_s + S_d}{s(s + i\lambda)} \qquad (2.270)$$

Differentiating with respect to $z$:

$$\frac{d}{dz}\mathcal{L}\{w_e\} = \sqrt{\frac{s + i\lambda}{\nu}}\left[A\sinh\left(\sqrt{\frac{s + i\lambda}{\nu}}z\right) + B\cosh\left(\sqrt{\frac{s + i\lambda}{\nu}}z\right)\right] \qquad (2.271)$$

Using the boundary conditions at $z = 0$, we obtain

$$B = \frac{T - \rho_\omega S_w}{\rho_\omega\sqrt{\nu s(s + i\lambda)}} \qquad (2.272)$$

and using the bottom boundary condition at $z = -h$,

$$A\cosh\left(\sqrt{\frac{s + i\lambda}{\nu}}h\right) - B\sinh\left(\sqrt{\frac{s + i\lambda}{\nu}}h\right) = \frac{i\lambda w_s + S_d}{s(s + i\lambda)} \qquad (2.273)$$

Solving for $A$,

$$A = \frac{(T - \rho_\omega S_w)\sinh\left(\sqrt{\frac{s + i\lambda}{\nu}}h\right)}{\rho_\omega\sqrt{\nu s(s + i\lambda)}\cosh\left(\sqrt{\frac{s + i\lambda}{\nu}}h\right)} + \frac{i\lambda w_s + S_d}{s(s + i\lambda)\cosh\left(\sqrt{\frac{s + i\lambda}{\nu}}h\right)} \qquad (2.274)$$

Therefore, the transformed solution becomes

$$\mathcal{L}\{w_e\} = \frac{(T - \rho_\omega S_w)}{\rho_\omega s\sqrt{\nu(s + i\lambda)}}\frac{\sinh\left(\sqrt{\frac{s + i\lambda}{\nu}}(z + h)\right)}{\cosh\left(\sqrt{\frac{s + i\lambda}{\nu}}h\right)}$$

$$+ \frac{(i\lambda w_s + S_d)\cosh\left(\sqrt{\frac{s + i\lambda}{\nu}}z\right)}{s(s + i\lambda)\cosh\left(\sqrt{\frac{s + i\lambda}{\nu}}h\right)} - \frac{i\lambda w_s + S_d}{s(s + i\lambda)} \qquad (2.275)$$

Taking the Laplace inverse of Eq. (2.275)

$$
w_e(z, t) = \left(\frac{T - \rho_\omega S_w}{\rho_\omega \sqrt{\nu}}\right) e^{-\lambda t} \mathcal{L}^{-1}\left\{\frac{\sinh\left(\sqrt{\frac{s}{\nu}}(z + h)\right)}{(s - i\lambda)\sqrt{s}\cosh\left(\sqrt{\frac{s}{\nu}}h\right)}\right\}
$$

$$
+ (i\lambda w_s + S_d) e^{-i\lambda t} \mathcal{L}^{-1}\left\{\frac{\cosh\left(\sqrt{\frac{s}{\nu}}z\right)}{s(s - i\lambda)\cosh\left(\sqrt{\frac{s}{\nu}}h\right)}\right\}
$$

$$
- (i\lambda w_s + S_d)\mathcal{L}^{-1}\left\{\frac{1}{s(s + i\lambda)}\right\} \tag{2.276}
$$

We know from the Laplace inverse table that

$$
\mathcal{L}^{-1}\left\{\frac{\sinh x\sqrt{s}}{\sqrt{s}\cosh a\sqrt{s}}\right\} = \frac{2}{a}\sum_{n=1}^{\infty}(-1)^{n-1}e^{-\frac{(2n-1)^2\pi^2 t}{4a^2}}\sin\left(\frac{(2n-1)\pi x}{2a}\right)
$$

$$
\mathcal{L}^{-1}\left\{\frac{\cosh x\sqrt{s}}{s\cosh a\sqrt{s}}\right\} = 1 + \frac{4}{\pi}\sum_{n=1}^{\infty}\frac{(-1)^n}{(2n-1)}e^{-\frac{(2n-1)^2\pi^2 t}{4a^2}}\cos\left(\frac{(2n-1)\pi x}{2a}\right)
$$

Using this information in conjunction with the convolution integral, we obtain

$$
w_e(z, t) = \frac{2(T - \rho_\omega S_w)}{\rho_\omega}\int_0^t e^{-i\lambda\tau}\left[\sum_{n=1}^{\infty}(-1)^{n-1}e^{-\frac{(2n-1)^2\pi^2\nu\tau}{4h^2}}\sin\left(\frac{(2n-1)\pi(z+h)}{2h}\right)\right]d\tau
$$

$$
+ (i\lambda w_s + S_d)\int_0^t e^{-i\lambda\tau}\left[1 + \frac{4}{\pi}\sum_{n=1}^{\infty}\frac{(-1)^n}{(2n-1)}e^{-\frac{(2n-1)^2\pi^2\nu\tau}{4h^2}}\cos\left(\frac{(2n-1)\pi z}{2h}\right)\right]d\tau
$$

$$
- \left(\frac{(i\lambda w_s + S_d)}{i\lambda}\right)[1 - e^{-i\lambda t}] \tag{2.277}
$$

This is the solution for $w_e(z, t)$ in complex integral form. Equating real and imaginary parts:

$$
u_e(z, t) = \left(\frac{2\gamma}{\rho_\omega h}\right)\int_0^t \cos(\lambda\tau - \theta)f(z, \tau)d\tau
$$

$$
+ \gamma_1\int_0^t \cos(\lambda\tau - \theta_1)g(z, \tau)d\tau
$$

$$
+ \frac{1}{\lambda}[-(\lambda u_s + S_{dy})(1 - \cos\lambda t) + (\lambda v_s - S_{dx})\sin\lambda t] \tag{2.278}
$$

$$v_e(z, t) = \left(-\frac{2\gamma}{\rho_\omega h}\right)\int_0^t \sin(\lambda\tau - \theta)f(z, \tau)d\tau$$

$$-\gamma_1\int_0^t \sin(\lambda\tau - \theta_1)g(z, \tau)d\tau$$

$$-\frac{1}{\lambda}[(\lambda v_s - S_{dx})(1 - \cos\lambda t) + (\lambda u_s + S_{dy})\sin\lambda t]  \qquad (2.279)$$

where

$$\left.\begin{array}{l} \gamma = \sqrt{(\tau_x - \rho_\omega S_{wx})^2 + (\tau_y - \rho_\omega S_{wy})^2} \\[2mm] \theta = \tan^{-1}\left(\dfrac{\tau_y - \rho_\omega S_{wy}}{\tau_x - \rho_\omega S_{wx}}\right) \\[4mm] \gamma_1 = \sqrt{(-\lambda w_{sy} + S_{dx})^2 + (\lambda w_{sx} + S_{dy})^2} \\[2mm] \theta_1 = \tan^{-1}\left(\dfrac{\lambda w_{sx} + S_{dy}}{-\lambda w_{sy} + S_{dx}}\right) \end{array}\right\} \qquad (2.280)$$

and

$$\left.\begin{array}{l} f(z, t) = \displaystyle\sum_{n=1}^{\infty}(-1)^{n-1}e^{-\frac{(2n-1)^2\pi^2\nu t}{4h^2}}\sin\left(\dfrac{(2n-1)\pi(z+h)}{2h}\right) \\[6mm] g(z, t) = 1 + \dfrac{4}{\pi}\displaystyle\sum_{n=1}^{\infty}\dfrac{(-1)^n}{(2n-1)}e^{-\frac{(2n-1)^2\pi^2\nu t}{4h^2}}\cos\left(\dfrac{(2n-1)\pi z}{2h}\right) \end{array}\right\} \qquad (2.281)$$

The solutions of these integrals are very easily obtained by numerical or analytical procedures.

Note:

$$\int_0^t \cos\lambda\tau e^{-\frac{(2n-1)^2\pi^2\nu\tau}{4h^2}}\,d\tau = \left[\frac{e^{-\frac{(2n-1)^2\pi^2\nu\tau}{4h^2}}}{\lambda^2 + \left(\frac{(2n-1)^2\pi^2\nu}{4h^2}\right)^2}\left(-\left[\frac{(2n-1)^2\pi^2\nu\tau}{4h^2}\right]\cos\lambda\tau + \lambda\sin\lambda t\right)\right]_0^t$$

$$= \frac{e^{-\frac{(2n-1)^2\pi^2\nu t}{4h^2}}}{\lambda^2 + \left(\frac{(2n-1)^2\pi^2\nu}{4h^2}\right)^2}\left[-\frac{(2n-1)^2\pi^2\nu}{4h^2}\cos\lambda\tau + \lambda\sin\lambda\tau\right]$$

$$+ \frac{(2n-1)^2\pi^2\nu}{\lambda^2 + \left\{\frac{(2n-1)^2\pi^2\nu}{4h^2}\right\}^2}\left(\frac{1}{4h^2}\right).$$

## Case III: Shallow Water Ocean

In this case $h \to 0$, $z \to 0$, or $(z + h) \to 0$ and so Eq. (2.276) reduces to (up to $O(h)$)

$$
w_e(z, t) \approx \left(\frac{T - \rho_\omega S_w}{\rho_\omega \nu}\right) e^{-i\lambda t} \mathcal{L}^{-1}\left\{\frac{(z + h)}{(s - i\lambda)}\right\}
$$

$$
+ (i\lambda w_s + S_d)\left\{e^{-i\lambda t} \mathcal{L}^{-1}\left\{\frac{1}{s(s - i\lambda)}\right\}\right\}
$$

$$
- (i\lambda w_s + S_d)\left\{\mathcal{L}^{-1}\left\{\frac{1}{s(s + i\lambda)}\right\}\right\}
$$

$$
= \frac{(T - \rho_\omega S_w)(z + h)}{\rho_\omega \nu} + (i\lambda w_s + S_d)\left[\frac{1}{i\lambda}(1 - e^{-i\lambda t}) - \frac{1}{i\lambda}(1 - e^{-i\lambda t})\right]
$$

$$
= \frac{(T - \rho_\omega S_w)(z + h)}{\rho_\omega \nu}
$$

So, we have

$$
\left. \begin{aligned}
u_e(z, t) &= \frac{\tau_x - \rho_\omega S_{wx}}{\rho_\omega \nu}(z + h) \\
v_e(z, t) &= \frac{\tau_y - \rho_\omega S_{wy}}{\rho_\omega \nu}(z + h)
\end{aligned} \right\} \tag{2.282}
$$

This approximation does not include any effects of Stokes drift and wave dispersion on the Eulerian currents. Use the following approximation:

Let $\cosh\left(\sqrt{\frac{s + i\lambda}{\nu}}h\right) \approx 1 + \frac{s + i\lambda}{2\nu}h^2$ and $\cosh\left(\sqrt{\frac{s + i\lambda}{\nu}}z\right) \approx 1 + \frac{s + i\lambda}{2\nu}z^2$. Then

$$
\mathcal{L}\{w_e\} = \frac{2(T - \rho_\omega S_w)}{\rho_\omega h^2}\left[\frac{z + h}{s\left(s + i\lambda + \frac{2\nu}{h^2}\right)}\right]
$$

$$
+ \frac{i\lambda w_s + S_d}{h^2}\left[\frac{z^2 - h^2}{s\left(s + i\lambda + \frac{2\nu}{h^2}\right)}\right]
$$

The inverse is given by

$$
w_e(z, t) = \frac{2(T - \rho_\omega S_w)(z + h)}{\rho_\omega h^2\left(i\lambda + \frac{2\nu}{h^2}\right)}\left[1 - e^{-\left(i\lambda + \frac{2\nu}{h^2}\right)t}\right] + \frac{(i\lambda w_s + S_d)(z^2 - h^2)}{h^2\left(i\lambda + \frac{2\nu}{h^2}\right)}\left[1 - e^{-\left(i\lambda + \frac{2\nu}{h^2}\right)t}\right]
$$

$$
= \frac{2(T - \rho_\omega S_w)(z + h)}{\rho_\omega(i\lambda h^2 + 2\nu)}\left[1 - e^{-\left(\frac{2\nu}{h^2} + i\lambda\right)t}\right] + \frac{(i\lambda w_s + S_d)(z^2 - h^2)}{i\lambda h^2 + 2\nu}\left[1 - e^{-\left(\frac{2\nu}{h^2} + i\lambda\right)t}\right]
$$

$$
\tag{2.283}
$$

Equating real and imaginary parts:

$$
\begin{aligned}
u_e(z,t) = \frac{2(z+h)}{\rho_\omega(4\nu^2 + \lambda^2 h^4)} & \left[ \{2\nu(\tau_x - \rho_\omega S_{wx}) + \lambda h^2(\tau_y - \rho_\omega S_{wy})\} \left(1 - \cos\lambda t\, e^{-\frac{2\nu t}{h^2}}\right) \right. \\
& \left. - \{2\nu(\tau_y - \rho_\omega S_{wy}) - \lambda h^2(\tau_x - \rho_\omega S_{wx})\}\sin\lambda t\, e^{-\frac{2\nu t}{h^2}} \right]
\end{aligned}
$$

$$
\begin{aligned}
+ \frac{(z^2 - h^2)}{4\nu^2 + \lambda^2 h^4} & \left[ \{2\nu(-\lambda w_{sy} + S_{dx}) + \lambda h^2(\lambda w_{sx} + S_{dy})\} \left(1 - \cos\lambda t\, e^{-\frac{2\nu t}{h^2}}\right) \right. \\
& \left. - \{2\nu(\lambda w_{sx} + S_{dy}) - \lambda h^2(-\lambda w_{sy} + S_{dx})\}\sin\lambda t\, e^{-\frac{2\nu t}{h^2}} \right]
\end{aligned}
$$

$$(2.284)$$

$$
\begin{aligned}
v_e(z,t) = \frac{2(z+h)}{\rho_\omega(4\nu^2 + \lambda^2 h^4)} & \left[ \{2\nu(\tau_y - \rho_\omega S_{wy}) - \lambda h^2(\tau_x - \rho_\omega S_{wx})\} \left(1 - \cos\lambda t\, e^{-\frac{2\nu t}{h^2}}\right) \right. \\
& \left. + \{2\nu(\tau_x - \rho_\omega S_{wx}) + \lambda h^2(\tau_y - \rho_\omega S_{wx})\}\sin\lambda t\, e^{-\frac{2\nu t}{h^2}} \right]
\end{aligned}
$$

$$
\begin{aligned}
+ \frac{(z^2 - h^2)}{4\nu^2 + \lambda^2 h^4} & \left[ \{2\nu(-\lambda w_{sy} + S_{dx}) + \lambda h^2(\lambda w_{sx} + S_{dy})\}\sin\lambda t\, e^{-\frac{2\nu t}{h^2}} \right. \\
& \left. + \{2\nu(\lambda w_{sx} + S_{dy}) - \lambda h^2(-\lambda w_{sy} + S_{dx})\} \left(1 - \cos\lambda t\, e^{-\frac{2\nu t}{h^2}}\right) \right]
\end{aligned}
$$

$$(2.285)$$

It can be easily seen that up to the $O(h)$ and $O(z)$:

$$
u_e(z,t) = \frac{\tau_x - \rho_\omega S_{wx}}{\rho_\omega \nu}(z + h)
$$

$$
v_e(z,t) = \frac{\tau_y - \rho_\omega S_{wx}}{\rho_\omega \nu}(z + h)
$$

which are the same as Eq. (2.282).

Note: The solutions obtained for infinite depth and finite depth oceans satisfy the governing equations plus all the boundary and initial conditions of the problem. However, shallow water solutions satisfy only the bottom boundary conditions. To obtain the correct solutions for the shallow water ocean case, the whole system of Navier-Stokes equations must be modified as demonstrated by Rahman [18]. Interested readers are also directed to Lighthill's *An Informal Introduction to Theoretical Fluid Mechanics*, published by Oxford University Press, Oxford, 1978. This study is motivated partly because Eulerian currents play a very important role in the study of ice floe drift.

## Acknowledgments

The author is grateful to the Natural Sciences and Engineering Research Council of Canada for its financial support. This report was completed while the author was on sabbatical leave 1 July 1999 to 30 June 2000 at the Bedford Institute of Oceanography at Dartmouth, McGill University in Montreal, and Trent University at Peterborough. Thanks are extended to Isaac Mulolani who helped in its preparation. I am grateful to Ms. Rhonda Sutherland for her excellent drafting of the figures. Thanks are also extended to CRC Press LLC for its interest in my work.

## References

1. Ekmar, V. W., On the influence of the earth's rotation on ocean currents, *Arkiv. F. Matematik...*, ii, 1905.
2. Hasselmann, K., On the nonlinear energy transfer in a gravity wave spectrum, Part 1: General theory, *J. Fluid Mech.*, 12, 481–500, 1962.
3. Hasselmann, K., On the nonlinear energy transfer in gravity wave spectrum, Part 2: Conservation theorems, wave-particle analogy, irreversibility, *J. Fluid Mech.*, 15, 273–281. Part 3. *ibid. 15*, 385–398, 1963.
4. Hasselmann, S. and Hasselmann, K., A symmetrical method of computing the nonlinear transfer in a gravity-wave spectrum, *Hamb. Geophys. Einzelschriften Reihn A Wiss. Abhand.*, 52, 138, 1981.
5. Hasselmann, S., Hasselmann, K., Komen, G. K., Janssen, P., Ewing, A. J., and Cardone, V., The WAM model—A Third generation ocean wave prediction model, *J. Phys. Oceanogra.*, 18, 1775–1810, 1988.
6. Huang, N. E., Derivation of Stokes drift for a deep water random gravity wave field, *Deep Sea Res.*, 18, 255–259, 1971.
7. Isaacson, M. de St. Q., Fixed and floating axisymmetric structures in waves, *J. Waterw., Port, Coastal and Ocean Div.*, ASCE, WW2, 180–199, 1982.
8. Jenkins, A. D., The use of a wave prediction model for driving a near surface current model, *Dtsch. Hydrogr. Z.*, 42, 134–149, 1989.
9. Komen, G. J., Hasselmann, S., and Hasselmann, K., On existence of a fully developed windsea spectrum, *J. Phys. Oceanogra.*, 14, 1271–1285, 1984.
10. Komen, G. J., Cavaler, L., Donelan, M., Hasselmann, K., Hasselmann, S., and Jenssen, P. A. E. M., *Dynamic and Modelling of Ocean Waves*, Cambridge University Press, Cambridge, 1994.
11. Lamb, H., *Hydrodynamics*, 6th edition, Cambridge University Press, Cambridge, 1945.
12. Lighthill, M. J., *An Informal Introduction to Theoretical Fluid Mechanics*, Oxford University Press, Oxford, 1986.
13. Longuet-Higgins, M. S., On the nonlinear transfer of energy in the peak of a gravity wave spectrum: a simplified model, *Proc. Roy. Soc. London*, A347, 311–328, 1976.
14. Masson, D. and LeBlond, P., Spectral evolution of wind generated surface gravity waves in a dispersed ice-field, *J. Fluid Mechanics*, 202, 43–81, 1989.
15. Perrie, W. and Hu, Y., Air-ice-ocean momentum exchange. Part II: Ice drift, *J. Phys. Oceanogra.*, 27, 1976–1996, 1997.
16. Phillips, O. M., *The Dynamics of Upper Oceans*, Cambridge University Press, Cambridge, 1960.
17. Rahman, M., *Applied Differential Equations for Scientists and Engineers*, Vol. 1 & Vol. 2, Computational Mechanics Publications, Southampton U.K., Boston, 1994.
18. Rahman, M., *Water Waves: Relating Modern Theory to Advanced Engineering Applications*, Oxford University Press, Oxford, 1995.
19. Webb, D. J., Nonlinear transfer between sea waves, *Deep-Sea Res.*, 25, 279–298, 1978.

# 2.3 System Approaches to Heave Compensation in Marine Dynamic Applications

*Ferial El-Hawary*

Compensating for underwater motion effects is required in areas such as control and operations of autonomous remotely operated vehicles (ROV), underwater seismic exploration, and buoy wave data analysis. In all these applications received signals contain an undesirable component due to the dynamics of both source and receiver. This is commonly known as the heave phenomenon.

The vehicle's position is estimated both at the surface and at the vehicle in underwater remotely operated vehicles [1]. As detailed in Jackson and Ferguson [2], a probing pulse is emitted from the surface, and the vehicle answers with a second pulse, thereafter the transponder array responds, and the time delays between the signals are measured separately at the surface and at the vehicle. Accuracy is essential in navigating the vehicle, and can be refined by compensating for motion effects. According to Smith [3], this aspect is also essential in controlling dynamic loads through the design of motion-compensated handling systems. The operation of manipulator controls in underwater robotics depends on compensation for motion effects (for details see Collins [4]). The application to floats of freely drifting acoustic sensors to measure signal propagation and ambient noise in the 1 to 20 Hz band is given in Culver and Hodgkiss [5].

The deployment of several freely drifting floats forms an array of sensors whose outputs can be combined after the experiment with a beam former. Float locations must be known to within one-tenth of a wavelength at the highest frequency of interest in order to effectively beam-form their outputs. The floats generate and receive acoustic pulses and thereby measure float-to-surface and float-to-float travel times. Estimating float positions and float position uncertainties from these travel-time measurements is the main task in this application.

Underwater seismic/acoustic exploration is adopted in evaluating the structure of the underwater layered media, and classification of sediments in terms of properties that aid in carrying out an exploration task. Identifying hydrocarbon formations is an important phase of the marine seismic process. Classical acoustic techniques have gained increasing use over the past several years for marine layer identification and classification of sediments (see Robinson and Durrani [6], El-Hawary [7], and Vetter [8]).

A deep-towed acoustic signal source and hydrophone receiver array are used. The source imparts energy to the water and underlying media. The source signal then undergoes multiple transmissions and reflections at the layers' boundaries. The ship's dynamics, coupled to the towed body (fish) through the towing cable, and the hydrodynamics of the towed body, cause vertical motions of the source and sensor. These components have the outcome of a varying acoustic wave travel path to the sea floor and to the sub-bottom reflectors between successive pings of the source.

The motion's effects appear on the reflection records along the ship track as additional undulations of the sea floor and of the sub-bottom reflectors (the underlying media). Removal of the heave component is an important preprocessing task for improving displays of the raw and filtered reflection data, and for extracting media parameters such as reflection coefficients and reflector depths. The design of the compensating filter facilitates reducing the residual heave effects, i.e., for delaying and advancing the recording trigger on successive firings so as to effect a smoothing or removal of such undulations. On the basis of the frequency response of the heave record, a model for the heave dynamics consistent with those found in the area of hydrodynamics can be postulated. This provides the basis for formulating the heave extraction problem as one of optimal linear estimation.

In all applications, filtering this component at the receiver side is required to enhance the quality of the received records for further processing to extract information. This demands an accurate model of the heave dynamics, and to identify its parameters based on short records. A number of advanced techniques are available to respond to this challenge.

## Heave Motion Modeling

Details of modeling the heave process can be found in marine hydrodynamics references (see Bhattacharya [9]). On the basis of the frequency content of the heave record, a model for the heave dynamics can be postulated. This model provides the basis for formulating the heave extraction problem as one of optimal linear estimation.

Field records of the return signal display a periodic random component that can be extracted to model the source heave motion effects making it possible to postulate expressions that describe the process. The frequency spectrum $Z(f)$ of the extracted heave component can be assumed to be the result of a purely random excitation (white noise) caused by current and wave effects on the towed assembly and ship, float, or the underwater vehicle.

We assume that the heave process is represented by a continuous linear time invariant (LTI) dynamic system. The simplest model is the second order model, which has been utilized in practical applications. From an accuracy point of view, higher order models can be expected to provide less error in modeling the process. In general we assume that the heave process is modeled using the following transfer function

$$T(s) = K\frac{N(s)}{D(s)} \tag{2.286}$$

The system gain is denoted by $K$. The numerator function $N(s)$ has $M$ zeros $z_i$, and is therefore given by

$$N(s) = (s - z_1)\,(s - z_2)...(s - z_M) \tag{2.286a}$$

The denominator function has $N$ zeros $p_i$, and is given by

$$D(s) = (s - p_1)\,(s - p_2)...(s - p_N) \tag{2.286b}$$

The modeling task resolves to finding the optimal values of poles and zeros using, for instance, least squares model parameter estimation techniques. Ideally, we should attempt to determine, *a priori*, the optimal model order, but this is quite an involved process. We process records assuming a series of $M$ and $N$ values and the resulting optimal models are compared in terms of accuracy.

## A Spectral Analysis Approach

Spectral analysis can be instrumental as the basis for heave compensation in underwater applications. A major part of signal processing is based on spectral analysis, typically for distinguishing and tracking signals of interest, and for extracting information from the relevant data. Given a finite number of noisy measurements of a discrete-time stochastic process, or its first few covariance lags, the classical spectral estimation problem involves estimating the shape of its continuous power spectrum. For some modern applications of signal processing, such as radar, sonar, and phased arrays, the spectrum of interest is a line spectrum, and the modern spectral estimation problem involves estimating the locations of these spectral lines.

The frequency resolution in conventional Fourier transform methods is roughly equal to the reciprocal of the data record length. As a result, additional constraints (or prior information) are included to enhance the resolution capability. This is achieved in modern methods where the data are modeled as the output of a linear system driven by white noise. With a properly selected model, these methods will lead to enhanced performance. In underwater processing applications, the spectral estimates have to be based on short data records and yet low-bias, low-variance, high-resolution estimates are desired.

In this section, we treat source dynamic motion evaluation in underwater applications, and discuss a parametric modeling approach based on power spectral density estimation using Burg's maximum entropy methodology. Attention is focused on the auto-regressive model as an efficient tool for carrying out the estimation task. The section is based on work reported in [10] where results using field data were presented.

### Parametric PSD Models

The power spectral density (PSD) is defined as the discrete-time Fourier transform of an infinite auto-correlation sequence (ACS). This transform relationship between the PSD and ACS may be considered as a nonparametric description of the second-order statistics of a random process (see Marple [11], and Kay and Marple [12]).

A parametric description of the second-order statistics may also be conceived by assuming a time series model of the random process. The PSD of the time-series model will then be a function of the model parameters rather than the ACS. A special case of models, driven by white noise, processes and possessing rational system functions, is used in this approach. This class includes the Auto-Regressive (AR) process model, the Moving Average (MA) process model, and the Auto-Regressive-Moving Average (ARMA) process model. A major reference is Box and Jenkins [13]. The output processes of this class of model have power spectral densities that are totally described in terms of the model parameters and the variance of the white noise process.

The motive for parametric models of random processes is the capability to attain better PSD estimators based upon the model than produced by conventional spectral estimators. Better spectral resolution is one key advancement area. Both the periodogram and correlogram methods generated PSD estimates from a windowed set of data or ACS estimates. The unavailable data or unestimated ACS values outside the window are implicitly zero. This is, as a rule, an unrealistic assumption that leads to distortions in the spectral estimate. Some knowledge about the process from which the data samples are taken is often available. This information may be used to construct a model that approximates the process that generated the observed time sequence. Such models will make more realistic assumptions about the data outside the window other than the zero assumption. Therefore, the requirement for window functions can be avoided, along with their distorting effect. The degree of enhancement in resolution and spectral fidelity, if any, is determined by the appropriateness of the selected model and the ability to fit the measured data or the ACS (known or estimated from the data) with a few model parameters.

**Auto-regressive PSD Estimation**

The auto-regressive (AR) time-series model approximates many discrete-time deterministic and stochastic processes. The sequence $x[n]$ is assumed to be the output of a casual filter that models the observed data as the response to a driving sequence $e_K[n]$, which is a white noise process of zero mean and variance $\sigma_{ep}^2$. For a Kth order AR model we have:

$$x[n] = -\sum_{k=1}^{K} h_k[k]x[n-k] + e_{KK}[n] \qquad (2.287)$$

This can be written as:

$$e_K[n] = \sum_{k=0}^{K} h_k[k]x[n-k] \qquad (2.288)$$

Here, we assume that $h_K[0] = 1$. Equation (2.288) reveals that $e_K[n]$ is obtained as the convolution of the filter sequence $h_K[n]$ and the input $x[n]$. As a result, we have the following using the $Z$-transform [14, 15]

$$E_K(z) = H(z)X(z) \qquad (2.289)$$

$H(z)$ is the filter $Z$-transfer function. As a result, the $Z$-transform of the correlation sequences is given by:

$$P_{e_K e_K}(z) = P_{xx}(z)H(z)H^a\left(\frac{1}{z^a}\right) \qquad (2.290)$$

To obtain the auto-regressive power spectral estimate of $x[n]$, we substitute $z = e^{j\omega T}$, to obtain:

$$P_{xx}(\omega) = \frac{\sigma_{eK}^2}{\varepsilon_{h_K h_K}(\omega)} \qquad (2.291)$$

Here, we take $T = 1$. The denominator function is given by:

$$\varepsilon_{h_K h_K}(\omega) = \left( \sum_{k=0}^{K} h_k[k] e^{-j\omega k} \right)^2 \tag{2.292}$$

It is clear from Eq. (2.292), that estimating the PSD is equivalent to estimating the AR model parameters $h_K[n]$. The parameters $h_K[K]$ of the AR $(K)$ model can be found from the autocorrelation sequence for lags 0 to $K$ by using the AR Yule-Walker normal equations or the discrete-time Wiener-Hopf equations given by

$$\begin{bmatrix} r_{xx}[0] & r_{xx}[-1] & \cdots & r_{xx}[-K] \\ r_{xx}[1] & r_{xx}[0] & \cdots & r_{xx}[-K+1] \\ \vdots & \vdots & \vdots & \vdots \\ \vdots & \vdots & \vdots & \vdots \\ \vdots & \vdots & \vdots & \vdots \\ \vdots & \vdots & \vdots & \vdots \\ r_{xx}[K] & r_{xx}[K-1] & \cdots & r_{xx}[0] \end{bmatrix} \begin{bmatrix} 1 \\ h_K[1] \\ \vdots \\ \vdots \\ \vdots \\ h_K(K) \end{bmatrix} = \begin{bmatrix} \sigma_{eK}^2 \\ 0 \\ 0 \\ 0 \\ 0 \\ 0 \\ 0 \end{bmatrix} \tag{2.293}$$

Thus, AR parameter estimation involves the solution of a Hermitian Toeplitz system, for $\sigma_{eK}^2$, $h_K[1]$, ..., $h_K[p]$, given the ACS of $x[n]$ over 0 to $K$ lags. This can be computed very efficiently using the Levinson algorithm. The linear prediction coefficients $h_1[1]$, ..., $h_K[K]$ are often termed the reflection coefficients. A special symbol $k_K = a_K[K]$ is typically applied to distinguish these particular linear prediction coefficients from the remaining coefficients.

The Levinson recursive solution to the Yule-Walker equations relates the order $K$ parameters to the order $K - 1$ parameters as:

$$h_K[n] = h_{K-1}[n] + k_K h_{K-1}^a[K-n] \tag{2.294}$$

for $n = 1$ to $n = K - 1$. The reflection coefficient $K_K$ is obtained from the known autocorrelations for lags 0 to $K - 1$

$$k_K = h_K[K] = \frac{-\sum_{n=0}^{K-1} h_{K-1}[n] r_{xx}[K-n]}{\sigma_{a(K-1)}^2} \tag{2.295}$$

The recursions for the driving white noise variance are given by:

$$\sigma_{eK}^2 = \sigma_{e(K-1)}^2 [1 - |k_K|^2] \tag{2.296}$$

where

$$\sigma_0^2 = r_{xx}[0] \tag{2.297}$$

## Application Results

We explored the performance of the auto-regressive PSD estimation technique when applied to a series of heave records used earlier with Kalman filtering as the benchmark. Reference [10] considered heave records taken from a cruise off the coast of Newfoundland, and concentrated on the influence of model

**TABLE 2.1**   Summary of Results

| K | Dominant Peak Value | Σ | Number of Peaks |
|---|---|---|---|
| 10 | 1.82 | 117.28 | 1 |
| 20 | 1.66 | 1163.96 | 2 |
| 30 | 2.58 | 116.95 | 4 |
| 40 | 14.76 | 118.09 | 6 |
| 50 | 10.20 | 115.99 | 6 |
| 60 | 19.51 | 112.32 | 5 |
| 70 | 7.43 | 43.87 | 12 |
| 80 | 8.52 | 45.64 | 12 |
| 90 | 5.60 | 34.20 | 11 |
| 100 | 5.33 | 29.74 | 18 |

order on the algorithm's results. All models exhibited a peak at about 45 kHz. The model of order $K = 40$ appears to represent the spectrum more accurately from an error point of view. The dominant peak value is 14.76, with a Σ of 118.09, the corresponding number of peaks is 6.

## A Kalman Filtering Approach

Kalman filter applications are well known for some different physical situations as reported in [16–19]. Heave component extraction details are presented in [20, 21] for the marine seismic application. Compensation for source heave involves two steps. First, a model of the heave phenomenon is obtained on the basis of available data. The type of model, as well as its order, are important considerations. In early applications [21] a linear, second-order model of the phenomenon was found satisfactory. Estimation of the model parameters is an important aspect. An iterative procedure has been used [22]. Once a model and its parameters have been identified, the process of Kalman filtering is performed. We have done this in two ways. In the early implementation, conventional Kalman filtering [20–22] was used satisfactorily in a majority of cases. More recently, parallel Kalman filters were applied to the exploration problem using a second-order model [23]. Investigations concerning higher order model implementations are given in [24].

The transfer function model of the heave motion can be written in the following standard discrete state space form suitable for applying Kalman filtering:

$$x(k + 1) = \phi(k + 1, k)x(k) + \Gamma(k + 1, k)w(k) \qquad (2.298)$$

The state transition matrix $\phi(k + 1, k)$ and the matrix $\Gamma(k + 1, k)$ are constants since the heave dynamics are described by a constant parameter model. The elements of the matrices are derived from the parameters of the frequency response model discussed above and we therefore write:

$$x(k + 1) = \phi x(k) + \Gamma w(k) \qquad (2.299)$$

The input sequence $w(k)$ is assumed to be a Gaussian white sequence with zero mean and covariance matrix $Q(k)$, which is positive semi-definite. The initial state is assumed to be a Gaussian random vector with zero mean and known covariance matrix $P(0)$. It is further assumed that $w(k)$ is independent of $x(k)$. The record of the heave component is assumed to be the basis for the measurement model given by

$$z(k + 1) = Hz(k + 1) + v(k + 1) \qquad (2.300)$$

The measurement error sequence v is assumed to be Gaussian with zero mean and covariance matrix $R(k)$. Assume that measurements $z(1), z(2), ..., z(j)$ are available, from which we like to estimate $x(k)$, denoted by $x(k|j)$. In filtering we have $j = k$, and we therefore wish to find $x(k|k)$. For prediction we have $k$ larger than $j$. In earlier applications one utilizes the standard predictor-corrector form of a Kalman filter (see Meditch [16]).

**Predictor**

In the predictor stage, the Kalman filter predicts the $k$th state vector based on the previous optimal estimate:

$$\hat{x}_-(k) = \phi(k-1)x_+(k-1) \tag{2.301}$$

In addition, the Kalman filter requires updating the error covariance matrix

$$P_-(k) = \phi(k-1)P_+(k-1)\phi^T(k-1) + \Gamma Q(k-1)\Gamma^T \tag{2.302}$$

**Corrector**

In the corrector stage, the Kalman filter produces an updated state estimate

$$\hat{x}_+(k) = \hat{x}_-(k) + K(k)r(k) \tag{2.303}$$

Here, $r$ is the innovation sequence given by:

$$r(k) = z(k) - Hx_-(k) \tag{2.304}$$

The Kalman filter update of the covariance matrix is:

$$P_+(k) = (1 - K(k)H(k))P_-(k) \tag{2.305}$$

The Kalman gain matrix $K$ is given by:

$$K(k) = P_-(k)H^T(K)A^{-1}(k) \tag{2.306}$$

where:

$$A(k) = R(k) + H(k)P_-(k)H^T(k) \tag{2.307}$$

For details of the application and computational results consult [22].

## A Recursive Weighted Least Squares Estimation Approach

This section discusses solving the heave compensation problem using the discrete Auto-Regressive Moving Average with external variable (ARMAX) model, which is an alternative model of the process, and proposes use of the recursive weighted least squares algorithm [25-28] for its solution. The algorithm is simpler than Kalman filtering in terms of the required knowledge of noise statistics and thus provides an attractive alternative to Kalman filtering. Computational results pertaining to the performance of the RWLS technique are given and the influence of the method's weighting functions on the filtering outcome is discussed.

Recently, attention has been drawn to a significant category of estimation techniques that can be interpreted as special cases of Kalman filtering. In this category one is interested in minimizing the following cost function

$$J = \sum_{j=1}^{k} \beta(k, j)\varepsilon^T(j) \wedge^{-1}(j)\varepsilon(j) \tag{2.308}$$

where the weighting function $\beta(k, j)$ is introduced to shape the memory of the filter and satisfies the following relationships [28, 29]:

$$\beta(k, j) = \pi(k)\beta(k - 1, j) \quad 1 \le j \le k - 1$$
$$\beta(k, k) = 1 \tag{2.309}$$

Note that these relations imply that

$$\beta(k, j) = \prod_{i=j+1}^{k} \lambda(i) \tag{2.310}$$

The matrix $\Gamma$ is diagonal whose entries signify the relative importance of the error components, but is usually taken as unity. This category of estimation techniques is useful in identifying time-varying parameters and has been discussed by, for example, Astrom and Wittenmark [25], Ljung and Soderstrom [26], Ljung [27], and Young [28].

The recursive estimation algorithm can be written in a number of different equivalent forms. Here we adopt the following Kalman-like form of the RWLS algorithm, which has many computational advantages:

$$\hat{x}(k) = \hat{x}(k - 1) + K(k)r(k) \tag{2.311}$$

Here, $r(k)$ is the innovations sequence defined by:

$$r(k) = z(k) - H(k)\hat{x}(k - 1) \tag{2.312}$$

The gain matrix $K(k)$ is defined by:

$$K(k) = P(k - 1)H^T(k)A^{-1}(k) \tag{2.313}$$

where $A(k)$ is defined by:

$$A(k) = \lambda(k) \wedge (k) + H(k)P(k - 1)H^T(k) \tag{2.314}$$

The equivalent of the state error covariance matrix is given by

$$P(k) = \frac{1}{\lambda(k)}[1 - K(k)H(k)]P(k - 1) \tag{2.315}$$

The algorithm presented here bears many resemblances to the Kalman filtering algorithm as discussed in detail in [29].

## Choice of Weighting Function

The choice of the weighting function $\beta(k, j)$ controls the way in which each measurement is incorporated relative to other measurements. The choice should clearly be such that measurements that are relevant to current system properties are included. If one chooses to assign less weight to older measurements such as in the case of variable system parameters, then a popular choice is given by:

$$\lambda(i) = \lambda$$

This leads to:

$$\beta(k, j) = \prod_{i=j+1}^{k} \lambda = \lambda^{k-j} \qquad (2.316)$$

This weighting function is referred to as an Exponential Weighting into the Past (EWP). It should be noted that $\lambda$ is a forgetting factor that shapes the estimator's memory. We choose $\lambda$ to be slightly less than one. With this weighting, measurements older than $T_o$ are included in the criterion with a weight of less than 36% of the most recent measurements. Here $T_o$ is referred to as the filter's or criterion memory time constant and is given by:

$$T_o = \frac{1}{1 - \lambda}$$

A second possible choice [27] is such that the forgetting factor is time varying and in this case we define:

$$\beta(k, j) = [\alpha_k]^{k-j} \qquad (2.317)$$

Here, we have $\alpha_k$ defined by the first order discrete filter:

$$\alpha_k = \lambda_0 \alpha_{k-1} + (1 - \lambda_0)\alpha$$

Typically, $\alpha_0 = 0.95$, $\lambda_0 = 0.99$ and $0 < \alpha < 1$. The closed form solution is:

$$\alpha_k = \alpha + [\alpha_0 - \alpha]\lambda_0^k$$

The forgetting factor starts at $\alpha_0$ and reaches steady state value of $\alpha$. For a reasonably large $k$ we have:

$$\lambda(k) = \alpha_{k-1} = \lambda_0[\lambda(k-1) - \alpha] + \alpha$$

Ljung and Soderstrom [26] take $\alpha$ to be unity to obtain:

$$\lambda(k) = \lambda_0[\lambda(k-1) - 1] + 1$$

The convergence of the filter is influenced by the choice of the weighting function.

**Computational Results**

Reference [29] reports on an extensive computational experiment that was conducted to explore the performance of the RWLS technique when applied to a series of heave records used earlier, with Kalman filtering as the benchmark. The order parameters of the ARMAX model considered are: $N = 2$ and $L = 3$. The author concentrated on the influence of weighting functions on the algorithm's results. Emphasis in results reported is on the choice of $\lambda_o$ in the recommended range of 0.7 to 0.9. It was found that the value $\lambda_o = 0.7$ involves less errors than higher choices of the weighting factor. Lower values of $\lambda_o$ did not appreciably improve the situation.

The filtering results compared favorably with those acquired using the conventional Kalman filter for a second-order heave dynamic model as outlined in Appendix A, with covariance matrices of $Q = [1.5\ Z_{av}]^2$ and $R = [0.1\ Z_{av}]^2$, where $Z_{av}$ is the average value of the heave record. The computational time requirements of the RWLS algorithm as tested are lower than those for the Kalman filter. The latter requires about 15 to 20 % more CPU time. The significant difference between the results of the two approaches is that the improvements in the filtering error using the RWLS technique take place later in the time horizon. It is noted that the KF results will (by their very nature) give the optimal estimates, in a least squares sense, weighted by the noise covariance matrices.

## A Generalized Kalman Filtering Approach

In this application area, an optimum solution that simultaneously accounts for the accuracy of the estimates and stability (lateral continuity) is achieved. One of the problems associated with KF is the difficulties in maintaining the lateral continuity, since each trace is analyzed separately (single channel operation).

The proposed generalized Kalman Filter (GKF) is based on the formulation developed by Lagarias and Aminzadeh [30], establishing a trade-off between the cost associated with estimation error and the cost related to the lateral discontinuity of the estimates. By assigning proper weights for accuracy and stability (WA and WS) in the objective function, the desired balance between accuracy and stability is achieved. This method has proven useful in multistage planning and economic forecasting. In this section we introduce the application of the Generalized Kalman filtering (GKF) method to heave compensation for underwater applications.

### The Generalized Kalman Filter

Consider the problem of finding forecasts of the state variables to optimize the production of a planning process. The process performance is defined by specifying an (expected) loss function. In a single-period process, the loss function is generally some measure of the forecast error, and one is interested in obtaining the most accurate forecast. Lagarias and Aminzadeh [30] recognize that in multistage processes, there may be an additional loss associated with unstable forecasts, i.e., forecasts for $x(k)$ in which the forecasts $x(k|k_1)$ and $x(k|k_2)$ made at different time periods $k_1$ and $k_2$ vary a good deal from each other. Consequently the optimal forecast for such multistage processes incorporates a cost trade-off between accuracy and stability of successive forecasts.

The usual perception of a forecast $\tilde{x}(k|k-j)$ for a quantity $x(k)$ generally postulates that the goal is to minimize some measure of the forecast error $x(k) - \tilde{x}(k|k-j)$. The loss function in the problems treated by Lagarias and Aminzadeh also includes terms related to the stability of successive forecasts. As a result forecasts $\tilde{x}(k|k-j)$ chosen to minimize the loss functions considered by Lagarias and Aminzadeh are called projections.

Optimal projections $\tilde{x}(k|k-j)$ are obtained for the case where the evolution of the state variables $x(k)$ over time is governed by a general discrete time linear system with the structure:

$$x(k+1) = \phi(k+1, k)x(k) + \Gamma(k+1, k)w(k) \tag{2.318}$$

Linear measurements $z(k)$ (which may include measurement errors) are made of this process:

$$z(k+1) = H(k+1)x(k) + v(k+1) \tag{2.319}$$

where $x(k)$ is a vector-valued process to be estimated and $z(k)$ is a vector of measurements made at time $k$. In Eq. (2.318), $\phi(k+1, k)$ and $\Gamma(k+1, k)$ are transition matrices, and $w(k)$ is a vector-valued white noise process describing uncertainties in the system driving process. Equation (2.319) relates the measurements $z(k+1)$ to the states $x(k+1)$ via an observation matrix $H(k+1)$, and $v(k+1)$ is a white noise process describing the uncertainties of observation. This model is not completely determined until the stochastic processes $v(k)$ and $w(k)$ have been specified. The special case where $v(k)$ and $w(k)$ are Gaussian white noise processes is called a Gaussian linear system.

It is well known that for the Gaussian linear system, the Kalman filter generates predictions $\hat{x}(k|k-j)$ for $x(k)$ made at time $k-j$ which are optimal predictions in the sense of minimizing the expected value of $L[x(k) - \hat{x}(k|k-j)]$ for a broad spectrum of loss functions. More generally the Kalman filtering algorithm yields the optimal linear predictors for the general discrete time linear system for many loss functions. The recursive structure of the Kalman filter makes it appropriate for rapid numerical computation. The computational efficiency in turn is the basis for the development of a large variety of suboptimal algorithms with a recursive structure for more complicated, nonlinear problems.

The loss functions studied are quadratic plus linear with respect to the final projection error $\varepsilon(k|k)$, and the projection changes $\varepsilon(k|k - j)$ for $j = 1, 2, ..., M-1$.

$$\varepsilon(k|k) = x(k) - \tilde{x}(k|k - 1) \tag{2.320}$$

$$\varepsilon(k|k - j) = \tilde{x}(k|k - i) - \tilde{x}(k|k - j - 1) \tag{2.321}$$

The loss function is defined by:

$$L = \frac{1}{2}\xi^T Q \xi + c^T \xi \tag{2.322}$$

where

$$\xi^T = [\varepsilon(k|k), \varepsilon(k|k - 1), ..., \varepsilon(k|k - M + 1)] \tag{2.323}$$

$$c^T = [c_1, c_2, ..., c_M] \tag{2.324}$$

and

$$Q = \begin{bmatrix} Q_{11} & Q_{12} & Q_{13} & \cdot & \cdot & Q_{1M} \\ Q_{21} & Q_{22} & Q_{23} & \cdot & \cdot & Q_{2M} \\ Q_{31} & Q_{32} & Q_{33} & \cdot & \cdot & Q_{3M} \\ \cdot & \cdot & \cdot & \cdot & \cdot & \cdot \\ \cdot & \cdot & \cdot & \cdot & \cdot & \cdot \\ Q_{M1} & Q_{M2} & Q_{M3} & \cdot & \cdot & Q_{MM} \end{bmatrix} \tag{2.325}$$

Here, $\varepsilon(k|k - i)$ and $c_i$ are $s \times 1$ column vectors, the $Q_{ij}$ are $s \times s$ matrices, and $c^T$ denotes the transpose of $c$. The matrix $Q$ is assumed to be positive definite symmetric. The projection changes $\varepsilon(k|k - j)$ measure instability of the projections. This loss function does not allow interaction of costs between plans aimed at different final periods.

Lagarias and Aminzadeh's [30] general result states that the optimal solution to this multistage problem has a simple expression in terms of the Kalman filter predictors for the state variables in different stages. A linear projection is defined to be one that is a matrix linear combination of the observation $z(j)$ and the initialization values $\hat{x}(0)$.

### General Result

Suppose that the state variables $x(k)$ are generated by a Gaussian linear process and the loss function $L$ is quadratic plus linear in the prediction errors and projection changes $\varepsilon(k|j)$. Then the optimal projections $\tilde{x}(k|k - j)$ for $0 \leq j \geq M - 1$ have the form

$$\tilde{x}(k|k - M + j) = \sum_{i=0}^{j} F_{ij}\hat{x}(k|k - M + i) + W_j \tag{2.326}$$

where $\hat{x}(k|k - M + i)$ are the Kalman filter estimates available at time $k - M + i$, and $F_{ij}$, $W_j$ are matrix weights that depend only on the loss function $L$ and not on the observations. Furthermore, the matrix weights satisfy:

$$\sum_{i=o}^{j} F_{ij} = I \quad \forall j \tag{2.327}$$

where $I$ is the identity matrix. Equation (2.326) also provides the optimal linear projections for the general discrete time linear process (not necessarily Gaussian) with the same loss function $L$.

A consequence of this result is that in computing the optimal projections, one needs only to retain the Kalman filter estimates $\hat{x}(k|k - M + j)$ for $0 \le i \le j$ instead of all the data $z(0)$, $z(1)$, ..., $z(k - M + j)$. In addition the matrices $F_{ij}$ and $W_j$ can be computed in advance from the loss function. This result also shows that the Kalman filter aspect of the solution is completely decoupled from the multi-period aspect realized in the matrix weights $F_{ij}$, $W_j$.

## Special Case

Consider the special case when the loss function model's cost effects, which are due to changes in successive updated projections, in which there is no cost interaction between projections two or more periods apart. That is, the form $Q$ in Eq. (2.304) has the block-diagonal form:

$$Q = \begin{bmatrix} Q_1 & 0 & 0 & . & . & 0 \\ 0 & Q_2 & 0 & . & . & 0 \\ 0 & 0 & Q_3 & . & . & 0 \\ 0 & 0 & 0 & . & . & 0 \\ 0 & 0 & 0 & . & . & 0 \\ 0 & 0 & 0 & . & . & Q_M \end{bmatrix} \tag{2.328}$$

This case corresponds to the setting where, when a new projection $\tilde{x}(k|k - j)$ is available, modifications are immediately made whose cost is a function of $\varepsilon(k|k - j)$ alone. In this case it is shown that the optimal projections satisfy a recursive relationship.

Suppose that the state variables $x(k)$ are generated by a Gaussian linear process, and the loss function is of the form in Eq. (2.322), where $Q$ is the block diagonal given by Eq. (2.328). Then the optimal projections are given by the following recursive relationship.

For $1 \le j \le M - 1$, we have:

$$\tilde{x}(k|k - j) = [I - G_j]\hat{x}(k|k - j) + G_j\tilde{x}(k|k - j - 1) + V_j \tag{2.329}$$

and for $j = M$, we have:

$$\tilde{x}(k|k - M) = \hat{x}(k|k - M) + V_M \tag{2.330}$$

Here, the $\hat{x}(k|k - j)$ are the Kalman filter estimates, and the $G_j$ and $V_j$ are matrix weights depending only on the loss function $L$. The weights $G_j$ satisfy the recursive relation:

$$G_j = N_j^{-1}Q_{j+1} \tag{2.331}$$

where

$$N_j = Q_{j+1} + Q_j - Q_jG_{j-1} \tag{2.332}$$

with initial value given by:

$$N_1 = Q_1 + Q_2 \tag{2.333}$$

The weights $V_j$ can be computed using the recursions:

$$V_j = \sum_{i=1}^{j} H_{j,i}(c_i - c_{i+1}) \tag{2.334}$$

where $c_{M+1} = 0$

$$H_{j,i} = N_j^{-1} Q_j H_{j-1,i} \tag{2.335}$$

for $i \neq j$ and

$$H_{j,j} = N_j^{-1} \tag{2.336}$$

Conclusions similar to this result can be proved for matrices more general than that in Eq. (2.328). For example, if the matrix $Q$ is a block tri-diagonal matrix, i.e., if $Q_{ij} = 0$ whenever $|i - j| \geq 2$, then one can show that the optimal projections satisfy a recursion of the form

$$\tilde{x}(k|k-j) = [I - G_j^{(1)} - G_j^{(2)}]\hat{x}(k|k-j) + G_j^{(1)}\tilde{x}(k|k-j-1)$$
$$+ G_j^{(2)}\tilde{x}(k|k-j-2) + V_j$$

In this case the formulae for the matrix weights $G^{(1)}$ and $G^{(2)}$ become more complicated than Eqs. (2.331) through (2.336).

The recursive Eqs. (2.331) through (2.336) of this result are convenient for fast computation. The recursions in Eqs. (2.331) and (2.332) have some similarity to those of the Kalman gain and error covariance matrix updates of the Kalman filter itself. Note, however, that the Kalman filter estimates do not depend at all on the loss function $L$. In contrast, the weights $G_{ij}$ rely only on the loss function and not on the observations.

## Application in Heave Compensation

For simplicity and ease of illustration, and without loss of generality, we will also assume $x$ to be a scalar. We define the final projection error by Eq. (2.320), and the projection changes are measured by Eq. (2.321).

The objective is to minimize Eq. (2.322) with $c^T = 0$ where $\xi^T$ is defined by Eq. (2.323) and

$$Q = \begin{bmatrix} Q_{11} & 0 & 0 & . & . & 0 \\ 0 & Q_{22} & 0 & . & . & 0 \\ 0 & 0 & Q_{33} & . & . & 0 \\ 0 & 0 & 0 & . & . & 0 \\ 0 & 0 & 0 & . & . & 0 \\ 0 & 0 & 0 & . & . & Q_{MM} \end{bmatrix} \tag{2.337}$$

In underwater applications, $k$ stands for time and $j$ would stand for trace number. Clearly $Q_{11}$ corresponds to the weight associated with the accuracy and $Q_{22}, Q_{33}, \ldots, Q_{MM}$, relate to the stability of the estimates in different stages.

The Ocean Engineering Handbook

If $Q_{11}$ is set to zero, then the solution will involve constant estimates (traces) in all stages. If $Q_{22}$, $Q_{33}$, ..., $Q_{MM}$ are set to zero, the solution will reduce to the conventional Kalman filtering estimates.

In the general case the solution will be the scalar version of Eq. (2.329):

$$\tilde{x}(k|k - j) = (1 - G_j)\hat{x}(k|k - j) + G_j\tilde{x}(k|k - j - 1) \qquad \text{for} \quad 1 \le j \le M - 1 \quad (2.338)$$

and for $j = M$, we have

$$\tilde{x}(k|k - M) = \hat{x}(k|k - M) \qquad (2.339)$$

where $\hat{x}$ is the conventional Kalman estimates, and the scalar version of Eqs. (2.331) through (2.333) is given by:

$$G_j = \frac{Q_{j+1}}{Q_{j+1} + Q_j(1 - G_{j-1})} \qquad (2.340a)$$

and

$$G_1 = \frac{Q_2}{Q_1 + Q_2} \qquad (2.340b)$$

or

$$G_j = \frac{1}{1 + \dfrac{Q_j}{Q_{j+1}}[1 - G_{j-1}]} \qquad (2.341a)$$

and

$$G_1 = \frac{1}{1 + \dfrac{Q_1}{Q_2}} \qquad (2.341b)$$

We can also express $\tilde{x}(k|k - j)$ in terms of conventional Kalman estimates only using the expressions of Eqs. (2.326) and (2.327) repeated for this case as:

$$\tilde{x}(k|k - M + j) = \sum_{i=0}^{j} F_{ij}\hat{x}(k|k - M = i) \qquad M - 1 > j \ge 0 \qquad (2.342)$$

where

$$\sum_{i=0}^{j} F_{ij} = 1 \qquad \forall j \qquad (2.343)$$

Starting with a heave record, application of KF or GKF with WS = 0 to individual records gives the best estimate for the heave sequence for each record. Applying a tracking algorithm using KF or GKF with WA = 0 will result in laterally continuous events. Clearly, in the first case, continuity (stability of estimates) is not achieved and in the second case the convolutional model assumption is violated with necessary substitutions of parameters and variables to account for this particular application.

Using the objective function described by Eq. (2.322), and with proper choice of *WS* and *WA* (or diagonal elements of *Q* matrix), an optimal solution will be achieved. This solution will involve a compromise between the above-mentioned two extreme cases.

In the more general case, different weight functions (other elements of *Q* matrix) may be considered for costs (penalties) associated with changes in the values of the estimates across a number of records. A reasonable choice for the weight factors is the one that varies with the inverse of the distances between any two records. This enables the analyst to incorporate all the *a priori* information, at least indirectly, in the formulation of the problem.

## Conclusions

The issue of compensating for underwater motion effects arises in a number of areas of current interest such as control and operations of autonomous remotely operated vehicles, underwater seismic exploration, and buoy wave data analysis. In this section we treated the problem of source dynamic motion evaluation in underwater applications using a number of advanced system theoretic approaches including spectral analysis, Kalman filtering, recursive weighted least squares, and the Generalized Kalman Filter.

## References

1. McFarlane, J. R., Frisbie, F. R., and Mullin, M., 1987 The evolution of deep ROV technology in Canada, *Oceans '87 Proceedings*, 1260–1266, 1987.
2. Jackson, E. and Ferguson, J., Design of ARC—Autonomous Remotely Controlled submersible, Oceans '84 Proceedings, 365–368, 1984.
3. Smith, G. R., Development of a 5000 meter remote operated vehicle for marine research, *Oceans '87 Proceedings*, 1254–1259, 1987.
4. Collins, J. S., Advanced marine robotics as a strategic technology for Canada and observations of a related large scale project in Japan, *Oceans '87 Proceedings*, 1246–1253, 1987.
5. Culver, R. L. and Hodgkiss, Jr., W. S., Comparison of Kalman and least squares filters for locating autonomous very low frequency acoustic sensors, *IEEE Journal of Oceans Engineering*, 13, No. 4, 282–290, 1988.
6. Robinson, E. A. and Durrani, T. S., *Geophysical Signal Processing*, Prentice-Hall, Englewood Cliffs, NJ, 1986.
7. El-Hawary, F. and Vetter, W. J., Spatial parameter estimation for ocean subsurface layered media, *Canadian Electrical Engineering Journal*, 5, 1, 28–31, 1980.
8. El-Hawary, F. and Vetter, W. J., Event enhancement on reflections from subsurface layered media, *IEEE Journal of Oceanic Engineering*, OE-7, 1, 51–58, 1982.
9. Bhattacharya, R., *Dynamics of Marine Vehicles*, Wiley-Interscience, New York, 1978.
10. El-Hawary, F., Application of Burg's spectral analysis to heave response modeling in underwater applications, in P. F. Fougere, Ed., *Maximum Entropy and Bayesian Methods*, Kluwer Publishers, 1990, 409–417.
11. Marple, S. L., *Digital Spectral Analysis with Applications*, Prentice-Hall Inc., Englewood Cliffs, NJ, 1987.
12. Kay, S. M. and Marple, S. L., Spectrum analysis—a modern perspective, *Proceedings of the IEEE*, 69, 1380–1418, 1981.
13. Box, G. E. P. and Jenkins, G. M., *Time Series Analysis: Forecasting and Control*, Holden-Day, San Francisco, CA, 1970.
14. Oppenheim, A. V. and Schafer, R. W., *Digital Signal Processing*, Prentice-Hall Inc., Englewood Cliffs, NJ, 1975.
15. Oppenheim, A. V. and Wilsky, A. S., *Signals and Systems*, Prentice-Hall Inc., Englewood Cliffs, NJ, 1983.
16. Meditch, J. S., *Stochastic Optimal Linear Estimation and Control*, McGraw-Hill, New York, 1969.

17. Sage, A. P. and Melsa, J. L., *Estimation Theory with Applications to Communications and Control*, McGraw-Hill, New York, 1971.

18. Gelb, A., *Applied Optimal Estimation*, MIT Press, Cambridge, MA, 1974.

19. Maybeck. P. S., *Stochastic Models, Estimation and Control*, Vol. 1, Academic Press, New York, 1979.

20. El-Hawary, F. and Vetter, W. J., Heave compensation of shallow marine seismic reflection records by Kalman filtering, *IEEE Oceans*, '81, Boston, MA, September 1981.

21. El-Hawary, F., Compensation for source heave by use of Kalman filter, *IEEE Journal of Oceanic Engineering*, OE-7, 2, 89–96, 1982.

22. El-Hawary, F., An approach to extract the parameters of source heave dynamics, *Canadian Electrical Engineering Journal*, 19–23, January 1987.

23. El-Hawary, F. and Ravindranath, K. M., Application of array processing for parallel linear recursive Kalman filtering in underwater acoustic exploration, in *Proceedings of IEEE Oceans '86*, Washington, D.C., Vol. 1, 336–340, September 1986.

24. El-Hawary, F., and Richards, T., Heave response modeling using higher order models, *International Symposium on Simulation and Modeling*, Santa Barbara, CA, May 1987.

25. Astrom, K. J. and Wittenmark, B., *Computer Controlled Systems: Theory and Design*, Prentice-Hall, Englewood Cliffs, NJ, 1984.

26. Ljung, L. and Soderstrom, T., *Theory and Practice of Recursive Identification*, MIT Press, Cambridge, MA, 1983.

27. Ljung, L., *System Identification: Theory for the User*, Prentice-Hall, Englewood Cliffs, NJ, 1987.

28. Young, P., *Recursive Estimation and Time-Series Analysis*, Springer Verlag, Berlin, 1986.

29. El-Hawary, F., A comparison of recursive weighted least squares estimation and Kalman filtering for source dynamic motion evaluation, *Canadian J. Electrical & Computer Engineering*, 17, 3, 136–145, 1992.

30. Lagarias, J. C. and Aminzadeh, F., Multistage planning and extended linear quadratic gaussian control problem, *Mathematics of Operations Research*, 8, 1, 42–63, 1983.

# 2.4   Approaches to Marine Seismic Extraction

## *Ferial El-Hawary*

Underwater seismic exploration using acoustic arrays seeks to determine the structure of underwater layered media through knowledge of significant geometrical and material properties. Identifying hydrocarbon formations is a main goal of the marine seismic method. The process involves three stages of image (or data) acquisition, processing, and pattern recognition (or interpretation). Classical acoustic techniques have been the major tools for marine layer identification and classification of sediments.

In conducting shallow marine seismic explorations to determine the structure of the media underwater, a deep towed acoustical signal source and a hydrophone receiver are used. High-pressure air is suddenly introduced into the water creating a pressure pulse. The resulting pressure waveform is non-periodic and may be approximated by a damped sinusoidal wave. The wave hits the seabed and propagates through the ocean floor and underlying structures, undergoing multiple transmissions and reflections at the interfaces. The marine seismic method uses information from reflected waves collected by arrays of acoustic receivers (hydrophones) for processing to extract required information about the properties of the media underwater. The parameters of the media are related to characteristic points on the received arrays, which are pre-filtered to compensate for source and sensor heave components. The filtered image is then processed to find amplitude, attenuation, and delay parameters in each individual record.

A successful signal-processing algorithm relies on the availability of appropriate models of the Earth's response to acoustic or seismic excitation inputs. The chapter begins with a comprehensive review of a number of important approaches to the modeling task. Models of this nature are known as synthetic seismograms and there is a wealth of literature in the area. The mathematical form of the input excitation function is known.

The modeling task can be generally stated as that of tracing the input wave as it travels through the media to the receiving equipment. Several modeling assumptions are common to all models discussed here. A flat completely elastic layered medium structure with plane acoustic wave propagation is assumed. Normal wave incidence is assumed. This is appropriate for the available field results where the source and the down-looking receivers are in close proximity. The models considered differ in the degree of detail and mathematical refinement. Models of acoustic wave propagation in lossless layered media systems can be classified into two broad categories, according to the degree of detail involved. Models in the first category utilize a parametric representation involving two signals for each layer. The basis for this type of model is the decomposition of the solution to the wave particle displacement (particle velocity and pressure) into forward and backward traveling components in each layer. The second category includes models derived assuming that the system is described by the lossless wave equation and boundary conditions and the nature of the solution of the wave equation requiring that the output be the sum of time shifted and scaled replicas of the source signals. The model in this case is a convolution summation, which is sometimes referred to as a nonparametric representation, Mendel et al. (1980). The models reviewed differ mainly in their approach to modeling the reflection coefficients at the interfaces.

The chapter continues with a discussion of a selection of techniques that allow the analyst to extract time domain attributes of the underwater media in a lossless environment. The following are discussed: Sequential Correlation Based Detection, Linearized Delay Model Detection, and Event Enhancement Filtering. The ability to incorporate losses in the formulation is important as it adds a more realistic element to the detection outcome. The final section deals with time domain attribute extraction accounting for losses. Here we begin with a frequency domain model that leads us to an iterative process for layer parameter estimation in lossy media.

## Parametric Models of Earth Response

This category of models gives rise to cascade-type matrix model structure, which is a feature common to all parametric-type models reviewed. The models differ in the choice of the two signals associated with each interface. Another difference is in the choice of representation, either in discrete-time (Z-transform), continuous time (Laplace transform), or the frequency domain.

Wuenschel's model employs the pressure and particle velocity in arriving at a transmission matrix type model in terms of the Laplace transform. The model due to Goupillaud, which uses the upward and downward propagating pressure wave amplitudes as the model variables is discussed. A scattering matrix type representation in the frequency domain is the feature of this model. Robinson's model is among the most widely accepted models. It is a scattering matrix type model developed in the Z-domain for equal time delay. Robinson's model is the basis for the predictive convolution method. The quasi-state space model of Mendel is also discussed. This is a time-domain representation in terms of a dynamical equation with multiple time delays referred to as a casual functional equation.

### Wuenschel's Model

Wuenschel (1960) approaches the modeling problem starting with the wave equation in terms of displacement $\zeta$ given in terms of the Laplace operators by

$$\frac{\partial^2 \zeta(S, X)}{\partial X^2} = \frac{S^2}{C^2} \zeta(S, X) \tag{2.344}$$

The solution of the above equation is given by

$$\zeta(S, X) = \zeta_i e^{-SX/C} + \zeta_r e^{SX/C}$$

The incident and reflected displacement amplitudes are denoted by $\zeta_i$ and $\zeta_r$ respectively. As a result the particle velocity $U(S, X)$ and pressure $P(S, X)$ are given by

$$U(S, K) = (\zeta_i e^{-SX/C} + \zeta_r e^{SX/C})s \tag{2.345}$$

$$P(S, K) = (-\zeta_i e^{-SX/C} + \zeta_r e^{SX/C})zs \tag{2.346}$$

Wuenschel considered two layers $(K - 1)$ and $K$ with interface at $X = 0$. The depth of layer $K$ is denoted by $d_K$. The particle velocity and pressure at $X = 0$, are thus obtained as:

$$U_{K-1} = S(\zeta_i + \zeta_r) \tag{2.347}$$

$$P_{K-1} = Z_K S(\zeta_i - \zeta_r) \tag{2.348}$$

At $X = d_K$, we obtain

$$U_K = S[\zeta_i e^{-Sd_K/C_K} + \zeta_r e^{Sd_K/C_K}] \tag{2.349}$$

$$P_K = Z_K S[-\zeta_i e^{-Sd_K/C_K} + \zeta_r e^{Sd_K/C_K}] \tag{2.350}$$

Eliminating $\zeta_i$ and $\zeta_r$, one obtains

$$\begin{bmatrix} U_K \\ P_K \end{bmatrix} = \underline{\alpha}_K \begin{bmatrix} U_{K-1} \\ P_{K-1} \end{bmatrix} \tag{2.351}$$

The matrix $\underline{\alpha}_K$ is the Wuenschel's layer matrix given by

$$\underline{\alpha}_K = \begin{bmatrix} \left( \dfrac{e^{(Sd_K/C_K)} + e^{-(Sd_K/C_K)}}{2} \right) & \left( \dfrac{e^{(Sd_K/C_K)} - e^{-(Sd_K/C_K)}}{2Z_K} \right) \\ Z_K \left( \dfrac{e^{(Sd_K/C_K)} - e^{-(Sd_K/C_K)}}{2} \right) & \left( \dfrac{e^{(Sd_K/C_K)} + e^{-(Sd_K/C_K)}}{2} \right) \end{bmatrix} \tag{2.352}$$

For an $N$-layer medium, assuming continuity of velocity and pressure along the interfaces, one can write

$$\begin{bmatrix} U_N \\ P_N \end{bmatrix} = \underline{\alpha}_N \underline{\alpha}_{N-1} \cdots \alpha_1 \begin{bmatrix} U_0 \\ P_0 \end{bmatrix} \tag{2.353}$$

Wuenschel's formulation can be used to express the particle velocity in terms of the forcing function of the source as follows. Assume that the $(N + 1)$th layer has an acoustic impedance of $Z_{N+1}$. The variables emerging from the $N$th layer are thus related by

$$P_N = -Z_{N+1} U_N$$

As a result the fundamental expression of Eq. (2.353) reduces to

$$\begin{bmatrix} U_N \\ -Z_{N+1} U_N \end{bmatrix} = \underline{A} \begin{bmatrix} U_0 \\ P_0 \end{bmatrix} \tag{2.354}$$

where

$$A = \begin{bmatrix} A_{11} & A_{12} \\ A_{21} & A_{22} \end{bmatrix} = \underline{\alpha}_N \underline{\alpha}_{N-1} \dots \underline{\alpha}_1 \tag{2.355}$$

It is easy to verify that

$$U_0 = \frac{(A_{22}/Z_{N+1}) + A_{12}}{(A_{21}/Z_{N+1}) + A_{11}}(-P_0) \tag{2.356}$$

Thus, the particle velocity at the source is expressed in terms of the forcing function of the source with a single factor of proportionality, which is a function of the characteristic impedances of the individual layers and the transit time across the layers.

## Goupillaud's Model

The variables used in Goupillaud's (1961) model are down and up-going wave amplitudes $D$ and $U$ as shown in Fig. (2.24). It can be shown that by defining the matrix

$$M_K = \frac{1}{T_K} \begin{bmatrix} e^{j\omega\zeta_K} & R_K e^{j\omega\zeta_K} \\ R_K e^{-j\omega\zeta_K} & e^{-j\omega\zeta_K} \end{bmatrix} \tag{2.357}$$

The layer equation is given by

$$\begin{bmatrix} D_K \\ U_K \end{bmatrix} = M_K \begin{bmatrix} D_{K+1} \\ U_{K+1} \end{bmatrix} \tag{2.358}$$

The transmission and reflection coefficients in the $K$th layer are $T_K$ and $R_K$, with

$$T_K = 1 + R_K \tag{2.359}$$

The travel time in layer $K$ is denoted by $T_{K'}$.
    For a system of $N$ interfaces, one obtains

$$\begin{bmatrix} D_0 \\ U_0 \end{bmatrix} = M_0^{(N-1)} \begin{bmatrix} D_N \\ U_N \end{bmatrix} \tag{2.360}$$

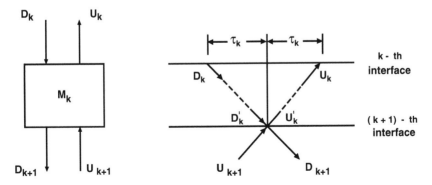

**FIGURE 2.24** Variable definitions for Goupillaud's model.

where the matrix product is defined by

$$M_0^{(N-1)} = M_0 M_1 \ldots M_{N-1} \tag{2.361}$$

The reflection response $U_0$ for a unit input and for a completely absorptive Nth interface $U_N = 0$ is given by

$$U_0 = \frac{\left[ \sum_{i=0}^{N} R_i \exp \cdot \left\{ -2j\omega \sum_{0}^{N} \tau_i \right\} + \sum_{i=0}^{N-2} \left( \sum_{\substack{j=1 \\ j \neq i}}^{N-1} \sum_{\substack{k=2 \\ k \neq i}}^{N} R_i R_j R_k \right) \exp \cdot \left\{ -2j\omega \left( \sum_{0}^{K} \tau_K - \sum_{0}^{j} \tau_j + \sum_{0}^{i} \tau_i \right) \right\} + \ldots \right]}{(1 + \Delta_1 + \Delta_2 + \ldots)} \tag{2.362}$$

It should be noted that $U_0$ is a frequency response expression and in a manner similar to that mentioned in Wuenschel's treatment, the total response is obtained through the Fourier integral.

### Robinson's Model

Robinson's Earth model of 1967 is in terms of the $Z$ or shift operator and is obtained in terms of the basic relation

$$\begin{bmatrix} D_{K+1}(Z) \\ U_{K+1}(Z) \end{bmatrix} = \frac{Z^{-1/2}}{T_K} \underline{N}_K \begin{bmatrix} D_K(Z) \\ U_K(Z) \end{bmatrix} \tag{2.363}$$

where

$$\underline{N}_K = \begin{bmatrix} Z & -R_K \\ -R_K Z & 1 \end{bmatrix} \tag{2.364}$$

As a result, the overall model is obtained as

$$\begin{bmatrix} D_{K+1}(Z) \\ U_{K+1}(Z) \end{bmatrix} = \frac{Z^{-K/2}}{T_K' \ldots T_2' T_1'} \underline{N}^{(K)} \begin{bmatrix} D_1(Z) \\ U_1(Z) \end{bmatrix} \tag{2.365}$$

where the matrix $\underline{N}^{(K)}$ denotes the product

$$\underline{N}^{(K)} = \underline{N}_K \ldots \underline{N}_2 \underline{N}_1 \tag{2.366}$$

An important feature of Robinson's formulation is that the prevailing matrix properties are exploited to arrive at an efficient implementation. It can be shown that

$$\underline{N}^{(K)} = \underline{N}_K \underline{N}_{K-1} \ldots \underline{N}_1 = \begin{bmatrix} Z^K n_{22}^{(K)}(Z^{-1}) & n_{12}^{(K)}(Z) \\ Z^K n_{12}^{(K)}(Z^{-1}) & n_{22}^{(K)}(Z) \end{bmatrix} \tag{2.367}$$

Thus, only $n_{22}$ and $n_{12}$ are needed, and can be generated recursively using

$$n_{12}^{(K)}(Z) = Z n_{12}^{(K-1)}(Z) - R_K n_{22}^{(K-1)}(Z) \tag{2.368}$$

$$n_{22}^{(K)}(Z) = n_{22}^{(K-1)}(Z) - R_K Z n_{12}^{(K-1)}(Z) \tag{2.369}$$

If the $(N + 1)$ layer is apsorptive, then the impulse response transfer function of the Earth model can be obtained as

$$G^{(N)}(Z) = \frac{E^{(N)}(Z)}{A^{(N)}(Z)} \tag{2.370}$$

where

$$A^{(N)}(Z) = n_{22}^{(N)}(Z) - R_0 Z^N n_{12}^{(N)}(Z^{-1}) \tag{2.371}$$

$$E^{(N)}(Z) = -Z^N n_{12}^{(N)}(Z^{-1}) \tag{2.372}$$

with

$$n_{12}^{(1)}(Z) = -R_1$$

$$n_{12}^{(1)}(Z) = 1$$

$$G^{(0)}(Z) = 0$$

The numerator and denominator $E^N(Z)$ and $A^N(Z)$ can be expressed in polynomial forms as

$$E^{(N)}(Z) = \sum_{j=1}^{N} \varepsilon_j^{(N)} Z^j \tag{2.373}$$

$$A^{(N)}(Z) = \sum_{j=0}^{N} a_j^{(N)} Z^j \tag{2.374}$$

with

$$a_0^{(N)} = 1$$

If one denotes the output by $Y(K)$ and the input by $M(K)$, then the signal time domain representation is

$$Y(K) + \sum_{j=1}^{N} a_j^{(N)} Y(K - j) = \sum_{j=1}^{N} \varepsilon_j^{(N)} M(K - j) \tag{2.375}$$

Using the defining relations, the coefficients of the characteristic polynomial $a_j^{(N)}$ can be obtained. Silvia (1977), gives the values in terms of the serial correlation defined by

$$\phi_j^{(N)} = \sum_{i=0}^{N} R_i R_{i+j} \qquad N \geq j \geq 1 \tag{2.376}$$

$$\phi_0^{(N)} = 1 \tag{2.377}$$

Table 2.2 gives the expressions corresponding to $N = 1,\dots, 4$. Note that residual terms such as $R_0$, $R_1$, $R_2$, $R_3$ are present. If one assumes a small reflection coefficient approximation, then these residuals may be neglected giving

$$a_j^{(N)} \cong \phi_j^{(N)} \tag{2.378}$$

$$a_j^{(N)} \cong \sum_{i=0}^{N} R_i R_{i+j} \tag{2.379}$$

**TABLE 2.2**

| N | $a_1^{(N)}$ | $a_2^{(N)}$ | $a_3^{(N)}$ | $a_4^{(N)}$ |
|---|---|---|---|---|
| 1 | $\phi_1^{(1)}$ | — | — | — |
| 2 | $\phi_1^{(2)}$ | $\phi_2^{(2)}$ | — | — |
| 3 | $\phi_1^{(3)}$ | $\phi_2^{(3)} + R_0 R_1 R_2 R_3$ | $\phi_2^{(3)}$ | — |
| 4 | $\phi_1^{(4)}$ | $\phi_2^{(4)} + R_4 R_3 R_2 R_1 + R_4 R_3 R_1 R_0 + R_3 R_2 R_1 R_0$ | $\phi_2^{(4)} + R_4 R_3 R_2 R_0 + R_4 R_2 R_1 R_0$ | $\phi_4^{(4)}$ |

**TABLE 2.3**

| N | $\Sigma_1^{(N)}$ | $\Sigma_2^{(N)}$ | $\Sigma_3^{(N)}$ | $\Sigma_4^{(N)}$ |
|---|---|---|---|---|
| 1 | $R_1$ | — | — | — |
| 2 | $R_1$ | $R_2$ | — | — |
| 3 | $R_1$ | $R_2 + R_3 R_2 R_1$ | $R_3$ | — |
| 4 | $R_1$ | $R_2 + R_4 R_3 R_1 + R_3 R_2 R_1$ | $R_3 + R_4 R_2 R_1 + R_4 R_3 R_2$ | $R_4$ |

In a similar fashion, the coefficients of the reflection polynomial $E^{(Z)}$ are found, as given in Table 2.3. If one uses the same small reflection approximation, one has

$$\varepsilon_j^{(N)} \cong R_j \tag{2.380}$$

It can thus be seen that polynomial $E^{(N)}(Z)$ contains the reflection information required for seismic data processing.

**Mendel's Model**

Mendel et al. (1980) developed a quasi-state space model for layered media that uses the ideas presented in the models discussed above. The model form introduced by Mendel is given by the layer-ordered relations

$$\tilde{d}_1(t + \tau_1) = - R_0 u_1(t) + (1 + R_0) m(t)$$
$$u_1(t + \tau_1) = - R_1 d_1(t) + (1 - R_1) u_2(t)$$
$$\vdots$$
$$\left. \begin{aligned} \tilde{d}_K(t + \tau_K) &= (1 + R_{K-1}) \tilde{d}_{K-1}(t) - R_{K-1} u_K(t) \\ u_K(t + \tau_K) &= R_K \tilde{d}_K(t) + (1 - R_K) u_{K-1}(t) \end{aligned} \right\} \quad K = 2, 3, \ldots, N - 1$$
$$\tilde{d}_N(t + \tau_N) = (1 + R_{N-1}) \tilde{d}_{N-1}(t) - R_{N-1} u_N(t)$$
$$u_N(t + \tau_N) = R_N \tilde{d}_N(t)$$

Note that

$$\tilde{d}_K(t) \triangleq d_K(t - \tau_K)$$

The above set of equations are dynamic with multiple time delays (generally incremental non-equal). The equations are referred to as casual functional equations.

A quasi-state space representation is obtained by defining

$$\underline{X}(t) = \text{col.}[u_1(t), \tilde{d}_1(t), \ldots, u_N(t), \tilde{d}_N(t)] \tag{2.381}$$

The $2N \times 2N$ matrix operator $\tilde{\underline{Z}}$ is defined by

$$\tilde{\underline{Z}} \triangleq \text{diag.}[Z_1, Z_1, Z_2, Z_2, \ldots, Z_K, Z_K] \tag{2.382}$$

where $\tilde{\underline{Z}}_i$ is a scalar operator denoting a $\tau_i$ sec time delay

$$\tilde{\tilde{Z}}_i \ f(t) = f(t - \tau_i)$$

As a result, one has

$$\tilde{\underline{Z}}^{-1} \underline{X}(t) = \underline{A} \underline{X}(t) + \underline{b} \underline{m}(t) \tag{2.383}$$

$$y(t) = \underline{C}^T \underline{X}(t) + R_0 \underline{m}(t) \tag{2.384}$$

This equation provides a conceptual solution given by

$$\underline{X}(t) = [\tilde{\underline{Z}}^{-1} - \underline{A}]^{-1} \underline{b} m(t) \tag{2.385}$$

and

$$y(t) = \left[ \underline{C}^T [\tilde{\underline{Z}}^{-1} - \underline{A}]^{-1} \underline{b} + R_0 \right] m(t) \tag{2.386}$$

The transfer function of the system will be a ratio of two polynomials in $Z_1, Z_2, \ldots, Z_N$, which should reduce to Robinson's formula when $Z_1 = Z_2 = \ldots = Z_N$. As an example, the two-layer case of Nahi, et al. (1978) gives

$$\frac{Y(\tilde{Z})}{M(\tilde{Z})} = \frac{R_0 + R_0 R_1 R_2 Z_2^2 + R_1 Z_1^2 + R_2 Z_1^2 Z_2}{1 + R_1 R_2 Z_2^2 + R_0 R_1 Z_1^2 + R_0 R_2 Z_1^2 Z_2^2} \tag{2.387}$$

for the case of equal $Z_i$s, i.e., $\tau_1 = \tau_1 = \ldots \tau_N \triangleq \tau$, as state space solution is possible, since the model reduces to

$$\underline{X} = [(K+1)\tau] = \underline{A} \underline{X}(K\tau) + \underline{b} m(K\tau) \tag{2.388}$$

whose solution is

$$\underline{X} = [(K+1)\tau] = \underline{A}^K \underline{X}(0) + \sum_{i=0}^{K} \underline{A}^{K-1} \underline{b} m(i\tau) \tag{2.389}$$

It appears that no solution to the general case has been documented in the literature.

## Nonparametric Models

Synthetic seismograms of the nonparametric type are developed with emphasis on the nature of the reflection coefficients. Three modeling approaches are reviewed here.

## PFC Approach

A nonparametric Earth model based on an approximation of the reflection coefficients is proposed in the work of Peterson, Fillippone, and Croker in (1955). For normal incidence, plane wave propagation between two media, the reflection coefficient is given in terms of acoustic impedences by

$$R = Z_2 - Z_1 / Z_2 + Z_1 \tag{2.390}$$

Assuming incremental changes, one sets

$$\Delta Z = Z_2 - Z_1 \tag{2.391}$$

$$Z \cong Z_1 \cong Z_2 \tag{2.392}$$

thus

$$R = \Delta Z / 2Z \tag{2.393}$$

As a result an approximate expression for the reflection coefficient between the two media is given by

$$R = \frac{1}{2}\Delta[\ln(Z)] \tag{2.394}$$

In multilayer media, the $i$th reflection coefficient is given by

$$R_i = \frac{1}{2}\Delta[\ln(Z_{i+1}) - \ln(Z_i)] \tag{2.395}$$

The seismograph representing the signal $y(t)$ received by the hydrophones is given by the summation

$$y(t) = m(t) + \sum_{i=1}^{N} R_i m(t - \tau_i) \tag{2.396}$$

where $m(t)$ is the input signal and the delays $\tau_i$ are given by the recursive relation

$$\tau_i = \tau_{i-1} + 2C_i^{-1}[d_i - d_{i-1}] \tag{2.397}$$

The velocity of sound propagation in medium $i$ is denoted by $C_i$. The vertical distance of the $i$th interface from reference is denoted by $d_i$.

## SLM Approach

In addition to the PFC approximation, Sengbush, Lawrence, and McDonald (1961) assume that the layer's density is velocity dependent according to this relation

$$\rho = KC^m \tag{2.398}$$

As a result, the reflection coefficient in terms of velocity of propagation is obtained as

$$R = \frac{m+1}{2}\Delta[\ln C] \tag{2.399}$$

SLM obtains an expression for a reflectivity function $\tilde{r}(t)$ by extending the concept to the continuous case, which is given by

$$\tilde{r}(t) = \frac{d}{dt} \ln C(t) \tag{2.400}$$

A synthetic seismogram can be obtained from a velocity log $C(t)$ as the output of a filter due to an input $\tilde{r}(t)$. The filter's impulse response is simply that of the shot pulse $m(t)$.

## BGW Approach

A method for computing the synthetic seismogram assuming variable sound propagation velocity profiles is due to Berryman, Goupillaud, and Waters (1958). In their paper, transformation of variables is introduced into the particle displacement equation.

$$\frac{\partial^2 \zeta(x,t)}{\partial t^2} = c^2 \frac{\partial^2 \zeta(x,t)}{\partial x^2} + 2c \frac{\partial c}{\partial x} \frac{\partial \zeta(x,t)}{\partial x} \tag{2.401}$$

The main BGW equation is given by

$$C^2 \frac{d^2 \pi}{dx^2} + \omega^2 \pi = 0 \tag{2.402}$$

The new variable $\pi(x)$ is given by

$$\pi(x) = C^2 d\zeta(x,t) / dx \tag{2.403}$$

The displacement is assumed to be in sinusoidal form. The pressure $p(x,t)$ can be obtained in terms of the new variable as

$$p(x,t) = \rho \pi(x) e^{j\omega t} \tag{2.404}$$

A closed-form solution to Eq. (2.402) can be obtained if the velocity profile in a layer $K$ is assumed to vary linearly with $x$ according to

$$C(x) = C_K + b_K(x - x_K) \tag{2.405}$$

The constants $C_K$ and $b_K$ are characteristics of the medium. The solution is of the form

$$\pi_K(x) = \pi_K^+ \exp.\left\{ \frac{1}{2}(1 - \beta_K)\gamma_K \right\} + \pi_K^- \exp.\left\{ \frac{1}{2}(1 + \beta_K)\gamma_K \right\} \tag{2.406}$$

The constant $\beta_K$ is given by the expression

$$\beta_K^2 = 1 - (4\omega^2 / b_K) \tag{2.407}$$

The new variable $\gamma_K$ is defined as

$$\gamma_K = \ln[1 + \{b_K(x - x_K) / C_K\}] \tag{2.408}$$

A layer reflection coefficient denoted by $R_K$ is the ratio

$$R_K = \pi_K^- / \pi_K^+ \tag{2.409}$$

BGW shows that a recursive form to obtain this coefficient is given by:

$$R_{K-1} = \frac{b_{K-1}\beta_{K-1} + (b_K - b_{K-1}) - b_K\beta_K(1 - R_K/1 + R_K)}{b_{K-1}\beta_{K-1} - (b_K - \beta_{K-1}) + b_K\beta_K(1 - R_K/1 + R_K)} e^{-\beta_{K-1}\ln(C_K/C_{K-1})} \tag{2.410}$$

To produce a synthetic seismogram, the bottom layer is assumed to have zero reflection and thus

$$R_N = 0$$

The computations are then carried out backward to obtain $R_{N-1}, R_{N-2}, \ldots, R_0$ for a specific frequency $\omega$. The overall synthetic seismogram is obtained through the use of the Fourier integral.

## Time Domain Attribute Extraction for Lossless Media

Some aspects of trace feature extraction from the reflected signal waveforms subsequent to heave compensation via time domain representations are discussed below. The acoustic pressure wave travels through the subsurface layered media and undergoes multiple reflections as it impinges on boundaries between successive layers. The signal received at the sensors (hydrophones) contains replicas of the original source signal that are corrupted by noise components due to various sources and modulated by the transmission media.

Assuming a linear lossless model of the wave propagation, one can express the received signal $y(t)$ as the sum of $N_L$ terms, each consisting of a version of the source signal $x(t)$ delayed by a cumulative two-way travel time $\tau_i$ with an amplitude scale factor $a_i$ as well as an additive noise term. The amplitude scale factors are related to the layers' reflection coefficients and the delay times are related to layer depth and the velocity of sound propagation in the given media. Thus

$$y(t) = \sum_{i=1}^{N_L} \alpha_i x(t - \tau_i) + v(t) \tag{2.411}$$

According to de Figueiredo and Shaw (1987), the model Eq. (2.411) is called a Tauberian approximation of the actual trace. In essence, the required information about the media is extracted in a time domain-based approach using peak and delay parameter detection techniques performed with the aid of Eq. (2.411). Figure 2.25 shows in graphic form how the Tauberian model arises as a natural consequence of wave propagation through media with varying properties such as the acoustic impedance. Performing the required information extraction task may appear to be straightforward. Many detection procedures are available, and we review three schemes in this section.

### Sequential Correlation-Based Detection

The estimation of delay and amplitude parameters can be carried out effectively using cross-correlation processing combined with minimum variance analysis (El-Hawary, 1985). Two steps are involved in this procedure.

The first step uses the fact that a signal such as $x(t)$, when cross-correlated with the signal $y(t)$, which is the sum of delayed replicas of $x(t)$ plus a random waveform will yield a cross-correlation function exhibiting large peaks at the values of the respective time delays. Peak and delay times are estimated using the cross-correlation between source and received signals. The result is a cross-correlation function with peaks at the unknown delay times. A direct search for the extrema results in an unrefined estimate of

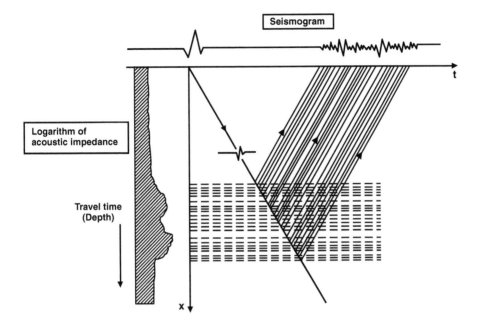

**FIGURE 2.25**  Elements of a marine seismic trace.

the delay parameters. In this step it is important that a threshold be established on the amplitude of the extrema considered. In practical implementations, past experience is relied on in a trial-and-error procedure to set a threshold that is appropriate for the area surveyed. A database of prior knowledge is an important component of an automated implementation of this fast procedure.

The second step refines the estimates using minimum variance information. For this purpose, the problem resolves to finding an optimal parameter estimate of the amplitude scale factors with optimality defined in the minimum of error variance sense. The optimal estimates of the amplitude scale factors are given by

$$\hat{a}_i = \frac{[(\sigma_v^2 / \sigma_{a_i}^2)\bar{a}_i + \phi_{xy}(\hat{\tau}_i)]}{[(\sigma_v^2 / \sigma_{a_i}^2) + \phi_{xx}(0)]} \tag{2.412}$$

Note that prior knowledge of the statistics of the noise process $v(t)$ is needed. This knowledge can also be achieved through the use of a matched filter implementation, where the impulse response of the filter is a reverse time replica of the source waveform. The matched filter simply maximizes the signal-to-noise ratio at the filter output.

Without *a priori* knowledge an estimate of peak amplitude is given by the ratio of cross-correlation of source and receiver signals to the autocorrelation of the source signal. Thus

$$\hat{a}_1 = \frac{\phi_{xy}(\hat{\tau}_i)}{\phi_{xx}(0)} \tag{2.413}$$

### Linearized Delay Model Detection

In this procedure (El-Hawary and Vetter, 1980), it is assumed that each delayed replica of the source signal is approximated by a first-order Taylor expansion about an estimate of each time delay.

$$\Delta\tau_i = \hat{\tau}_i - \tau_i \tag{2.414}$$

A further assumption is made that a rough estimate of the delays involved is available. The delayed replica is then represented by the values of source signal and its derivative, both evaluated at the estimate of the delay.

$$x(t - \tau_i) = x(t - \hat{\tau}_i) - \Delta\tau_i x'(t - \hat{\tau}_i) \tag{2.415}$$

As a result an approximation to the Tauberian expansion can be written as

$$y(t) = \sum_{i=1}^{N_L} a_i x(t - \hat{\tau}_i) + b_i x'(t - \hat{\tau}_i) + v(t) \tag{2.416}$$

The new variables $b_i$ are given by the products

$$b_i = a_i \Delta\tau_i \tag{2.417}$$

In discrete for Eq. (2.416) is written in the familiar linear measurement model given by

$$Y = HX + V \tag{2.418}$$

The problem, therefore, is one of linear parameter estimation, which is then solved efficiently using recursive algorithms.

It is important to realize that the choice of the estimate of the delays plays a central role in determining the success of this procedure. Aids in choosing these values include the requirement of consistency among successive records and in some instances (especially in the initial processing phase), it may be necessary to use results of a sequential correlation detector.

### Event Enhancement Filtering

Event enhancement filtering is a heuristic-based procedure (El-Hawary, 1982) that has been applied to field data successfully for enhancing image reflections from subsurface layered media. The effect of amplitude parameters is amplified by a squaring process. The positive and negative parts of the reflected image play a main role in event enhancement. The filtering process is essentially one of prewhitening through differentiation, followed by shifting and multiplication to sharpen the signal.

Note that if the noise term in the Tauberian expression is neglected, a more direct approach can be employed to extract the amplitude scale factors and delays. This method, suggested by de Figueiredo and Shaw (1987), involves use of the frequency domain transfer function evaluated at equidistant frequencies and obtaining a least squares estimate of an intermediate unknown parameter vector. This optimal estimate is then used to define the coefficients of a polynomial equation in a complex variable related to the amplitude scale factors. Solution of this polynomial equation yields the required delay time results. With the delays available, the amplitude scale factors are obtained from a set of linear equations. This method is referred to as the Prony method.

Note that the outcome of an algorithm may not provide satisfactory results that conform with other verification results, depending on the source of the given data. Figure 2.26 shows a successful enhancement of an underwater layer's image shown on the left side, while an expansion of a number of individual traces is shown on the right side to illustrate the sharpening effect of the filtering process, showing consistency between records. It is, therefore, important early in the analysis phase to determine which scheme to use. A knowledge-based system recognizing the selection criteria and limitation of each scheme is, therefore, of extreme importance. This part of the knowledge-based system can be enabled as a production rule. Substantial information is available in the seismic waveform beyond that reflected in amplitude and delay time, which is currently unexploited yet available to describe the layers traversed by the interrogating medium. This part of the knowledge-based system can be encoded as pattern

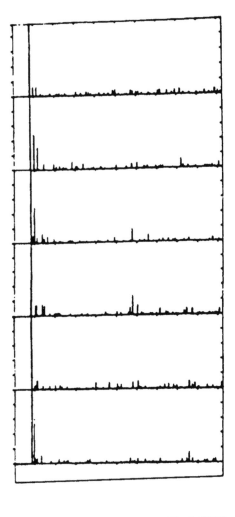

**FIGURE 2.26** Typical enhanced underwater layers' image.

classifiers. Thus, an extremely novel artificial intelligence system can be designed as a hybrid system containing both expert system and pattern recognition technologies.

## Time Domain Attribute Extraction Accounting for Losses

A discussion of attribute extraction in the time domain by identifying media properties accounting for attenuation effects, and a review of the basic problem and techniques available for processing the image follow. The inclusion of attenuation effects is discussed in terms of the added complexity. An algorithm

**FIGURE 2.27**  Typical illustration of signal decay due to attenuation.

to extract physically related information for amplitude, attenuation, and delay parameters contained in the reflected image records using an iterative inversion process based on Newton's method is described.

Attenuation causes the amplitude of a plane sine wave to decay exponentially with the distance traveled in the medium as shown graphically in Figure 2.27. There are at least three possible mechanisms that cause attenuation (Hamilton, 1971, 1972). These include anelastic losses in the frame of sediment particles, viscous losses due to relative movement of sediments and pore water, and losses due to volume scattering.

Assuming propagation in one layer at a given frequency $\omega$, the amplitude of the sound pressure wave at distance $d$ from the source is given by

$$M(d) = M_0 e^{-\alpha(\omega)d} \tag{2.419}$$

The attenuation rate $\alpha(w)$ is a function of the frequency of the propagating pressure wave.

The product of the attenuation rate and distance traveled is assumed to be proportional to the radian frequency $\omega$. The medium damping ratio $D_i$, which depends on the thickness of the layer and the attenuation rate $\alpha$, is defined. Thus, an expression for the medium amplitude scale factor $A_i$ in terms of frequency and damping ratio is

$$A_i = a_i e^{-\omega D_i} \tag{2.420}$$

The new parameter $D_i$ is assumed to be frequency independent. The attenuation model assumed here requires that a frequency domain approach be used in the image processing algorithm.

## A Frequency Domain Model

The basic model adopted to represent the return pressure signal for each channel $y(t)$ assumes that $y(t)$ is the sum of delayed replicas of the source signal $x(t)$. The replicas are scaled and delayed versions of the original signal. The number of replicas is equal to the number of layers in the model $N_L$. The amplitude scale factors are denoted by $A_i$ and vary with layer thickness and density, they are also frequency dependent as indicated in Eq. (2.420). The delay parameters are denoted by $\tau_i$ and are related to layer thickness and sound propagation velocity.

$$y(t) = \sum_{i=1}^{N_L} A_i x(t - \tau_i) \tag{2.421}$$

Equation (2.421) must be written in the frequency domain by finding the Fourier transform of both sides. An approximation to the actual expression can be written as

$$Y(\omega) = \sum_{i=1}^{N_L} A_i e^{-j\omega\tau_i} X(\omega) \tag{2.422}$$

The exact evaluation of the frequency response of Eq. (2.422) employs the modulation property of the Fourier transform. The approximation involved here assumes that the amplitudes are time invariant.

Now a frequency domain transfer function can be obtained and is defined by

$$Z(\omega) = \frac{Y(\omega)}{X(\omega)} \tag{2.423}$$

As a result, the complex-valued transfer function is expressed as

$$Z(\omega) = \sum_{i=1}^{N_L} A_i e^{-j\omega\tau_i} \tag{2.424}$$

The interpretation task involves finding the real parameters $a_i$, $D_i$, and $\tau_i$ for all layers considered $i = 1, 2, \ldots, N_L$, given the transfer function $Z(\omega)$. As a preparatory step, we separate real and imaginary portions of the transfer function to obtain

$$Z_r(\omega) = \sum_{i=1}^{N_L} A_i \cos\omega\tau_i \tag{2.425}$$

and

$$-Z_{\text{Im}}(\omega) = \sum_{i=1}^{N_L} A_i \sin\omega\tau_i \tag{2.426}$$

Note that the Fourier transform of $Z$ is obtained at the discrete frequencies $\omega_j$ over the pass band of the acoustic signal. Assuming that the frequency index $j$ ranges from 1 to $N_\omega$

$$Z_r(\omega) = \sum_{i=1}^{N_L} A_i \cos\omega\tau_i + \eta_r \tag{2.427}$$

and

$$-Z_{\text{Im}}(\omega) = \sum_{i=1}^{N_L} A_i \sin\omega\tau_i + \eta_{\text{Im}} \tag{2.428}$$

Equations (2.427) and (2.428) are the basis for a nonlinear parameter estimation algorithm. Note that noise components $\eta_r$ and $\eta_{\text{Im}}$ are present in the real and imaginary portions of the transfer function. Our search is for the parameters $a_i$, $D_i$, and $\tau_i$ for all layers considered $i = 1, 2, \ldots, N_L$.

**Layer Parameter Estimation**

The aim of parameter estimation technique is to evaluate the parameters to minimize the noise components in the sum of the squares sense

$$I = \sum_{j=1}^{N_\omega} \eta_r^2(\omega_j) + \eta_{\text{Im}}^2(\omega_j) \tag{2.429}$$

The minimization is carried out with respect to the unknown layer parameters. The vector $U$ of the unknown parameters is defined by

$$U = \begin{bmatrix} u_1 \\ u_2 \\ \vdots \\ u_{N_L} \end{bmatrix} \tag{2.430}$$

Each individual subvector is given by

$$u_i = \begin{bmatrix} a_i \\ D_i \\ \tau_i \end{bmatrix} \tag{2.431}$$

The condition for least square parameter estimation is

$$f = \left[ \frac{\partial I}{\partial U} \right] = 0 \tag{2.432}$$

Equation (2.432) represents $3N_L$ individual equations, which are written as

$$f_{lk} = \sum_{j=1}^{N_\omega} \eta_r \left[ \frac{\partial \eta_r}{\omega u_{lk}} \right] + \eta_{\mathrm{Im}} \left[ \frac{\partial \eta_r}{\omega u_{lk}} \right] = 0 \tag{2.433}$$

Here, Eqs. (2.427) and (2.428) are rewritten as

$$\eta_r(\omega_j) = Z_r(\omega_j) - \sum_{i=1}^{N_L} A_i \cos \omega_j \tau_i \tag{2.434}$$

$$\eta_{\mathrm{Im}}(\omega_j) = -Z_{\mathrm{Im}}(\omega_j) - \sum_{i=1}^{N_L} A_i \sin \omega_j \tau_i \tag{2.435}$$

Next, obtain the partial derivatives required and expand Eq. (2.433) to the following form:

$$f_{l1} = \sum_{j=1}^{N_\omega} -e^{-\omega_j D_l}[\eta_r(\omega_j) \cos \omega_j \tau_l + \eta_{\mathrm{Im}}(\omega_j) \sin \omega_j \tau_l] = 0 \tag{2.436}$$

$$f_{l2} = \sum_{j=1}^{N_\omega} a_l \omega_j e^{-\omega_j D_l}[\eta_r(\omega_j) \cos \omega_j \tau_l + \eta_{\mathrm{Im}}(\omega_j) \sin \omega_j \tau_l] = 0 \tag{2.437}$$

and

$$f_{l3} = \sum_{j=1}^{N_\omega} a_l \omega_j e^{-\omega_j D_l}[\eta_r(\omega_j) \sin \omega_j \tau_l + \eta_{\mathrm{Im}}(\omega_j) \sin \omega_j \tau_l] = 0 \tag{2.438}$$

These equations completely specify the required estimates of the parameters.

## An Iterative Inversion Process

In contrast to the problem that neglects attenuation, the parameter estimation Eqs. (2.436), (2.437), and (2.438) are nonlinear and require an iterative solution technique. The Newton-Raphson iterative technique may be employed to perform the inversion process. Starting with the initial guess of the unknown parameters, denoted by $U^{(0)}$, we obtain successively (hopefully) improved estimates of the unknowns according to Newton's iterative formula

$$U^{k+1} = U^k + \Delta^k \tag{2.439}$$

The vector of the incremental improvements $\Delta_k$ on the $k$th iteration is obtained as the solution of the following system of linear equations that involves the Jacobian matrix $J$ evaluated at the current values of the unknowns. Thus

$$f(u^i) + J(u^i)\Delta^i = 0 \tag{2.440}$$

The Newton-Raphson method is a powerful solution technique that requires the evaluation of the Jacobian matrix formed by taking the partial derivative of each equation with respect to the unknown variables of interest. For the given structure of the layered media, constructing the Jacobian with reference to partitioned submatrices that correspond to each layer and to the layers' associated unknowns is more convenient. The required elements of each submatrix are detailed in El-Hawary (1988). Newton's method is known for its quadratic convergence provided that an initial guess, which is reasonably close to the actual solution, is available.

The complexity of this iterative inversion process is due to the requirement of evaluating the Jacobian matrix J whose dimensions are $3N_L \times 3N_L$ at each iteration. The elements involve multiple summations and trigonometric function evaluations as indicated earlier. A second complicated factor arises since the evaluation of the incremental improvements requires solving a set of linear equations that is not sparse. Finally, a good initial guess of the solution must be available to guarantee convergence of the iterations.

In order to avoid these complexities, a number of improvements can be introduced. Use of an initial guess of the unknown amplitudes and delays based on a prior solution for the same data record, neglecting attenuation, is possible. A second possibility involves the evaluation of the Jacobian only on the first few iterations and subsequently fixing it in a quasi-Newton implementation. Third, for a given area, start by assuming a minimum number of layers involved and increase the number if warranted.

## Conclusions

This chapter offered a comprehensive review of some approaches to modeling underwater acoustic wave propagation known as synthetic seismograms. A model involves tracing the input wave as it travels through the media to the receiving equipment. Modeling assumptions taken include a flat completely elastic layered medium structure with plane acoustic wave propagation, and normal wave incidence. The models considered differ in the degree of detail and mathematical refinement. Models of acoustic wave propagation in lossless layered media systems were classified according to the degree of detail involved. These include a parametric representation involving two signals for each layer based on decomposing the solution to the wave particle displacement, (particle velocity, and pressure) into forward and backward traveling components in each layer. The second category is a convolution summation, which is sometimes referred to as a nonparametric representation. The models reviewed differ mainly in their approach to modeling the reflection coefficients at the interfaces.

The chapter continued with a discussion of a selection of techniques that allow the analyst to extract time domain attributes of the underwater media in a lossless environment. Here we discussed sequential correlation based detection, linearized delay model detection, and event enhancement filtering. We concluded with time domain attribute extraction accounting for losses, based on a frequency domain model that leads to an iterative process for layer parameter estimation in lossy media.

# References

Aminzadeh, F. and Chatterjee, S., Application of clustering in exploration seismology, in *Pattern Recognition and Image Processing*, F. Aminzadeh, (Ed.): Geophysical Press., London, 1987, 372–387.

Berryman, L. H., Goupillaud, P. L., and Waters, K. H., Reflections from Multiple Transition Layers, *Geophysics* 23, 223–243, 1958,

Bhattacharya, R., *Dynamics of Marine Vehicles*, Wiley-Interscience, New York, 1978.

Bois, P., Autoregressive pattern recognition applied to the delimitation of oil and gas reservoirs: *Geophys. Prosp.*, 28, 572–591, 1980.

Bois, P., Determination of nature of reservoirs by use of pattern recognition algorithm with prior learning, *Geophys. Prosp.*, 29, 687–701, 1981.

Bonamini, R., De Mori, R., Lettera, A., and Rogerro, R., An electro–cardiographic signal understanding system, in *Pattern Recognition, Theory and Applications*, J. Kittler, K.S. Fu, and L.F. Pau, (Eds.), D. Reidel, Dordrecht, 1981, 443–464.

Chen, C. H., Recognition of underwater transient patterns, *Pattern Recognition*, 18, 6, 485–490, 1985.

DeFigueiredo, R. J. P. and Shaw, S., Spectral and artificial intelligence methods for seismic stratigraphic analysis, in *Pattern Recognition and Image Processing*, F. Aminzadeh, (Ed.): Geophysical Press, London, 1987, 426–445.

Dodds, D. J., *Bottom-Interacting Ocean Acoustics*, Vol. IV:5, Plenum Press, New York, 1980.

El-Hawary, F., (1982). Seismic signal processing for geotechnical properties applied to marine sediments, in *Associate Committee for Research on Shoreline Erosion and Sedimentation, National Research Council of Canada*, Proceedings of Workshop on Atlantic Coastal Erosion and Sedimentation, 1982.

El-Hawary, F. and Vetter, W. J., estimation of subsurface layer parameters by use of a multiple-reflection model for layered media, in *POAC* (Proceedings of the Fourth International Conference on Port and Ocean Engineering Under Arctic Conditions held in St. John's, Newfoundland, 1977), 1087–1099.

El-Hawary, F. and Vetter, W. J., Subsurface layered media parameter estimation using a linearized multiple reflection model, in *Oceans '78* (Proceedings of IEEE Oceans Conference held in Washington, D.C., 1978).

El-Hawary, F. and Vetter, W. J., Spatial parameter estimation for ocean subsurface layered media, *Canadian Electrical Engineering Journal* 5, 28–31, 1980.

El-Hawary, F., Compensation for source heave by use of Kalman filter, *IEEE J. Oceanic Eng.* OE-7, 89–96, 1982.

El-Hawary, F., Accounting for attenuation in sub-surface layered media parameter estimation: A frequency domain approach, presented at *IEEE Oceans '83 Conference*, 45–49, 1983.

El-Hawary, F., An approach to seismic information extraction, in *Time Series Analysis: Theory and Practice*, Vol. 6, O.D. Anderson, J. K. Ord, and E. A. Robinson (Eds.): Elsevier, Amsterdam, 1985, 223–238.

El-Hawary, F., An approach to extract the parameters of source heave dynamics, *Can. Elec. Eng. J.* 12, 1, 19–23, 1987a.

El-Hawary, F., Image processing of underwater acoustic arrays geophysical exploration including attenuation effects, presented at *International Conference on Digital Signal Processing*, Florence, Italy, 623–627, 1987b.

El-Hawary, F., Parallel Kalman filtering for heave compensation using higher order models, presented at *IEEE Pacific Rim Conference on Communications, Computer, and Signal Processing*, Victoria, B.C., 355–359, 1987c.

El-Hawary, F., Image analysis methods from seabed reflections and multiple reflections, *Int. J. Pattern Recognition Artificial Intelligence* 1, 2, 261–272, 1987d.

El-Hawary, F., Role of parameter estimation and correlation techniques in remote imaging with emphasis on signal attenuation in varied media, presented at SPIE Wave Propagation and Scattering in Varied Media Conference, 927, 112–121, 1988.

El-Hawary, F. and Ravindranath, K. M., Application of array processing for parallel linear recursive Kalman filtering in underwater acoustic exploration, presented at *IEEE Oceans '86*, 1, 336–340, 1986.

El-Hawary, F. and Richards, T., Heave response modeling using higher order models, presented at *Int. Symp. Simulation and Modeling*, Santa Barbara, CA, 60–64, 1987.

El-Hawary, F. and Vetter, W. J., Event enhancement on reflections from subsurface layered media, *IEEE J. Oceanic Eng.* OE-7, 1, 51–58, 1981.

El-Hawary, F. and Vetter, W. J., Heave compensation of shallow marine seismic reflection records by Kalman filtering, presented at *IEEE Oceans '81*, 1981.

Fu., K. S., *Syntactic Pattern Recognition and Applications*, Prentice-Hall, Englewood Cliffs, NJ, 1982.

Gelb, A., *Applied Optimal Estimation*, MIT Press, Cambridge, MA, 1974.

Goupillaud, P. L., An approach to inverse filtering of near-surface layer effects from seismic records. *Geophysics* 26, 754–760, 1961.

Hagen, D. C., The application of principal component analysis to seismic data sets, presented at *Second Int. Symp. Computer-Aided Seismic Analysis and Discrimination*, 98–109, 1981.

Hamilton, E. L., Elastic properties of marine sediments, *J. Geophys. Res.* 76, 2, 1971.

Hamilton, E. L., Compressional wave attenuation in marine sediments, *J. Geophys.* 37, 4, 1972.

Hassab, J. C. and Chen, C. H., On constructing an expert system for contact localization and tracking, *Pattern Recognition*, 18, 6, 465–474, 1985.

Huang, K. Y. and Fu, K. S., Detection of bright spots in seismic signal using tree classifiers, *Geoexploration*, 23, 121–145, 1984.

Hutchins, R. W., Removal of tow fish motion noise from high resolution seismic profiles, presented at *SEG-U.S. Navy Symposium on Acoustic Imaging Technology and On Board Data Recording and Processing Equipment*, National Space Technology Lab., Bay St. Louis, MI, 1978.

Hutchins, R. W., Dodds, D. J., Parrot, R., and King, L. H., Characterization of sea floor sediments by geo-acoustic scattering models using high resolution seismic data, presented at *Oceanology Int. Conference*, Brighton, 1982.

Justice, J. H., Hawkins, D. J., and Wong, G., Multidimensional attribute analysis and pattern recognition for seismic interpretation, *Pattern Recognition*, 18, 6, 391–408, 1985.

Khattri, K. and Gir, R., A study of seismic signatures of sedimentation models using synthetic seisomograms, *Geophys. Prosp.*, 24, 454–477, 1976.

Khattri, K., Sinvhal, A., and Awashti, K., Seismic discriminants of stratigraphy derived from Monte Carlo simulation of sedimentary formations, *Geophys. Prosp.*, 27, 168–195, 1979.

Kung, S. Y., VLSI signal processing: From transversal filtering to concurrent array processing, in *VLSI and Modern Signal Processing*, S. Y. Kung, H. J. Whitehouse, and T. Kailath (Eds.): Prentice-Hall, Englewood Cliffs, NJ, 127–152, 1985.

Love, P. L. and Simaan, M., Segmentation of seismic section via image processing and AI techniques, *Pattern Recognition* 18, 6, 409–419, 1985.

Mathieu, P. G. and Rice, G. W., Multivariate analysis used in the detection of stratigraphic anomalies from seismic data, *Geophysics* 34, 507–515, 1969.

McCormick, M. E., *Ocean Engineering Wave Mechanics*, Wiley-Interscience, New York, 1973.

Medich, J. S., *Stochastic Optimal Linear Estimation and Control*, McGraw-Hill, New York, 1969.

Mendel, J. M. and Ashrafi, F. H., A survey of approaches to solving inverse problems for lossless layered media systems, *IEEE Transactions on Geoscience and Remote Sensing*, GE-18, 320–330, 1980.

Mendel, J. M., *Optimal Seismic Deconvolution: An Estimation-Based Approach*, Academic Press, New York, 1983.

Mendel, J. M., Some modeling problems in reflection seismology, *IEEE ASSP* 3, 2, 4–17, 1986.

Nahi, N. E., Mendel, J. M., and Silverman, L. M., Recursive Derivation of Reflection Coefficients from Noisy Seismic Data. Presented at IEEE Int. Conf. Acoustics, Speech, and Signal Processing, Tulsa, OK, 1978.

Peterson, R. A., Fillippone, W. R. and Croker, F. B., The Synthesis of Seismograms from Well Log Data, *Geophysics XX*, 516–538, 1955.

Price, W. G. and Bishops, R. E. D., Probabilistic Theory of Ship Dynamics, Wiley-Interscience, New York, 1974.

Robinson, E. A., *Multichannel Time Series Analysis with Digital Computer Programs*, Holden Day, San Francisco, 1967.

Robinson, E. A. and Durrani, T. S., *Geophysical Signal Processing*, Prentice-Hall, Englewood Cliffs, NJ, 1986.

Sengbush, R. L., Lawrence, P. L., and Mcdonald, F. J., Interpretation of Synthetic Seismograms, *Geophysics XXIV*, 138–157, 1961.

Severance, R. W., Optimum filtering and smoothing of buoy wave data, *J. Hydronaut.* 9, 69–74, 1975.

Silvia, M. T., Deconvolution of Geophysical Time Series, Doctoral Dissertation, Northeastern University, Boston, MA, 1977.

Sinvhal, A. and Khattri, K., Application of seismic reflection data to discriminate subsurface lithostratigraphy, *Geophysics* 48, 1498–1513, 1983.

Sinvhal, A., Khattri, K., Sinvhal, H., and Awashti, A. K., Seismic indicators of stratigraphy, *Geophysics* 49, 1196–1212, 1984.

Stremler, F. G., *Introduction to Communication Systems*, 2nd ed., Addison-Wesley, New York, 1977, 1982.

Travassos, R. H., Application of systolic array technology to recursive filtering, in *VLSI and Modern Signal Processing*, S. Y. Kung, H. J. Whitehouse, and T. Kailath, (Eds.), Prentice-Hall, Englewood Cliffs, NJ, 1985, 375–388.

Trorey, A. W., Theoretical seismograms with frequency and depth dependent absorption, *Geophysics* 27, 766–785, 1962.

White, J. E., *Seismic Waves: Radiation, Transmission, and Attenuation*, McGraw-Hill, New York, 1965.

Wuenschel, P. E., Seismogram Synthesis Including Multiples and Transmission Coefficients, *Geophysics XXV*, 106–129, 1960.

# 3

# Position Control
# Systems for Offshore
# Vessels

**Jann Peter Strand**
*ABB Industri AS*

**Asgeir J. Sørensen**
*The Norwegian University of
Science and Technology*

**Marit Rønæss**
*The Norwegian University of
Science and Technology*

**Thor I. Fossen**
*The Norwegian University of
Science and Control*

## 3.1  Marine Positioning Systems

*Jann Peter Strand and Asgeir J. Sørensen*

The history of automated closed-loop ship control starts with Elmer Sperry (1860–1930), who constructed the first automatic ship steering mechanism in 1911 for course keeping (Allensworth [2] and Bennet [6]). This device is referred to as the "Metal Mike," and it functioned much as a skilled pilot or a helmsman. "Metal Mike" did compensate for varying sea states by using feedback control and automatic gain adjustments. Later in 1922, Nicholas Minorsky (1885–1970) presented a detailed analysis of a position feedback control system where he formulated a three-term control law. Today it is referred to as Proportional-Integral-Derivative (PID) control, Minorsky, [19]. These three different behaviors were motivated by observing the way in which a helmsman steered a ship.

**FIGURE 3.1**   Marine operations in offshore oil and gas exploration.

In the 1960s systems for automatic control of horizontal position and course were developed. Such systems for simultaneous control of three horizontal motions (surge, sway, and yaw motions) are today commonly known as *dynamic positioning (DP) systems*. A dynamically positioned vessel is by the class societies, e.g., Det norske Veritas [8], American Bureau of Shipping [1], and Lloyd's Register of Shipping [18], defined as a vessel that maintains its position and heading (fixed location or predetermined track) exclusively by means of active thrusters. This is accomplished either by installing tunnel thrusters in addition to the main screw(s), or by using azimuthing thrusters, which can produce thrust in different directions.

While in DP operated ships the thrusters are the sole source of station keeping, the assistance of thrusters are only complementary to the mooring system in the case of *thruster-assisted position mooring (PM) systems*. Here, most of the position keeping is provided by a deployed anchor system. In severe environmental conditions, the thrust assistance is used to minimize the vessel excursions and line tension by mainly increasing the damping in terms of velocity feedback control. For turret-anchored ships (see Figs. 3.1 and 3.2) without natural weather-vaning properties, the thrusters are also used for automatic control of the heading, similar to DP operated vessels.

In this text, a *marine positioning system* is either defined as a *dynamic positioning* system or a *thruster-assisted position mooring* system. The offshore oil and gas industry has been a driving force in the development of these systems. Today positioning systems are used for a wide range of vessel types and marine operations. The different applications using marine positioning systems are:

- **Offshore oil and gas industry:** Typical applications in the offshore market are offshore service vessels, drilling rigs and drilling ships, shuttle tankers, cable and pipe layers, floating production off-loading and storage units (FPSOs), crane and heavy lift vessels, geological survey vessels, and multipurpose vessels. Cable and pipe laying are typical operations that also need tracking functionality.

- **Shipping:** Currently there is a trend toward more automatic control of marine/merchant vessels, beyond the conventional autopilot. This involves guidance systems coupled to automatic tracking control systems, either at high or low speed. In addition, more sophisticated weather routing and weather planning systems are expected. Automatic docking systems, and a need for precise positioning using DP systems when operating in confined waters will be used more in the future.

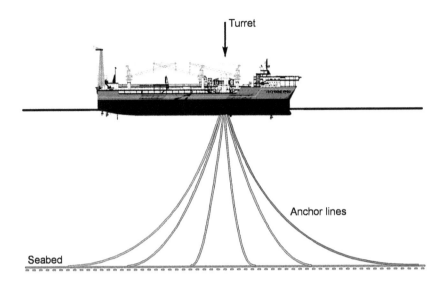

**FIGURE 3.2** Turret-anchored FPSO.

- **Cruise and yacht:** The cruise and yacht markets also make use of more automatic positioning control. In areas where anchors are not allowed due to vulnerable coral reefs, DP systems are used for station keeping. Precise positioning is also required for operating in harbors and confined waters.
- **Fisheries:** Application of more sophicsticated guidance, navigation, and control systems for positioning of ships during fishing are motivated by the need for precise positioning, reduced fuel consumption, and intelligent selective fishing.

The marine positioning system's functionality, design methods, sensors, and total reliability have improved over the years and are still a topic for research. The introduction of global satellite navigation systems, such as GPS (U.S. system) and GLONASS (Russian system), have increased the availability and reduced the cost of obtaining reliable position measurements, which of course are crucial for maintaining position. This has further increased the popularity of dynamic positioning systems beyond the oil and gas market such as the more traditional merchant vessels, cruise ships, yachts, and in fisheries.

DP systems have traditionally been a *low-speed* application, where the basic DP functionality is either to keep a fixed position and heading or to move slowly from one location to another. In addition, specialized tracking functions for cable and pipe layers, and remote operated vehicle (ROV) operations have been available. The traditional *autopilot* functionality has, over the years, become more sophisticated. Often a course correction function is available for correction of course set-point due to environmental disturbances and drifting, such that the vessel follows a straight line. Way-point tracking is used when a vessel is supposed to follow a predefined track, e.g., defined by several way-point coordinates. The trend today is for typical high-speed operation functionality to merge with the DP functionality, giving *one unified system* for all speed ranges and types of operations.

The first DP systems introduced in the beginning of 1960s were designed using conventional PID controllers in cascade with low pass and/or notch filters to suppress the wave-induced motion components. From the middle of the 1970s, more advanced control techniques based on linear optimal control and Kalman filter theory were proposed by Balchen et al. [4]. This work was later modified and extended by Balchen et al. [5], Grimble et al. [13] and [14], Fung and Grimble [12], Sælid et al. [23], and more lately by Sørensen et al. [25], Sørensen and Strand [26], and Strand [22].

*Thruster-assisted position mooring (PM) systems* have been commercially available since the 1980s. They provide a flexible solution for floating structures for drilling and oil and gas exploitation on the smaller and marginal fields. In PM systems the thrusters are complementary to the mooring system.

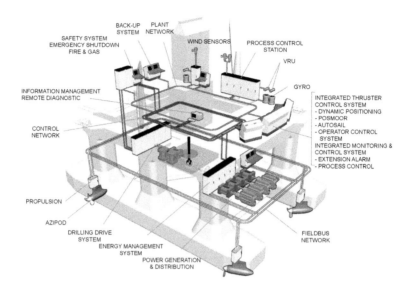

**FIGURE 3.3** Illustration of different systems and components in a large, integrated plant (printed with kind permission from ABB Industri AS).

Modeling and control of turret-moored ships are treated in Strand et al. [21]. Further extension of controller design using nonlinear theory of both DP and PM systems is addressed by Strand [22].

## System Overview

DP and PM systems may comprise the following subsystems (Hansen [16]).

- Power system. This comprises all units necessary to supply the DP system with power.
- Control system. This comprises:
  1. Computer/joystick systems.
  2. Sensor and position reference systems. Hardware, software, and sensors to supply information and/or corrections necessary to give accurate position and heading references.
  3. Display system and operator panels.
  4. Associated cabling and cable routing.
- Thruster system. This involves all components and systems necessary to supply the DP system with thrust force and direction. The thruster system includes thrusters with drive units and necessary auxiliary systems, including piping and main propellers and rudders, that are under the control of the DP system.
- Mooring System (applicable for moored vessels only).

### Power Plant

The power generationand distribution systems are divided into the following main parts:

- A *power plant* with prime mover and generators, the latter either a diesel engine or turbine driven.
- A *power distribution* system with main switchboards, usually split in two, three, or four sections. The normal operation mode for DP vessels is to split the network in at least two sections in order to be tolerant of a worst-case single failure, knocking out one power section.
- Voltage conditioners or filters for reducing harmonic interference.
- Transformers for feeding of alternate voltage levels.
- Low-voltage switchboards and motor control centers.

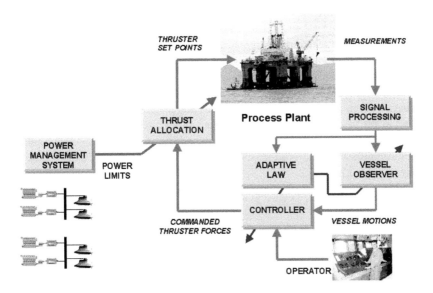

**FIGURE 3.4** Illustration of the functional modules in a positioning system, with close integration with the power management system.

- Rotating converters for frequency conversion and clean power supply.
- Uninterruptible power supply of sensitive equipment and automation systems.

## Automation System

The merging of software and hardware platforms in automation systems has enabled totally integrated automation systems comprising:

- DP/PM control systems.
- Autosailing system (including autopilot).
- Power management system for handling of generators, blackout prevention, power limitation, load sharing, and load shedding. For advanced vessels more sophisticated energy management systems are used for intelligent power planning and allocation.
- Vessel automation and HVAC control systems.
- Cargo and ballast control.
- Process automation system.
- Emergency shutdown, and fire and gas detection systems.
- Off-loading control system.

## Sensor and Position Reference Systems

It is common to divide the different measurements used by a positioning system into position reference systems and sensor systems. The most commonly used position reference systems are:

1. **Satellite Navigation Systems:** The most commonly used navigation system for marine vessels is Navstar GPS, which is a U.S. satellite navigation system with world coverage. An alternative system is the Russian system GLONASS, which only covers certain regions of the globe (see Parkinson and Spiker [20]). A European system, Gallileo, is currently under construction. For local area operations it is now possible to achieve meter accuracy by using *Differential* GPS (DGPS), and sub-decimeter accuracy by using *Carrier Differential* GPS (CDGPS). The development of *wide area augmentation systems* (WAAS) is expected to give meter accuracy across entire continents. When using a satellite navigation system, at least four satellites must be visible in order to compute

a reliable position estimate (three for sea level navigation). If the ship is entering a shadowed region, and redundant signals are not available, there will be a loss of position measurements. Other causes for degraded position measurements (or in the worst case, loss of measurement) are ionospheric disturbances, reflections in the water surface, etc. For DGPS the accuracy is typically within a 1-meter radius with 95% accuracy.

2. **Hydroacoustic Position Reference (HPR) Systems:** By using one or several transponders located on fixed positions on the seabed, and one or several transducers mounted under the hull, the position of the vessel is measured. The accuracy of such systems depends on the water depth and the horizontal distance between the transponder and the transducer. The most common principles for performing measurements include super short baseline (SBL), super/ultra short baseline (S/USBL), or long baseline (LBL).

3. **Taut Wire:** A *taut wire* system is used to measure the relative position of a floating vessel at rest. This system consists of a heavy load located at the sea floor. The load is connected to the vessel by a wire cord or heavy metal chain that is held in constant tension by using a winch on board the vessel. The angles at the top and bottom of the wire are measured, along with the length of the wire. Hence, the relative position $(x,y,z)$ can be computed by solving three geometric equations with three unknowns.

Other position reference systems are microwave systems (ARTEMIS, MICRORAGER, MICRO FIX), radiowave systems (SYLEDIS), optical systems (laser beams), and riser angle position measurement systems. The latter system is based on instrumentation on a marine riser attached to the vessel when performing drilling operations.

The sensor systems may comprise:

1. Gyrocompass and/or magnetic compass, which measures the heading of the vessel.
2. Vertical reference unit (VRU), which at minimum measures the vessel heave, roll, and pitch motions. Angular velocities are also often available. One of the main functions for the VRU is to adjust the position measurements provided by GPS, hydroacoustic position reference (HPR) systems, etc. for roll and pitch motions. For deep water DP operations, the accuracy of the roll and pitch signals must be high providing accurate HPR position measurements.
3. An inertial motion unit (IMU) typically contains gyros and accelerometers in 3 axes that can be used to measure the body-fixed accelerations in surge, sway, and heave; the angular rates in roll, pitch, and yaw; and the corresponding Euler angles (Britting [7], Titterton and Weston [27]). The IMU can be integrated in a filter (observer) with DGPS and HPR measurements, for instance, in order to produce accurate velocity estimates. In most cases only IMU Euler angles are used in conjunction with DGPS. This is a minimum configuration since the Euler angles are needed in order to transform the measured GPS position corresponding to the GPS antenna down to the vessel fixed coordinate system usually located in the centerline of the ship.
4. Wind sensors, which measure the wind velocity and direction relative to the vessel.
5. Draft sensors (used for vessels operated over a wide range of drafts).
6. Environmental sensors: wave sensors (significant wave height, direction, peak frequency), current sensors (velocity and direction at sea surface and at different depths). Environmental sensors are not a class requirement. These sensors are quite common on the most sophisticated offshore installations.
7. Other sensors depending on type of operation, e.g., for pipe layers, pipe tension are also measured and utilized by the DP system.

In many installations redundant measurements are available, and the number and types of measurements required are specified by certain class rules, see for example DnV [8]. Redundant measurements increase the safety and availability of the positioning system.

The heading of the vessel is usually measured by one or several gyrocompasses, which dynamically are accurate. During rapid turning operations, however, the gyros will produce a steady-state offset that gradually decreases to zero when the course is constant again. By using two GPS antennas, the heading can be measured even more accurately, without the drifting effect. In the case of anchored vessels, measurements of line length and tension are usually interfaced to the PM control system. *Any* positioning system requires measurement of the vessel position and heading. Wind sensors, measuring the wind speed and relative direction, are commonly used for wind feedforward control. In most commercial systems, measurements of surge and sway velocities are *not* available with sufficient accuracy and must be estimated online. A guidance rule by Det norske Veritas [8] is that the accuracy of the position reference data is generally to be within a radius of 2% of the water depth for bottom-based systems, and within a radius of 3 meters for surface-based systems.

A minimum sensor and navigation configuration for a DP system typically consists of at least one position reference system, one gyro compass, one VRU for roll and pitch measurement, and one wind sensor. The redundancy of the DP system can of course be increased by multiple measurement devices. However, one should also consider using systems based on different measurement principles, giving full redundancy, not only in hardware configuration.

The trend today is integration between sensor systems and position reference systems. Over recent years the prices of accurate inertial motion units (IMUs) have been reduced. This trend enables integration between inertial navigation systems (which are based on the IMU sensor units) and other position reference systems, e.g., GPS.

## Propulsion System

The propulsion system consists typically of prime movers such as diesel engines, generators, transmissions, and thrusters. A *thruster* is defined in this document as the general expression for a propeller unit. A ship can be equipped with several types of thrusters. Conventional ships typically have a *main* propulsion unit located aft of the ship. Another common type is the *tunnel* thruster, which is a propeller inside a tunnel that goes through the hull and produces a fixed-direction transverse force. A third type is the *azimuth* (rotatable) thruster, which can produce thrust in any direction. While the main and the tunnel thrusters have fixed force directions, the direction of the azimuth thruster can be changed, either manually by the operator or automatically by the positioning control system. Rudders in combination with main propellers can also be used actively in the positioning system to produce a transverse force acting on the stern. In positioning systems, the main objective is to control the thruster's force. However, since this resulting force cannot be measured directly, it is common to control the thruster revolution speed (RPM control) or the pitching of the propeller blades (pitch control). The servomechanisms for the propulsion devices must be designed to give accurate and fast response. This is often referred to as *low-level* thrust control, which will be treated in more detail in the last section.

## Data Network and Process Stations

On larger systems the power, automation, and the different parts of the positioning system are connected by data networks, and the positioning control systems are implemented in the local process stations. It is important that these systems meet the necessary redundancy level and communication real-time requirements, such that the positioning system is reliable. Today it is common to integrate all automation systems into a common plant network.

## Operator Stations and MMI

The operator's link to the positioning system is through the Man-Machine Interface (MMI) on the process stations. On the operating stations, it is common to arrange a joystick system, giving the operator the opportunity to directly, control the thruster forces in surge and sway and moment in yaw. The MMI should be user-friendly, and sufficient information should be presented for the operator so that correct decisions can be made.

## Mooring System (for Anchored Vessels)

Generally, a mooring system consists of *n* lines connected to the structure and horizontally spread out in a certain pallern. At the seabed the lines are connected to anchors. *Spread mooring* systems are used both for semisubmersibles and turret-anchored ships. In the latter case the anchor lines are connected to a *turret* on the ship, which can be rotated relative to the ship. The number of anchor lines may vary, typically from 6 to 12 lines. The length of the anchor lines are adjusted by winches and determines the pretension and thus the stiffness of the mooring system.

## Positioning System Functionality and Modules

The positioning systems provided by the different vendors in the market may be designed using different methods, but the basic positioning functionality is more or less based on the same principles, as outlined in the following.

### Signal Processing

All signals from external sensors should be thoroughly analyzed and checked in a separate signal processing module. This comprises testing of the individual signals and sensor signal voting and weighting when redundant measurements are available. The individual signal quality verification often includes tests for signal range and variance, frozen signals, and signal wild points. If an erroneous signal is detected, the measurement is not used by the positioning system. The weighted signals from each sensor group should not contain any steps or discontinuities when utilized further in the system to ensure safe operation.

### Vessel Observer

Filtering and state estimation are important features of a positioning system. In most cases today, accurate measurements of vessel velocities are not available. Hence, estimates of the velocities must be computed from noisy position and heading measurements through a state *observer*. The position and heading measurements are corrupted with colored noise, mainly caused by wind, waves, and ocean currents. However, only the slowly varying disturbances should be counteracted by the propulsion system, whereas the oscillatory motion due to the waves (first-order wave loads) should *not* enter the feedback loop. The so-called wave-frequency modulation of the thrusters will cause unnecessary wear and tear on the propulsion equipment. In the observer, *wave filtering* techniques are used, which separate the position and heading measurements into low-frequency (LF) and wave-frequency (WF) parts. The estimated LF position, heading, and velocities are utilized by the feedback controller. The observer is also needed when the position or heading measurements are temporarily unavailable. This situation is called *dead reckoning*, and in this case the *predicted* estimates from the observer are used in the control loop. Another feature of the observer is that it estimates the unmodeled and unmeasured slowly varying forces and moments, mainly due to second-order wave loads and ocean current.

### Controller Logic

The positioning system can be operated in different modes of operation (see section entitled "Modes of Operation"). Many kinds of internal system status handling and mode transitions, model adaptation, etc. are governed by the controller logic. Smooth transitions are needed between the different modes of operation such as issue alarm and warnings, operator interaction, etc.

### Feedback Control Law

The positioning controllers are often of the PD type (multivariable or decoupled in surge, sway, and yaw), where feedback is produced from the estimated LF position and heading deviations and estimated LF velocities. The underlying control methods may, however, vary. Traditionally, decoupled controllers and linear quadratic controllers have been popular. In addition to the PD part, integral action is needed to compensate for the static (or slowly varying) part of the environmental loads. The controller should be optimized with respect to positioning accuracy, fuel consumption, and wear and tear on the propulsion system. In designing the positioning controller, the control inputs are forces in surge, sway, and moment in yaw. In the context of positioning systems, this may be regarded as *high-level control*, since the actual control inputs are shaft speed (RPM control) or the pitching of the propeller blades (pitch control), which indirectly controls the developed force. In the case of azimuth thrusters, the directions of each thrust device are additional control inputs.

## Guidance System and Reference Trajectories

In tracking operations, where the ship moves from one position and heading to another, a reference model is needed for achieving a smooth transition. In the most basic case the operator specifies a new desired position and heading, and a reference model generates smooth reference trajectories/paths for the vessel to follow. A more advanced guidance system involves way-point tracking functionality, optimal path planning, and weather routing for long distance sailing. The guidance system could be interfaced with electric map systems.

## Feedforward Control Laws

The most common feedforward control term is *wind feedforward*. Based on measurements of wind speed and direction, estimates of wind forces and moment acting on the ship are computed. As a consequence, a fast disturbance rejection with respect to varying wind loads can be obtained. In order to improve the performance of the system during tracking operations, a *reference feedforward* is computed. This is done by using a model of the ship dynamics, and reference accelerations and velocities given by the reference model. In PM systems, wind feedforward is normally enabled in yaw only, since the mooring system should compensate for stationary wind loads in surge and sway. In addition, a line break detection algorithm monitors the line tension signals and the corresponding vessel motions, in order to automatically detect breaks in the anchor lines. When a line break is detected, the line break feedforward controller is activated in order for the thrusters to compensate for the lost forces and moment produced by the broken line. This will ease the load in the surrounding lines and thus prevent yet another line break.

## Thrust Allocation

The high-level feedback and feedforward controllers compute commanded forces in surge, sway, and moment in yaw. The *thrust allocation* module computes the corresponding force and direction commands to each thrust device. The thurst allocation algorithm should be optimized for fuel consumption, wear and tear on the thruster devices, and for obtaining the commanded thrust in surge, sway, and yaw. In addition, the function should take into account saturation of the RPM/pitch inputs and forbidden directional sectors. The thrust allocation module is also the main link between the positioning system and the power management system (PMS). The positioning system has a very high priority as a power consumer. In any case, the thrust allocation must handle the power limitation of the thrusters in order to avoid power system overload or blackout. The thrust allocation problem is treated by Durham [10] and Sørdalen [24]. The thrust allocation module receives continuously updated input from the power management system about available power and prevailing power plant configuration with status on the bus ties and generators. This should prevent power blackout and undesired load shedding of other important power consumers.

## Model Adaptation

The parameters in the mathematical model describing the vessel dynamics will vary with different operational and environmental conditions. In a model-based observer and controller design, the positioning system should automatically provide the necessary corrections to the vessel model and controller gains subject to changes in vessel draught, wind area, and variations in the sea state. This can be obtained either by gain-scheduling techniques or continuously by using nonlinear and adaptive formulations. In addition, other adaptive control and estimation methods may be applied, either run in batches or processed online.

## Advisory and Surveillance Systems

Use of advisory systems for diagnostics, simulation, and analysis of future operational requirements, subject to varying environmental and operational conditions, becomes of increasing importance for optimal operational planning. Such systems are integrated with the positioning system and typical features are described below:

1. **DP and PM Vessel Motion Simulators.** Such simulators include mathematical models of the environment and the vessel motion. The operator can simulate the performance of the positioning system,

either using the prevailing environmental conditions or by specifying any environmental condition. In addition, different types of failure scenarios can be simulated such as power or thruster failure on drive- and drift-off. In PM systems, failures in one or several lines can be simulated as well.

2. **Consequence Analysis.** This is an advanced version of a position capability analysis that continuously verifies that the vessel is capable of keeping position and heading for different failure scenarios during the prevailing conditions. This can be loss of one or several thrusters, one engine room, or mooring lines (if applicable). In many cases a similar off-line version is also available, where *any* environmental condition, operation, or failure situation can be simulated, at the request of the operator.

## Modes of Operation

The DP and PM control functionality is closely connected to the surge, sway, and yaw degrees of freedom (DOF), which can be viewed as independent of the actual thruster configuration, as long as there is enough thrust capacity to fulfill the force and moment demands. The surge, sway, and yaw degrees of freedom (DOFs) can individually be controlled in the following modes of operation:

1. **Manual Control:** By using a joystick and a rotation knob, the operator of the ship can generate force set points in surge and sway and a moment set point in yaw for manual control of the ship.
2. **Damping Control:** Feedback is produced from estimated low-frequency vessel velocities and the objective is to regulate the vessel in the specific axis toward zero. This mode is especially applicable for anchored vessels, where effective damping will reduce possible large oscillatory motions, experienced about the reasonance period of the anchored vessel, and thus reduce the stress on the mooring system. Damping is also useful in DP for obtaining a smooth transition between transit speed and fixed-position operations.
3. **Set-Point Control:** Feedback is produced from both estimated low-frequency velocities and position/ yaw angle. The objective is to keep the actual axis at the specified set point position or heading.
4. **Tracking Control:** The vessel tracks a reference trajectory that is computed from the old to the new position or heading set point.

The most common DP operation is set-point control in all three axes, often referred to as *station keeping*. Other fully automatic modes of operation can be station keeping with weather optimal positioning, and roll-pitch damping. In the first case the vessel automatically tends to the heading, where the effect of the environmental loads are minimized. In the latter case, the roll and pitch motion is suppressed by proper action of the horizontal thrust components. A combination of different modes is common, e.g., semiautomatic mode where a joystick is used for manual surge and sway control, while the heading is automatically controlled. Other tailor-made DP functions exist for tracking operations such as cable and pipe laying, ROV operations, etc.

The combination of damping control in surge and sway and set-point control of heading is often used in PM systems, especially in bad weather. Moreover, the vessel will tend to an equilibrium position where the mean environmental forces are balanced by the mooring forces. Assuming a proper heading set point, this equilibrium position will be optimal with respect to the thrust usage and the fuel consumption. For turret-moored ships, automatic heading control is often the most important function. By maintaining an optimal heading against the weather, the effect of the environmental loads, and thereby the stress on the mooring system, will be minimized.

## Classification and Guidelines

In many cases it is up to the owners to decide the type of classification to be sought (or to seek class approval at all) for the positioning system installed. The following class societies all have rules for classification of DP systems: Det norske Veritas (DnV) [8], The American Bureau of Shipping (ABS) [1], Lloyd's Register of Shipping (LRS) [18], etc. The International Marine Organization (IMO) has developed Guidelines for Dynamic Positioning, in order to provide an international standard for DP systems on all types of new vessels. The purpose of the guidelines and class rules is to recommend design criteria, necessary equipment, operating requirements, and a test and documentation system for DP systems to reduce the risk to personnel, the vessel itself, other vessels and structures, subsea installations, and the environment, while operating under dynamic positioning control. Taking into consideration that DP-operated vessels often operate in different parts of the world, such standardization provides a useful tool for the different Costal states to specify local rules and regulations, defining levels of safety requirements, and requirements for redundancy and operations for DP vessels. For mooring vessels using automatic thruster assistance, IMO does not have any guidelines. Det norske Veritas has the class notation called POSMOOR ATA, DnV [9] and Lloyd's Register of Shipping has the notation PM or PM T1, LRS. A more general document for analysis, design, and evaluation of moored, floating units is published by the American Petroleum Institute [3] containing some guidelines for DP systems.

The requirements for hardware and software in DP systems are closely connected to the level of redundancy, which is defined as:

**Definition 3.1 (Redundancy)** the ability of a component or system to maintain or restore its function when a single fault has occurred. This property can be obtained by installation of multiple components, systems, or alternative means of performing a function.

A DP system consists of components and systems acting together to achieve sufficiently reliable position-keeping capability. The necessary reliability of such systems is determined by the degree of the consequences associated with a loss of position-keeping capability. The larger the consequence, the more reliable the DP system should be. Reflecting this philosophy, the requirements have been grouped into three different equipment classes. The equipment class depends on the specific DP operation, which may be governed by Costal state rules and regulations or by agreement between the DP operator company and its customers. Short descriptions of the different classes are given below.

- **Class 1:** For equipment class 1, loss of position may occur in the event of a single fault, e.g., the DP control system need *not* be redundant.
- **Class 2:** For equipment class 2, loss of position must not occur in the event of a single fault in any active component or system. The DP control system must have redundancy in all active components, e.g., the hardware must consist of at least two independent computer systems with self-checking routines and redundant data transfer arrangements and plant interfaces. It must also have at least three independent position reference systems and three sensor systems for vertical motion measurement, three gyrocompasses, and three wind sensors.
- **Class 3:** Same as DP class 2, with additional requirements with respect to redundancy in technical design and physical arrangement.

Class 2 or 3 systems should include a "Consequence analysis" function that continuously verifies that the vessel will remain in position even if the worst single failure occurs.

The IMO Guidelines also specify the relationship between equipment class and type of operation. DP drilling operations and production of hydrocarbons, for instance, require equipment class 3, according to the IMO.

## Safety Aspects

The main purpose of a DP sytem is to keep position (and heading) within certain excursion limits within a specified weather window, or so-called design environment. In order to meet the designed positioning capability, the system components should be reliable and the necessary redundancy requirements should be met. To ensure this, several design methods are available as described in the following.

### Capability and Simulations

During the design phase it is important to verify that the amount of power and thrust capacity installed on a vessel will provide the necessary holding capacity. This can be done either by static or dynamic analysis. In a static analysis, only the mean slowly varying forces due to wind, current, and waves are considered.

The items of data required in such a study include:

1. Main vessel particulars such as displacement, length, breadth, and operating draught.
2. Direction-dependent wind, current, and wave-drift coefficients, from which the corresponding forces and moments can be computed.
3. The maximum environmental conditions in which the vessel should operate in dynamic position-ing (wind speed, significant wave height, and current speed). Important parameters are dominat-ing wave period and the statistical description of the waves, usually described by wave spectrum formulations such as the Bretcheider spectrum, the Pierson-Moskowitz spectrum, or the Joint North Sea Wave Project (JONSWAP) spectrum.

A rule to thumb is that the most loaded thruster should not use more than 80% of the maximum thrust in the design environment to compensate for static loads, (API). The 20% margin is then left for compensation for dynamic variations. The results of such a static analysis can be presented as a *capability plot*. The static analysis with 20% margin is often too conservative. This suggests that the capability analysis should be complemented with dynamic (time domain) simulations, where the thrust allocation, the inherent thruster dynamics and dynamic thruster losses, forbidden azimuthing sectors, and the whole control loop are taken into account in the final verification. In the traditional capability simulations, the effect of power limitation is often neglected, which often is the most limiting factor in failure situations. A capability analysis with power limitation is also important in the design of the power plant; how many power busses, amount of power generation on each bus, which thruster to be connected to the different busses, etc.

## Reliability and Redundancy

From a safety perspective, a DP system can be viewed as four different subsystems. Each subsystem can then be further split recursively into sub-subsystems. For instance:

- Level 1: Power system. Level 2: Power generation, power distribution, drives, etc.
- Level 1: Propulsion system. Level 2: Main screw, tunnel thrusters, azimuth thrusters.
- Level 1: Positioning control system. Level 2: Computer and I/O, Operator MMI, UPS, Operator interaction.
- Level 1: Sensor system. Level 2: Gyros, position reference systems, wind sensors.

Starting from the bottom, the reliability of each subcomponent can be specified, and depending on the level of redundancy, the reliability and availability of the total system can be computed using statistical methods. Each component can be characterized by:

- Failure rate $\lambda$, defined as maximum number of failures per million hours.
- Mean time between failures (MTBF), for one component given by

$$MTBF = \frac{1}{l}$$

- Total down time, $T_d$, of a component, including mean time to repair (MTTR).
- Availability, $A$, defined as

$$A = \frac{MTBF}{MTBF + T_d}$$

The above characteristics are specified for each component or each level in a reliability analysis and will, at the top level, characterize the whole plant.

The cost of making design changes during the initial project phase is small, but as the project progresses the cost of design changes increases significantly. Minor changes in a system can cause project delays and huge additional costs during commissioning and sea trials. During the whole design phase, the reliability of the total system can be thoroughly investigated by different reliability methodologies in order to detect system design errors and thus minimize the risk. Such analyses of the DP system and its subsystems are not required by the class societies or the IMO Guidelines. However, the Costal states, oil companies, or customers, etc. may require it, in addition to classification approval. There are different methods available for assessing the reliability of such complex systems.

### Failure Analysis

One common methodology is the Failure Mode and Effect Analysis (FMEA). This is a qualitative reliability technique for systematically analyzing each possible failure mode within a system and identifying the resulting effect on that system, the mission, and personnel. This analysis can be extended by a criticality analysis (CA), a quantitative procedure that ranks failure modes by their probability and consequence.

## References

1. American Bureau of Shipping (ABS). *Guide for Thrusters and Dynamic Positioning Systems*, New York, 1994.
2. Allensworth, T., A Short History of Sperry Marine, http:\\www.sperry-marine.com/pages/history.html, Litton Marine Systems, 1999.
3. American Petroleum Institute (API). *Recommended Practice for Design and Analysis of Station keeping Systems for Floating Structures*, Exploration and Production Department, API Recommended Practice 2SK, 1st ed., 1995.
4. Balchen, J. G., Jenssen, N. A., and Sælid, S., Dynamic positioning using Kalman filtering and optimal control theory, *IFAC/IFIP Symp. on Aut. in Offshore Oil Field Operation*, Holland, Amsterdam, 183–186, 1976.
5. Balchen, J. G., Jenssen, N. A., Mathisen, E., and Sælid, S., Dynamic positioning system based on Kalman filtering and optimal control. *Modeling, Identification and Control*, 1, (3), 135–163, 1980.
6. Bennet, S., *A History of Control Engineering 1800–1930*, Peter Peregrinus, London, 1979.
7. Britting, K. R., *Inertial Navigation System Analysis*, Wiley Interscience, New York, 1971.
8. Det norske Veritas (DnV), *Rules and Regulations of Ships Newbuildings, Special Equipment and Systems*, Additional Class, Part 6, Chapter 7: Dynamic Positioning Systems (DP). Norway, 1990.
9. Det norske Veritas (DnV), *Rules and Regulations of Mobile Offshore Units, Special Equipment and Systems*, Additional Class, Part 6, Chapter 2: Position Mooring (POSMOOR). Norway, 1996.
10. Durham, W. C., Constrained control allocation, *Journal of Guidance, Control and Dynamics*, 16, 4, 1993.
11. Farrell, J. A. and Barth, M., *The Global Positioning System & Inertial Navigation*, McGraw-Hill, New York, 1998.
12. Fung, P. T-K. and Grimble, M., Dynamic ship positioning using self-tuning Kalman filter, *IEEE Transaction on Automatic Control*, 28, 3, 339–349, 1983.
13. Grimble, M. J., Patton, R. J., and Wise, D. A., The design of dynamic ship positioning control systems using stochastic optimal control theory. *Optimal Control Applicatons & Methods*, 1, 167–202, 1980a.
14. Grimble, M. J., Patton, R. J., and Wise, D. A., Use of Kalman filtering techniques in dynamic ship positioning systems. *IEE Proc.*, 127, 3, Pt. D, 93–102, 1980b.
15. Grimble, M. J. and Johnson, M. A., *Optimal Control and Stochastic Estimation: Theory and Applications*, Vol. 2. John Wiley & Sons, New York, 1988.
16. Hansen, J. F., Optimal Power Control of Marine Systems, Ph. D. thesis, Dept. of Engineering Cybernetics, Norwegian University of Science and Technology, Trondheim, Norway, 2000.
17. Lloyd's Register of Shipping (LRS). *Rules and Regulations for the Classification of Ships*, Part 7, Chapter 8: Positional Mooring and Thruster Assisted Positional Mooring Systems. U.K., 1997.
18. Lloyd's Register of Shipping (LRS). *Rules and Regulations for the Classification of Ships*, Part 7, Chapter 4: Rules for the Construction and Classification of Dynamic Positioning Systems Installed in Ships, U. K., 1997.

19. Minorsky, N., Directional stability of automatic steered bodies, *J. Amer. Soc. of Naval Engineers,* 34, 2, 280–309, 1922.
20. Parkinson, B. W. and Spiker, J. J., (eds.), *The Global Positioning System: Theory and Applications,* American Institute of Aeronautics and Astronautics, Reston, VA, 1996.
21. Strand, J. P., Sørensen, A. J., and Fossen, T. I., Design of automatic thruster assisted position mooring systems for ships, *Modeling, Identification and Control,* MIC, 19, 2, 1998.
22. Strand, J. P., Nonlinear Position Control Systems Design for Marine Vessels. Ph. D. thesis, Dept. of Engineering Cybernetics, Norwegian University of Science and Technology, Trondheim, Norway, 1999.
23. Sælid, S., Jenssen, N. A., and Balchen, J. G., Design and analysis of a dynamic positioning system based on Kalman filtering and optimal control, *IEEE Transactions on Automatic Control,* 28, 3, 331–339, 1983.
24. Sørdalen, O. J., Optimal thrust allocation for marine vessels. *IFAC Journal of Control Engineering Practice,* 5, 9, 1223–1231, 1997.
25. Sørensen, A. J., Sagatun, S. I., and Fossen, T. I., Design of a dynamic positioning system using model-based control. *IFAC Journal of Control Engineering Practice,* 4, 3, 359–368, 1996.
26. Sørensen, A. J. and Strand, J. P., Positioning of low-water-plane-area marine constructions with roll and pitch damping. *IFAC Journal of Control Engineering in Practice,* 8, 2, 205–213, 2000.
27. Titterton, D. H. and Weston, J. L., *Strapdown Inertial Navigation Technology,* IEE, London, U. K., 1997.

# 3.2 Mathematical Modeling of Dynamically Positioned and Thruster-Assisted Anchored Marine Vessels

*Asgeir J. Sørensen and Marit Ronæss*

## Introduction

Dynamically positioned (DP) vessels or thruster-assisted position mooring (PM) of anchored marine vessels can, in general, be regarded as a low velocity or Froude number application. This assumption will particularly be used in the formulation of mathematical models used in conjunction with the controller design. Easy access to computer capacity, and the presence of efficient control system design toolboxes, such as Matlab, Simulink, etc. have motivated more extensive use of numerical simulations for design and verification of control systems. Essential to success is the ability to make sufficiently detailed mathematical models of the actual process. From an industrial point of view, the same tendency is also driven by the fact that numerical simulations contribute to a reduction in the time needed for configuration and tuning of the control plant during commissioning and sea trials. For the purpose of model-based observer and controller design, it is sufficient to derive a simplified mathematical model, which nevertheless is detailed enough to describe the main physical characteristics of the dynamic system. Then, structural information on the physical properties of the dynamic system are incorporated in the controller design in order to achieve a better performance and robustness compared to the conventional PID control design methods. For processes dominated by nonlinearities, model-based control design simplifies the overall control algorithm software since linearizations about different working points is avoided.

Hence, the mathematical models may be formulated at two complexity levels. The first level consists of a simplified mathematical description containing only the main physical properties of the process. This model may constitute a part of the controller and is denoted as the *control plant model.* The second modeling complexity level may be a comprehensive description of the actual process. The main purpose of this model, denoted as the *process plant model,* simulates the real plant dynamics including process disturbance, sensor outputs, and control inputs.

This section will focus on the formulation of *process plant models.* In this context this means mathematical models of the vessel dynamics and external forces and moments, in terms of environmental loads,

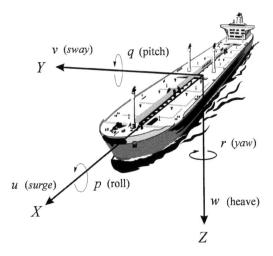

**FIGURE 3.5** Definition of surge, sway, heave, roll, pitch, and yaw modes of motion in body-fixed frame.

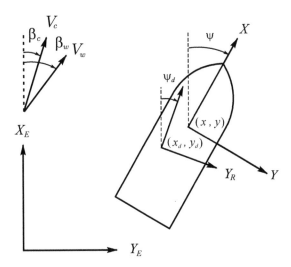

**FIGURE 3.6** Definition of frames: Earth-fixed, reference-parallel, and body-fixed.

thruster/propeller forces, and mooring forces (if any), acting on the vessel. Descriptions of the *control plant models* will be presented in the controller design sections.

## Kinematics

The different reference frames used are illustrated in Fig. 3.6 and described below:

1. The Earth-fixed reference frame is denoted as the $X_E Y_E Z_E$-frame. Measurement of the vessel's position and orientation coordinates are done in this frame relative to a defined origin. One should notice that each position reference system has its own local coordinate system, which has to be transformed into the common Earth-fixed reference frame.
2. The reference-parallel $X_R Y_R Z_R$-frame is also Earth-fixed, but rotated to the desired heading angle $\psi_d$. The origin is translated to the desired $x_d$ and $y_d$ position coordinates. Assuming small amplitudes of motion, it is convenient to use this frame in the development of the control schemes.
3. The body-fixed $XYZ$-frame is fixed to the vessel body with the $x$-axis positive forward, $y$-axis positive to the starboard, and $z$-axis positive downward. For ships it is common to assume that

the center of gravity is located in the center line of the vessel, and that the submerged part of the vessel is symmetric about the $xz$-plane (port/starboard). Here, it is assumed that the origin is located in the mean oscillatory position in the average water plane. Hence, the center of gravity is located at $(x_G, 0, z_G)$ in body coordinates. The motion and the loads acting on the vessel are calculated in this frame.

## 6 DOF Kinematics

The linear and angular velocities of the vessel in the Earth-fixed frame is given by the transformation:

$$\dot{h} = \begin{bmatrix} \dot{h}_1 \\ \dot{h}_2 \end{bmatrix} = \begin{bmatrix} J_1(h_2) & \mathbf{0}_{3 \times 3} \\ \mathbf{0}_{3 \times 3} & J_2(h_2) \end{bmatrix} \begin{bmatrix} n_1 \\ n_2 \end{bmatrix} = J(h_2)n. \tag{3.1}$$

The vectors defining the vessel's Earth-fixed position and orientation, and the body-fixed translation and rotation velocities use SNAME [10] notation given by:

$$\begin{aligned} h_1 &= [x, y, z]^T, & h_2 &= [f, u, c]^T, \\ n_1 &= [u, v, w]^T, & n_2 &= [p, q, r]^T. \end{aligned} \tag{3.2}$$

Here, $\boldsymbol{\eta}_1$ denotes the position vector in the Earth-fixed frame, and $\boldsymbol{\eta}_2$ is a vector of Euler angles. $\boldsymbol{\nu}_1$ denotes the body-fixed linear surge, sway, and heave velocity vector, and $\boldsymbol{\nu}_2$ denotes the body-fixed angular roll, pitch, and yaw velocity vector. For surface vessels, the orientation is normally represented in terms of Euler angles. Hence, the rotation matrix $J(\boldsymbol{\eta}_2)$ [ **SO** (3) $\times \mathbb{R}^{3 \times 3}$ is defined as:

$$J_1(h_2) = \begin{bmatrix} cccu & -sccf + ccsusf & scsf + cccfsu \\ sccu & cccf + sfsusc & -ccsf + susccf \\ -su & cusf & cucf \end{bmatrix}, \tag{3.3}$$

$$J_2(h_2) = \begin{bmatrix} 1 & sftu & cftu \\ 0 & cf & -sf \\ 0 & sf/cu & cf/cu \end{bmatrix}, \quad cu \neq 0, \tag{3.4}$$

where and $c \cdot = \cos(\cdot)$, $s \cdot = \sin(\cdot)$ and $t \cdot = \tan(\cdot)$.

## 6 DOF Kinematics (Small Roll and Pitch Angle Representation)

A frequently used kinematic approximation for metacentric stable vessels is:

$$\dot{h} = J(h_2)n, \tag{3.5}$$

$$\Downarrow \quad h_2 = [0, 0, c]^T,$$

$$\dot{h} = J(c)n, \tag{3.6}$$

where

$$J(c) = \begin{bmatrix} R(c) & \mathbf{0}_{3 \times 3} \\ \mathbf{0}_{3 \times 3} & I_{3 \times 3} \end{bmatrix}, \quad \text{where} \quad R(c) = \begin{bmatrix} cc & -sc & 0 \\ sc & cc & 0 \\ 0 & 0 & 1 \end{bmatrix}. \tag{3.7}$$

### 3 DOF Kinematics (Horizontal Motion)

For surge, sway, and yaw, the 6 DOF kinematics reduce to:

$$\dot{h} = \mathbf{R}(c)n, \qquad \mathbf{R}^{-1}(c) = \mathbf{R}^T(c) \tag{3.8}$$

where we have redefined the state vectors according to: $\boldsymbol{\eta} = [x, y, \psi]^T$ and $\boldsymbol{\nu} = [u, v, r]^T$.

In a reference-parallel formulation we can define

$$\dot{h}_R = \mathbf{R}(c - c_d)n. \tag{3.9}$$

**Remark 3.1**  When considering moored structures, such as turret-moored tankers or moored semisubmersibles, it is common to locate the origin of the Earth-fixed frame in the natural equilibrium point for the mooring system. This position is often referred to as the *field zero point (FZP)*. The body-fixed frame is located in the geometrical center of the mooring system on the structure; for turret-moored ships this will be the center of turret (COT).

**Remark 3.2**  In a reference-parallel frame formulation assuming small amplitudes of motion in yaw $(|c - c_d|)$, we have $\dot{h}_R = \mathbf{R}(c - c_d)n \Rightarrow \dot{h}_R = \mathbf{I}n$ , where $\mathbf{I}$ is the identity matrix. Then $\dot{h} = \mathbf{R}(c_d)\dot{h}_R$ .

## Vessel Dynamics

In mathematical modeling of marine vessel dynamics, it is common to separate the total model into a low-frequency (LF) model and a wave-frequency (WF) model by superposition. Hence, the total motion is a sum of the corresponding LF and WF components, see Fig. 3.7. The WF motions are assumed to be caused by first-order wave loads. Assuming small amplitudes, these motions will be well represented by a linear model. The LF motions are assumed to be caused by second-order mean and slowly varying wave loads, current loads, wind loads, and mooring and thrust forces. These motions are generally nonlinear, but linear approximations about certain operating points can be found.

One should notice that for surface vessels, only the horizontal motions, i.e., the surge, sway, and yaw degrees-of-freedom (DOF), are subject to control. In the design of positioning control systems, it is sufficient to only consider the horizontal-plane dynamics in the LF model. This is normally an appropriate assumption, where the effect of vertical-plane dynamics, i.e., the heave, roll, and pitch DOF, will be of minor practical interest for the control problem. However, according to Sørensen and Strand [11], under certain circumstances this assumption may cause unacceptable reduction in the control performance. In their work it was suggested that the control problem be characterized into two categories, whether the natural periods in roll and pitch are within or outside the bandwidth of the positioning controller. Hence, the modeling problem can either be regarded as a three or a six DOF problem, where the following prerequisites can be made:

- **3 DOF:** For conventional ship and catamaran hulls in the low-frequency model, only the three horizontal-plane surge, sway, and yaw DOF are of practical interest for the controller design. For these vessels it can be assumed that the LF vertical-plane dynamics and the thruster action will not have any influence on each other.

- **6 DOF:** For marine structures with a small waterplane area and low metacentric height, which results in relatively low hydrostatic restoring compared to the inertial forces, an unintentional coupling phenomenon between the vertical and the horizontal planes through the thruster action can be invoked. Examples are found in semisubmersibles and SWATHs, which typically have natural periods in roll and pitch in the range of 35–65 seconds. These natural periods are normally within the bandwidth of the positioning controllers, and should therefore be considered in the controller design. If the inherent vertical damping properties are small, the amplitudes of roll and pitch may be emphasized by the thruster's induction by up to two degrees in the resonance range.

These oscillations have caused discomfort in the vessel's crew and have in some cases limited operation. Hence, both the horizontal and vertical planes' DOF should be considered.

In the WF model it is normal to include all 6 DOF no matter how the LF model is formulated. The vertical-plane WF motions must be used to adjust the acquired position measurements provided by the vessel's installed position reference systems, such as GPS and hydroacoustic position reference systems.

## Nonlinear Low-Frequency Vessel Model

The nonlinear 6 DOF body-fixed coupled equations of the LF motions in surge, sway, heave, roll, pitch, and yaw are written as follows:

$$\mathbf{M}\dot{n} + \mathbf{C}_{RB}(n)n + \mathbf{C}_A(n_r)n_r + \mathbf{D}(n_r) + \mathbf{G}(h) = t_{env} + t_{moor} + t. \tag{3.10}$$

The right-hand expression of Eq. (3.10) represents generalized external forces acting on the vessel and is treated later in this section. Forces in surge, sway, and heave and moments in roll, pitch, and yaw are referred to as generalized forces. $t_{env}$ [ $\mathbf{R}^6$ represents the slowly varying environmental loads with the exception of current loads acting on the vessel. The effect of current is already included on the left-hand side of Eq. (3.10) by the introduction of the relative velocity vector. $t$ [ $\mathbf{R}^6$ represents the generalized forces generated by the propulsion system. Even if only the horizontal-plane surge, sway, and yaw DOF are subject to control, geometrical coupling to the vertical-plane heave, roll, and pitch DOF will be invoked due to the actual location of the thrusters. As discussed above, the produced thrust components in the vertical-plane will be important to consider for marine structures with small waterplane areas. If the vessel is attached to a mooring system, the effect of this is represented by $t_{moor}$ [ $\mathbf{R}^6$.

**Generalized inertial forces, M$\dot{n}$:**   The system inertia matrix  $\mathbf{M}$ [ $\mathbf{R}^{636}$ including added mass is defined as:

$$\mathbf{M} 5 \begin{bmatrix} m - X_{\dot{u}} & 0 & -X_{\dot{w}} & 0 & mz_G - X_{\dot{q}} & 0 \\ 0 & m - Y_{\dot{v}} & 0 & -mz_G - Y_{\dot{p}} & 0 & mx_G - Y_{\dot{r}} \\ -Z_{\dot{w}} & 0 & m - Z_{\dot{w}} & 0 & -mx_G - Z_{\dot{q}} & 0 \\ 0 & -mz_G - K_{\dot{v}} & 0 & I_x - K_{\dot{p}} & 0 & -I_{xz} - K_{\dot{r}} \\ mz_G - M_{\dot{u}} & 0 & -mx_G - Z_{\dot{q}} & 0 & I_y - M_{\dot{q}} & 0 \\ 0 & mx_G N_{\dot{v}} & 0 & -I_{zx} - N_{\dot{p}} & 0 & I_z - N_{\dot{r}} \end{bmatrix}, \tag{3.11}$$

where $m$ is the vessel mass, $I_x$, $I_y$, and $I_z$ are the moments of inertia about the x-, y-, and z-axes and $I_{xz} = I_{zx}$ are the products of inertia. The zero-frequency added mass coefficients $X_{\dot{u}}$, $X_{\dot{w}}$, $X_{\dot{q}}$, $X_{\dot{v}}$, and so on at low speed in surge, sway, heave, roll, pitch, and yaw due to accelerations along the corresponding and the coupled axes are defined as in Faltinsen [1]. Hence, it can be shown that the system inertia matrix is symmetrical and positive definite, i.e., $\mathbf{M} = \mathbf{M}^T > 0$ and $\dot{\mathbf{M}} = \mathbf{0}$.

**Generalized Coriolis and centripetal forces, $\mathbf{C}_{RB}(\boldsymbol{v})\boldsymbol{v} + \mathbf{C}_A(\boldsymbol{v}_r)\boldsymbol{v}_r$:**   The matrix $\mathbf{C}_{RB}(\boldsymbol{v}) \in \mathbb{R}^{6\times6}$ is the skew-symmetric Coriolis and centripetal matrix written (Fossen [4]):

$$\mathbf{C}_{RB}(n) = \begin{bmatrix} 0 & 0 & 0 & c_{41} & -c_{51} & -c_{61} \\ 0 & 0 & 0 & -c_{42} & c_{52} & -c_{62} \\ 0 & 0 & 0 & -c_{43} & -c_{53} & c_{63} \\ -c_{41} & c_{42} & c_{43} & 0 & -c_{54} & -c_{64} \\ c_{51} & -c_{52} & c_{53} & c_{54} & 0 & -c_{65} \\ c_{61} & c_{62} & -c_{63} & c_{64} & c_{65} & 0 \end{bmatrix}, \tag{3.12}$$

where

$$
\begin{aligned}
&c_{41} = m z_G r && c_{42} = m w && c_{43} = m(z_G p - v) && \\
&c_{51} = m(x_G q - w) && c_{52} = m(z_G r + x_G p) && c_{53} = m(z_G q + u) && c_{54} = I_{xz} p - I_z r \\
&c_{61} = m(v + x_G r) && c_{62} = -m u && c_{63} = m x_G p && c_{64} = I_y q \\
&c_{65} = I_x p + I_{xz} r.
\end{aligned}
\tag{3.13}
$$

Wichers [14] divided the effect of current into two parts: the potential part and the viscous part. The potential part is formulated according to Fossen [4]:

$$
\mathbf{C}_A(n_r) =
\begin{bmatrix}
0 & 0 & 0 & 0 & -c_{a51} & -c_{a61} \\
0 & 0 & 0 & -c_{a42} & 0 & -c_{a62} \\
0 & 0 & 0 & -c_{a43} & -c_{a53} & 0 \\
0 & c_{a42} & c_{a43} & 0 & -c_{a54} & -c_{a64} \\
c_{a51} & 0 & c_{a53} & c_{a54} & 0 & -c_{a65} \\
c_{a61} & c_{a62} & 0 & c_{a64} & c_{a65} & 0
\end{bmatrix},
\tag{3.14}
$$

where

$$
\begin{aligned}
&c_{a42} = 2 Z_{\dot{w}} w - X_{\dot{u}} u_r - Z_{\dot{q}} q && c_{a43} = Y_{\dot{p}} p + Y_{\dot{v}} v_r + Y_{\dot{r}} r && \\
&c_{a51} = Z_{\dot{q}} q + Z_{\dot{w}} w + X_{\dot{w}} u r && c_{a53} = X_{\dot{q}} q - X_{\dot{u}} u_r - X_{\dot{w}} w && c_{a54} = Y_{\dot{r}} v_r + K_{\dot{p}} p + N_{\dot{r}} r \\
&c_{a61} = Y_{\dot{v}} v_r - Y_{\dot{p}} p - Y_{\dot{r}} r && c_{a62} = X_{\dot{u}} u_r + X_{\dot{w}} w + X_{\dot{q}} q && c_{a64} = X_{\dot{q}} u_r + Z_{\dot{q}} w + M_{\dot{q}} q \\
&c_{a65} = Y_{\dot{p}} v_r + K_{\dot{p}} p + K_{\dot{r}} r
\end{aligned}
\tag{3.15}
$$

Notice that the so-called Munk moments appear from the expression $\mathbf{C}_A(\boldsymbol{\nu}_r)\boldsymbol{\nu}_r$.

**Generalized damping and current forces, $\mathbf{D}(\boldsymbol{\nu}_r)$:** The damping vector may be divided into nonlinear and linear components according to:

$$
\mathbf{D}(n_r) = \mathbf{D}_L n + \mathbf{d}_{NL}(n_r, g_r).
\tag{3.16}
$$

Furthermore, the effect of current load is normally included in the nonlinear damping term by the definition of the relative velocity vector according to:

$$
n_r = [u - u_c \quad v - v_c \quad w \quad p \quad q \quad r]^T.
\tag{3.17}
$$

The horizontal current components are defined as:

$$
u_c = V_c \cos(b_c - c), \qquad v_c = V_c \sin(b_c - c),
\tag{3.18}
$$

where $V_c$ and $\beta_c$ are the current velocity and direction, respectively (see Fig. 3.6). Notice that the current velocity components in heave, roll, pitch, and yaw are not considered. The total relative current vector is then defined as for $u_r = u - u_c$ and $v_r = v - v_c$:

$$
U_{cr} = \sqrt{u_r^2 + v_r^2}.
\tag{3.19}
$$

The relative drag angle is found from the following relation:

$$
g_r = \operatorname{atan2}(v_r, u_r),
\tag{3.20}
$$

where $atan2$ is the four-quadrant arctangent function of the real parts of the elements of $X$ and $Y$, such that $-\pi \le atan2(Y, X) \le \pi$. The nonlinear damping is assumed to be caused by turbulent skin friction and viscous eddy-making, also denoted as vortex shedding, Faltinsen [1] and Faltinsen and Sortland [2]. Assuming small vertical motions, the 6-dimensional nonlinear damping vector is formulated as:

$$
\mathbf{d}_{NL}(n_r, g_r) = 0.5 r_w L_{pp}
\begin{bmatrix}
DC_{cx}(g_r)|U_{cr}|U_{cr} \\
DC_{cy}(g_r)|U_{cr}|U_{cr} \\
BC_{cz}(g_r)|w|w \\
B^2 C_{cf}(g_r)|p|p \\
L_{pp} BC_{cu}(g_r)|q|q \\
L_{pp} DC_{cc}(g_r)|U_{cr}|U_{cr}
\end{bmatrix},
\tag{3.21}
$$

where $C_{cx}(\gamma_r)$, $C_{cy}(\gamma_r)$, $C_{cz}(\gamma_r)$, $C_{c\phi}(\gamma_r)$, $C_{c\theta}(\gamma_r)$, and $C_{c\psi}(\gamma_r)$ are the nondimensional drag coefficients found by model tests for the particular vessel with some defined location of origin, $\rho_w$ is the density of water, $L_{pp}$ is the length between the perpendiculars, $D$ is the draft, and $B$ is the breadth. Here, for the last element in Eq. (3.21), it is assumed that:

$$
0.5 r_w \#_{L_{pp}} (D(x) C_{cy}^{2D}(g_r, x) x (v_r + rx)|v_r + rx|) dx < 0.5 r_w L_{pp}^2 DC_{cc}(\gamma_r)|U_{cr}|U_{cr},
\tag{3.22}
$$

implying that the effect of yaw angular velocity, $r$, is small relative to effect of $v_r$.

It is important to notice that for decreasing velocities, linear damping becomes more significant than nonlinear damping. The strictly positive linear damping matrix $\mathbf{D}_L [ \mathbb{R}^{6\times6}$ caused by linear wave drift damping and the laminar skin friction is written as:

$$
\mathbf{D}_L = -
\begin{bmatrix}
X_u & 0 & X_w & 0 & X_q & 0 \\
0 & Y_v & 0 & Y_p & 0 & Y_r \\
Z_u & 0 & Z_w & 0 & Z_q & 0 \\
0 & K_v & 0 & K_p & 0 & K_r \\
M_u & 0 & M_w & 0 & M_q & 0 \\
0 & N_v & 0 & N_p & 0 & N_r
\end{bmatrix}.
\tag{3.23}
$$

The most important contribution to linear damping in the LF model is viscous skin friction and wave drift damping. This kind of damping must not be confused with the frequency-dependent wave radiation damping used in the wave-frequency model. The effect of wave radiation damping can be neglected in the LF model due to the low frequency of oscillation. Wave drift damping in surge can be interpreted as added resistance for the vessel advancing in waves and is proportional to the square of the significant wave height.

For most bodies it is hard to calculate the damping coefficients. A combination of empirical formulas, model tests, and computational fluid dynamics (CFD) are normally used to find the damping coefficients. In Table 3.1 the dominating damping effects are described for the horizontal plane DOF. The heave, roll, and pitch will normally follow the same tendency as the sway and yaw.

The Keulegan-Carpenter number is defined as $KC = UT/D$, where $U$ is the free stream velocity, $T$ is the oscillation period, and $D$ is the characteristic length of the body. For sway and yaw the nonlinear damping due to eddy-making will dominate the damping until a very low velocity, where the damping behavior is seen to be linear, suggesting that laminar skin friction is present.

The coefficients can be calculated by special software or found by model tests. In the last years the importance of wave-drift damping has been more appreciated, and some effort has been made to include wave-drift damping predictions in hydrodynamic software programs, see Finne and Grue [3] and

**TABLE 3.1** Dominating Damping Effects

| Dominating Damping | High Sea State | Low Sea State |
|---|---|---|
| Surge | Linear wave drift. Nonlinear turbulent skin friction. | Nonlinear turbulent skin friction, when $|u_r| > 0$. Linear laminar skin friction, for low KC number and $u_r \to 0$. |
| Sway | Nonlinear eddy-making. Linear wave drift. | Nonlinear turbulent skin friction, when $|u_r| > 0$. Linear laminar skin friction, for low KC number and $u_r \to 0$. |
| Yaw | Nonlinear eddy-making. Linear wave drift. | Nonlinear turbulent skin friction, when $|u_r| > 0$. Linear laminar skin friction, for low KC number and $u_r \to 0$. |

the references therein. The cross-flow principle and strip theory can be used to calculate viscous damping in sway and yaw. The cross-flow principle assumes that the flow separates due to the cross-flow past the ship, and that the transverse forces on a cross-section is mainly due to separated flow effects on the pressure distribution around the ship. The method is semi-empirical in the sense that empirical drag coefficients are employed. For conventional ships, data on nonlinear drag coefficients for the horizontal plane modes are available, see OCIMF [9]. Instead of applying semi-empirical methods, model tests are often used. When using model tests it should be kept in mind that scale effects may be important for the viscous forces. The transition from laminar to turbulent boundary layer is dependent on the Reynolds number, which is different in model and in full scale. This may also affect the separation point and thus the eddy-making damping. In cases with a clearly defined separation point, such as a bilge keel, scale effects are not supposed to be significant for the eddy-making damping. Wave drift damping is considered to be unaffected by scale effects.

If data already exists from other ships, based on model tests or numerical calculations, experience has shown that by using proper scaling techniques these coefficients may represent a reasonably good estimate for ships with similar geometrical hull shapes. For further details about damping, the reader is referred to Faltinsen [1], Faltinsen and Sortland [2], and Newman [7, 8].

**Generalized restoring forces, $G(\eta)$:** Here, small roll and pitch angles are assumed, such that the restoring vector can be linearized to $G\eta$, where $G \in \mathbb{R}^{6\times 6}$ is a matrix of linear generalized gravitation and buoyancy force coefficients and is written as follows for $xz$-plane symmetry.

$$
G = -\begin{bmatrix}
0 & 0 & 0 & 0 & 0 & 0 \\
0 & 0 & 0 & 0 & 0 & 0 \\
0 & 0 & Z_z & 0 & Z_u & 0 \\
0 & 0 & 0 & K_f & 0 & 0 \\
0 & 0 & M_z & 0 & M_u & 0 \\
0 & 0 & 0 & 0 & 0 & 0
\end{bmatrix},
\tag{3.24}
$$

where the coefficients are defined as:

$$
Z_z \triangleq -r_w g A_{WP} \tag{3.25}
$$

$$
Z_u = M_z \triangleq r_w g \iint_{A_{WP}} x \, dA \tag{3.26}
$$

$$
K_f \triangleq -r_w g \nabla (z_B - z_G) - r_w g \iint_{A_{WP}} y^2 \, dA = -r_w g \nabla \overline{GM_T} \tag{3.27}
$$

$$
M_u \triangleq -r_w g \nabla (z_B - z_G) - r_w g \iint_{A_{WP}} x^2 \, dA = -r_w g \nabla \overline{GM_L}. \tag{3.28}
$$

Here, $g$ is the acceleration of gravity, $A_{WP}$ is the waterplane area, $\nabla$ is the displaced volume of water, and $GM_T$ and $GM_L$ are the transverse and longitudinal metacentric heights, respectively.

## Linear Wave-Frequency Model

In a reference-parallel formulation we can define

$$\dot{h}_R = \mathbf{R}(c - c_d)n. \tag{3.29}$$

$\dot{h}_R = \mathbf{R}(c - c_d)n \Rightarrow \dot{h}_\mathbf{R} = \mathbf{I}n,$ where $\mathbf{I}$ is the identity matrix $\mathbf{I}$. Then $\dot{h} = \mathbf{R}(c_d)\dot{h}_\mathbf{R}$.

Linear theory assuming small waves and amplitudes of motion are assumed. The WF motion is calculated in the Earth-fixed frame in the literature. Since it is possible to obtain results in irregular seas by linearly superposing results from regular wave components, it is sufficient to analyze the vessel in regular sinusoidal waves of small steepness. The hydrodynamic problem in regular waves is solved as two subproblems added together to give the total linear wave-induced loads [1]. Potential theory is assumed, neglecting viscous effects.

- Wave Reaction: Forces and moments on the vessel when the vessel is forced to oscillate with the wave excitation frequency. The hydrodynamic loads are identified as added mass and wave radiation damping terms.
- Wave Excitation: Forces and moments on the vessel when the vessel is restrained from oscillating and there are incident waves. This gives the wave excitation loads which are composed of so-called Froude-Kriloff and diffraction forces and moments.

The coupled equations of WF motions in surge, sway, heave, roll, pitch, and yaw are assumed to be linear, and can be formulated in the body-fixed frame as:

$$\mathbf{M}(v)\dot{n}_w + \mathbf{D}_p(v)n_w + \mathbf{G}h_w = t_{\text{wave1}}, \tag{3.30}$$

where $\boldsymbol{\eta}_w = [\eta_{w1}, \eta_{w2}, \eta_{w3}, \eta_{w4}, \eta_{w5}, \eta_{w6}]^T$ is the WF motion vector in the Earth-fixed frame. $\tau_{\text{wave1}}$ [ $\mathbb{R}^6$ is the first order wave excitation vector, which will be modified for varying vessel headings relative to the incident wave direction. $\mathbf{M}(\omega)$ [ $\mathbb{R}^{6\times 6}$ is the system inertia matrix containing frequency dependent added mass coefficients in addition to the vessel's mass and moment of inertia. $\mathbf{D}_p(\omega)$ [ $\mathbb{R}^{6\times 6}$ is the wave radiation (potential) damping matrix. The linearized restoring coefficient matrix $\mathbf{G}$ [ $\mathbb{R}^{6\times 6}$ is the same as in Eq. (3.24). The structure of the mass and damping matrices are the same as in Eqs. (3.11) and (3.23). It is assumed that the mooring system will not influence the WF motions, see Triantafyllou [12]. For small yaw motions in the reference-parallel frame, the WF motion vector becomes:

$$\dot{h}_{Rw} = \mathbf{J}(c - c_d)\dot{n}_w, \tag{3.31}$$

$$\Downarrow$$

$$\dot{h}_{Rw} = n_w. \tag{3.32}$$

Then, Eq. (3.30) can be reformulated to the reference-parallel and Earth-fixed frames according to:

$$\mathbf{M}(v)\ddot{h}_{Rw} + \mathbf{D}_p(v)\dot{h}_{Rw} + \mathbf{G}h_{Rw} = t_{\text{wave1}}. \tag{3.33}$$

$$\dot{h}_w = \mathbf{J}(c_d)\dot{h}_{Rw} \tag{3.34}$$

An important feature of the added mass terms and the wave radiation damping terms are the memory effects, which in particular are important to consider for nonstationary cases, i.e., rapid changes of heading angle. Memory effects can be taken into account by introducing a convolution integral or a so-called retardation function [7, 8].

## Environmental Loads

The slowly varying environmental loads acting on the vessel are composed of:

$$t_{\text{env}} = t_{\text{wind}} + t_{\text{wave2}}. \tag{3.35}$$

Remember that the current load vector is already included in the relative velocity vector in the nonlinear damping term.

### Wind Load Model

The effect of wind may be divided into mean, slowly varying, and rapidly varying wind loads. The components of the wind velocities are defined according to:

$$u_w = V_w \cos(b_w - c), \qquad v_w = V_w \sin(b_w - c), \tag{3.36}$$

where $V_w$ and $\beta_w$ are the wind velocity and direction, respectively (see Figure 3.6). The wind velocity is assumed to be much larger than the vessel velocity, such that the wind load vector is formulated:

$$t_{\text{wind}} = 0.5 r_a \begin{bmatrix} A_x C_{wx}(g_w)|u_w|u_w \\ A_y C_{wy}(g_w)|v_w|v_w \\ 0 \\ -A_x L_{xz} C_{wx}(g_w)|u_w|u_w \\ A_y L_{yz} C_{wy}(g_w)|v_w|v_w \\ A_y L_{oa} C_{wc}(g_w)|v_w|v_w \end{bmatrix}. \tag{3.37}$$

Here, $\rho_a$ is the density of air, $L_{oa}$ is the overall length of the vessel, $L_{xz}$ and $L_{yz}$ are the vertical distances between transverse and longitudinal origin and the wind load point of attack, $A_x$ and $A_y$ are the lateral and longitudinal areas of the nonsubmerged part of the ship projected on the $xz$-plane and $yz$-plane. $C_{wx}(\gamma_w)$, $C_{wy}(\gamma_w)$, and $C_{w\psi}(\gamma_w)$ are the nondimensional wind coefficients in surge, sway, and yaw, respectively. These coefficients are often found by model testing or by employing semi-empirical formulas as presented in Isherwood [6]. The relative wind angle is defined as:

$$g_w = b_w - c. \tag{3.38}$$

### Wave Load Model

The wave drift loads contribute to a significant part of the total excitation force in the low-frequency model. The second-order wave effects are divided into mean, slowly varying (difference frequencies), and rapidly varying (sum frequencies), wave loads. For the applications considered here, the effect of the rapidly varying second-order wave loads can be neglected. The determination of the second-order wave effects can be done by means of quadratic transfer functions [1, 7]:

$$t_{\text{wave2}}^i = t_{\text{wm}}^i + t_{\text{wsv}}^i, \qquad i = 1..6$$

$$= \bigwedge_{j51}^{N} \bigwedge_{k51}^{N} A_j A_k [T_{jk}^{ic} \cos((v_k - v_j)t + {}'_k - {}'_j) + T_{jk}^{is} \sin((v_k - v_j)t + {}'_k - {}'_j)], \tag{3.39}$$

where $\omega_j$ is the wave frequency, $A_j$ is the wave amplitude and $\varepsilon_j$ is a random phase angle. The quadratic transferfunctions $T_{jk}$ are dependent on both the first- and second-order velocity potentials, which require a nonlinear panel methodology. In addition, it is very time consuming to calculate the $T_{jk}$ for all combinations of $\omega_k$ and $\omega_j$. Hence, this motivates us to derive some simplifications. One should notice that the transfer functions, when $k = j$, $T_{jj}$ represents the mean wave loads, and can be calculated from the first-order velocity potential only. The most interesting slowly varying wave loads are those where $\omega_k - \omega_j$ is small and the loads are truly slowly varying. Normally, $T_{jk}$ will not vary significantly with the frequency. Then, the following approximation by Newman [7] will give satisfactory results:

$$T_{jk}^{ic} = T_{kj}^{ic} = \frac{1}{2}(T_{jj}^{ic} + T_{kk}^{ic}) \tag{3.40}$$

$$T_{jk}^{is} = T_{kj}^{is} = 0 \tag{3.41}$$

The slowly varying loads are approximated by the mean drift loads, and hence, the computation becomes much simpler and less time consuming. This approximation, based on frequency-dependent wave drift coefficients, will then be applied by dividing the sea wave spectrum (usually of Pierson-Moskowitz type) into N equal frequency intervals with corresponding wave frequency, $\omega_j$, and amplitude, $A_j$, and the wave drift loads are found to be:

$$\begin{aligned} t_{wave2}^{i} &= \bar{t}_{wm}^{i} + t_{wsv}^{i} \\ &= 2\left( \bigwedge_{j51}^{N} A_j (T_{jj}^{i}(\nu_j, b_{wave} - c))^{1/2} \cos(\nu_j t + \acute{}_j) \right)^2, \end{aligned} \tag{3.42}$$

where $T_{jj}^{i} > 0$ is the frequency-dependent wave drift function and $\beta_{wave}$ is the mean wave direction (assumed to follow the same sign convention as wind and current). A disadvantage with this approximation is the numerical generation of high-frequency components with no physical meaning. By numerical filtering this can be avoided. Equation (3.42) can also be extended to include wave spreading. In general, the second-order wave loads are much smaller than the first-order wave loads. The second-order wave loads are proportional to the square of the wave amplitude, whereas the first-order wave loads are proportional to the wave amplitude. This means that the second-order wave loads have an increased importance for increasing sea states.

## Mooring System

Generally, a mooring system consists of $n_m$ lines connected to the structure and horizontally spread out in a certain pattern. The anchor lines are composed of chain, wire lines, or synthetic material, often partitioned into several segments of different types and properties. For turret-moored ships, when the turret is rotatable, the relative angle between the turret and the body-fixed frame is given by the *turret angle* $\alpha_{tu}$. The length of each anchor line is adjusted by winches and determines the pretension and thus the stiffness of the mooring system. The anchor lines enter the turret through fairleads below the hull and the coordinates are defined as *terminal points* (TP). Mooring lines are subjected to three types of excitation [12]: large amplitude LF motions; medium amplitude WF motions; and small amplitude, very high-frequency, vortex-induced vibrations. For the purpose of Posmoor control system design, it is appropriate to consider the mooring lines' influence on the low-frequency vessel model.

A horizontal-plane spread mooring model can be formulated as:

$$t_{moor} = 2\mathbf{R}^{\mathsf{T}}(c)\mathbf{g}_{mo}(h) - \mathbf{d}_{mo}(n), \tag{3.43}$$

where $\boldsymbol{\tau}_{mo} \in \mathbb{R}^3$ is the vector of generalized mooring forces, $\mathbf{d}_{mo} \in \mathbb{R}^3$ represents the additional damping in the system due to the mooring system, and $\mathbf{g}_{mo} \in \mathbb{R}^3$ is the Earth-fixed *restoring term*:

$$\mathbf{g}_{mo} = \bigwedge_{i51}^{n_m} \begin{bmatrix} H_i \cos b_i \\ H_i \sin b_i \\ H_i \bar{x}_i \sin b_i - H_i \bar{y}_i \cos b_i \end{bmatrix}, \tag{3.44}$$

which is the vectorial sum of the force contribution from each line. $H_i$ is the horizontal force at the attachment point of the ship along the direction of line $i$, and $\beta_i$ is the Earth-fixed direction of the line. $\bar{x}_i$ and $\bar{y}_i$ are the corresponding moment arms. In a *quasi-static* approach, by disregarding the dynamic effects in the mooring lines, the restoring forces $\mathbf{g}_{mo}$ are treated as functions of the low-frequency ship position and heading $\boldsymbol{\eta}$ only according to:

$$\mathbf{g}_{mo} \overset{q.s.}{<} \bar{\mathbf{g}}_{mo}(h; a_{tu}), \tag{3.45}$$

where the horizontal force contributions $H_i$ in Eq. (3.44) are replaced by the *static line characteristics* (distance/force relationships) for each line $i$ by:

$$\bar{H}_i = f_{Hi}(h_i), \tag{3.46}$$

which is a function of the horizontal distance $h_i$ between TP and the anchor of each line. About a working point, the line characteristics in Eq. (3.46) can be linearized by:

$$\bar{H}_i = \bar{H}_{oi} + \frac{df_{Hi}}{dh_i}\bigg|_{h_i 5 h_{io}} \Delta h_i, \tag{3.47}$$

where $\bar{H}_{oi}$ is the average horizontal force in the working point $h_{io}$, and $\frac{df_{Hi}}{dh_i}\big|_{h_i 5 h_{io}}$ is the slope of the line characteristics in Eq. (3.46) at $h_{io}$. By assuming a fixed anchor line length and neglecting the influence of the current field along the line profile, the generalized mooring forces in Eq. (3.43) in a working point can be approximated by a 1st-order Taylor expansion of the static restoring mooring forces and the mooring damping about the working points $\boldsymbol{\eta} = \boldsymbol{\eta}_o$ and $\boldsymbol{\nu} = 0$ according to:

$$\bar{\mathbf{g}}_{mo}(h) = \bar{\mathbf{g}}_{mo}(h_o) + \mathbf{G}_{mo}(h - h_o) + h.o.t. \tag{3.48}$$

$$\mathbf{d}_{mo}(n) = \mathbf{D}_{mo}n + h.o.t., \tag{3.49}$$

where h.o.t. denotes higher order terms and:

$$\mathbf{G}_{mo} = \frac{\partial \bar{\mathbf{g}}_{mo}}{\partial h}\bigg|_{h 5 h_o}, \quad \mathbf{D}_{mo} = \frac{\partial \mathbf{d}_{mo}}{\partial n}\bigg|_{n 5 (0)}. \tag{3.50}$$

For simplicity, the Earth-fixed frame is often placed in the natural equilibrium point of the mooring system, i.e., $\bar{\mathbf{g}}_{mo}(h_o = 0) = 0$. $\mathbf{D}_{mo}$ and $\mathbf{G}_{mo}$ are the linearized mooring damping and stiffness matrices assumed to only contribute in the horizontal plane. They can, for symmetrical mooring patterns about the $xz$- and $yz$-planes, be formulated as:

$$\mathbf{G}_{mo} = \text{diag}\{g_{m11} \ g_{m22} \ 0 \ 0 \ 0 \ g_{m66}\}, \tag{3.51}$$

$$\mathbf{D}_{mo} = \text{diag}\{d_{m11} \ d_{m22} \ 0 \ 0 \ 0 \ d_{m66}\}. \tag{3.52}$$

Hence, the *quasi-static* mooring model can be written:

$$t_{\text{moor}} = 2\mathbf{R}^{\text{T}}(c)\mathbf{G}_{\text{mo}}h - \mathbf{D}_{\text{mo}}n. \tag{3.53}$$

For further details on modeling, see References [1], [5], and [13]. For fully dynamically positioned vessels with no anchor system, $\tau_{\text{moor}}$ is equal to zero.

## Thrust Model

Usually, perfect control action without imposing thruster dynamics is assumed. Unfortunately, this will not be true in a real system. The thrust response is affected by the dynamics in the actuators and the drive system. This will cause reduced command-following capabilities such as phase lag and amplitude reduction when the frequency increases. There will also be a loss of thrust efficiency due to disturbances in the water inflow to the thruster blades, caused by thrust-to-thrust and thrust-to-hull interactions, current, and vessel velocities. In addition, influence from the free surface will affect the thrust efficiency. Reduced thruster efficiency caused by disturbances in the water inflow is compensated for in the thrust allocation algorithm. The dynamics in the actuator and drive systems will be accounted for in the controller. Experience obtained from full-scale experiments indicates that a first-order model is well suited to this purpose. Hence, thruster dynamics are represented by the following model:

$$\dot{t} = -A_{\text{thr}}(t + t_{\text{c}}), \tag{3.54}$$

where $\tau_{\text{c}} \in \mathbb{R}^3$ is the commanded thrust vector in surge, sway, and yaw produced by the controller. $A_{\text{thr}} = \text{diag}(1/T_1, 1/T_2, 1/T_3)$ is the diagonal thruster dynamics matrix. Determination of the commanded thrust vector will be treated in more detail in the following sections.

## References

1. Faltinsen, O. M., *Sea Loads on Ships and Offshore Structures,* Cambridge University Press, Cambridge, U.K., 1990.
2. Faltinsen, O. M. and Sortland, B., Slow drift eddy making damping of a ship, *Applied Ocean Reserach,* 9, 1, 37–46, 1987.
3. Finne, S. and Grue, J., On the complete radiation-diffraction problem and wave-drift damping of marine bodies in the yaw mode of motion, *J. Fluid Mechanics,* 357, 289–320, 1998.
4. Fossen, T. I., *Guidance and Control of Ocean Vehicles,* John Wiley & Sons, Ltd., Chichester, England, 1994.
5. Huse, E. and Matsumoto, K., Mooring line damping due to first- and second-order vessel motion, *Proc. 21 Annual Offshore Technology Conference,* Houston, Texas, 4, 135–148, 1989.
6. Isherwood, M. A., Wind resistance of merchant ships, *Trans. Inst. Naval Arch.,* RINA, 115, 327–338, 1972.
7. Newman, J. N., Second order, slowly varying forces on vessels in irregular waves, *Proc. Int. Symp. Dynamics of Marine Vehicles and Structures in Waves,* Mechanical Engineering Publication Ltd., London, 1974, 182–186.
8. Newman, J. N., *Marine Hydrodynamics,* MIT Press, Cambridge, MA, 1977.
9. *Prediction of Wind and Current Loads on VLCCs,* Oil Companies International Marine Forum, Witherby & Co. Ltd., London, UK, Second Edition, 1994.
10. SNAME. The Society of Naval Architects and Marine Engineers. Nomenclature for Treating the Motion of a Submerged Body Through a Fluid, *Technical and Research Bulletin,* (1–5), 1950.
11. Sørensen, A. J. and Strand, J. P., Positioning of small-waterplane-area marine constructions with roll and pitch damping. *IFAC Journal of Control Engineering Practice,* 3, 2, 205–213, 2000.
12. Triantafyllou, M. S., Cable mechanics for moored floating systems, BOSS '94, Boston, MA, 67–77, 1994.

13. Triantafyllou, M. S. and Yeu, D. K. P., Damping of moored floating structures, *Proc. 26th Annual Offshore Technology Conference*, Houston, Texas, 3, 215–224, 1994.
14. Wichers, J. E. W., *Position Control—From Anchoring to DP System*, UETP, Marine Science and Technology, Trondheim, Norway, 1993, Section 18, 1–23.

## 3.3   Position and Velocity Observer Design

### *Thor I. Fossen and Jann Peter Strand*

Filtering and state estimation are important features of both dynamic (DP) and positioning mooring (PM) systems, see Fossen [6] and Strand [15]. The main purpose of the state estimator (observer) is to reconstruct unmeasured signals and perform filtering before the signals are used in a feedback control system (see Fig. 3.8). The input to the state estimator is sensor data, e.g., from an inertial measurement, unit (IMU), compass measurements, and a positioning reference system (see the section entitled "System Overview").

The main purpose of the state estimator is to estimate velocities and current and wave drift forces from position measurements. In addition, first-order wave disturbances should be filtered out, e.g., by using a notch filter suppressing wave-induced disturbances close to the peak frequency of the wave spectrum. Even though accurate measurements of the vessel velocities are available when using differential

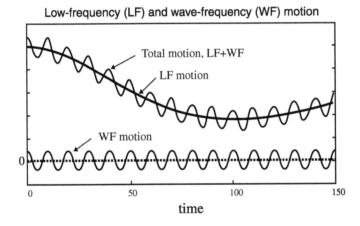

**FIGURE 3.7**   The total motion of a ship is modeled as a LF response with the WF response added as an output disturbance.

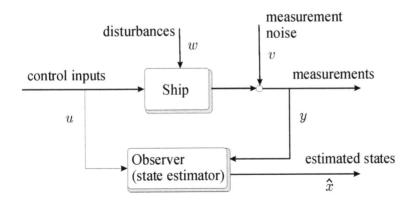

**FIGURE 3.8**   Block diagram showing the principle of state estimation.

GPS systems or Doppler log, a velocity state estimator must be designed in order to satisfy the classification rules for DP and PM systems. In the case of temporary loss of position and heading measurements, the observer must be able to operate in a *dead reckoning* mode implying that *predicted* observer velocity, position, and heading are used for feedback. A temporary loss of these measurements will not affect the positioning accuracy. When the necessary signals reappear, the estimated values will give a smooth transition back to the true position and heading.

In addition to DGPS, taut wire and hydroaccoustic reference systems are used to obtain measurement redundancy. All these systems, however, require that the estimates of the velocities are reconstructed from noisy position and heading measurements through a state observer. The position and heading measurements are corrupted with colored noise caused by wind, waves, and ocean currents. However, only the slowly varying disturbances should be counteracted by the propulsion system, whereas the oscillatory motion due to the waves (first-order wave-induced disturbances) should *not* enter the feedback loop. This is done by using so-called wave filtering techniques, which separate the position and heading measurements into low-frequency (LF) and wave-frequency (WF) position and heading parts (see Fig. 3.7).

In this section we discuss three different methods for position and velocity state estimation:

1. *Extended Kalman filter design (1976–present):* The traditional Kalman filter based estimators are linearized about a set of predefined constant yaw angles, typically 36 operating points in steps of 10 degrees, to cover the whole heading envelope between 0 and 360 degrees [2], [3], [4], [9], [10], [17]. When this estimator is used in conjunction with a linear quadratic Gaussian (LQG), PID, or $H_\infty$ controller in conjunction with a separation principle, there is no guarantee of global stability of the total nonlinear system. However, these systems have been used by several DP producers since the 1970s. The price for using linear theory is that linearization of the kinematic equations may degrade the performance of the system. Kalman filter based DP and PM systems are commercially available from several companies, e.g., ABB Industri, Alstom/Cegelec, Kongsberg Simrad, Nautronix, etc.

2. *Nonlinear observer design (1998–present):* The nonlinear observer is motivated by passivity arguments [7]. Also the nonlinear observer includes wave filtering, velocity, and bias state estimation. In addition it is proven to be global asymptotic stable (GES), through a passivation design. Compared to the Kalman filter, the number of tuning parameters is significantly reduced, and the tuning parameters are coupled more directly to the physics of the system. By using a nonlinear formulation, the software algorithms in a practical implementation are simplified. The observer of Fossen and Strand [7] has been used by Aarset, Strand and Fossen [1], Strand and Fossen [14] and Loria, Fossen and Panteley [12] in output-feedback controller design. The nonlinear observer has been implemented and tested by ABB Industri in 1998, and is now a commercial product.

3. *Adaptive and nonlinear observer design (1999–present):* The nonlinear and passive observer is further extended to adaptive wave filtering by an augmentation design technique (see Strand and Fossen [16]). This implies that the wave frequency model can be estimated recursively *online* such that accurate filtering is obtained for different sea states.

## Objectives

The objective of this section is to present the state-of-the-art and most recent observer design for conventional surface ships and rigs operating at *low speed*. The vessels can either be free-floating or anchored. In the latter case we assume a symmetrical, spread mooring system, with linear response of the mooring system. It is assumed that the vessels are *metacentric stable*, which implies that restoring forces exist in heave, roll, and pitch, such that these motions can be modeled as damped oscillators with zero mean and limited amplitude. Hence, the horizontal motions (surge, sway, and yaw) are considered in the modeling and observer design. The objectives for the observer are (see Strand [15]):

- *Velocity estimation.* It is assumed that only position and heading measurements are available. Hence, one objective is to produce velocity estimates for feedback control.
- *Bias estimation.* By estimating a bias term, accounting for slowly varying environmental loads and unmodeled effects, there will be no steady-state offsets in the velocity estimates. Moreover, the bias estimates can be used as a feedforward term in the positioning controller.
- *Wave filtering.* The position and heading signals used in the feedback controller should *not* contain the WF part of the motion. By including a synthetic wave-induced motion model in the observer, wave filtering is obtained.

**Definition 3.2 (wave filtering):** Wave filtering can be defined as the reconstruction of the LF motion components from noisy measurements of position and heading by means of an observer. In addition to this, noise-free estimates of the LF velocities should be produced. This is crucial in ship motion control systems since the WF part of the motion should *not* be compensated for by the positioning system. If the WF part of the motion enters the feedback loop, this will cause unnecessary wear and tear on the actuators and increase the fuel consumption.

## Ship Model

### Kinematics

Let the position $(x, y)$ and the orientation (heading) $\psi$ of the ship in the horizontal plane relative to an Earth-fixed (EF) frame be represented by the vector $\boldsymbol{\eta} = [x, y, \psi]^T$.

Velocities decomposed in the body-fixed frame are denoted by the vector $\boldsymbol{v} = [u, v, r]^T$ where $u$ is the alongship velocity (surge), $v$ is the athwartship velocity (sway) and $r$ is the rotational velocity (yaw).

### Vessel Dynamics

It is common to separate the modeling of marine vessels into a LF model and WF model [6]. The nonlinear LF equations of motion are driven by second-order mean and slowly varying wave, current, and wind loads as well as thruster forces. The WF motion of the ship is due to first-order wave-induced loads (see Figure 3.7). For the purpose of model-based observer and controller design, it is sufficient to derive a simplified mathematical model, which nevertheless is detailed enough to describe the main physical characteristics of the dynamic system.

### Nonlinear Low-Frequency Control Plant Model

The following *station keeping* LF ship model is used for the horizontal motion (surge, sway, and yaw):

$$\dot{h} = \mathbf{R}(c)n, \tag{3.55}$$

$$\mathbf{M}\dot{n} + \mathbf{D}v + \mathbf{R}^T(c)\mathbf{G}h = t + \mathbf{R}^T(c)\mathbf{b}, \tag{3.56}$$

where the rotation matrix $\mathbf{R}(a): \mathfrak{R} \rightarrow SO(3)$ in yaw is defined as:

$$\mathbf{R}(a) = \begin{bmatrix} \cos a & -\sin a & 0 \\ \sin a & \cos a & 0 \\ 0 & 0 & 1 \end{bmatrix}. \tag{3.57}$$

Notice that $\mathbf{R}^{-1}(a) = \mathbf{R}^T(a)$. Furthermore $\tau \in \mathfrak{R}^3$ is a control vector of forces and moment provided by the propulsion system (thrusters). $\mathbf{M} \in \mathfrak{R}^{3 \times 3}$ is the inertia matrix including hydrodynamic added inertia, $\mathbf{D} \in \mathfrak{R}^{3 \times 3}$ is a damping matrix, $\mathbf{G} \in \mathfrak{R}^{3 \times 3}$ is a stiffness matrix, due to a mooring system and $\mathbf{b} \in \mathfrak{R}^3$ is a bias term, accounting for external forces and moment due wind, waves, and currents. A symmetrical, spread mooring system is assumed where we, for simplicity and without loss of generality, have placed the Earth-fixed frame in the natural equilibrium point of the mooring system. For low-speed applications, the different matrices are defined according to Fossen [6] and Strand [15]:

$$
\mathbf{M} = \begin{bmatrix} m - X_{\dot u} & 0 & 0 \\ 0 & m - Y_{\dot v} & mx_G - Y_{\dot r} \\ 0 & mx_G - N_{\dot v} & I_z - N_{\dot r} \end{bmatrix},
$$

$$
\mathbf{D} = \begin{bmatrix} -X_u & 0 & 0 \\ 0 & -Y_v & -Y_r \\ 0 & -N_v & -N_r \end{bmatrix}, \quad \mathbf{G} = \begin{bmatrix} -X_x & 0 & 0 \\ 0 & -Y_y & 0 \\ 0 & 0 & -N_c \end{bmatrix}.
$$

We have here assumed that **G** is a constant, diagonal matrix.

**Linear Wave Frequency Model for Control Plant**

A linear second-order WF model is considered to be sufficient for representing the WF-induced motions and is formulated as:

$$
\dot j = \mathbf{A}vj + \mathbf{E}_w\mathbf{w}_w, \tag{3.58a}
$$

$$
h_w = \mathbf{C}vj, \tag{3.58b}
$$

where $\eta_w = [x_w, y_w, \psi_w]^T, \xi \in \Re^6$ and $\mathbf{w}_w \in \Re^3$ is a zero-mean Gaussian white noise vector and:

$$
\mathbf{A}_w = \begin{bmatrix} 0 & \mathbf{I} \\ -\Omega^2 & -2\Lambda\Omega \end{bmatrix}, \quad \mathbf{C}_w = \begin{bmatrix} 0 & \mathbf{I} \end{bmatrix}, \quad \mathbf{E}_w = \begin{bmatrix} 0 \\ \mathbf{K}_w \end{bmatrix}. \tag{3.59}
$$

where $\Omega = \mathrm{diag}\{\omega_1, \omega_2, \omega_3\}$, $\Lambda = \mathrm{diag}\{\zeta_1, \zeta_2, \zeta_3\}$ and $\mathbf{K}_w = \mathrm{diag}\{K_{w1}, K_{w2}, K_{w3}\}$. This model corresponds to three decoupled WF models:

$$
\frac{h_{w_i}}{w_{w_i}}(s) = \frac{K_{w_i}s}{s^2 + 2z_iv_is + v_i^2}, \quad (i = 1, 2, 3) \tag{3.60}
$$

representing the first-order wave-induced disturbances on the vessel. From a practical point of view, the WF model parameters are slowly varying quantities, depending on the prevailing sea state. Typically, the periods $T_i$ of the dominating waves are in the range of 5 to 20 seconds in the North Sea corresponding to a wave frequency $\omega_i = 2\pi/T_i$. The relative damping ration $\zeta_i$ will typically be in the range 0.05–0.10.

**Bias Model**

A frequently used bias model $\mathbf{b} \in \Re^3$ for marine control applications is:

$$
\dot{\mathbf{b}} = -\mathbf{T}_b^{-1}\mathbf{b} + \mathbf{E}_b\mathbf{w}_b, \tag{3.61}
$$

where $\mathbf{w}_b \in \Re^3$ is a zero-mean bounded disturbance vector, $\mathbf{T}_b \in \Re^{3 \times 3}$ is a diagonal matrix of bias time constants, and $\mathbf{E}_b$ a diagonal matrix scaling the amplitude of $\mathbf{w}_b$. The bias model accounts for slowly varying forces and moment due to second-order wave loads, ocean currents, and wind. In addition, a bias model will account for errors in modeling of the constantmooring loads, actuator thrust losses, and other unmodeled slowly varying dynamics.

## Measurements

The measurement equation is written:

$$\mathbf{y} = h + h_w + \mathbf{v}, \tag{3.62}$$

where the measurement $\mathbf{y}$ is treated as the sum of the vessel LF motion $\boldsymbol{\eta}$ and the WF motion $\boldsymbol{\eta}_w$ while $\mathbf{v} \in \mathfrak{R}^3$ is Gaussian measurement noise.

## Resulting Model

The resulting model is:

$$\dot{j} = \mathbf{A}_w j + \mathbf{E}_w \mathbf{w}_w, \tag{3.63a}$$

$$\dot{h} = \mathbf{R}(c)n, \tag{3.63b}$$

$$\dot{\mathbf{b}} = -\mathbf{T}_b^{-1}\mathbf{b} + \mathbf{E}_b \mathbf{w}_b, \tag{3.63c}$$

$$\mathbf{M}\dot{n} = -\mathbf{D}n - \mathbf{R}^T(c)\mathbf{G}h + \mathbf{R}^T(c)\mathbf{b} + t, \tag{3.63d}$$

$$\mathbf{y} = h + \mathbf{C}_w j + \mathbf{v}. \tag{3.63e}$$

The state-space model is of dimension $\mathbf{x} \in \mathfrak{R}^{15}$, $\boldsymbol{\tau} \in \mathfrak{R}^3$, and $\mathbf{y} \in \mathfrak{R}^3$. In addition, it is common to augment two additional states to the state-space model if wind speed and direction are available as measurements. If not, these are treated as slowly varying disturbances to be included in the bias term $\mathbf{b}$. In the next section, we will show how all these states can be estimated by using only three measurements.

## Extended Kalman Filter Design

The extended Kalman (EKF) filter is based on the nonlinear model:

$$\dot{\mathbf{x}} = \mathbf{f}(\mathbf{x}) + \mathbf{Bu} + \mathbf{Ew}, \tag{3.64a}$$

$$\mathbf{y} = \mathbf{Hx} + \mathbf{v}, \tag{3.64b}$$

where $\mathbf{f}(\mathbf{x})$, $\mathbf{B}$, $\mathbf{E}$, and $\mathbf{H}$ are given by Eqs. (3.63a) through (3.63e). Moreover:

$$\mathbf{f}(\mathbf{x}) = \begin{bmatrix} \mathbf{A}_w j \\ \mathbf{R}(c)n \\ -\mathbf{T}_b^{-1}\mathbf{b} \\ -\mathbf{M}^{-1}\mathbf{D}n - \mathbf{M}^{-1}\mathbf{R}^T(c)\mathbf{G}h + \mathbf{M}^{-1}\mathbf{R}^T(c)\mathbf{b} \end{bmatrix}, \quad \mathbf{B} = \begin{bmatrix} 0 \\ 0 \\ 0 \\ \mathbf{M}^{-1} \end{bmatrix}, \tag{3.65}$$

$$\mathbf{E} = \begin{bmatrix} \mathbf{E}_w \\ 0 \\ \mathbf{E}_b \\ 0 \end{bmatrix}, \quad \mathbf{H} = \begin{bmatrix} 0 & \mathbf{I} & 0 & \mathbf{C}_w \end{bmatrix}, \tag{3.66}$$

where $\mathbf{x} = [\boldsymbol{\xi}^T, \boldsymbol{\eta}^T, \mathbf{b}^T, \boldsymbol{\nu}^T]^T$, $[\mathbf{w} = \mathbf{w}_w^T, \mathbf{w}_v^T, ]^T$ and $\mathbf{u} = \boldsymbol{\tau}$. Hence, for a commercial system with $n = 15$ states the covariance weight matrices $\mathbf{Q} = E(\mathbf{w}^T \mathbf{w}) \in \mathfrak{R}^{n \times n}$ (process noise) and $\mathbf{R} = E(\mathbf{v}^T\mathbf{v}) \in \mathfrak{R}^{3k \times 3k}$ (position measurement noise) where $k$ denotes the number of reference systems in use. This again implies that $n + n(n + 1)/2 = 135$ ODEs must be integrated online. In order to simplify the tuning procedure these two matrices are usually treated as two diagonal design matrices, which can be chosen by applying *Bryson's inverse square method* [5], for instance.

### Discrete-Time EKF Equations

The discrete-time EKF equations are given by [8]:

**Initial values:**

$$\bar{\mathbf{x}}_{k50} = \mathbf{x}_0 \tag{3.67a}$$

$$\bar{\mathbf{P}}_{k50} = \mathbf{P}_0 \tag{3.67b}$$

**Corrector:**

$$\mathbf{K}_k = \mathbf{P}_k\mathbf{H}^T[\mathbf{H}\bar{\mathbf{P}}_k\mathbf{H}^T + \mathbf{R}]^{-1} \tag{3.68a}$$

$$\hat{\mathbf{P}}_k = (\mathbf{I} - \mathbf{K}_k\mathbf{H})\bar{\mathbf{P}}_k(\mathbf{I} - \mathbf{K}_k\mathbf{H})^T + \mathbf{K}_k\mathbf{R}\mathbf{K}_k^T \tag{3.68b}$$

$$\hat{\mathbf{x}}_k = \bar{\mathbf{x}}_k + \mathbf{K}_k(\mathbf{y}_k - \mathbf{H}\bar{\mathbf{x}}_k) \tag{3.68c}$$

**Predictor:**

$$\hat{\mathbf{P}}_{k+1} = \Phi_k\bar{\mathbf{P}}_k\Phi_k^T + \Gamma_k\mathbf{Q}\Gamma_k^T \tag{3.69a}$$

$$\bar{\mathbf{x}}_{k+1} = \mathbf{f}_k(\hat{\mathbf{x}}_k, \mathbf{u}_k) \tag{3.69b}$$

where $\mathbf{f}_k(\hat{\mathbf{x}}_k, \mathbf{u}_k)$, $\Phi_k$ and $\Gamma_k$ can be found by using *forward Euler* for instance. Moreover:

$$\mathbf{f}_k(\hat{\mathbf{x}}_k, \mathbf{u}_k) = \hat{\mathbf{x}}_k + h[\mathbf{f}((\hat{\mathbf{x}}_k) + \mathbf{B}\mathbf{u}_k)], \tag{3.70a}$$

$$\Phi_k = \left.\frac{\neq\mathbf{f}_k(\mathbf{x}_k, \mathbf{u}_k)}{\neq\mathbf{x}_k}\right|_{\mathbf{x}_k5\bar{\mathbf{x}}_k}, \tag{3.70b}$$

$$\Gamma_k = h\mathbf{E}, \tag{3.70c}$$

where $h > 0$. The EKF has been used in most industrial ship control systems. It should, however, be noted that the there is no proof of global asymptotic stability when the system is linearized. In particular, it is difficult to obtain asymptotic convergence of the bias estimates $\hat{\mathbf{b}}$ when using the EKF algorithm in DP and PM. In the next section, we will demonstrate how a nonlinear observer can be designed to meet the requirement of global exponential stability (GES) through a passivation design. The nonlinear observer has excellent convergence properties and it is easy to tune since the covariance equations are not needed.

## Nonlinear Observer Design

Two different nonlinear observers will be presented. The first one is similar to the observer of Fossen and Strand [7] for dynamically positioned (free-floating) ships with extension to spread mooring-vessel systems [15]. The second representation of the nonlinear observer is an augmented design where the filtered state of the innovation signals are used to obtain better filtering, see Strand and Fossen [16]. By using feedback from the high-pass filtered innovation in the WF part of the observer there will be no steady-state offsets in the WF estimates. Another advantage is that by using the low-pass filtered innovation for bias estimation, these estimates will be less noisy and can thus be used directly as a feedforward term in the control law.

The adaptive observer proposed in the section entitled "Adaptive and Nonlinear Observer Design" is an extension of the augmented observer. SPR-Lyapunov analysis is used to prove passivity and stability of the nonlinear observers.

### Observer Equations in the Earth-Fixed Frame

When designing the observer, the following assumptions are made in the Lyapunov analysis:

**A1** Position and heading sensor noise are neglected, that is $\mathbf{v} = \mathbf{0}$, since this term is negligible compared to the wave-induced motion.

**A2** The amplitude of the wave-induced yaw motion $\psi_w$ is assumed to be small, that is less than 1 degree during normal operation of the vessel and less than 5 degrees in extreme weather conditions. Hence, $\mathbf{R}(\psi) \approx R(\psi + \psi_w)$. From **A1** this implies that $\mathbf{R}(\psi) \approx R(\psi_y)$, where $\psi_y \triangleq \psi + \psi_w$ denotes the measured heading.

These assumptions are used only as a matter of convenience so that Lyapunov stability theory can be used to derive the structure of the observer updating mechanism. As it turns out, the SPR-Lyapunov based observer equations are robust for white Gaussian white noise, so these assumptions can be relaxed when implementing the observer.

We will also exploit the following model properties of the inertia and damping matrices in the passivation design:

$$\mathbf{M} = \mathbf{M}^T > 0, \dot{\mathbf{M}} = 0, \mathbf{D} > 0.$$

### Observer Equations (Representation 1)

A nonlinear observer copying the vessel and environmental models in Eqs. (3.63a) through (3.63e) is:

$$\dot{\hat{j}} = \mathbf{A}_w\hat{j} + \mathbf{K}_1\tilde{\mathbf{y}} \tag{3.71a}$$

$$\dot{\hat{h}} = \mathbf{R}(c_y)\hat{n} + \mathbf{K}_2\tilde{\mathbf{y}} \tag{3.71b}$$

$$\dot{\hat{b}} = -\mathbf{T}_b^{-1}\hat{b} + \mathbf{K}_3\tilde{\mathbf{y}} \tag{3.71c}$$

$$\mathbf{M}\dot{\hat{n}} = -\mathbf{D}\hat{n} - \mathbf{R}^T(c_y)\mathbf{G}\hat{h} + \mathbf{R}^T(c_y)\hat{b} + t + \mathbf{R}^T(c_y)\mathbf{K}_4\tilde{\mathbf{y}} \tag{3.71d}$$

$$\hat{\mathbf{y}} = \hat{h} + \mathbf{C}_w\hat{j}, \tag{3.71e}$$

where $\tilde{\mathbf{y}} = \mathbf{y} - \hat{\mathbf{y}}$ is the innovation vector and $\mathbf{K}_1 \in \Re^{6 \times 3}$, and $\mathbf{K}_2, \mathbf{K}_3, \mathbf{K}_4 \in \Re^{3 \times 3}$ are observer gain matrices to be determined later. The nonlinear observer is implemented as 15 ODEs with no covariance updates.

**Observer Estimation Errors**

The observer estimation errors are defined as $\tilde{j} = j - \hat{j}$, $h = h - \hat{h}$, $\tilde{\mathbf{b}} = \mathbf{b} - \hat{\mathbf{b}}$, and $\tilde{n} = n - \hat{n}$. Hence, from Eqs. (3.63a) through (3.63e) and Eqs. (3.71a) through (3.71e) the observer error dynamics is:

$$\dot{\tilde{j}} = \mathbf{A}_w \tilde{j} - \mathbf{K}_1 \tilde{\mathbf{y}} + \mathbf{E}_w \mathbf{w}_w \tag{3.72a}$$

$$\dot{\tilde{h}} = \mathbf{R}(c_y)\tilde{n} - \mathbf{K}_2 \tilde{\mathbf{y}} \tag{3.72b}$$

$$\dot{\tilde{\mathbf{b}}} = -\mathbf{T}_b^{-1}\tilde{\mathbf{b}} - \mathbf{K}_3 \tilde{\mathbf{y}} + \mathbf{E}_b \mathbf{w}_b \tag{3.72c}$$

$$\mathbf{M}\dot{\tilde{n}} = -\mathbf{D}\tilde{n} - \mathbf{R}^T(c_y)\mathbf{G}\tilde{h} + \mathbf{R}^T(c_y)\tilde{\mathbf{b}} - \mathbf{R}^T(c_y)\mathbf{K}_4\tilde{\mathbf{y}} \tag{3.72d}$$

$$\tilde{\mathbf{y}} = \tilde{h} + \mathbf{C}_w \tilde{\xi}. \tag{3.72e}$$

By defining a new output

$$\tilde{\mathbf{z}}_o \overset{\mathrm{D}}{5} \mathbf{K}_4 \tilde{\mathbf{y}} + \mathbf{G}\tilde{h} - \tilde{\mathbf{b}} \overset{\mathrm{D}}{5} \mathbf{C}_o \tilde{\mathbf{x}}_o, \tag{3.73}$$

and the vectors

$$\tilde{\mathbf{x}}_o \overset{\mathrm{D}}{5} \begin{bmatrix} \tilde{j} \\ \tilde{h} \\ \tilde{\mathbf{b}} \end{bmatrix}, \quad \mathbf{w} \overset{\mathrm{D}}{5} \begin{bmatrix} \mathbf{w}_w \\ \mathbf{w}_b \end{bmatrix}, \tag{3.74}$$

the *error dynamics* in Eqs. (3.72a) through (3.72d) can be written in compact form as:

$$\mathbf{M}\dot{\tilde{n}} = -\mathbf{D}\tilde{n} - \mathbf{R}^T(c_y)\mathbf{C}_o \tilde{\mathbf{x}}_o \tag{3.75a}$$

$$\dot{\tilde{\mathbf{x}}}_o = \mathbf{A}_o \tilde{\mathbf{x}}_o + \mathbf{B}_o \mathbf{R}(\psi_y)\tilde{\nu} + \mathbf{E}_o \mathbf{w}, \tag{3.75b}$$

where

$$\mathbf{A}_o = \begin{bmatrix} \mathbf{A}_w - \mathbf{K}_1\mathbf{C}_w & -\mathbf{K}_1 & 0 \\ -\mathbf{K}_2\mathbf{C}_w & -\mathbf{K}_2 & 0 \\ -\mathbf{K}_3\mathbf{C}_w & -\mathbf{K}_3 & -\mathbf{T}_b^{-1} \end{bmatrix},$$

$$\mathbf{C}_o = [\mathbf{K}_4\mathbf{C}_w \quad \mathbf{K}_4 + \mathbf{G} \quad -\mathbf{I}],$$

$$\mathbf{B}_o = \begin{bmatrix} 0 \\ \mathbf{I} \\ 0 \end{bmatrix}, \mathbf{E}_o = \begin{bmatrix} \mathbf{E}_w & 0 \\ 0 & 0 \\ 0 & \mathbf{E}_b \end{bmatrix}.$$

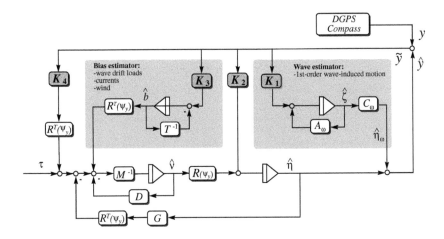

**FIGURE 3.9**   Block description of the observer.

The observer gain matrices can be chosen such that the error dynamics is passive and GES.

It is convenient to prove passivity and stability by using an SPR-Lyapunov approach. In the error dynamics in Fig. 3.9 two new error terms $\varepsilon_z$ and $\varepsilon_v$ are defined according to:

$$\varepsilon_z \triangleq -\mathbf{R}^T(c_y)\tilde{z}_o, \qquad \varepsilon_v \triangleq \mathbf{R}(c_y)\tilde{v}. \tag{3.76}$$

Thus, the observer error system can be viewed as two linear blocks $\mathcal{H}_1$ and $\mathcal{H}_2$, interconnected through the bounded transformation matrix $\mathbf{R}(\psi_y)$. Based on the physical properties of the ship dynamics, we can make the following statement:

**Proposition 3.1** The mapping $\varepsilon_z \mapsto \tilde{n}$ is *state strictly passive*, (system $\mathcal{H}_1$) in Fig. 3.9 is strictly passive.
**Proof.** Let

$$S_1 = \frac{1}{2}\tilde{n}^T\mathbf{M}\tilde{n} > 0 \tag{3.77}$$

be a positive definite storage function. From Eq. (3.75a) we have:

$$\dot{S}_1 = -\frac{1}{2}\tilde{n}^T(\mathbf{D} + \mathbf{D}^T)\tilde{n} + \tilde{n}^T\varepsilon_z \tag{3.78}$$

$$\Downarrow$$

$$\tilde{n}^T\varepsilon_z \geq \dot{S}_1 + \beta\tilde{n}^T\tilde{n}, \tag{3.79}$$

where $\beta = \frac{1}{2}\lambda_{\min}(\mathbf{D} + \mathbf{D}^T) > 0$ *and* $\lambda_{\min}(\cdot)$ denotes the minimum eigenvalue. Thus, Eq. (3.79) proves that $\varepsilon_z \mapsto \tilde{n}$ is state strictly passive (see, e.g., Khalil [11]). Moreover, since this mapping is strictly passive, post-multiplication with the bounded transformation matrix $\mathbf{R}(\psi_y)$ and pre-multiplication by its transpose will not affect the passivity properties. Hence, the block $\mathcal{H}_1$ is strictly passive. $\square$

Passivity and stability of the total system will be provided if the observer gain matrices $\mathbf{K}_1,\ldots,\mathbf{K}_4$ can be chosen such that the mapping $\varepsilon_v \mapsto \tilde{Z}_o$ is strictly positive real (SPR). This is obtained if the matrices $\mathbf{A}_o$, $\mathbf{B}_o$, $\mathbf{C}_o$ in Eqs. (3.75a) through (3.75b) satisfy the *Kalman-Yakubovich-Popov (KYP) Lemma*:

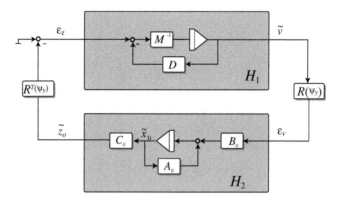

**FIGURE 3.10**    Block diagram of the observer error dynamics.

**Lemma 3.1 (Kalman-Yakubovich-Popov)** Let $Z(s) = C(sI − A)^{-1}B$ be an $m \times m$ transfer function matrix, where $A$ is Hurwitz, $(A, B)$ is controllable, and $(A, C)$ is observable. Then $Z(s)$ is strictly positive real (SPR), if and only if there exist positive definite matrices $P = P^T$ and $Q = Q^T$ such that:

$$PA + A^TP = -Q \qquad (3.80)$$

$$B^TP = C \qquad (3.81)$$

**Proof.** See Yakubovich [18] or Khalil [11].

**Theorem 3.1 (Main result: passive observer error dynamics)** The nonlinear observer error dynamics in Eqs. (3.71a) through (3.71d) is passive if the observer gain matrices $K_i$ ($i = 1,...,4$) are chosen such that Eq. (3.75b) satisfies the KYP-Lemma.

**Proof.** Since it is established that $\mathcal{H}_1$ is strictly passive and $\mathcal{H}_2$, which is given by $A = A_o$, $B = B_o$ *and* $C = -C_o$, can be made SPR by choosing the gain matrice*s* $K_i$ ($i = 1,..., 4$) according to the KYP lemma, the nonlinear observer error in Eqs. (3.71a) through (3.71d) is passive. In addition the observer error dynamics is GES (see Fig. 3.10), see Fossen and Strand [7]. □

In practice it is easy to find a set of gain matrices $K_i$ ($i = 1,..., 4$) satisfying the KYP lemma. Since the mooring stiffness matrix $G$ is assumed to be diagonal, the mapping $\varepsilon_v \mapsto Z_o$ will, by choosing a diagonal structure of the observer, gain matrices

$$K_1 = \begin{bmatrix} \text{diag}\{k_1, k_2, k_3\} \\ \text{diag}\{k_4, k_5, k_6\} \end{bmatrix} \qquad (3.82a)$$

$$K_2 = \text{diag}\{k_7, k_8, k_9\} \qquad (3.82b)$$

$$K_3 = \text{diag}\{k_{10}, k_{11}, k_{12}\} \qquad (3.82c)$$

$$K_4 = \text{diag}\{k_{13}, k_{14}, k_{15}\} \qquad (3.82d)$$

be described by three decoupled transfer functions:

$$\frac{\tilde{z}_o}{n}(s) = \text{diag}\{h_{z1}(s), h_{z2}(s), h_{z3}(s)\} \qquad (3.83)$$

where a typical transfer function $h_{zi}(s)$ is presented in Fig. 3.11. The wave filtering properties are clearly seen in the notch effect in the frequency range of the wave motion. In order to meet the SPR requirement,

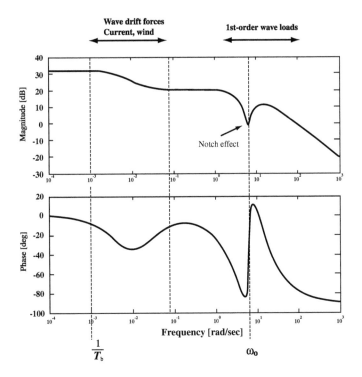

**FIGURE 3.11** A typical bode plot of the transfer function $h_{zi}(s)$.

one necessary condition is that the transfer functions $h_{zi}(s)$, $(i = 1, 2, 3)$ have phase greater than -90° and less than +90°. Regarding the choice of observer gain matrices, the tuning procedure can be similar to that for the observer for free-floating ships in Fossen and Strand [7], where loop-shaping techniques are used.

## Augmented Observer (Representation 2)

The proposed observer in the section entitled "Nonlinear Observer Design" can be further refined by augmenting a new state. The augmented design provides more flexibility, better filtering, and it is the basis for the adaptive observer in the section entitled "Adaptive and Nonlinear Observer Design." We start by adding a new state, $\mathbf{x}_f$, in the observer, which is the low-pass filtered innovation $\tilde{\mathbf{y}}$. Moreover:

$$\dot{\mathbf{x}}_f = -\mathbf{T}_f^{-1}\mathbf{x}_f + \tilde{\mathbf{y}} = -\mathbf{T}_f^{-1}\mathbf{x}_f + \tilde{h} + \mathbf{C}_w\tilde{j}, \qquad (3.84)$$

where $\mathbf{x}_f \in \Re^3$, and $\mathbf{T}_f = \text{diag}\{\mathbf{T}_{f1}, \mathbf{T}_{f2}, \mathbf{T}_{f3}\}$ contains positive filter constants. High-pass filtered innovation signals can be derived from $\mathbf{x}_f$ by:

$$\tilde{\mathbf{y}}_f = -\mathbf{T}_f^{-1}\mathbf{x}_f + \tilde{\mathbf{y}} = -\mathbf{T}_f^{-1}\mathbf{x}_f + \tilde{h} + \mathbf{C}_w\tilde{j}. \qquad (3.85)$$

Thus, both the low-pass and high-pass filtered innovations are available for feedback. Moreover,

$$\left.\begin{array}{l} \mathbf{x}_f^{\{i\}}(s) = \dfrac{1}{11T_{fi}s}\tilde{\mathbf{y}}^{\{i\}}(s) \\[1.5em] \tilde{\mathbf{y}}_f^{\{i\}}(s) = \dfrac{T_{fi}s}{11T_{fi}s}\tilde{\mathbf{y}}^{\{i\}}(s) \end{array}\right\}, \quad (i = 1, 2, 3). \qquad (3.86)$$

The cut-off frequency in the filters should be below the frequencies of the dominating waves in the WF model in Eqs. (3.58a) through (3.58b). The augmented observer equations are:

$$\dot{\hat{j}} = \mathbf{A}_w\hat{j} + \mathbf{K}_{1h}\tilde{\mathbf{y}}_f \tag{3.87a}$$

$$\dot{\hat{h}} = \mathbf{R}(c_y)\hat{n} + \mathbf{K}_2\tilde{\mathbf{y}} + \mathbf{K}_{21}\mathbf{x}_f + \mathbf{K}_{2h}\tilde{\mathbf{y}}_f \tag{3.87b}$$

$$\dot{\hat{b}} = -\mathbf{T}_b^{-1}\hat{b} + \mathbf{K}_3\tilde{\mathbf{y}} + \mathbf{K}_{31}\mathbf{x}_f \tag{3.87c}$$

$$\mathbf{M}\dot{\hat{n}} = -\mathbf{D}\hat{n} - \mathbf{R}^T(c_y)\mathbf{G}\hat{h} + \mathbf{R}^T(c_y)\hat{b} + \tau$$
$$+ \mathbf{R}^T(c_y)(\mathbf{K}_4\tilde{\mathbf{y}} + \mathbf{K}_{41}\mathbf{x}_f + \mathbf{K}_{4h}\tilde{\mathbf{y}}_f) \tag{3.87d}$$

$$\hat{\mathbf{y}} = \hat{h} + \mathbf{C}_w\hat{j}, \tag{3.87e}$$

where $\mathbf{x}_f$ is the low-pass filtered innovation vector and $\tilde{\mathbf{y}}_f$ is the high-pass filtered innovation given by Eqs. (3.84) and (3.85), respectively. Here $\mathbf{K}_{1h} \in \mathbb{R}^{6\times3}$ and $\mathbf{K}_{21}, \mathbf{K}_{2h}, \mathbf{K}_{31}, \mathbf{K}_{41}, \mathbf{K}_{4h} \in \mathbb{R}^{3\times3}$ are new observer gain matrices to be determined.

### Augmented Observer Error Equations

The augmented observer error dynamics can be written compactly as:

$$\mathbf{M}\dot{\tilde{n}} = -\mathbf{D}\tilde{n} - \mathbf{R}^T(c_y)\mathbf{C}_a\tilde{\mathbf{x}}_a \tag{3.88a}$$

$$\dot{\tilde{\mathbf{x}}} = \mathbf{A}_a\tilde{\mathbf{x}}_a + \mathbf{B}_a\mathbf{R}(c_y)\tilde{n} + \mathbf{E}_a\mathbf{w}, \tag{3.88b}$$

where

$$\tilde{\mathbf{x}}_a \triangleq \begin{bmatrix} \tilde{j}^T & \tilde{h}^T & \mathbf{x}_f^T & \tilde{b}^T \end{bmatrix}^T, \tag{3.89}$$

$$\tilde{\mathbf{z}}_a \triangleq \mathbf{K}_4\tilde{\mathbf{y}} + \mathbf{K}_{41}\mathbf{x}_f + \mathbf{K}_{4h}\tilde{\mathbf{y}}_f + \mathbf{G}\tilde{h} - \tilde{b} \triangleq \mathbf{C}_a\tilde{\mathbf{x}}_a, \tag{3.90}$$

and

$$\mathbf{A}_a = \begin{bmatrix} \mathbf{A}_w - \mathbf{K}_{1h}\mathbf{C}_w & -\mathbf{K}_{1h} & \mathbf{K}_{1h}\mathbf{T}_f^{21} & 0 \\ -(\mathbf{K}_2+\mathbf{K}_{2h})\mathbf{C}_w & -(\mathbf{K}_2+\mathbf{K}_{2h}) & \mathbf{K}_{21}\mathbf{T}_f^{-1}-\mathbf{K}_{21} & 0 \\ \mathbf{C}_w & \mathbf{I} & -\mathbf{T}_f^{-1} & 0 \\ -\mathbf{K}_3\mathbf{C}_w & -\mathbf{K}_3 & -\mathbf{K}_{31} & -\mathbf{K}_b^{-1} \end{bmatrix},$$

$$\mathbf{B}_a = \begin{bmatrix} 0 \\ \mathbf{I} \\ 0 \\ 0 \end{bmatrix}, \quad \mathbf{E}_a = \begin{bmatrix} \mathbf{E}_w & 0 \\ 0 & 0 \\ 0 & 0 \\ 0 & \mathbf{E}_b \end{bmatrix},$$

$$\mathbf{C}_a = \begin{bmatrix} (\mathbf{K}_4+\mathbf{K}_{4h})\mathbf{C}_w & (\mathbf{K}_4+\mathbf{K}_{4h})+\mathbf{G} & -\mathbf{K}_{4h}\mathbf{T}_f^{-1}+\mathbf{K}_{41} & -\mathbf{I} \end{bmatrix}.$$

The signals $\tilde{\mathbf{y}}_f$ and $\mathbf{x}_f$ are extracted from $\tilde{\mathbf{x}}_a$ by $\tilde{\mathbf{y}}_f = \mathbf{C}_h\tilde{\mathbf{x}}_a$ and $\mathbf{x}_f = \mathbf{C}_l\tilde{\mathbf{x}}_a$ where

$$\mathbf{C}_h = \begin{bmatrix} \mathbf{C}_w & \mathbf{I} & -\mathbf{T}_f^{-1} & 0 \end{bmatrix}, \qquad \mathbf{C}_l = \begin{bmatrix} 0 & 0 & \mathbf{I} & 0 \end{bmatrix}. \tag{3.91}$$

Again, the gain matrices should be chosen such that $(\mathbf{A}_a, \mathbf{B}_a, \mathbf{C}_a$, satisfies the KYP lemma in order to obtain passivity and GES, see [16] for more details.

## Adaptive and Nonlinear Observer Design

In this section we treat the case where the parameters of $\mathbf{A}_w$ in the WF model in Eq. (3.58a) are *not* known [16]. The parameters vary with the different sea states in which the ship is operating. Gain-scheduling techniques, using off-line batch processing frequency trackers and external sensors such as wind velocity, wave radars, and roll, pitch angle measurements can also be used to adjust the WF model parameters to varying sea states [6]. Additional sensor units can, however, be avoided by using an adaptive observer design. Since the wave models are assumed to be decoupled in surge, sway, and yaw, $\Lambda$ and $\Omega$ in $\mathbf{A}_w$ are diagonal matrices given by:

$$\mathbf{A}_w(u) = \begin{bmatrix} 0 & \mathbf{I} \\ -\Omega^2 & -2\Lambda\Omega \end{bmatrix} ? \begin{bmatrix} 0 & \mathbf{I} \\ -\mathrm{diag}(u_1) & -\mathrm{diag}(u_2) \end{bmatrix}, \tag{3.92}$$

where $\boldsymbol{\theta} = [u_1^T, u_2^T]^T$, and $\boldsymbol{\theta}_1, \boldsymbol{\theta}_2 \in \mathfrak{R}^3$ contain the unknown wave model parameters to be estimated. We start with the following assumption:

**A3** (Constant environmental parameters). It is assumed that the unknown parameters $\Omega$ and $\Lambda$ in $\mathbf{A}_w$ are constant or at least slowly varying compared to the states of the system.

In addition, the wave spectrum parameters are limited by:

$$\left. \begin{array}{l} 0 < \omega_{o,\,min} < \omega_{oi} < \omega_{o,\,max} \\ 0 < \zeta_{min} < \zeta_i < \zeta_{max} \end{array} \right\} \quad i = 1, 2, 3 \tag{3.93}$$

such that $\mathbf{A}_w$ is Hurwitz. Hence, the unknown wave model parameters are treated as constants in the analysis, such that:

$$\dot{u} = 0. \tag{3.94}$$

### Adaptive Observer Equations

The adaptive version of the observer is based on the augmented observer Eqs. (3.87a) through (3.87e), except in the WF part where we now propose to use the estimated WF parameters, $\hat{u}$, such that:

$$\dot{\hat{j}} = \mathbf{A}_w(\hat{u})\hat{j} + \mathbf{K}_{1h}\tilde{\mathbf{y}}_f. \tag{3.95a}$$

The parameter update law is:

$$\begin{aligned} \dot{\hat{u}} &= -\Gamma_w\Phi(\hat{j})\mathbf{C}_h\tilde{\mathbf{x}}_a \\ &= -\Gamma_w\Phi(\hat{j})\tilde{\mathbf{y}}_f, \qquad \Gamma_w > 0 \end{aligned} \tag{3.96}$$

where $\Phi(\hat{j}) \in \mathfrak{R}^{6 \times 3}$ is the regressor matrix. The regressor matrix is further investigated by considering the error dynamics.

**Adaptive Observer Error Dynamics**

The adaptive WF observer error dynamics is:

$$\dot{\tilde{j}} = \mathbf{A}_w j - \mathbf{A}_w(\hat{u})\hat{j} - \mathbf{K}_{1h}\tilde{\mathbf{y}}_f + \mathbf{E}_w \mathbf{w}_w. \tag{3.97}$$

By adding and subtracting $\mathbf{A}_w \hat{j}$, defining

$$\mathbf{B}_w \Phi^T(\hat{j})\tilde{u} \triangleq (\mathbf{A}_w - \mathbf{A}_w(\hat{u}))\hat{j}, \tag{3.98}$$

where $\tilde{u} = \hat{u} - u$ denotes the estimation error,

$$\Phi^T(\hat{j}) \triangleq [\text{diag}(\hat{j}_1) \quad \text{diag}(\hat{j}_2)], \tag{3.99}$$

$$\mathbf{B}_w \triangleq [0 \quad \mathbf{I}]^T, \tag{3.100}$$

where $\hat{j} = [\hat{j}_1^T, \hat{j}_2^T]^T$, $\hat{j}_1, \hat{j}_2 \in \mathbb{R}^3$, and by using Eq. (3.85), then Eq. (3.97) can be rewritten as:

$$\dot{\tilde{j}} = (\mathbf{A}_w - \mathbf{K}_{1f}\mathbf{C}_w)\tilde{j} - \mathbf{K}_{1f}\tilde{h} + \mathbf{B}_w \Phi^T(\hat{j})\tilde{u}$$
$$+ \mathbf{K}_{1f}\mathbf{T}_f^{-1}\mathbf{x}_f + \mathbf{E}_w \mathbf{w}_w. \tag{3.101}$$

The observer error dynamics can be written compactly as:

$$\mathbf{M}\dot{\tilde{n}} = -\mathbf{D}\tilde{n} - \mathbf{R}^T(c_y)\mathbf{C}_a\tilde{\mathbf{x}}_a \tag{3.102a}$$

$$\dot{\tilde{\mathbf{x}}} = \mathbf{A}_a\tilde{\mathbf{x}}_a + \mathbf{B}_a\mathbf{R}(c_y)\tilde{n} + \mathbf{H}_a\Phi^T(\hat{j})\tilde{u} + \mathbf{E}_a\mathbf{w}, \tag{3.102b}$$

where

$$\mathbf{H}_a = [\mathbf{B}_w^T \quad 0 \quad 0 \quad 0]^T. \tag{3.103}$$

In the adaptive case, we want the WF adaptive law to be updated by the high-pass filtered innovations signals. Hence, it is required that the observer gain matrices are chosen such that:

$$\mathbf{A}_a^T\mathbf{P}_a + \mathbf{P}_a\mathbf{A}_a = -\mathbf{Q}_a, \tag{3.104a}$$

$$\mathbf{B}_a^T\mathbf{P}_a = \mathbf{C}_a, \tag{3.104b}$$

$$\mathbf{H}_a^T\mathbf{P}_a = \mathbf{C}_h. \tag{3.104c}$$

in order to obtain passivity, see Strand and Fossen [16]. It should be noted that GES cannot be guaranteed for this case since this requires persistency of excitation. However, global convergence of all state estimation errors to zero can be guaranteed at the same time as the parameter estimation error is bounded. We will, however, see from the experimental results that the parameters also converge to their true values when considering a ship at rest but exposed to waves.

## Experimental Results

Both the augmented and the adaptive observer have been implemented and tested at the Guidance, Navigation and Control (GNC) Laboratory at the Department of Engineering Cybernetics, NTNU. A detailed description of the laboratory is found in Strand [15]. In the experiments, CyberShip I was used (see Fig. 3.12). A nonlinear PID controller is used for maintaining the ship at the desired position $(x_d, y_d)$ and heading $\psi_d$.

An illustration of the experimental setup is given in Fig. 3.13. The experimental results are transformed to full scale by requiring that the *Froude number* $F_n = U/\sqrt{L_g}$ = constant. Here, $U$ is the vessel speed, $L$ is the length of the ship, and $g$ is the acceleration of gravity. The scaling factors are given in Table 3.2 where $m$ is the mass and the subscripts $m$ and $s$ denote the model and the full-scale ship, respectively. The length of the model ship is $L_m = 1.19$ meters and the mass is $m_m = 17.6$ kg. A full-scale ship similar to Cybership I typically has a length of 70 to 90 meters and mass of 4000 to 5000 tons.

The experiment can be divided in three phases:

- *Phase I (No waves)*. Initially the ship is maintaining the desired position and heading with no environmental loads acting on the ship (calm water). The reference heading is −140 degrees. When the data acquisition starts, a wind fan is switched on. There is no adaptive wave filtering and the observer is identical to the augmented design (Representation 2). The effect of the wind loads are reflected in the bias estimates in Fig. 3.14.

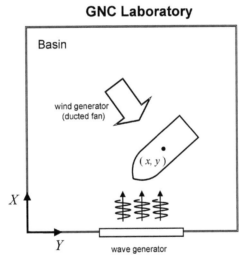

**FIGURE 3.12**    Experimental setup in model basin.

**TABLE 3.2**    Scaling Factors Used in the Experiments (Bis Scaling)

| | |
|---|---|
| Position | $L_s/L_m$ |
| Linear velocity | $\sqrt{L_s/L_m}$ |
| Angular velocity | $\sqrt{L_m/L_s}$ |
| Linear acceleration | 1 |
| Angular acceleration | $L_m/L_s$ |
| Force | $m_s/m_m$ |
| Moment | $\dfrac{m_s L_s}{m_m L_m}$ |
| Time | $\sqrt{L_s/L_m}$ |

**FIGURE 3.13**     CyberShip I: Model ship scale 1:70.

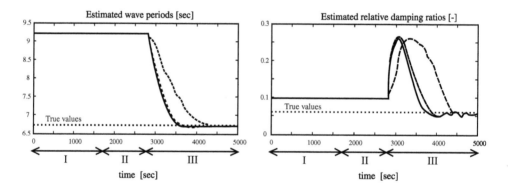

**FIGURE 3.14**     Estimated wave periods (left) and estimated relative damping ratios (right) for surge (solid), sway (dashed), and yaw (dotted). The adaptive wave filter is activated after 2800 seconds.

- *Phase II (Waves, adaptive wave filter is off).* After 1700 seconds the wave generator is started. In this phase we can see the performance of the observer without the adaptive wave filter. In the wave model we are assuming that the dominating wave period is 9.2 seconds and the relative damping ratio is 0.1 (see Fig. 3.15).
- *Phase III (Waves, adaptive wave filter is on).* After 2800 seconds the adaptive wave filter is activated. The estimates of dominating wave period and relative damping are plotted in Fig. 3.15 for surge, sway, and yaw.

A spectrum analysis of the position and heading measurements shows that the estimated wave periods converge to their true values, that is wave periods of approximately 7.8 seconds and relative damping ratios of 0.07 (see Fig. 3.15). In Fig. 3.14 the measured position deviation and heading are plotted together with the corresponding LF estimates. The effect of the adaptive wave filtering is clearly seen in Fig. 3.16, where the innovation signals are significantly reduced during phase III, when the adaptation is active and the wave model parameters start converging to their true values. The effect of *bad* wave filtering is reflected by noisy control action by the thrusters during phase II (see Fig. 3.16). A zoom-in of the heading

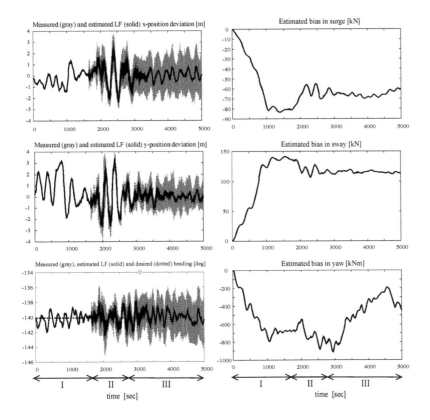

**FIGURE 3.15** Left column: Measured position and heading (gray) together with corresponding LF estimates (solid). Right column: Estimated bias in surge, sway, and yaw.

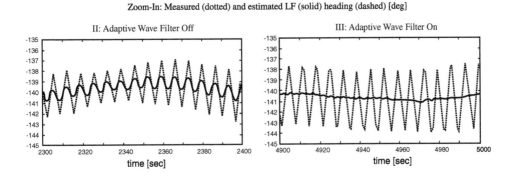

**FIGURE 3.16** Zoom-in of measured and estimated LF heading. Left: Observer without adaptive wave filtering. Right: Observer with adaptive wave filtering.

measurement together with the LF estimate is given is Fig. 3.17, both for phase II and III. Here, we see that the LF estimates have a significant WF contribution when the adaptive wave filter is off. This is the reason for the noisy control action in phase II. The other zoom-in shows excellent LF estimation when the adaptive wave filter is active and the wave model parameters have converged to their true values. Hence, it can be concluded that adaptive wave filtering yields a significant improvement in performance compared to filters with fixed WF model parameters operating in varying sea states.

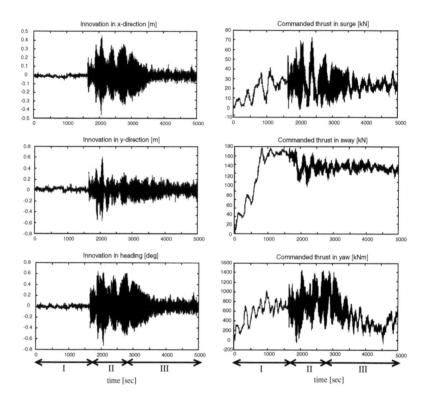

**FIGURE 3.17**    Left column: Innovation in position and heading. Right column: Commanded thrust in surge, sway, and yaw.

## References

1. Aarset, M. F., Strand, J. P., and Fossen, T. I., Nonlinear vectorial observer backstepping with integral action and wave filtering for ships, *Proc. of the IFAC Conf. on Control Appl. in Marine Systems (CAMS '98)*, Fukuoka, Japan, October 27–30, 1998, 83–89.

2. Balchen, J. G., Jenssen, N. A., and Sælid, S., Dynamic positioning using Kalman filtering and optimal control theory, *IFAC/IFIP Symposium on Automation in Offshore Oil Field Operation*, Amsterdam, The Netherlands, 183–186. 1976.

3. Balchen, J. G., Jenssen, N. A., and Sælid, S., Dynamic positioning of floating vessels based on Kalman filtering and optimal control, *Proc. of the 19th IEEE Conf. on Decision and Control*, New York, NY, 852–864, 1980.

4. Balchen, J. G., Jenssen, N. A., Mathisen, E., and Sælid, S., Dynamic positioning system based on Kalman filtering and optimal control. *Modeling, Identification and Control MIC*-1(3): 135–163, 1980.

5. Bryson, A. E. and Bryson, Y. C., *Applied Optimal Control*, Ginn and Co., Boston, MA 1969.

6. Fossen, T. I., *Guidance and Control of Ocean Vehicles,* John Wiley & Sons, Ltd., Chichester, England, 1994.

7. Fossen, T. I. and Strand, J. P., Passive nonlinear observer design for ships using Lyapunov methods: experimental results with a supply vessel, (Regular Paper) *Automatica* AUT-35(1): 3–16, January 1999.

8. Gelb, A. (Ed.), *Applied Optimal Estimation*, MIT Press, Boston, MA, 1988.

9. Grimble, M. J., Patton, R. J., and Wise, D. A., The design of dynamic positioning control systems using stochastic optimal control theory, *Optimal Control Applications and Methods* 1, 167–202, 1980.

10. Grimble, M. J., Patton, R. J., and Wise, D. A., Use of Kalman filtering techniques in dynamic ship positioning systems, *IEE Proceedings Vol. 127*, Pt. D, No. 3, 93–102, 1980.

11. Khalil, H. K., *Nonlinear Systems*, Prentice-Hall, Englewood Cliffs, NJ, 1996.

12. Loria, A., Fossen, T. I., and Panteley, E., A cascaded approach to a separation principle for dynamic ship positioning, *IEEE Transactions of Control Systems Technology*, to appear in 2000.

13. Savage, P. G., Strapdown Inertial Navigation, *Lecture Notes. Strapdown Associates Inc*, Minnetonka, MN, 1990.

14. Strand, J. P. and Fossen, T. I., Nonlinear output feedback and locally optimal control of dynamically positioned ships: experimental results, *Proc. of the IFAC Conference on Control Applications in Marine Systems (CAMS '98)*, Fukuoka, Japan, October 27–30, 1998, 89–95.

15. Strand, J. P., Nonlinear Position Control Systems Design for Marine Vessels, Doctoral Dissertation, Department of Engineering Cybernetics, Norwegian University of Science and Technology, June 1999.

16. Strand, J. P. and Fossen, T. I., Nonlinear Passive Observer for Ships with Adaptive Wave Filtering, in *New Directions in Nonlinear Observer Design*, H. Nijmeijer and T. I. Fossen, Eds., Springer-Verlag London Ltd., 113–134, 1999.

17. Sørensen, A. J., Sagatun, S. I., and Fossen, T. I., Design of a dynamic positioning system using model based control, *Journal of Control Engineering Practice* CEP-4, 3, 359–368, 1996.

18. Yakubovich, V. A., A frequency theorem in control theory, *Siberian Mathematical Journal* 14, 2, 384–420, 1973.

## 3.4 Design of Controllers for Positioning of Marine Vessels

*Asgeir J. Sørensen, Thor I. Fossen, and Jann Peter Strand*

In the design of dynamic positioning (DP) systems and thruster-assisted position mooring (PM) systems it has been adequate to regard the control objective as a three degrees-of-freedom (DOF) problem in the horizontal plane, in surge, sway, and yaw, respectively (see Balchen et al. [1], Balchen et al. [2], Grimble et al. [5] and [6], Fung and Grimble [4], Sælid et al. [10], Grimble and Johnson [7], Fossen [3], Sørensen et al. [12], Strand et al. [8] and Strand [9]). However, for certain marine structures, especially those with a small waterplane area and low metacentric height, which results in relatively low hydrostatic restoring compared to the inertia forces, an unintentional coupling phenomenon between the vertical and the horizontal planes through the thruster action can be invoked. An example of this is found in semisubmersibles, which typically have natural periods in roll and pitch in the range of 35 to 65 seconds. These natural periods are within the bandwidth of the positioning controller. In Sørensen and Strand [14] it was shown that roll and pitch may be unintentionally excited by the thruster system, which is only supposed to act in the horizontal plane. If the inherent vertical damping properties are small, the amplitudes of roll and pitch may be emphasized by the thruster's induction by up to 2 to 5° in the resonance range. These oscillations have caused discomfort in the vessel's crew and have in some cases limited operation.

In this section different model-based multivariable control strategies accounting for both horizontal and vertical motions, with the exception of heave, are described. Since it is undesirable to counteract the wave-frequency (WF) motions caused by first-order wave loads, the control action of the propulsion system is produced by the low-frequency (LF) part of the vessel's motion, which is caused by current, wind, and second-order mean and slowly varying wave loads. The computation of feedback signals to be used in the controller are based on nonlinear observer theory as presented in the previous section.

### Control Plant Model

As noted in the mathematical modeling section, we may formulate the models into two complexity levels. The first level consists of a simplified mathematical description containing only the main physical properties of the process. This model may constitute a part of the controller and is here denoted as the *control plant model*. The second modeling complexity level is a comprehensive description of the actual process. The main purpose of this model, denoted as the *process plant model*, is to simulate the real plant dynamics including process disturbance, sensor outputs, and control inputs.

### Linear Low-Frequency Model

For the purpose of controller design and analysis, it is convenient to derive a linear LF control plant model about zero vessel velocity. By assuming small roll and pitch amplitudes, and that the yaw angle is defined with respect to the desired heading angle, the rotation matrix $\mathbf{J}(h_2)$ can be approximated by the identity matrix. The measured position and heading signals are transformed into the reference-parallel frame before the estimator and the controller process them.

In the new control strategy for small waterplane area marine constructions, the conventional 3 DOF multivariable controller in surge, sway, and yaw will be extended to account for couplings to roll and pitch. In order to derive the new controller it is appropriate to define the model reduction matrices:

$$
\mathbf{H}_{5 \times 6} = \begin{bmatrix} 1 & 0 & 0 & 0 & 0 & 0 \\ 0 & 1 & 0 & 0 & 0 & 0 \\ 0 & 0 & 0 & 1 & 0 & 0 \\ 0 & 0 & 0 & 0 & 1 & 0 \\ 0 & 0 & 0 & 0 & 0 & 1 \end{bmatrix},
$$

$$
\mathbf{H}_{5 \times 3} = \begin{bmatrix} 1 & 0 & 0 \\ 0 & 1 & 0 \\ 0 & 0 & 0 \\ 0 & 0 & 0 \\ 0 & 0 & 1 \end{bmatrix}, \tag{3.105}
$$

$$
\mathbf{H}_{3 \times 6} = \begin{bmatrix} 1 & 0 & 0 & 0 & 0 & 0 \\ 0 & 1 & 0 & 0 & 0 & 0 \\ 0 & 0 & 0 & 0 & 0 & 1 \end{bmatrix},
$$

$$
\mathbf{H}_{3 \times 3} = \mathbf{I}_{3 \times 3}, \quad \mathbf{H}_{6 \times 6} = \mathbf{I}_{6 \times 6}.
$$

Hence, the $i \times i$ dimensional mass, damping, and restoring matrices can be written in reduced order form according to:

$$
\begin{aligned}
\mathbf{M}_i &= \mathbf{H}_{i \times 6} \mathbf{M} \mathbf{H}_{i \times 6}^T, \\
\mathbf{D}_i &= \mathbf{H}_{i \times 6} (\mathbf{D}_L + \mathbf{D}_m) \mathbf{H}_{i \times 6}^T, \\
\mathbf{G}_i &= \mathbf{H}_{i \times 6} (\mathbf{G}_B + \mathbf{G}_m) \mathbf{H}_{i \times 6}^T,
\end{aligned} \tag{3.106}
$$

where $i = 3$ describes the conventional horizontal model matrices in surge, sway, and yaw, and $i = 5$ will represent the 5 DOF model of surge, sway, roll, pitch, and yaw. The corresponding linear LF state-space model can be formulated as:

$$
\begin{aligned}
\dot{\mathbf{x}}_i &= \mathbf{A}_i \mathbf{x}_i + \mathbf{B}_i t_{ic} + \mathbf{E}_i \mathbf{w}_i \\
\mathbf{y}_i &= \mathbf{C}_i \mathbf{x}_i + \mathbf{v}_i.
\end{aligned} \tag{3.107}
$$

For $i = 3$ and $i = 5$, the state-space vectors become:

$$
\mathbf{x}_3 = [u, v, r, x, y, c]^T, \tag{3.108}
$$

$$
\mathbf{x}_5 = [u, v, q, p, r, x, y, f, u, c]^T. \tag{3.109}
$$

$t_{ic}$ [ $\mathbb{R}^3$ is the commanded control vector. $\mathbf{w}_i$, $\mathbf{y}_i$, $\mathbf{v}_i$, [ $\mathbb{R}^i$ are the disturbance, measurement, and noise vectors, respectively. The system matrix $\mathbf{A}_i$ [ $\mathbb{R}^{2i \times 2i}$ is then written:

$$\mathbf{A}_i = \begin{bmatrix} -\mathbf{M}_i^{-1}\mathbf{D}_i & -\mathbf{M}_i^{-1}\mathbf{G}_i \\ \mathbf{I}_{i \times i} & \mathbf{0}_{i \times i} \end{bmatrix}, \tag{3.110}$$

where $\mathbf{I}_{i \times i}$ [ $\mathbb{R}^{i \times i}$ is the identity matrix and $\mathbf{0}_{i \times i}$ [ $\mathbb{R}^{i \times i}$ is the zero matrix. The control input matrix $\mathbf{B}_i$ [ $\mathbb{R}^{2i \times 3}$ is defined to be:

$$\mathbf{B}_i \begin{bmatrix} \mathbf{M}_i^{-1}\mathbf{H}_{i \times 3} \\ \mathbf{0}_{i \times i}\mathbf{H}_{i \times 3} \end{bmatrix}. \tag{3.111}$$

The disturbance matrix $\mathbf{E}_i$ [ $\mathbb{R}^{2i \times i}$ becomes:

$$\mathbf{E}_i = \begin{bmatrix} \mathbf{M}_i^{-1} \\ \mathbf{0}_{i \times i} \end{bmatrix}. \tag{3.112}$$

In this text it is assumed that only the position and the orientation are measured. Converting the position and heading measurements to the reference-parallel frame, the LF measurement matrix $\mathbf{C}_i$ [ $\mathbb{R}^{i \times 2i}$ becomes:

$$\mathbf{C}_i = [\mathbf{0}_{i \times i} \quad \mathbf{I}_{i \times i}]. \tag{3.113}$$

### Linear Wave-Frequency Model

In the controller design, synthetic white-noise-driven processes consisting of uncoupled harmonic oscillators with damping will be used to model the WF motions. Applying model reduction as for the LF model, the synthetic WF model can be written in state-space form for $i = 3$ and $i = 5$ according to:

$$\begin{aligned} \dot{\mathbf{x}}_{iw} &= \mathbf{A}_{iw}\mathbf{x}_{iw} + \mathbf{E}_{iw}\mathbf{n}_i \\ \mathbf{y}_{iw} &= \mathbf{C}_{iw}\mathbf{x}_{iw}, \end{aligned} \tag{3.114}$$

where the state-space vector is defined as being:

$$\mathbf{x}_{iw} = [h_{iw}^T, \quad j_{iw}^T]^T. \tag{3115}$$

$\mathbf{y}_{iw}$ [ $\mathbb{R}^i$ is the measurement vector of the wave-frequency motion, and $\mathbf{n}_i$ [ $\mathbb{R}^i$ is the white noise vector. The system matrix $\mathbf{A}_{iw}$ [ $\mathbb{R}^{2i \times 2i}$, the disturbance matrix $\mathbf{E}_{iw}$ [ $\mathbb{R}^{2i \times i}$ and the measurement matrix $\mathbf{C}_{iw}$ [ $\mathbb{R}^{i \times 2i}$ are formulated as in Fossen [3].

## Horizontal Plane Controller

The horizontal plane positioning controller consists of feedback and feedforward controller terms, which may be linear or nonlinear. Usually, linear controllers have been used. However, extensive research in this field has contributed to the introduction of nonlinear control theory (see Strand [9] and the references therein).

### Resulting Control Law

The resulting horizontal plane station-keeping control law, including both feedback and feedforward terms is written:

$$t_{3c} = t_w + t_t + t_i + t_{pd}, \tag{3.116}$$

where the different terms are defined below.

## Horizontal Plane PD Control Law

A conventional linear multivariable proportional and derivative (PD) type controller based on the horizontal plane surge, sway, and yaw DOF can be written:

$$t_{pd} = 2G_p e - G_d \dot{e}, \tag{3.117}$$

where the error vector is decomposed in the reference-parallel frame according to:

$$e = R^T(c_d)[\hat{h} - h_d]^T, \tag{3.118}$$

with $\hat{h} = [\hat{x}, \hat{y}, \hat{c}]^T$, $\hat{h}_d = [x_d, y_d, c_d]^T$ and $G_p, G_d$ [ $\mathbb{R}^{3 \times 3}$ are the nonnegative controller gain matrices found by appropriate control synthesis methods, e.g., linear quadratic performance index, etc.

## Integral Action

The integral control law is defined as in Sørensen et al. [12] according to:

$$\dot{t}_i = A_{wi} t_i + G_i e, \tag{3.119}$$

where $A_{wi}$ [ $\mathbb{R}^{3 \times 3}$ is the anti-windup precaution matrix, and $-G_i$ [ $\mathbb{R}^{3 \times 3}$ is the nonnegative integrator gain matrix.

## Wind Feedforward Control Action

In order to obtain fast disturbance rejection with respect to varying wind loads, it is desirable to introduce a feedforward wind controller. The feedforward wind control law is taken to be:

$$t_w = -G_w \hat{t}_{wind}, \tag{3.120}$$

where $\hat{t}_{wind}$ [ $\mathbb{R}^3$ is the vector of estimated wind forces and moment in surge, sway, and yaw, respectively. It is assumed that the coupling in wind loads is covered by the wind coefficients, so that the gain matrix $G_w$ [ $\mathbb{R}^{3 \times 3}$ is a nonnegative diagonal matrix.

## Model Reference Feedforward Control Action

In order to improve the performance of the controller during tracking operations, a feedforward control action based on input from the reference model is included. The feedforward control action is written:

$$t_t = M_3 a_d + D_3 v_d + d_3(v_d) + C(v_d)v_d, \tag{3.121}$$

where $a_d$ and $v_d$ [ $\mathbb{R}^3$ are the desired generalized reference acceleration and velocity vectors, respectively computed by appropriate reference models.

## Horizontal Plane Controller with Roll-Pitch Damping

The horizontal plane positioning control law, Eq. (3.116), can be extended to also include roll and pitch damping. This is motivated by the fact the hydrodynamic couplings the surge and sway feedback loops will be extended to, incorporate feedback from the low frequency estimated pitch and roll angular velocities, denoted as $\hat{p}$ and $\hat{q}$, respectively.

### Roll-Pitch Control Law

By using a linear formulation, the roll-pitch control law is formulated according to:

$$t_{rpd} = 2G_{rpd} \begin{bmatrix} \hat{p} \\ \hat{q} \end{bmatrix}, \tag{3.122}$$

where the roll-pitch controller gain matrix $\mathbf{G}_{rpd} \in \mathbb{R}^{3 \times 2}$ is defined as:

$$\mathbf{G}_{rpd} = \begin{bmatrix} 0 & g_{xq} \\ g_{yp} & 0 \\ g_{cp} & 0 \end{bmatrix}, \tag{3.123}$$

and $g_{xq}$, $g_{yp}$, and $g_{\psi p}$ are the corresponding nonnegative roll-pitch controller gains.

### Resulting Control Law

The resulting positioning control law including roll-pitch damping becomes:

$$t_{5c} = t_{3c} + t_{rpd}. \tag{3.124}$$

## Controller Analysis

The effect of the roll-pitch damping controller can be shown by analyzing the impact of the extended control law given by Eq. (3.124) on the linearized model given in Eq. (3.107). The coupled linearized low-frequency surge-pitch and sway-roll-yaw models will be considered. In the analysis it is assumed that the wind feedforward controller and the integral control action will compensate for the wind load, the current load, and the mean second-order wave drift loads. Hence, the linearized coupled surge-pitch model can be written:

$$m_{11}\dot{u} + m_{15}\dot{q} + d_{11}u + d_{15}q + g_{11}x = t_{5\,surge,} \tag{3.125}$$

$$m_{51}\dot{u} + m_{55}\dot{q} + d_{51}u + d_{55}q + g_{55}u = 0, \tag{3.126}$$

where subscript $ij$ reflects the element of the matrices in Eq. (3.106) for $i = 6$. By inserting Eq. (3.125) in Eq. (3.126), the pitch dynamics can be reformulated to:

$$\left( m_{55} - \frac{m_{51}m_{15}}{m_{11}} \right)\dot{q} + \left( d_{55} - \frac{m_{51}d_{15}}{m_{11}} \right)q + g_{55}u$$

$$+ \left( d_{51} - \frac{m_{51}d_{11}}{m_{11}} \right)u - \frac{m_{51}g_{11}}{m_{11}}x = \frac{m_{51}}{m_{11}}t_{5\,surge}. \tag{3.127}$$

The pure surge part of the multivariable control law in Eq. (3.124), assuming only PD and the roll-pitch control actions is written as:

$$t_{5\,surge} = -g_x x - g_u u - \frac{m_{11}}{m_{51}}g_{xq}q, \tag{3.128}$$

where $g_x$, $g_u \geq 0$ are the PD controller gains. Substituting Eq. (3.128) in Eq. (3.127) gives the following closed-loop pitch dynamics:

$$\left( m_{55} - \frac{m_{51}m_{15}}{m_{11}} \right)\dot{q} + \left( d_{55} - \frac{m_{51}d_{15}}{m_{11}} + g_{xq} \right)q + g_{55}u$$

$$+ \left( d_{51} - \frac{m_{51}d_{11}}{m_{11}} + g_u\frac{m_{51}}{m_{11}} \right)u - (g_{11} + g_x)\frac{m_{51}}{m_{11}}x = 0. \tag{3.129}$$

Assume that the horizontal plane controller will force the surge velocity $u$ and position deviation $x$ to zero. Hence, the closed-loop pitch dynamics in Eq. (3.129) can be simplified to:

$$\left(m_{55} - \frac{m_{51}m_{15}}{m_{11}}\right)\dot{q} + \left(d_{55} - \frac{m_{51}d_{15}}{m_{11}} + g_{xq}\right)q + g_{55}u = 0. \tag{3.130}$$

Equation (3.130) can be recognized as a second-order mass-damper-spring system, where it is clearly shown that the effect of the new control law increases the damping in pitch.

A similar formulation can be found for the coupled sway-roll-yaw model. The coupled equations of motion in sway, roll, and yaw are:

$$m_{22}\dot{v} + m_{24}\dot{p} + m_{26}\dot{r} + d_{22}v + d_{24}p + d_{26}r + g_{22}y = t_{5\,\mathrm{sway}}, \tag{3.131}$$

$$m_{42}\dot{v} + m_{44}\dot{p} + m_{46}\dot{r} + d_{42}v + d_{44}p + d_{46}r + g_{44}f = 0, \tag{3.132}$$

$$m_{62}\dot{v} + m_{64}\dot{p} + m_{66}\dot{r} + d_{62}v + d_{64}p + d_{66}r + g_{66}c = t_{5\,\mathrm{yaw}}, \tag{3.133}$$

where $t_{\mathrm{sway}}^{5}$ and $t_{\mathrm{yaw}}^{5}$ are the control actions in sway and yaw, respectively. Thus, by including feedback from angular roll velocity in the sway and yaw control laws:

$$t_{5\,\mathrm{sway}} = -g_{y}y - g_{yc}c - g_{v}v - g_{vr}r - \frac{g_{yp}}{b_{y}}p, \tag{3.134}$$

$$t_{5\,\mathrm{yaw}} = -g_{cy}y - g_{c}c - g_{rv}v - g_{r}r - \frac{g_{cp}}{b_{c}}p, \tag{3.135}$$

a similar damping effect as in the pitch case can be obtained. Here, $g_{y}$, $g_{\psi}$, $g_{v}$, $g_{r} \geq 0$ are the diagonal gains of the PD part of the control law. Moreover, $g_{y\psi}$, $g_{\psi p}$, $g_{vp}$, and $g_{rv}$ are the off-diagonal controller gains with appropriate signs reflecting the hydrodynamic coupling between sway and yaw. As for surge, assume that the sway and yaw velocities, $v$ and $r$, position and angle, $y$ and $\psi$ are forced to zero by the horizontal-plane controller. Thus, the closed-loop roll dynamics becomes:

$$m_{p}\dot{p} + (d_{p} + g_{yp} + g_{cp})p + g_{44}f = 0, \tag{3.136}$$

where:

$$m_{p} = m_{44} - \frac{m_{42}m_{24}}{m_{22}} + \frac{m_{46} - \frac{m_{42}m_{26}}{m_{22}}}{1 - \frac{m_{62}}{m_{66}}\frac{m_{26}}{m_{22}}}\left(\frac{m_{62}}{m_{66}}\frac{m_{24}}{m_{22}} - \frac{m_{64}}{m_{66}}\right), \tag{3.137}$$

$$d_{p} = d_{44} - \frac{m_{42}d_{24}}{m_{22}} + \frac{m_{46} - \frac{m_{42}m_{26}}{m_{22}}}{1 - \frac{m_{62}}{m_{66}}\frac{m_{26}}{m_{22}}}\left(\frac{m_{62}}{m_{66}}\frac{d_{24}}{m_{22}} - \frac{d_{64}}{m_{66}}\right), \tag{3.138}$$

$$b_{y} = \frac{m_{42}}{m_{22}} + \frac{m_{62}}{m_{22}m_{66}}\frac{\frac{m_{42}m_{26}}{m_{22}} - m_{46}}{\left(1 - \frac{m_{62}}{m_{66}}\frac{m_{26}}{m_{22}}\right)}, \tag{3.139}$$

$$b_c = \frac{1}{m_{66}\left(1 - \frac{m_{62}}{m_{66}}\frac{m_{26}}{m_{22}}\right)} \cdot \frac{m_{46} - \frac{m_{42}m_{26}}{m_{22}}}{} \quad (3.140)$$

It is clearly shown here how the damping is increased in roll by applying the extended control law.

## Thrust Allocation

### Optimal Thrust Allocation

The relation between the control vector $\tau_{ic}$ [ $R^3$ and the produced thruster action $u_c$ [ $R^r$ is defined by Fossen [3]:

$$t_{ic} = T_{33r}(a)Ku_c, \quad (3.141)$$

where $T_{3\times r}(\alpha)$ [ $R^{3\times r}$ is the thrust configuration matrix, $\alpha$ [ $R^r$ is the thruster orientation vector, and $r$ is the number of thrusters. K [ $R^{r\times r}$ is the diagonal matrix of thrust force coefficients. For a fixed mounted propeller or thruster, the corresponding orientation angle is set to a fixed value reflecting the actual orientation of the device itself. In the case of an azimuthing thruster, $\alpha_i$, there is an additional control input to be determined by the thrust allocation algorithm. $u_c$ is the control vector of either pitch-controlled, revolution-controlled, or torque- and power-controlled propeller inputs [13]. The commanded control action and direction provided by the thrusters becomes [12]:

$$u_c = K^{-1}T^{+}_{33r}(a)t_{ic}, \quad (3.142)$$

where $T^{+}_{33r}(a)$ [ $R^{r\times 3}$ is the pseudo-inverse thrust configuration matrix.

### Geometrical Thrust Induction

The effect of the commanded control action provided by the thrusters in Eq. (3.142) on the 6 DOF vessel model defined in the mathematical modeling section can be calculated to be:

$$t = T_{6\times r}(a)Ku_c, \quad (3.143)$$

where $\tau$ [ $R^6$ is the corresponding actual control vector acting on the vessel, and $T_{6\times r}(\alpha)$ [ $R^{6\times r}$ is the thrust configuration matrix accounting for both the horizontal and the vertical contribution of the produced thruster actions.

## Case Study

A simulation study of a dynamically positioned semisubmersible is carried out to demonstrate the effect of the thruster-induced roll and pitch motions. The performance of the new control strategy, denoted as $\tau_{5c}$, and the conventional horizontal control law, denoted as $\tau_{3c}$, is compared. A semisubmersible (see Fig. 3.18), which is equipped with four azimuthing thrusters, each able to produce a force of 1000 $kN$ located at the four corners at the two pontoons, is used in the simulations. The operational draft is equal to 24 m, vessel mass at operational draft is 45,000 tons, the length is 110 m, and the breadth is 75 m. Radius of gyration in roll is 30 m, in pitch equal to 33 m, and in yaw 38 m. The undamped resonance periods in roll and pitch are found to be 55 s and 60 s, respectively.

### Frequency Domain Analysis

The linearized system given in Eq. (3.107) is analyzed in the frequency domain. The transfer functions of the coupled surge-pitch model subjected to disturbances are shown for the open-loop system, closed-loop system

**FIGURE 3.18**   Semisubmersible.

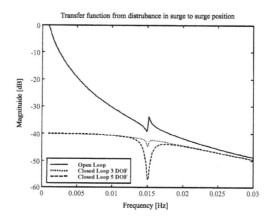

**FIGURE 3.19**   Normalized transfer function from disturbance in surge to surge position for the open-loop system (solid), applying conventional controller $\tau_{3c}$ (dotted) and new controller $\tau_{5c}$ (dashed).

with the conventional horizontal plane (3 DOF) controller, and the new extended roll-pitch (5 DOF) controller. In Fig. 3.19 it can be seen that a significant notch effect is achieved about the pitch resonance frequency with the new control law. Figures 3.20 and 3.21 show how the damping in pitch is amplified about the resonance frequency applying the new control law. Similar progress for the coupled sway-roll-yaw model can also be shown.

### Time Domain Numerical Simulations

The time domain numerical simulations are performed with a significant wave height of 6 m, wave peak period of 10 s, current velocity of 0.5 m/s and wind velocity of 10 m/s. The Earth-fixed directions of waves, wind, and current are colinear and equal to 225° in this simulation. The induced thrust components in roll and pitch are clearly shown in Fig. 3.22 and 3.23. Accordingly, it is important to control these components by the roll-pitch control law to ensure a safe and optimal operation.

By applying the proposed new control strategy, the roll and pitch amplitudes are significantly reduced as shown in the time series of Fig. 3.24 and 3.25. It is also evident that the new controller does not reduce the horizontal positioning accuracy (Fig. 3.24 and 3.25). The effect of the suppressed roll and pitch motions is, as expected, most effective about the resonance frequencies, as shown in the power spectra

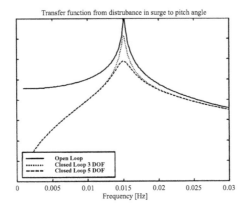

**FIGURE 3.20**    Normalized transfer function from disturbance in surge to pitch angle for the open-loop system (solid), applying conventional controller $\tau_{3c}$ (dotted) and new controller $\tau_{5c}$ (dashed).

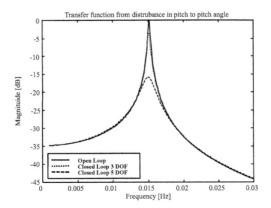

**FIGURE 3.21**    Normalized transfer function from disturbance in pitch to pitch angle for the open-loop system (solid), applying conventional controller $\tau_{3c}$ (dotted) and new controller $\tau_{5c}$ (dashed).

of Fig. 3.26. The simulations further indicate that, by maintaining the same level of positioning accuracy, the total energy consumption when applying the new controller will be lower than using the conventional design philosophy. Oscillations in roll and pitch will, through the hydrodynamic coupling, induce motions in surge, sway, and yaw, which the controller must compensate for. In the roll pitch control strategy these coupling effects are exploited in an optimal manner to damp the oscillations, and avoid unintentional roll and pitch induced surge, sway, and yaw motions.

## References

1. Balchen, J. G., Jenssen, N. A., and Saelid, S., Dynamic positioning using Kalman filtering and optimal control theory, *IFAC/IFIP Symp. on Aut. in Offshore Oil Field Operation*, Holland, Amsterdam, 183–186, 1976.
2. Balchen, J. G., Jenssen, N. A., Mathisen, E., and Sælid, S., A dynamic positioning system based on Kalman filtering and optimal control, *Modeling, Identification and Control*, 1, 3, 135–163, 1980.
3. Fossen, T. I., *Guidance and Control of Ocean Vehicles*, John Wiley & Sons, Ltd., Chichester, England, 1994.
4. Fung, P. T-K. and Grimble, M., Dynamic ship positioning using self-tuning Kalman filter, *IEEE Transaction on Automatic Control*, 28, 3, 339–349, 1983.

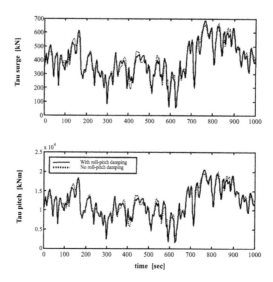

**FIGURE 3.22**  Produced thrust components in surge and pitch applying conventional controller $\tau_{3c}$ (dotted) and new controller $\tau_{5c}$ (solid).

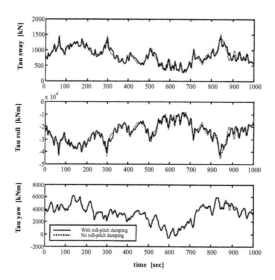

**FIGURE 3.23**  Produced thrust components in sway, roll, and yaw applying conventional controller $\tau_{3c}$ (dotted) and new controller $\tau_{5c}$ (solid).

5. Grimble, M. J., Patton, R. J., and Wise, D. A., The design of dynamic ship positioning control systems using stochastic optimal control theory, *Optimal Control Applications & Methods,* 1, 167–202, 1980a.
6. Grimble, M. J., Patton R. J., and Wise, D. A., Use of Kalman filtering techniques in dynamic ship positioning systems, *IEE Proc.,* 127, 3, Pt. D, 93–102, 1980b.
7. Grimble, M. J. and Johnson, M. A., *Optimal Control and Stochastic Estimation: Theory and Applications,* 2, John Wiley and Sons, New York, 1988.
8. Strand, J. P., Sørensen, A. J., and Fossen, T. I., Design of automatic thruster-assisted position mooring systems for ships, *Modeling, Identification and Control,* MIC, 19, 2, 1998.

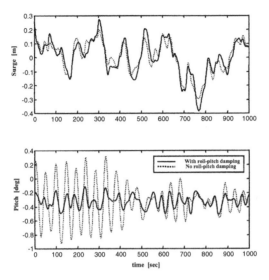

**FIGURE 3.24** Surge position and pitch angle applying conventional controller $\tau_{3c}$ (dotted) and new controller $\tau_{5c}$ (solid).

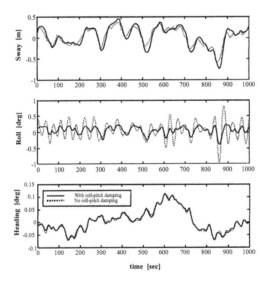

**FIGURE 3.25** Sway position, roll, and yaw angles applying conventional controller $\tau_{3c}$ (dotted) and new controller $\tau_{5c}$ (solid).

9. Strand, J. P., Nonlinear Position Control Systems Design for Marine Vessels, Ph.D. thesis, Norwegian University of Science and Technology, 1999.
10. Sælid, S., Jenssen, N. A., and Balchen, J. G., Design and analysis of a dynamic positioning system based on Kalman filtering and optimal control, *IEEE Transactions on Automatic Control*, 28, 3, 331–339, 1983.
11. Sørdalen, O. J., Optimal thrust allocation for marine vessels, *IFAC Journal of Control Engineering Practice*, 5, 9, 1223–1231, 1997.
12. Sørensen, A. J., Sagatun, S. I., and Fossen, T. I., Design of a dynamic positioning system using model-based control, *Journal of Control Engineering Practice*, 4, 3, 359–368, 1996.

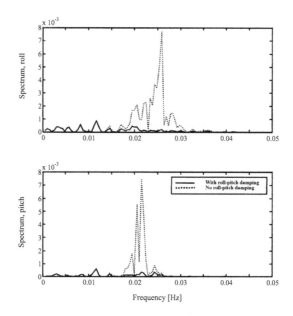

**FIGURE 3.26** Power spectra of roll and pitch applying conventional controller $\tau_{3c}$ (dotted) and new controller $\tau_{5c}$ (solid).

13. Sørensen, A. J., Ådnanes, A. K., Fossen, T. I., and Strand, J. P., A new method of thruster control in positioning of ships based on power control, in *Proc. 4th IFAC Conf. on Maneuvering and Control of Marine Craft (MCMC '97)*, Brijuni, Croatia, 172–179, 1997.
14. Sørensen, A. J. and Strand, J. P., Positioning of low-water plane-area marine constructions with roll and pitch damping, *IFAC Journal of Control Engineering in Practice*, 8, 2, 205–213, 2000.

## 3.5 Weather Optimal Positioning Control

*Thor I. Fossen and Jann Peter Strand*

Conventional DP systems for ships and free-floating rigs are usually designed for station keeping by specifying a desired constant position $(x_d, y_d)$ and a desired constant heading angle $\psi_d$. In order to minimize the ship's fuel consumption, the desired heading $\psi_d$ should, in many operations, be chosen such that the yaw moment is zero. For vessels with port/starboard symmetry, this means that the mean environmental forces due to wind, waves, and currents attack through the center line of the vessel. In this case the ship must be rotated until the yaw moment is zero.

Unfortunately, it is impossible to measure or compute the direction of the mean environmental force with sufficient accuracy. Hence, the desired heading $\psi_d$ is usually taken to be the measurement of the mean wind direction, which can be easily measured. In the North Sea, however, this can result in large offsets from the true mean direction of the total environmental force. The main reason for this is the unmeasured current force component and waves that do not coincide with the wind direction. Hence, the DP system can be operated under highly suboptimal conditions if saving fuel is the issue. A small offset in the optimal heading angle will result in a large use of thrust.

One attractive method for computing the weather optimal heading $\psi_d$ is to monitor the resulting thruster forces in the $x$- and $y$-directions. Hence, the bow of the ship can be turned in one direction until the thruster force in the $y$-direction approaches zero. This method is appealing, but the total resulting thruster forces in the $x$- and $y$-directions have to be computed since there are no sensors doing this job directly. The sensors only measure the angular speed and pitch angle of the propellers. Hence, the thrust for each propeller must be computed by using a model of the thruster characteristic resulting in a fairly rough estimate of the total thruster force in each direction.

Another principle, proposed by Pinkster and Nienhuis [6], is to control the $x$- and $y$ position using a PID feedback controller, in addition to feedback from the yaw velocity, such that the vessel tends toward the optimal heading. This principle, however, requires that the rotation point of the vessel be located in a certain distance *fore* of the center of gravity, or even fore of the bow, and it also puts restrictions on the thruster configuration and the number of thrusters installed.

This section describes the design of a new concept for *weather optimal positioning control* (WOPC) of ships and free-floating rigs (see Fossen and Strand [2] and Strand [7]). The proposed control strategy was patented by ABB Industri AS in cooperation with the authors in December 1998. The control objective is that the vessel heading should adjust automatically to the mean environmental disturbances (wind, waves, and currents) such that a minimum amount of energy is used in order to save fuel and reduce $NO_x/CO_x$-emissions without using any environmental sensors. This is particularly useful for shuttle tankers and FPSOs, which can be located at the same position for long time. Also DP operated supply vessels, which must keep their position for days in loading/off-loading operations, have a great WOPC fuel-saving potential.

Nonlinear and adaptive backstepping designs are used to derive the WOPC system. The concept of WOPC can also be implemented by using other control methods, e.g., feedback linearization.

The key feature of the WOPC is that no information about the environmental disturbances is required. This is important since the mean environmental disturbances acting on a floating vessel cannot be accurately measured or computed. We show that the ship can be exponentially stabilized on a circle arc with constant radius by letting the bow of the ship point toward the origin of the circle. In order to maintain a fixed position at the same time, a translatory circle center control law is designed. Moreover, the circle center is translated online such that the Cartesian position is constant while the bow of the ship is automatically turned up against the mean environmental force (weather-vaning). This approach is motivated by a pendulum in the gravity field where gravity is the unmeasured quantity. The circular motion of the controlled ship, where the mean environmental force can be interpreted as an unknown force field, copies the dynamics of a pendulum in the gravity field (see Fig. 3.27).

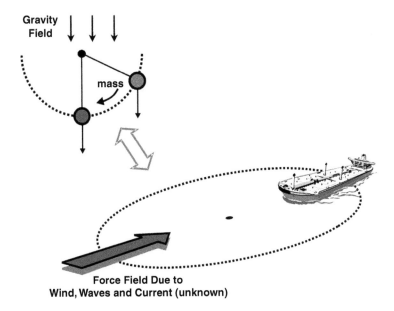

**FIGURE 3.27** Weather optimal positioning principle: Equivalent to a pendulum in the gravity field, where gravity is the unmeasured quantity.

## Mathematical Modeling

### Ship Model

In this section we consider a low-frequency, *low-speed* ship model in 3 DOF:

$$\dot{\eta} = \mathbf{R}(c)n, \tag{3.144}$$

$$\mathbf{M}\dot{n} + \mathbf{C}(n)n + \mathbf{D}(n)n = t + \mathbf{w}, \tag{3.145}$$

where the Earth-fixed position $(x, y)$ and heading $\psi$ is represented by $\boldsymbol{\eta} = [x, y, \psi]^T$ and the vessel-fixed vessel velocities are represented by $\boldsymbol{\nu} = [u, v, r]^T$. From the low-speed assumption, which is applied in the Lyapunov stability analysis, $\mathbf{M} = \mathbf{M}^T > 0$, $\dot{\mathbf{M}} = \mathbf{0}$, and $\mathbf{D}(\boldsymbol{\nu}) > 0$, $;n\,[\,\mathbf{R}^3$ . $t\,[\,\mathbf{R}^3$ is the control vector of forces in surge and sway and moment in yaw provided by the propulsion system. Unmodeled external forces and moment due to wind, currents, and waves are lumped together into an vessel-fixed disturbance vector $\mathbf{w}\,[\,\mathbf{R}^3$ . The rotation matrix $\mathbf{R}\,(\psi)$ in yaw is defined in the section entitled "Kinematics".

### Polar Coordinates

The Cartesian coordinates $(x, y)$ are related to the polar coordinates by:

$$x = x_0 + r\cos g, \tag{3.146}$$

$$y = y_0 + r\sin g, \tag{3.147}$$

where $(x_0, y_0)$ is the origin of a circle with radius $\rho$ and polar angle $\gamma$:

$$r = \sqrt{(x - x_0)^2 + (y - y_0)^2}, \tag{3.148}$$

$$g = \text{atan}((y - y_0), (x - x_0)). \tag{3.149}$$

Time differentiation of Eqs. (3.146) and (3.147) yields:

$$\dot{x} = \dot{x}_0 + \dot{r}\cos g - r\sin g\,\dot{g}, \tag{3.150}$$

$$\dot{y} = \dot{y}_0 + \dot{r}\sin g + r\cos g\,\dot{g}. \tag{3.151}$$

Define the state vectors

$$\mathbf{p}_0 \; \tilde{\texttt{5}} \; [x_0, y_0]^T, \qquad \mathbf{x} \; \tilde{\texttt{5}} \; [r, g, c]^T. \tag{3.152}$$

From Eqs. (3.150) and (3.151) a new kinematic relationship can be written in terms of the vectors $\mathbf{p}_0$ and $\mathbf{x}$ as:

$$\dot{\eta} = \mathbf{R}(g)\mathbf{H}(r)\dot{\mathbf{x}} \; 1 \; \mathbf{L}\dot{\mathbf{P}}_0, \tag{3.153}$$

where

$$\mathbf{H}(r) = \begin{bmatrix} 1 & 0 & 0 \\ 0 & r & 0 \\ 0 & 0 & 1 \end{bmatrix}, \qquad \mathbf{L} = \begin{bmatrix} 1 & 0 \\ 0 & 1 \\ 0 & 0 \end{bmatrix}. \tag{3.154}$$

From Eq. (3.153) the Cartesian kinematics in Eq. (3.144) can be replaced by a differential equation for the *polar coordinates*:

$$\dot{\mathbf{x}} = \mathbf{T}(\mathbf{x})n - \mathbf{T}(\mathbf{x})\mathbf{R}^T(c)\mathbf{L}\dot{\mathbf{p}}_0, \tag{3.155}$$

where

$$\mathbf{T}(\mathbf{x}) \stackrel{\Delta}{=} \mathbf{H}^{-1}(r)\mathbf{R}^T(g)\mathbf{R}(c)$$

$$= \mathbf{H}^{-1}(r)\mathbf{R}^T(g - c). \tag{3.156}$$

Note that the conversion between *Cartesian* and *polar* coordinates is only a *local diffeomorphism*, since the radius must be kept larger than a minimum value, i.e., $\rho > \rho_{min} > 0$, in order to avoid the singular point $\rho = 0$.

### Ship Model Transformation

The ship model, Eq. (3.145), can be represented by polar coordinates by using Eq. (3.155) and substituting

$$n = \mathbf{T}^{-1}(\mathbf{x})\dot{\mathbf{x}} + \mathbf{R}^T\mathbf{L}\dot{\mathbf{p}}_0 \tag{3.157}$$

$$\dot{n} = \mathbf{T}^{-1}(\mathbf{x})\ddot{\mathbf{x}} + \dot{\mathbf{T}}^{-1}(\mathbf{x})\dot{\mathbf{x}} + \mathbf{R}^T\mathbf{L}\ddot{\mathbf{p}}_0 + \dot{\mathbf{R}}^T\mathbf{L}\dot{\mathbf{p}}_0, \tag{3.158}$$

such that:

$$\mathbf{M}\dot{n} + \mathbf{C}(n)n + \mathbf{D}(n)n = t + \mathbf{w}$$
$$\Updownarrow \ r > 0$$
$$\mathbf{M}_x(\mathbf{x})\ddot{\mathbf{x}} + \mathbf{C}_x(n, \mathbf{x})\dot{\mathbf{x}} + \mathbf{D}_x(n, \mathbf{x})\dot{\mathbf{x}} = \mathbf{T}^{-T}\mathbf{q}(n, \mathbf{x}, \dot{\mathbf{p}}_0, \ddot{\mathbf{p}}_0)$$
$$+ \mathbf{T}^{-T}t + \mathbf{T}^{-T}\mathbf{w}, \tag{3.159}$$

where

$$\mathbf{M}_x(\mathbf{x}) = \mathbf{T}^{-T}(\mathbf{x})\mathbf{M}\mathbf{T}^{-1}(\mathbf{x}) \tag{3.160}$$

$$\mathbf{C}_x(n, \mathbf{x}) = \mathbf{T}^{-T}(\mathbf{x})(\mathbf{C}(n) - \mathbf{M}\mathbf{T}^{-1}(\mathbf{x})\dot{\mathbf{T}}(\mathbf{x}))\mathbf{T}^{-1}(\mathbf{x}) \tag{3.161}$$

$$\mathbf{D}_x(n, \mathbf{x}) = \mathbf{T}^{-T}(\mathbf{x})\mathbf{D}(n)\mathbf{T}^{-1}(\mathbf{x}) \tag{3.162}$$

$$\mathbf{q}(n, \mathbf{x}, \dot{\mathbf{p}}_0, \ddot{\mathbf{p}}_0) = \mathbf{M}\mathbf{R}^T(c)\mathbf{L}\ddot{\mathbf{p}}_0 + \mathbf{M}\dot{\mathbf{R}}^T(c)\mathbf{L}\dot{\mathbf{p}}_0$$
$$+ [\mathbf{C}(n) + \mathbf{D}(n)]\mathbf{R}^T(c)\mathbf{L}\dot{\mathbf{p}}_0. \tag{3.163}$$

Here, $\mathbf{M}_x(\mathbf{x})$, $\mathbf{C}_x(\boldsymbol{\nu}, \mathbf{x})$ and $\mathbf{D}_x(\boldsymbol{\nu}, \mathbf{x})$ can be shown to satisfy:

$$\mathbf{M}_x(\mathbf{x}) = \mathbf{M}_x^T(\mathbf{x}) > 0, \qquad \mathbf{D}_x(n, \mathbf{x}) > 0, \qquad \forall \mathbf{x}.$$

The ship dynamics also satisfy the skew-symmetric property:

$$\mathbf{z}^T(\dot{\mathbf{M}}_x - 2\mathbf{C}_x)\mathbf{z} = 0, \qquad \forall \mathbf{z}, \mathbf{x}. \tag{3.164}$$

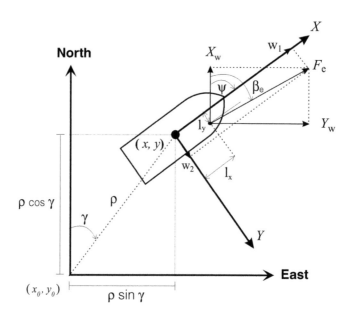

**FIGURE 3.28**  Environmental force $F_e$ decomposed into the components $w_1$ and $w_2$.

### Disturbance Modeling

The steady-state LF motion of the ship and also the ship's equilibrium position depend on the *unknown* environmental loads acting on the vessel. Let the environmental loads due to *wind*, *waves*, and *currents* be represented by:

- a slowly varying mean force $F_e$ that attacks the ship at a point $(l_x, l_y)$ in body-fixed coordinates.
- a slowly varying mean direction $\beta_e$, relative to the Earth-fixed frame (see Fig. 3.28).

The term *slowly varying* simply states that the mean environmental forces are slow compared to the vessel kinematics and dynamics. The slowly varying terms include second-order wave-induced disturbances (wave drift), currents, and mean wind forces. The WF motion is assumed to be filtered out of the measurements by using a *wave filter*.

Since there are no sensors that can be used to measure $(F_e, \beta_e)$ and $(l_x, l_y)$ with sufficient accuracy, it is impossible to use feedforward from the environmental disturbances. This motivates the following assumptions:

**A1:**  The unknown mean environmental force $F_e$ and its direction $\beta_e$ are assumed to be constant or at least slowly varying.

**A2:**  The unknown attack point $(l_x, l_y)$ is constant for each constant $F_e$.
**Discussion:**  These are good assumptions since the ship control system is only supposed to counteract the slowly varying motion components of the environmental disturbances.

From Figure 3.28 the body-fixed environmental load vector $\mathbf{w} \in R^3$ can be expressed as:

$$\mathbf{w} = \begin{bmatrix} w_1(c) \\ w_2(c) \\ w_2(c) \end{bmatrix} = \begin{bmatrix} F_e\cos(b_e - c) \\ F_e\sin(b_e - c) \\ l_x F_e\sin(b_e - c) - l_y F_e\cos(b_e - c) \end{bmatrix}. \tag{3.165}$$

Notice that the environmental loads vary with the heading angle $\psi$ of the ship. Moreover:

$$F_e = \sqrt{w_1^2 + w_2^2} \tag{3.166}$$

$$\beta_e = \psi + \tan^{-1}(w_2/w_1). \tag{3.167}$$

The environmental forces $X_w$ and $Y_w$ with attack point $(l_x, l_y)$ are shown in Fig. 3.28. It should be noted that the attach point $l_x = l_x(\psi)$ and $l_y = l_y(\psi)$ will also change with the yaw angle $\psi$. This relationship will be a complicated function of the hull and superstructure geometries.

## Weather Optimal Control Objectives

The weather optimal control objectives make use of the following definitions [2]:

**Definition 3.3 (Weather optimal heading)** The *weather optimal heading angle* $\psi_{opt}$ is given by the equilibrium state where the yaw moment $w_3(\psi_{opt}) = 0$ at the same time as the bow of the ship is turned against weather (mean environmental disturbances), that is, $w_2(\psi_{opt}) = 0$. This implies that the moment arms $l_x(\psi_{opt}) = $ constant and $l_y(\psi_{opt}) = 0$, and:

$$\mathbf{w}(c_{opt}) = \begin{bmatrix} w_1(c_{opt}) \\ w_2(c_{opt}) \\ w_3(c_{opt}) \end{bmatrix} = \begin{bmatrix} -F_e \\ 0 \\ 0 \end{bmatrix}.$$

Hence, the mean environmental force attacks the ship in the bow (minimum drag coefficient for water and wind loads).

**Definition 3.4 (Weather optimal positioning)** *Weather optimal positioning* (station keeping) is defined as the equilibrium state where

$$w_1(c_{opt}) \; 5 \; 2F_e, \qquad w_2(c_{opt}) \; 5 \; w_3(c_{opt}) \; 5 \; l_y(c_{opt}) \; 5 \; 0 \qquad (3.168)$$

(weather optimal heading) and the position $(x, y) = (x_d, y_d)$ is kept constant.

These definitions motivate the following two control objectives:

**O1: Weather Optimal Heading Control (WOHC):** This is obtained by restricting the ship to move on a circle with *constant* radius $\rho = \rho_d$ and, at the same time, force the ship's bow to point toward the center of the circle until the weather optimal heading angle $\psi = \psi_{opt}$ is reached (see Fig. 3.29). An analogy to this is a pendulum in a gravity field (see Fig. 3.27). The position $(x, y) = (x_0 + \rho \cos \gamma, y_0 + \rho \sin \gamma)$ will vary until the weather optimal heading angle is reached. This is obtained by specifying the control objective in polar coordinates according to:

$$\rho_d = \text{constant} \qquad (3.169)$$

$$\dot{g}_d = 0 \qquad (3.170)$$

$$\psi_d = \pi + \gamma. \qquad (3.171)$$

**Discussion:** The requirement $\rho_d = $ *constant* implies that the ship moves on a circle with constant radius. The second requirement $\dot{g}_d = 0$ implies that the tangential speed $\rho\dot{g}$ is kept small while the last requirement $\psi_d = \pi + \gamma$, ensures that the ship's bow points toward the center of the circle.

**O2: Weather Optimal Positioning Control (WOPC):** In order to maintain a fixed Earth-fixed position $(x, y) = (x_d, y_d)$, the *circle center* $\mathbf{p}_0 = [x_0, y_0]^T$ must be moved simultaneously as *Control Objective O1* is satisfied. This is referred to as *translatory circle center control*.

We will derive a nonlinear and adaptive backstepping controller that satisfies O1 and O2.

## Nonlinear and Adaptive Control Design

In this section a positioning controller is presented by using the polar coordinate representation of the ship dynamics. Backstepping is used to derive the feedback controller [4]. The control law will be derived in three successive steps:

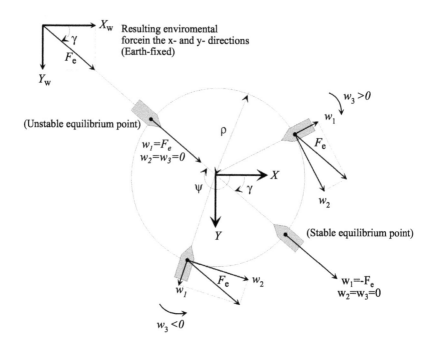

**FIGURE 3.29**   Principle for weather optimal heading control (WOHC).

1. *Nonlinear backstepping (PD-control):* the ship is forced to move on a circle arc with desired radius $\rho_d$, with minimum tangential velocity $r\dot{g}$ and desired heading $\psi_d$.
2. *Adaptive backstepping (PID-control):* this is necessary to compensate for the *unknown* environmental force $F_e$.
3. *Translatory circle center control:* the circle center $(x_0, y_0)$ is translated such that the ship maintains a constant position $(x_d, y_d)$ even though it is moving on a *virtual* circle arc. Hence, the captain of the ship will only notice that the ship is rotating a yaw angle $\psi$ about a constant position $(x_d, y_d)$ until the weather optimal heading $\psi_{opt}$ is reached.

## Nonlinear Backstepping (PD-Control)

A general positioning controller is derived by using *vectorial backstepping*. The tracking objective is specified in polar coordinates by the smooth reference trajectory $\mathbf{x}_d = [\rho_d, \gamma_d, \psi_d]^T \in C^3$ where:

$$\mathbf{x}_d, \dot{\mathbf{x}}_d, \ddot{\mathbf{x}}_d [ \mathcal{L}_\infty.$$

Since the transformed system in Eq. (3.159) is of order 2, backstepping is performed in two *vectorial steps* resulting in a nonlinear PD-control law. First, we define a *virtual reference trajectory* as:

$$\dot{\mathbf{x}}_r \overset{D}{5} \dot{\mathbf{x}}_d - \Lambda\mathbf{z}_1, \qquad (3.172)$$

where $\mathbf{z}_1 = \mathbf{x} - \mathbf{x}_d$ is the Earth-fixed tracking error and $\Lambda > 0$ is a diagonal design matrix. Furthermore, let $\mathbf{z}_2$ denote a measure of tracking defined according to:

$$\mathbf{z}_2 \overset{\text{5}}{3} \dot{\mathbf{x}} - \dot{\mathbf{x}}_r = \dot{\mathbf{z}}_1 + \Lambda\mathbf{z}_1. \qquad (3.173)$$

From Eq. (3.173), the following expressions are obtained:

$$\dot{\mathbf{x}} = \mathbf{z}_2 + \dot{\mathbf{x}}_r \tag{3.174}$$

$$\ddot{\mathbf{x}} = \dot{\mathbf{z}}_2 + \ddot{\mathbf{x}}_r. \tag{3.175}$$

This implies that the vessel model in Eq. (3.159) can be expressed in terms of $\mathbf{z}_2$, $\dot{\mathbf{x}}_r$, and $\ddot{\mathbf{x}}_r$ as:

$$\mathbf{M}_x \dot{\mathbf{z}}_2 + \mathbf{C}_x \mathbf{z}_2 + \mathbf{D}_x \mathbf{z}_2 = \mathbf{T}^{-T} t + \mathbf{T}^{-T} \mathbf{q}(\cdot) - \mathbf{M}_x \ddot{\mathbf{x}}_r - \mathbf{C}_x \dot{\mathbf{x}}_r \, 2 \, \mathbf{D}_x \dot{\mathbf{x}}_r \, 1 \, \mathbf{T}^{-T} \mathbf{w} \,. \tag{3.176}$$

**Step 1:** Let $\mathbf{z}_1$ be the first error variable, which from Eq. (3.173) has the dynamics:

$$\dot{\mathbf{z}}_1 \, 5 \, -\Lambda \mathbf{z}_1 + \mathbf{z}_2. \tag{3.177}$$

A *Lyapunov function candidate* (LFC) for the first step is:

$$V_1 \, 5 \, \frac{1}{2} \mathbf{z}_1^T \mathbf{K}_p \mathbf{z}_1 \tag{3.178}$$

$$\dot{V}_1 \, 5 \, -\mathbf{z}_1^T \mathbf{K}_p \Lambda \mathbf{z}_1 + \mathbf{z}_1^T \mathbf{K}_p \mathbf{z}_2, \tag{3.179}$$

where $\mathbf{K}_p = \mathbf{K}_p^T > 0$ is a constant design matrix.

**Step 2:** In the second step we choose a LFC motivated by the *"pseudo" kinetic energy*, that is:

$$V_2 = V_1 + \frac{1}{2} \mathbf{z}_2^T \mathbf{M}_x \mathbf{z}_2, \qquad \mathbf{M}_x = \mathbf{M}_x^T > 0. \tag{3.180}$$

Time differentiation of $V_2$ along the trajectories of $\mathbf{z}_1$ and $\mathbf{z}_2$ yields:

$$\dot{V}_2 = \dot{V}_1 + \mathbf{z}_2^T \mathbf{M}_x \dot{\mathbf{z}}_2 + \frac{1}{2} \mathbf{z}_2^T \dot{\mathbf{M}}_x \mathbf{z}_2, \tag{3.181}$$

which by substitution of Eqs. (3.179) and (3.176) yields

$$\dot{V}_2 \, 5 \, -\mathbf{z}_1^T \mathbf{K}_p \Lambda \mathbf{z}_1 + \frac{1}{2} \mathbf{z}_2^T (\dot{\mathbf{M}}_x - 2\mathbf{C}_x) \mathbf{z}_2 - \mathbf{z}_2^T \mathbf{D}_x \mathbf{z}_2 + \mathbf{z}_2^T \mathbf{T}^{-T} \mathbf{w}$$
$$+ \mathbf{z}_2^T (\mathbf{K}_p \mathbf{z}_1 + \mathbf{T}^{-T} t + \mathbf{T}^{-T} \mathbf{q}(\cdot) - \mathbf{M}_x \ddot{\mathbf{x}}_r - \mathbf{C}_x \dot{\mathbf{x}}_r \, 2 \, \mathbf{D}_x \dot{\mathbf{x}}_r). \tag{3.182}$$

By using the property of Eq. (3.164) and choosing the nonlinear PD-control law as:

$$\mathbf{T}^{-T} t = \mathbf{M}_x \ddot{\mathbf{x}}_r + \mathbf{C}_x \dot{\mathbf{x}}_r + \mathbf{D}_x \dot{\mathbf{x}}_r - \mathbf{K}_p \mathbf{z}_1 - \mathbf{K}_d \mathbf{z}_2 - \mathbf{T}^{-T} \mathbf{q}(\cdot), \tag{3.183}$$

where $\mathbf{K}_d > 0$ is a strictly positive design matrix, we finally get:

$$\dot{V}_2 = -\mathbf{z}_1^T \mathbf{K}_p \Lambda \mathbf{z}_1 - \mathbf{z}_2^T (\mathbf{K}_d + \mathbf{D}_x) \mathbf{z}_2 \, 1 \, \mathbf{z}^T \mathbf{T}^{-T} \mathbf{w}. \tag{3.184}$$

Notice that the dissipative term $\mathbf{z}_2^T \mathbf{D}_x \mathbf{z}_2 > 0$, $\forall \mathbf{z}_2 \neq 0$ is exploited in the design as it appears in the expression for $\dot{V}_2$. With the control law in Eq. (3.183), the closed-loop dynamics becomes:

$$\mathbf{M}_x \dot{\mathbf{z}}_2 + (\mathbf{C}_x + \mathbf{D}_x + \mathbf{K}_d)\mathbf{z}_2 + \mathbf{K}_p \mathbf{z}_1 = \mathbf{T}^{-T}\mathbf{w}. \qquad (3.185)$$

The error dynamics of the resulting system becomes *nonautonomous* since:

$$
\begin{bmatrix} \mathbf{K}_p & \mathbf{0}_{3 \times 3} \\ \mathbf{0}_{3 \times 3} & \mathbf{M}_x \end{bmatrix} \begin{bmatrix} \dot{\mathbf{z}}_1 \\ \dot{\mathbf{z}}_2 \end{bmatrix} = -\begin{bmatrix} \mathbf{K}_p \Lambda & \mathbf{0}_{3 \times 3} \\ \mathbf{0}_{3 \times 3} & \mathbf{C}_x + \mathbf{D}_x + \mathbf{K}_d \end{bmatrix} \begin{bmatrix} \mathbf{z}_1 \\ \mathbf{z}_2 \end{bmatrix}
$$

$$
+ \begin{bmatrix} \mathbf{0}_{3 \times 3} & \mathbf{K}_p \\ -\mathbf{K}_p & \mathbf{0}_{3 \times 3} \end{bmatrix} \begin{bmatrix} \mathbf{z}_1 \\ \mathbf{z}_2 \end{bmatrix} + \begin{bmatrix} \mathbf{0}_{3 \times 1} \\ \mathbf{T}^{-T} \end{bmatrix} \mathbf{w} \qquad (3.186)
$$

$$\Updownarrow$$

$$\mathcal{M}(\mathbf{x})\dot{\mathbf{z}} = -\mathcal{K}(\mathbf{x}, n)\mathbf{z} + S\mathbf{z} + \bar{b}(\mathbf{x})\mathbf{w}, \qquad (3.187)$$

where the different matrices are defined as:

$$\mathcal{M}(\mathbf{x}) = \mathcal{M}^T(\mathbf{x}) = \begin{bmatrix} \mathbf{K}_p & \mathbf{0}_{3 \times 3} \\ \mathbf{0}_{3 \times 3} & \mathbf{M}_x(\mathbf{x}) \end{bmatrix} \qquad (3.188)$$

$$\mathcal{K}(\mathbf{x}, n) = \begin{bmatrix} \mathbf{K}_p \Lambda & \mathbf{0}_{3 \times 3} \\ \mathbf{0}_{3 \times 3} & \mathbf{C}_x(\mathbf{x}, n) + \mathbf{D}_x(\mathbf{x}, n) + \mathbf{K}_d \end{bmatrix} > 0 \qquad (3.189)$$

$$S = -S^T = \begin{bmatrix} \mathbf{0}_{3 \times 3} & \mathbf{K}_p \\ -\mathbf{K}_p & \mathbf{0}_{3 \times 3} \end{bmatrix} \qquad (3.190)$$

$$\bar{\mathcal{B}}(\mathbf{x}) = \begin{bmatrix} \mathbf{0}_{3 \times 1} \\ \mathbf{T}^{-T}(\mathbf{x}) \end{bmatrix}. \qquad (3.191)$$

In the absence of disturbances, $\mathbf{w} ; 0$, the origin $\mathbf{z} = 0$ is uniformly locally exponentially stable (ULES) according to Lyapunov. Global results cannot be achieved due to the local diffeomorphism between the Cartesian and polar coordinates, that is the transformation matrix $\mathbf{T}(\mathbf{x})$ is singular for $\rho = 0$.

With disturbances $\mathbf{w} \neq 0$, the closed-loop system is input-to-state-stable (ISS). In the next section, we will show how adaptive backstepping (backstepping with integral action) can be used to obtain ULES for the case of a nonzero disturbance vector $\mathbf{w} \neq 0$.

**Adaptive Backstepping (PID-Control)**

If the disturbance vector $\mathbf{w}$ has a nonzero mean, this will result in a steady-state offset when using the nonlinear PD-controller. Since the ship is restricted to move on a circle arc where $\mathbf{w}$ can be viewed as a force field, there will be a stable and unstable equilibrium point on the circle arc (similar to a pendulum in the gravity field). The stable equilibrium point is (see Fig. 3.29):

$$\mathbf{w} = f F_e = \begin{bmatrix} -1 \\ 0 \\ 0 \end{bmatrix} F_e . \qquad (3.192)$$

Since the disturbance $F_e$ is assumed to be slowly varying, we can apply adaptive backstepping to obtain integral effect in the system. Thus, in the analysis it will be assumed that

$$\dot{F}_e = 0. \tag{3.193}$$

Let the estimate of $F_e$ be denoted as $\hat{F}_e$, and $\tilde{F} = \hat{F}_e - F_e$. An additional step in the derivation of the backstepping control must be performed in order to obtain an adaptive update law for $\hat{F}_e$. Moreover:

**Step 3:** The adaptive update law is found by adding the squared parameter estimation error to $V_2$. Moreover:

$$V_3 = V_2 + \frac{1}{2s}\tilde{F}_e^2, \qquad s > 0, \tag{3.194}$$

where

$$\dot{V}_3 = \dot{V}_2 + \frac{1}{s}\dot{\tilde{F}}_e\tilde{F}_e. \tag{3.195}$$

The nonlinear control law in Eq. (3.183) is modified to:

$$t = \mathbf{T}^T(\mathbf{M}_x\ddot{\mathbf{x}}_r + \mathbf{C}_x\dot{\mathbf{x}}_r + \mathbf{D}_x\dot{\mathbf{x}}_r - \mathbf{K}_p\mathbf{z}_1 - \mathbf{K}_d\mathbf{z}_2) - \mathbf{q}(\cdot) - f\hat{F}_e, \tag{3.196}$$

where the last term $f\hat{F}_e$ provides integral action. Hence, the $\mathbf{z}_2$-dynamics becomes:

$$\mathbf{M}_x\dot{\mathbf{z}}_2 + (\mathbf{C}_x + \mathbf{D}_x + \mathbf{K}_d)\mathbf{z}_2 + \mathbf{K}_p\mathbf{z}_1 = -\mathbf{T}^{2^T}f\tilde{F}_e. \tag{3.197}$$

This implies that:

$$\begin{aligned}
\dot{V}_3 &= -\mathbf{z}_1^T\mathbf{K}_p\Lambda\mathbf{z}_1 - \mathbf{z}_2^T(\mathbf{K}_d + \mathbf{D}_x)\mathbf{z}_2 - \mathbf{z}_2^T\mathbf{T}^{-T}f\tilde{F}_e + \frac{1}{\sigma}\dot{\tilde{F}}_e\tilde{F}_e \\
&= -\mathbf{z}_1^T\mathbf{K}_p\Lambda\mathbf{z}_1 - \mathbf{z}_2^T(\mathbf{K}_d + \mathbf{D}_x)\mathbf{z}_2 + \tilde{F}_e\left(-f^T\mathbf{T}^{21}\mathbf{z}_2 + \frac{1}{\sigma}\dot{\tilde{F}}_e\right).
\end{aligned} \tag{3.198}$$

The adaptive law $\dot{\tilde{F}}_e = \dot{\hat{F}}_e$ is chosen as:

$$\dot{\hat{F}}_e = sf^T\mathbf{T}^{21}\mathbf{z}_2, \qquad s > 0, \tag{3.199}$$

such that

$$\dot{V}_3 = -\mathbf{z}_1^T\mathbf{K}_p\Lambda\mathbf{z}_1 - \mathbf{z}_2^T(\mathbf{K}_d + \mathbf{D}_x)\mathbf{z}_2 \leq 0. \tag{3.200}$$

The *nonautonomous* error dynamics for the adaptive backstepping controller can be written:

$$\mathcal{M}(\mathbf{x})\dot{\mathbf{z}} = [-\mathcal{K}(\mathbf{x}, n) + S]\mathbf{z} + \mathcal{B}(\mathbf{x})\tilde{F}_e \tag{3.201}$$

$$\dot{\tilde{F}}_e = -s\mathcal{B}^T(\mathbf{x})\mathbf{z}, \tag{3.202}$$

where

$$\mathcal{B}(\mathbf{x}) = \begin{bmatrix} \mathbf{0}_{331} \\ -\mathbf{T}^{-T}(\mathbf{x})f \end{bmatrix}. \tag{3.203}$$

In order to satisfy *Control Objective O1* we must choose the controller gains according to:

$$\mathbf{K}_p = \begin{bmatrix} k_{p1} & 0 & 0 \\ 0 & 0 & 0 \\ 0 & 0 & k_{p3} \end{bmatrix}, \quad \mathbf{K}_d = \begin{bmatrix} k_{d1} & 0 & 0 \\ 0 & k_{d2} & 0 \\ 0 & 0 & k_{d3} \end{bmatrix}, \quad \Lambda = \begin{bmatrix} l_1 & 0 & 0 \\ 0 & 0 & 0 \\ 0 & 0 & l_3 \end{bmatrix}. \tag{3.204}$$

Notice that $k_{p2} = \lambda_2 = 0$. This implies that the ship is free to move on the circle arc with tangential velocity $r\dot{g}$. The gain $k_{d2} > 0$ is used to increase the tangential damping (D-control) while the radius $\rho$ and heading $\psi$ are stabilized by using PID-control.

**Semi-Definite Matrices**

Since the controller gains $k_{p2}$ and $\lambda_2$ are chosen to be zero, the matrices:

$$\mathbf{K}_p \geq 0, \qquad \Lambda \geq 0 \tag{3.205}$$

are only positive semi-definite resulting in a positive semi-definite $V_3$. *Uniform local asymptotic stability* (ULAS) of the equilibrium $(\mathbf{z}, \tilde{F}_e) = (\mathbf{0}, 0)$ can, however, be proven since the system is ISS. We therefore consider the error dynamics of the reduced order system $(\mathbf{z}_{1r}, \mathbf{z}_2)$ given by:

$$\mathbf{z}_{1r} = \mathbf{E}\mathbf{z}_1, \qquad \mathbf{E} = \begin{bmatrix} 1 & 0 & 0 \\ 0 & 0 & 1 \end{bmatrix}. \tag{3.206}$$

This implies that:

$$\begin{aligned} \dot{\mathbf{z}}_{1r} &= 2\mathbf{E}\Lambda\mathbf{z}_1 + \mathbf{E}\mathbf{z}_2 \\ &= 2(\mathbf{E}\Lambda\mathbf{E}^T)\mathbf{z}_{1r} + \mathbf{E}\mathbf{z}_2. \end{aligned} \tag{3.207}$$

Notice that the last step is possible since the diagonal matrices $\Lambda = \text{diag}\{\lambda_1, 0, \lambda_3\}$ satisfies:

$$\Lambda\mathbf{E}^T\mathbf{z}_{1r} = \Lambda\mathbf{z}_1. \tag{3.208}$$

Hence, the error dynamics in Eqs. (3.201) through (3.202) can be transformed to:

$$\mathcal{M}_r(\mathbf{x})\dot{\mathbf{z}}_r = [-\mathcal{K}_r(\mathbf{x}, n) + \mathcal{S}_r]\mathbf{z}_r + \mathcal{B}_r(\mathbf{x})\tilde{F}_e \tag{3.209}$$

$$\dot{\tilde{F}}_e = -s\mathcal{B}_r^T(\mathbf{x})\mathbf{z}_r, \tag{3.210}$$

where $\mathbf{z}_r = [\mathbf{z}_{1r}^T, \mathbf{z}_2^T]^T$ and:

$$\mathcal{M}_r(\mathbf{x}) = \mathcal{M}_r^T(\mathbf{x}) = \begin{bmatrix} (\mathbf{E}\mathbf{K}_p\mathbf{E}^T) & \mathbf{0}_{2\times3} \\ \mathbf{0}_{3\times2} & \mathbf{M}_x(\mathbf{x}) \end{bmatrix} \tag{3.211}$$

$$\mathcal{K}_r(\mathbf{x}, n) = \begin{bmatrix} (\mathbf{E}\mathbf{K}_p\mathbf{E}^T)(\mathbf{E}\Lambda\mathbf{E}^T) & \mathbf{0}_{2 \times 3} \\ \mathbf{0}_{3 \times 2} & \mathbf{C}_x(\mathbf{x}, n) + \mathbf{D}_x(\mathbf{x}, n) + \mathbf{K}_d \end{bmatrix} > 0 \qquad (3.212)$$

$$\mathcal{S}_r = -\mathcal{S}_r^T = \begin{bmatrix} \mathbf{0}_{2 \times 2} & \mathbf{E}\mathbf{K}_p \\ -\mathbf{K}_p\mathbf{E}^T & \mathbf{0}_{3 \times 3} \end{bmatrix} \qquad (3.213)$$

$$\mathcal{B}_r(\mathbf{x}) = \begin{bmatrix} \mathbf{0}_{2 \times 1} \\ \mathbf{T}^{-T}(\mathbf{x})f \end{bmatrix}. \qquad (3.214)$$

We have used the fact that $\mathbf{K}_p\mathbf{E}^T\mathbf{z}_{1r} = \mathbf{K}_p\mathbf{z}_{1r}$ for $\mathbf{K}_p = \mathrm{diag}\{k_{p1}, 0, k_{p3}\}$.

### Nonautonomous Lyapunov Analysis

Recall that the Lyapunov function $V_3$ is only semi-definite since $\mathbf{K}_p$ is positive semi-definite. Since the system is ISS, asymptotic tracking is guaranteed by:

$$V_{3r} = \frac{1}{2}\left[\mathbf{z}_{1r}^T(\mathbf{E}\mathbf{K}_p\mathbf{E}^T)\mathbf{z}_{1r} + \mathbf{z}_2^T\mathbf{M}_x\mathbf{z}_2 + \frac{1}{s}\tilde{F}_e^2\right] > 0, \qquad (3.215)$$

$$\dot{V}_{3r} = -\mathbf{z}_{1r}^T(\mathbf{E}\mathbf{K}_p\mathbf{E}^T)(\mathbf{E}\Lambda\mathbf{E}^T)\mathbf{z}_{1r} - \mathbf{z}_2^T(\mathbf{K}_d + \mathbf{D}_x)\mathbf{z}_2 \leq 0. \qquad (3.216)$$

where $\mathbf{E}\mathbf{K}_p\mathbf{E}^T > 0$ and $\mathbf{E}\Lambda\mathbf{E}^T > 0$. Hence, $\mathbf{z}_{1r}$, $\mathbf{z}_2$, $\tilde{F}_e$ [ $\mathcal{L}_\infty$ . ULES of the equilibrium point $(\mathbf{z}_{1r}, \mathbf{z}_2, \tilde{F}_e)$ = $(\mathbf{0}, \mathbf{0}, 0)$ follows by using the stability theorem of Fossen et al. [3] (see also Loria et al. [5]) for nonlinear *nonautonomous* systems where $V_3 > 0$ (*positive definite*) and $\dot{V}_3 \leq 0$ (*negative semi-definite*). The reason that $\dot{V}_3$ is only negative semi-definite is that a negative term proportional to $-\tilde{F}_e^2$ is missing in the expression for $\dot{V}_3$. We are now ready to state the main theorem of this section.

**Theorem 3.2 (Main result: ULES weather optimal position control)** The equilibrium point $(\mathbf{z}_1, \mathbf{z}_2, \tilde{F}_e)$ = $(\mathbf{0}, \mathbf{0}, 0)$ of the nonlinear system in Eqs. (3.209) and (3.210), with the control law in Eq. (3.196) and the parameter adaptation law in Eq. (3.199), is ULES.

**Proof.** see Fossen and Strand [2]. □

### Translatory Circle Center Controller

The adaptive backstepping controller of the previous section satisfies control objective O1, that is weather optimal heading control. Weather optimal position control, control objective O2, can be satisfied by moving the circle center $\mathbf{p}_0 = [x_0, y_0]^T$ such that the ship maintains a constant position $\mathbf{p} = [x, y]^T$.

In order to meet the fixed position control objective, an update law for the circle center $\mathbf{p}_0$ must be derived. Recall that the Cartesian Earth-fixed position of the ship is given by:

$$\mathbf{p} = \mathbf{L}^T h, \qquad (3.217)$$

where $\mathbf{L}$ is defined in Eq. (3.154). Let $\tilde{\mathbf{p}} \triangleq \mathbf{p} - \mathbf{p}_d$ denote the corresponding deviation from the desired position vector $\mathbf{p}_d \triangleq [x_d, y_d]^T$. The desired position can either be constant (regulation) or a smooth time-varying reference trajectory. The control law for translation of the circle center is derived by considering the following LFC:

$$V_p = \frac{1}{2}\tilde{\mathbf{p}}^T\tilde{\mathbf{p}}, \qquad (3.218)$$

where

$$\dot{V}_p = \tilde{\mathbf{p}}^T(\dot{\mathbf{p}} - \dot{\mathbf{p}}_d) = \tilde{\mathbf{p}}^T(\mathbf{L}^T\dot{h} - \dot{\mathbf{p}}_d). \tag{3.219}$$

By using Eq. (3.153), $\mathbf{L}^T\mathbf{L} = \mathbf{I}_{2\times2}$ and $\dot{\mathbf{x}} = \mathbf{z}_2 + \dot{\mathbf{x}}_r$ we get:

$$\begin{aligned} \dot{V}_p &= \tilde{\mathbf{p}}^T[\mathbf{L}^T(\mathbf{R}(g)\mathbf{H}(r)\dot{\mathbf{x}} + \mathbf{L}\dot{\mathbf{p}}_0) - \dot{\mathbf{p}}_d] \\ &= \tilde{\mathbf{p}}^T(\dot{\mathbf{p}}_0 - \dot{\mathbf{p}}_d + \mathbf{L}^T\mathbf{R}(g)\mathbf{H}(r)\dot{\mathbf{x}}_r) + \tilde{\mathbf{p}}^T\mathbf{L}^T\mathbf{R}(g)\mathbf{H}(r)\mathbf{z}_2. \end{aligned} \tag{3.220}$$

Now, by choosing the circle center update law as:

$$\dot{\mathbf{p}}_0 = \dot{\mathbf{p}}_d - \mathbf{L}^T\mathbf{R}(g)\mathbf{H}(r)\dot{\mathbf{x}}_r - k_0\tilde{\mathbf{p}}, \tag{3.221}$$

where $k_o > 0$, we get:

$$\dot{V}_p = -k_o\tilde{\mathbf{p}}^T\tilde{\mathbf{p}} + \tilde{\mathbf{p}}^T\mathbf{L}^T\mathbf{R}(g)\mathbf{H}(r)\mathbf{z}_2. \tag{3.222}$$

Unfortunately, a cross term in $\tilde{\mathbf{p}}$ and $\mathbf{z}_2$ will appear in the expression for $\dot{V}_{3r}$ if Eq. (3.221) is applied. In order to guarantee that $\dot{V}_{3r} \neq 0$, we must modify the weather optimal controller in Eq. (3.196) and add a negative term $-\tilde{\mathbf{p}}^T\tilde{\mathbf{p}}$ to the expression:

$$\dot{V}_{3r} = -\mathbf{z}_{1r}^T(\mathbf{E}\mathbf{K}_p\mathbf{E}^T)(\mathbf{E}\Lambda\mathbf{E}^T)\mathbf{z}_{1r} - \mathbf{z}_2^T(\mathbf{K}_d + \mathbf{D}_x)\mathbf{z}_2, \tag{3.223}$$

to remove the cross term.

### Weather Optimal Position Control (WOPC)

The cross terms involving $\tilde{\mathbf{p}}$ and $\mathbf{z}_2$ in $\dot{V}_p$ can be removed by modifying the nonlinear controller in Eq. (3.196) to:

$$\begin{aligned} t &= \mathbf{T}^T(\mathbf{M}_x\ddot{\mathbf{x}}_r + \mathbf{C}_x\dot{\mathbf{x}}_r + \mathbf{D}_x\dot{\mathbf{x}}_r - \mathbf{K}_p\mathbf{z}_1 - \mathbf{K}_d\mathbf{z}_2) - \mathbf{q}(\cdot) - f\hat{F}_e \\ &\quad - \mathbf{T}^T\mathbf{E}^T(r)\mathbf{R}^T(g)\mathbf{L}\tilde{\mathbf{p}}. \end{aligned} \tag{3.224}$$

The last term in $\tau$ implies that:

$$\dot{V}_{3r} = -\mathbf{z}_{1r}^T(\mathbf{E}\mathbf{K}_p\mathbf{E}^T)(\mathbf{E}\Lambda\mathbf{E}^T)\mathbf{z}_{1r} - \mathbf{z}_2^T(\mathbf{K}_d + \mathbf{D}_x)\mathbf{z}_2 - \tilde{\mathbf{p}}^T\mathbf{L}^T\mathbf{R}(g)\mathbf{H}(r)\mathbf{z}_2. \tag{3.225}$$

Consider the LFC:

$$V_{\text{wopc}} = V_{3r} + V_p. \tag{3.226}$$

$$\dot{V}_{\text{wopc}} = -\mathbf{z}_{1r}^T(\mathbf{E}\mathbf{K}_p\mathbf{E}^T)(\mathbf{E}\Lambda\mathbf{E}^T)\mathbf{z}_{1r} - \mathbf{z}_2^T(\mathbf{K}_d + \mathbf{D}_x)\mathbf{z}_2 - k_o\tilde{\mathbf{p}}^T\tilde{\mathbf{p}} \tag{3.227}$$

and therefore the equilibrium point $(\mathbf{z}_{1r}, \mathbf{z}_2, \tilde{F}_e, \tilde{\mathbf{p}}) = (\mathbf{0}, \mathbf{0}, 0, \mathbf{0})$ is ULES.

The term $\ddot{\mathbf{p}}_0$ is needed in the expression for $\mathbf{q}(\cdot)$. This term is computed from Eq. (3.221) as:

$$\begin{aligned} \ddot{\mathbf{p}}_0 &= \ddot{\mathbf{p}}_d - k_o(\dot{\mathbf{p}} - \dot{\mathbf{p}}_d) - \mathbf{L}^T\mathbf{R}(g)\mathbf{H}(r)\ddot{\mathbf{x}}_r \\ &\quad - \mathbf{L}^T\dot{\mathbf{R}}(g)\mathbf{H}(r)\dot{\mathbf{x}}_r - \mathbf{L}^T\mathbf{R}(g)\dot{\mathbf{H}}(r)\dot{\mathbf{x}}_r. \end{aligned} \tag{3.228}$$

## Experimental Results

The experimental setup is shown in Fig. 3.30. The experimental results are transformed to full scale according to Table 3.2. In the scaling we used $L_s = 70L_m$ meters and $m_s = 4500$ tons. A ducted fan is used to generate a slowly varying or constant wind disturbance.

### Experiment 1: Weather Optimal Heading Control (WOHC)

In the first experiment the ship was allowed to move on the circle arc (circle center controller was turned off). This is referred to as WOHC. The fixed origin and circle arc are shown in Figure 3.31. Notice that the initial heading is approximately 30 degrees, see Figure 3.33, while the position $(x, y) \approx (13, -43)$ (see Fig. 3.32). These values were obtained when the fan was initially directed in 210 degrees (opposite direction of the ship heading).

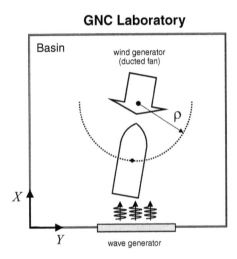

FIGURE 3.30  Experimental setup showing the directions of the wind and wave generators.

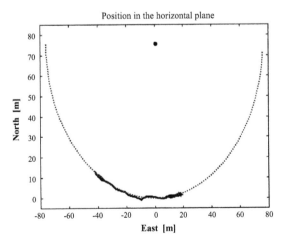

FIGURE 3.31  WOHC experiment showing the circular motion of the ship when the circle center controller is turned off (WOHC).

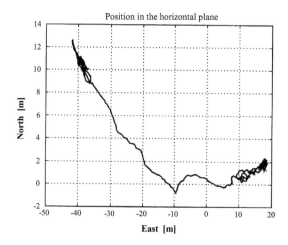

**FIGURE 3.32** WOHC experiment showing the $(x, y)$ position of the model ship (m) when the fan is rotated.

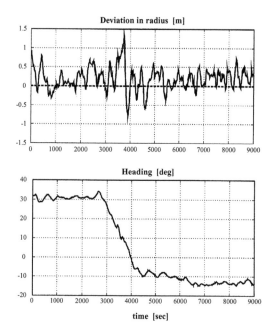

**FIGURE 3.33** WOHC experiment showing the performance of the radius regulator (upper plot) and weather optimal heading (lower plot) vs. time (s).

After 3000 seconds, the fan was slowly rotated to 165 degrees corresponding to a weather optimal heading of $-15°$ (see Fig. 3.33). During this process, the ship starts to move on the circle arc (with heading toward the circle center) until it is stabilized at its new heading, which is $-15°$. The new position on the circle arc is $(x, y) \approx (3, 20)$. This clearly demonstrates that the ship heading converges to the optimal value (copies the dynamics of a pendulum in the gravity field). This is done without using any external wind sensor.

In the next experiment, we show how the circle center can be translated online in order to obtain a constant position $(x, y)$.

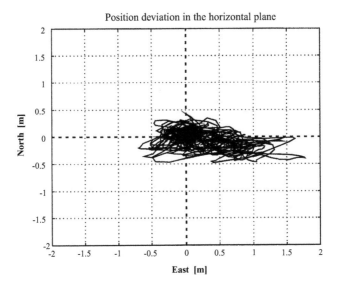

**FIGURE 3.34** WOPC experiment showing station keeping to $(x_d, y_d) = (0, 0)$.

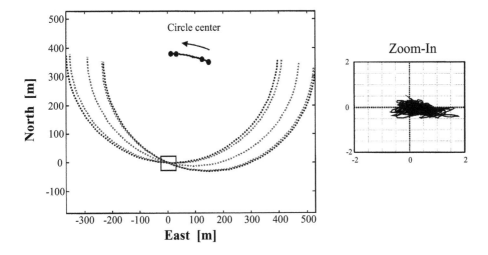

**FIGURE 3.35** WOPC experiment showing how the circle center is moving during station keeping.

## Experiment 2: Weather Optimal Position Control (WOPC)

In the second experiment the ship should maintain its position (circle center controller is turned on). This is referred to as WOPC. The performance during station keeping (dynamic positioning) is shown in Fig. 3.34, while the translation of the circle is shown in Fig. 3.35. The position controller works within an accuracy of $\pm 1$ meters, which is the accuracy of the position reference systems.

Again, the weather optimal heading is changed from approximately $23°$ to $2°$, but the position $(x, y)$ of the ship is not changed. The position deviations and the weather optimal heading are shown in Fig. 3.36. These values are obtained by moving the fan from an initial angle of 203 degrees to 182 degrees.

The last plots (Fig. 3.37) show the deviation for the radius regulator (upper plot) and how the circle center $(x_0, y_0)$ is changed online by the circle center control law (lower plots) in order to obtain a fixed

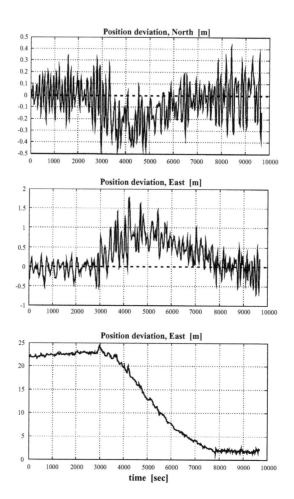

**FIGURE 3.36** WOPC experiment showing the North and East position accuracies (upper plots) and weather optimal heading (lower plot) vs. time (seconds). The position accuracy is within ±1 m while the heading changes from 23° to 2° as the fan is rotated.

position (x, y). The experiment shows that the ship will turn up against an unknown disturbance (wind) at the same time the ship maintains its position.

## References

1. Fossen, T. I., *Guidance and Control of Ocean Vehicles,* John Wiley & Sons, Ltd., Chichester, England, 1994.
2. Fossen, T. I. and Strand, J. P., Nonlinear passive weather optimal positioning control (WOPC) system for ships and rigs: experimental results, (Regular Paper) *Automatica,* to appear.
3. Fossen. T. I., Loria, A., and Teel, A., A theorem for UGAS and ULES of nonautonomous systems: robust control of mechanical systems and ships, *International Journal of Robust and Nonlinear Control,* to appear.
4. Krstić, M., Kanellakopoulos, I., and Kokotovic, P. V., *Nonlinear and Adaptive Control Design,* John Wiley & Sons, New York, 1995.
5. Loria, A., Fossen, T. I., and Teel, A., UGAS and ULES of Non-autonomous systems: applications to integral action control of ships and manipulators, *Proceedings of the 5th European Control Conference (ECC '99),* Karlsruhe, Germany, September 1999.

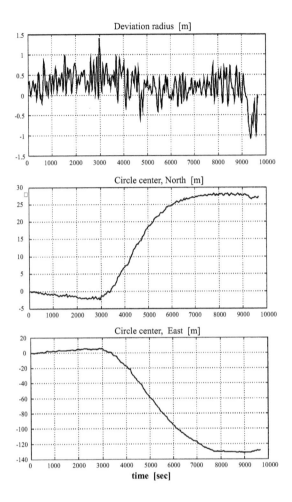

**FIGURE 3.37** WOPC experiment showing the deviation for the radius regulator (upper plot) and the translation of the circle center $(x_0, y_0)$ (lower plots) vs. time in seconds. The radius deviation is within $\pm 1$ m during the rotation of the fan.

6. Pinkster, J. A. and Nienhuis, U., Dynamic positioning of large tankers at sea, *Proc. of the Offshore Technology Conference (OTC'96)*, Houston, TX, 1996.
7. Strand, J. P., Nonlinear Position Control Systems Design for Marine Vessels, Doctoral Dissertation, Department of Engineering Cybernetics, Norwegian University of Science and Technology, June 1999.

# 3.6  Methods for Thrust Control

*Asgeir J. Sørensen*

## Introduction

Installed power capacity on marine vessels and offshore installations is normally limited. In addition, there is an increased focus on environmental aspects motivating technical solutions, which reduces the total energy consumption and emission of exhaust gases. The different subsystems/equipment installed on a vessel or offshore installation can be categorized into two parts, *producers* and *consumers* of energy. For dynamically positioned vessels, the thruster system normally represents one of the main consumers of energy, and is regarded as a critical system with respect to safety. On the contrary, the dynamic positioning system is only an auxiliary system enabling the vessel to do a profitable operation of one kind or another, such as drilling, oil production, loading, and so on. Hence, thruster usage should not cause a load-shedding of those productive consumers or, in the worst case, cause a total power blackout because of unintended power consumption. The strong requirements for vessel performance, operational availability, and overall safety have therefore resulted in increased focus on the total vessel concept and the interactions among the different equipment and systems installed. Flexibility in operation has enabled electrical power generation and distribution systems for propulsion, positioning, oil production, drilling, and loading, where all equipment and control systems are integrated into a common power plant and automation network. In fully integrated systems, functional in addition to physical integration of poower and automation systems combined with thorough marine process knowledge introduces new and far better opportunities to optimized the overall vessel mission objective at lower life cycle costs. In order to accomplish this, it is essential to properly address energy control by consumers and producers of electrical power on board the vessel. If the various consumers (thrusters, pumps, compressors, etc.) of power act separately, the power generation system must be dimensioned and operated with larger safety margins to account for the corresponding larger mean power demands and unintentional power peaks.

This section will illustrate the importance of focusing on low-level actuator control, exemplified in thruster control, in order to achieve a successful control result, which does not have negative impact on the other ship systems. In the ship positioning control literature, the focus has been on high-level controller design, where the control input has been regarded as forces in surge and sway and moment in yaw. Until recently, less attention has been paid to thruster allocation and low-level control of local thruster devices with a few exceptions (see Blanke [2], Grimble and Johnson [5], Healey et al. [7], Sørdalen [12], Sørensen et al. [13], Whitcomb and Yoerger [14] and [15], Bachmayer et al. [1], and Fossen and Blanke [6]).

## Problem Formulation

As seen in the previous sections, the positioning systems include different control functions for automatic positioning and guidance. A positioning controller computes commanded forces in surge and sway, and moment in yaw. The high-level positioning controller produces a commanded thrust vector $t_{ic}$ [ $R^3$ in surge, sway, and yaw. The problem of finding the corresponding force and direction of the thrusters that meets high-level thrust commands is called *thrust allocation*. By using singular value decomposition and geometrical filtering techniques, the optimal force and direction for each thrust device can be found, avoiding singular thrust behavior, with reduced wear and tear and energy consumption for any thrust configuration and type of thruster (see Sørdalen [12]).

The propeller pitch ratio is measured as the angle at a radius $0.7R$ of the propeller blade or as the ratio of traveled distance per revolution divided by the propeller diameter. Traditionally, the propeller pitch ratio or the shaft speed is used to indirectly control the propeller force toward the reference set point, $T_{ref}$ specified by the thrust allocation. In thrust devices, a local pitch or speed controller is present. A static mapping from the reference force to the actual control input, $u_{ref}$, which can be either pitch or propeller speed, is used according to:

$$u_{ref} = g(T_{ref}). \tag{3.229}$$

Since normally sensors are not available for measuring the actual force developed by the propeller, there are no guarantees for fulfilling the high-level thrust commands, and the mapping from commanded thruster force to actual propeller force can be viewed as an open-loop system. Thus, sophisticated control designs will be significantly degraded with respect to performance and stability margins, if high-level control inputs are not produced by local controllers. Traditionally, the propeller and thruster devices can be controllable pitch propeller (CPP) with fixed speed, controllable speed with fixed pitch propeller (FPP), or controllable pitch and speed in combination.

Conventionally, the resulting pitch or speed set point signals are determined from stationary propeller force-to-speed/pitch relations based on information about thruster characteristics found from model tests

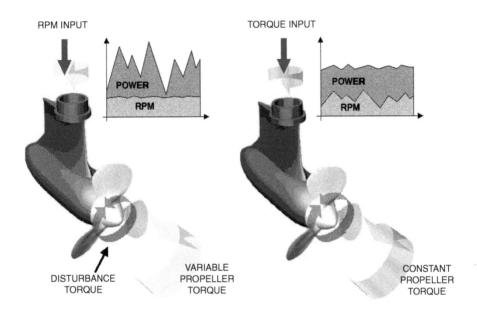

**FIGURE 3.38**   Torque and power control reduce the dynamic load variations and allow for lower spinning reserves.

and bollard pull tests provided by the thruster manufacturer. These relations may later be modified during sea trials. However, as shown later, they are strongly influenced by the local water flow around the propeller blades, hull design, operational philosophy, vessel motion, waves, and water current. In conventional positioning systems, variations in these relations are not accounted for in the control system resulting in reduced positioning performance with respect to accuracy and response time. In addition, the variations may also lead to deterioration of performance and stability in the electrical power plant network, due to unintentional peaks or power drops caused by load fluctuations on the propeller shafts, as shown in Fig. 3.38. The unpredictable load variations force the operator to have more available power than necessary. This implies that the diesel generators will get more running hours at lower loads in average, which creates more tear, wear, and maintenance. This motivates us to find methods for local thruster control.

In this section, the method of Sørensen et al. [13] on torque and power control of propellers is described using forces-to-torque/power mappings. Hence, the force-to-pitch/speed (RPM) mappings [see Eq. (3.229)] are replaced by force-to-torque/power mappings. This approach will yield more stable power loads, and the method is less sensitive to disturbances.

## Mathematical Modeling

### Propeller Characteristics

The propeller thrust $T_{th}$ and torque $Q_{th}$ are formulated according to:

$$T_{th} = rD^4 K_T |n| n, \tag{3.230}$$

$$Q_{th} = rD^5 K_Q |n| n, \tag{3.231}$$

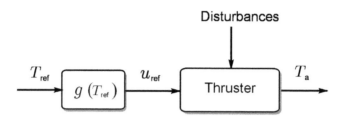

**FIGURE 3.39** Relationship between commanded thruster force, $T_{ref}$, and actual developed propeller force, $T_a$.

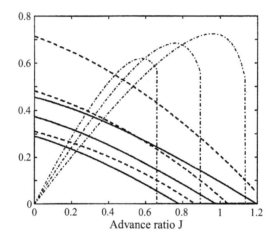

**FIGURE 3.40** Open water $K_T$ (solid), $10 \cdot K_Q$ (dash), and $\eta_o$ (dash-dot) as a function of advance ratio $J$ for $P/D = 0.7, 0.89,$ and 1.1.

where $\rho$ is the water density, $D$ is the propeller diameter, and $n$ is the propeller speed (revolutions per second). The expressions of $K_T$ and $K_Q$ are found through so-called *open water* tests, usually performed in a cavitation tunnel or a towing tank. These nondimensional thrust and torque coefficients are described by the following parameters [11]:

$$K_T = f_1\left(J, \frac{P}{D}, \frac{A_E}{A_o}, Z\right),$$

$\hspace{11cm}$ (3.232)

$$K_Q = f_2\left(J, \frac{P}{D}, \frac{A_E}{A_o}, Z, R_n, \frac{t}{c}\right),$$

$\hspace{11cm}$ (3.233)

where $P/D$ is the pitch ration, $A_E/A_o$ is the expanded-area ratio, $Z$ is the number of blades, $R_n$ is the Reynolds number, $t$ is the maximum thickness of the blade section, and $c$ is the chord length of the blade section. $J$ is the advance ratio defined as:

$$J = \frac{V_a}{nD},$$

$\hspace{11cm}$ (3.234)

where $V_a$ is the inflow velocity to the propeller. The open water propeller efficiency in undisturbed water is given as the ratio of the work done by the propeller in producing a thrust force divided by the work required to overcome the shaft torque according to:

$$h_o = \frac{T_{th}V_a}{2pnQ_{th}} = \frac{J}{2p}\frac{K_T}{K_Q}.$$

$\hspace{11cm}$ (3.235)

The curves for $K_T$, $K_Q$, $\eta_o$ are shown in Fig. 3.40 for different pitch ratios for a Wageningen B-screw series based on Table 5 in Oosterveld and van Oossanen [11], with $R_n = 2 \cdot 10^6$, $Z = 4$, $D = 3.1$ meters, and $A_E/A_o = 0.52$.

## Propulsion Efficiency

The difference between the ship speed $U$ and $V_a$ over the propeller disc is called the *wake*. For a ship moving forward, $V_a$ is less than $U$ since the after-body flow changes its magnitude between ship speed near the vessel and zero far from the vessel. The stationary relationship for axial water inflow can be modeled as [4]:

$$V_a = U(1 - (w_w + w_p + w_v)) = U(1 - w), \tag{3.236}$$

where $w$ is the *wake fraction number*, typically in the range of $0 < w < 0.4$. $w_w$ is the wake fraction caused by the wave motion of the water particles. $w_p$ is the wake fraction caused by so-called potential effects for a hull advancing forward in an ideal fluid, and $w_v$ is the wake fraction caused by viscous effects due to the effect of boundary layers. One should notice that Eq. (3.236) must only severe as a rough estimate, since it does not account for the interactions between the vessel motion and the actual wave particle flow. By applying a moment equation to a control volume surrounding the inlet flow, the thrust can be related to the rate of change of moment to the control volume. This will give a dynamic equation for the advance speed (see Healey et al. [7], Blanke et al. [3], and the references therein).

The suction of the propeller generally reduces the pressure at the stern resulting in increased drag $d_s$ according to:

$$d_s = T_{th}(1 - t_d), \tag{3.237}$$

where $t_d$ is the thrust-deduction coefficient, typically in the range of $0 < t_d < 0.2$ caused by pressure reduction due to potential effects, viscous effects, waves, and appendices. In some extreme cases $t_d$ may become negative.

The thrust coefficient behind the hull is normally assumed to be unchanged compared to open water, while the torque coefficient will be affected by the change in inflow at the stern, $K_{QB}$. This is accounted for by the relative rotative efficiency:

$$h_r = \frac{h_B}{h_o} = \frac{\frac{J}{2p}\frac{K_T}{K_{QB}}}{h_o} = \frac{K_Q}{K_{QB}}. \tag{3.238}$$

The overall propulsion efficiency can now be found as the ration between the useful work done by the product of drag and ship speed divided by the work required to overcome the shaft torque:

$$h_p = \frac{d_s U}{2 p n Q_{th}} = h_h h_o h_r h_m, \tag{3.239}$$

where $h_h = \frac{1-t_d}{1-w}$ is defined as the hull efficiency and is according to Newman [10] in the range of 1.0 to 1.2. $\eta_m$ is the mechanical efficiency typically in the range of 0.8 to 0.9.

## Propeller Losses

In addition to the modeled propeller losses caused by axial water inflow, several other effects contribute to reduction of propeller thrust and torque:

- Water inflow perpendicular to the propeller axis caused by current, vessel speed, or jets from other thrusters will introduce a force in the direction of the inflow due to deflection of the propeller race. This is often referred to as cross-coupling drag.
- For heavily loaded propellers, ventilation (air suction) caused by decreasing pressure on the propeller blades may occur, especially when the submergence of the propeller decreases due to wave motion.

- For extreme conditions with large vessel motions, the in-and-out-of-water effects will result in a sudden drop in thrust and torque following a hysteresis pattern.
- Both thrust reduction and change of thrust direction may occur due to thruster-hull interaction caused by frictional losses and pressure effects when the thruster race sweeps along the hull. The last is referred to as the Coanda effect.
- Thruster-thruster interaction, caused by the propeller race from one thruster on neighboring thrusters, may lead to significant thrust reduction if appropriate precautions are not taken in the thruster allocation algorithm.

Sensitivity to the different types of losses depends on what types of propellers and thrusters are used, application of skegs and fins, hull design, and operational philosphy. Main propellers are subject to large thrust losses due to air ventilation and in-and-out-of-water effects. Rotatable azimuth thrusters are subject to dominating losses caused by hull friction and interaction with other thrusters. Tunnel thrusters are subject to losses caused by non-axial inflow due to current, vessel speed, and ventilation phenomena in heavy weather.

The effects of the above-mentioned losses are modeled by the thrust and the torque reduction functions $h_T$ and $h_Q$ based on a combination of analytical and empirical data. In Fig. 3.50, thrust reduction functions [9] accounting for in-and-out-of-water effects are illustrated. Hence, the actual thrust $T_a$ and torque $Q_a$ experienced by the propeller can be expressed as:

$$T_a = h_T(n,\mathbf{x}_p,u_p)(1 - t_d)T \stackrel{\text{R}}{\text{S}} f_T(n, \mathbf{x}_p,u_p) \qquad (3.240)$$

$$Q_a = h_Q(n,\mathbf{x}_p,u_p)Q \stackrel{\text{R}}{\text{S}} f_Q(n, \mathbf{x}_p,u_p), \qquad (3.241)$$

where $\mathbf{x}_p$ represents dynamic states, such as vessel motion, propeller submergence, and environmental conditions. Propeller-dependent parameters are represented by the vector $\boldsymbol{\theta}_p$.

### Propeller Shaft Model

Let $Q_m$ denote the torque generated by the thruster motor. A torque balance (Fig. 3.41) for the propeller shaft is written:

$$I_s\dot{v} = Q_m - Q_a \qquad (3.242)$$

where $I_s$ is the moment of inertia of the shaft, and $\omega = 2\pi n$ is the angular shaft speed.

The power delivered by the motor and the actual propeller shaft power accounting for the effect of thrust losses are given by:

$$P_m = vQ_m = 2pnQ_m \qquad (3.243)$$

$$P_a = vQ_a = 2pnQ_a. \qquad (3.244)$$

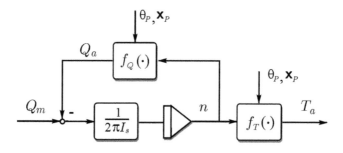

**FIGURE 3.41**   Propeller shaft model.

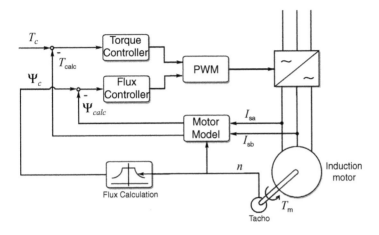

**FIGURE 3.42**  Torque loop in electrical motor drive.

By combining Eqs. (3.230), (3.231), and (3.244), it is possible to express the propeller thrust as a function of propeller power and torque according to:

$$T_a = \text{sgn}(n)\frac{r^{1/3}D^{2/3}K_T h_T(1-t_d)}{(2p\cdot K_Q\cdot h_Q)^{2/3}}\cdot P_a^{2/3} \tag{3.245}$$

$$T_a = \frac{h_T(?)(1-t_d)}{h_Q(?)}\frac{K_T}{K_Q D}?Q_a. \tag{3.246}$$

### Torque Loop in an Electrical Motor Drive

The torque control is inherent in the design of most applied control schemes for variable speed drive systems. The torque is controlled by means of motor currents and motor fluxes with high accuracy and bandwidth (see Fig. 3.42). Theoretically, the rise time of the torque in PWM (Pulse Width Modulated) drives is limited by the motor's inductance (in load commutated inverters, LCIs, also by the DC choke). However, in practice the controller limits the rate of change of torque in order to prevent damage to the mechanics. The closed loop of motor and torque controller may, for practical reasons, be assumed to be equivalent to the first-order model:

$$Q_m(s) = \frac{1}{1+T_T s}Q_c(s), \tag{3.247}$$

where the time constant $T_T$ is in the range of 20 to 200 milliseconds (ms).

## Local Thruster Controller Design

In this section, speed-controlled thruster drives are considered. The traditional speed control loop is treated and the new methods based on power or torque control are discussed. The sensitivity to thrust losses is analyzed for the following three schemes:

- Shaft speed feedback control
- Torque feedforward control
- Power feedback control

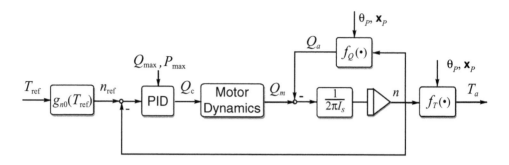

**FIGURE 3.43**   Shaft speed feedback control principle.

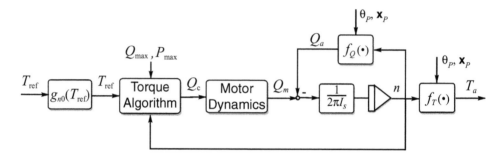

**FIGURE 3.44**   Torque feedforward control principle.

### Shaft Speed Feedback Control

In conventional FPP systems a speed controller is used to achieve the commanded propeller force, as shown in Fig. 3.43. Given a specified force command, $T_{ref}$, the corresponding reference (commanded) speed in the propeller, $n_{ref}$, is found by the stationary function:

$$n_{ref} = g_{n0}(T_{ref}) = \text{sgn}(T_{ref}) \sqrt{\left| \frac{T_{ref}}{rD^4 K_{T0}} \right|}, \qquad (3.248)$$

which is the inverse function of the nominal shaft-speed-versus-force characteristics given in Eq. (3.230) for typical $K_{T0} = K_T$ ($J = 0$).

The rated (nominal) torque and power are denoted $T_N$ and $P_N$. The speed controller is a PID controller with saturation limits, which calculates the necessary torque to obtain increased/decreased speed. The maximum torque is:

$$Q_{max} = aQ_N, \qquad (3.249)$$

where $\alpha$ is typically in the range of 1.1 to 1.2.

### Torque Feedforward Control

In the proposed torque control strategy for FPP drives, the outer speed control loop is removed, and the thruster is controlled by its inner torque control loop with a commanded torque, $T_c$, as the set point in Fig. 3.44. Based on Eq. (3.231), the torque reference is written:

$$Q_{ref} = rD^5 K_{Q0} |n_{ref}| n_{ref}, \qquad (3.250)$$

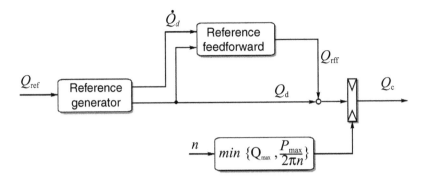

**FIGURE 3.45** Torque algorithm.

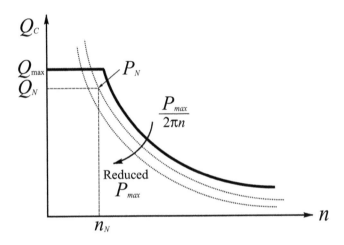

**FIGURE 3.46** Torque limitation.

where typically, $K_{Q0} = K_Q(J = 0)$. By combining Eqs. (3.248) and (3.250), the mapping between the commanded thrust force $T_{ref}$ and reference torque becomes:

$$Q_{ref} = \frac{DK_{Q0}}{K_{T0}} T_{ref} = g_{Q0}(T_{ref}).$$
$$(3.251)$$

To the existing torque loop a new function, *torque algorithm*, is added. The torque reference is filtered through a reference generator, which yields smooth bounded torque references and $\dot{Q}_d$, and $Q_d$. In order to speed up the response, a reference feedforward control action is computed. The commanded torque must also be limited by the maximum torque capability $Q_{max}$ and power capability $P_{max}$, see Fig. 3.45.

The torque-limiting function is described in Fig. 3.46. The torque must also be limited by the maximum power, which yields hyperbolic limit curves for the torque as a function of speed. Since maximum power is not limited by the converter and motor ratings only, but also by the available power in the generators, this limit will vary accordingly. By this method the power limitation will become fast and accurate, allowing utilization of the system's capability with built-in blackout prevention.

## Power Control

An alternative control strategy for propeller control, based on power control, is also possible. The torque control loop is maintained, but the commanded torque is found from a commanded power $P_{ref}$ (Fig. 3.47). This power reference is a signed value in order to determine the torque direction.

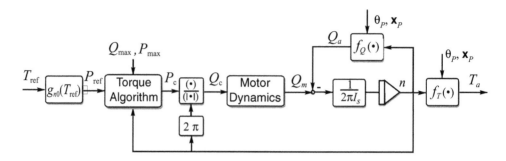

**FIGURE 3.47**   Power feedback control principle.

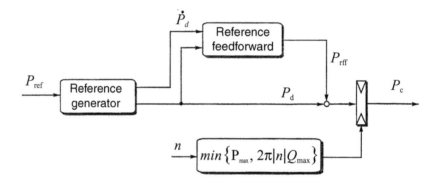

**FIGURE 3.48**   Commanded power algorithm.

By combining Eqs. (3.248) and (3.250), the mapping between the reference thrust force $T_{ref}$ and the power reference becomes:

$$P_{ref} = sgn(T_{ref})\frac{2pK_{Q0}}{\sqrt{r}DK_{T0}^{3/2}}|T_{ref}|^{3/2} = g_{P0}(T_{ref}). \qquad (3.252)$$

To the existing torque loop a new function, *power algorithm*, is added (Fig. 3.48). The power reference is filtered through a reference generator, which yields smooth desired torque references $P_d$ and $\dot{P}_d$. In order to speed up the response, a reference feedforward control action is computed. The commanded power is limited by the maximum power capability, $P_{max}$, and the maximum torque capability, $Q_{max}$, as before.

## Sensitivity to Thrust Losses

In order to compare the different control schemes, a simplified sensitivity analysis is carried out to address the most significant properties of speed control, torque control, and power control. Pitch control is not studied, but is expected in general to have the same behavior as speed control.

### Shaft Speed Feedback Control

Assume that the shaft speed equals the speed reference, $n_{ref}$, at steady state (Fig. 3.43). Using Eqs. (3.230), (3.240) and (3.248), the relationship between actual propeller thrust and the reference thrust becomes:

$$T_a = h_T(?)(1 - t_d)\frac{K_T}{K_{T0}}T_{ref} \overset{\text{D}}{=} s_n(?)T_{ref} \qquad (3.253)$$

## Torque Feedforward Control

From a similar consideration of the torque control scheme (Fig. 3.44), it is assumed that the motor torque and the propeller torque equal the torque reference, $Q_{\text{ref}}$, at steady state. Hence, combining Eqs. (3.230), (3.231), (3.240), (3.241), (3.248), and (3.250) yields:

$$T_a = \frac{h_T(?)(1 - t_d)K_{Q0}}{h_Q(?)} \frac{K_T}{K_Q K_{T0}} T_{\text{ref}} \overset{\text{\footnotesize?}}{\cdot} s_Q(?)T_{\text{ref}} \tag{3.254}$$

## Power Feedback Control

For the power control scheme (Fig. 3.47), the propeller and motor power tend toward the power reference, $P_{\text{ref}}$, at steady state. Hence, combining Eqs. (3.230), (3.231), (3.240), (3.241), (3.248), (3.250), and (3.252) yields:

$$T_a = \frac{h_T(?)(1 - t_d)}{h_Q^{2/3}(?)} \left(\frac{K_{Q0}}{K_Q}\right)^{2/3} \frac{K_T}{K_{T0}} T_{\text{ref}} \overset{\text{\footnotesize?}}{\cdot} s_P(?)T_{\text{ref}} \tag{3.255}$$

## Positioning Performance

The effect of positioning performance for the different control schemes is clearly seen from the two examples illustrated in Figs. 3.49 and 3.51.

## Example 3.1

Let $t_d = 0$ and $h_T = h_Q = 1$. By using Eqs. (3.232), (3.233), $K_{T0} = K_T(J = 0)$, and $K_{Q0} = K_Q(J = 0)$ with the same parameters as used in Fig. 3.40 for $P/D = 0.89$, the sensitivity functions $s_n$, $s_Q$, and $s_p$ in Eqs. (3.253) through (3.255) can be computed as a function of advance ratio $J$ (see Fig. 3.49). As expected, all three sensitivity functions decrease, resulting in increased thrust losses, for increasing $J$. The speed control scheme is less robust for variation in $J$, while the torque control scheme shows the best robustness.

## Example 3.2

Thrust losses (Fig. 3.50) caused by in-and-of-water effects can be modeled by the thrust and the torque reduction functions $h_F$ and $h_T$ in Eqs. (3.240) through (3.241) as a function of the submergence of the propeller shaft $h$ divided by the propeller radius $R$. The reduced disc area due to the reduced submergence is accounted for in the computation of Eqs. (3.232) and (3.233), according to Lehn [8]. Let $t_d = 0$, $K_{T0} = K_T(J = 0)$, with the same parameters as used in Fig. 3.40 for $P/D = 0.89$, then the sensitivity functions $s_n$, $s_T$, and $s_p$ in Eqs. (3.253) through (3.255) can be computed as a function of $h/R$ (see Fig. 3.51). As expected, all three sensitivity functions decrease, resulting in increased thrust losses, for decreasing $h/R$. The speed control scheme is less robust for reduced submergence, while the torque control scheme shows the best robustness.

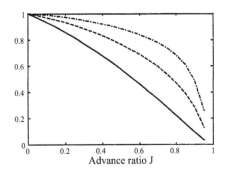

**FIGURE 3.49** Sensitivity functions for the different control schemes: $s_n$ (solid), $s_Q$ (dash-dot), and $s_p$ (dash) as function of advance ratio $J$ for $P/D = 0.89$.

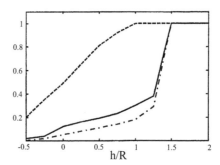

**FIGURE 3.50** Mean thrust losses $h_T$ (dash-dot), $h_Q$ (solid), and reduced disk area (dash) due to reduced submergence in calm water of a heavily loaded propeller as a function $h/R$ for $P/D = 0.89$.

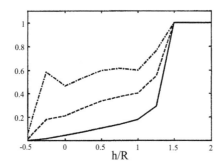

**FIGURE 3.51** Sensitivity functions for the different control schemes: $s_n$ (solid), $s_Q$ (dash-dot), and $s_p$ (dash) as function of $h/R$, $h_T$, and $h_Q$.

## Electrical Power Plant Network Performance and Stability

The power plant on ships are composed of several gas turbine or diesel engines driving electrical generators. For safe operation there must, in all load conditions, be enough spinning reserves with sufficient available power for unpredictable variations in load in order to prevent blackout. Blackout prevention by means of reducing load on heavy consumers must typically respond faster than 500 ms to be effective. With torque and power control, the propeller load is less sensitive to variations in the surroundings, creating less power disturbances in the network and improved voltage and frequency quality. Additionally, the maximum power consumption may easily be limited to the available power in both schemes, since the power limitation is explicit in the torque and power control algorithm. This is in contrast to speed-controlled and pitch-controlled propellers, where the actual power load must be measured as a feedback signal with an inherent time lag that controls the blackout prevention response time. The accurate and fast control of power and power limitation in torque and power control gives less unpredictable load changes, and less need for power limitation in torque and power control gives less unpredictable load changes, and less need for available power. Hence, there will be a reduced probablity of blackout due to overload, since unintentional power peaks will be suppressed. In traditional speed-controlled drives, there can be overshoot in the commanded torque/power up to 5%. If this occurs for several thrusters simultaneously, significantly more power than requested by the positioning system is consumed. With torque and power control, this transient overshoot problem is removed. The number of running generators can be reduced, such that the average loading can be higher. This creates less wear and tear, and maintenance of the prime movers.

## Discussion

Torque control has been shown to have the lowest sensitivity in obtained thrust with respect to disturbances and thrust losses and is thus the most robust control strategy among the evaluated alternatives. The variance in power is somewhat higher than for power control, giving more disturbances in the power network. Where power variation and quality if essential, the power control may be an alternative. However, the performance at low speeds and thrust demand is expected to be poor since small variations in power reference gives large variations in thrust. Hence, the algorithm will be highly sensitive to modeling errors. For such applications, a combination of these two strategies, where torque control is dominating at low speed and power control at higher speed, should be applied. Speed control and pitch control are assumed to have similar characteristics with respect to sensitivity to disturbances and thruster losses.

## References

1. Bachmayer, R., Whitcomb, L. L., and Grosenbaugh, M. A., An accurate four-quadrant nonlinear dynamical model for marine thrusters: theory and experimental validation, *IEEE Journal of Oceanic Engineering*, 25, 1, 146–159, 2000.
2. Blanke, M., Ship Propulsion Losses Related to Automatic Steering and Prime Mover Control, Ph. D. Thesis, Servolaboratory, Technical University of Denmark, Lyngby, Denmark, 1981.
3. Blanke, M., Lindegaard, K. P., and Fossen, T. I., Dynamic model for thrust generation of marine propellers, in *Proceedings of the IFAC Conference on Maneuvering and Control of Marine Craft (MCMC '2000)*, Aalborg, Denmark, 2000.
4. Carlton, J. S., *Marine Propellers and Propulsion,* Butterworth-Heinemann, Oxford, U. K., 1994.
5. Grimble, M. J. and Johnson, M. A., *Optimal Control and Stochastic Estimation: Theory and Applications,* 2, John Wiley Sons, New York, 1988.
6. Fossen., T. I. and Blanke, M., Nonlinear output feedback control of underwater vehicle propellers using feedback from estimated axial flow velocity, *IEEE Journal of Oceanic Engineering*, 25, 2, April, 2000.
7. Healey, A. J., Rock, S. M., Cody, S., Miles, D., and Brown, J. P., Toward an improved understanding of thrust dynamics for underwater vehicles, *IEEE Journal of Oceanic Engineering*, 20, 4, 354–360, 1995.
8. Lehn, E., *Practical Methods for Estimation of Thrust Losses,* Report MT51A92-003, 513003.00.06, Marintek, Trondheim, Norway, 1992.
9. Minsaas, K. J., Thon, H. J., and Kauczynski, W., Influence of ocean environment on thruster performance, in *Int. Symp. Propeller and Cavitation,* supplementary volume, 124–142. Shanghai: The Editorial Office of Shipbuilding of China, 1986.
10. Newman, J. N., *Marine Hydrodynamics,* MIT Press, Cambridge, MA, 1977.
11. Oosterveld, M. W. C. and van Oossanen, P., Further computer-analyzed data of the Wageningen B-screw series, *Int. Shipbuilding Progress,* 22, 251–262, 1975.
12. Sørdalen, O. J., Optimal thrust allocation for marine vessels, IFAC *Journal of Control Engineering Practice,* 5, 9, 1223–1231, 1997.
13. Sørensen, A. J., Ådnanes, A. K., Fossen, T. I., and Strand, J. P., A new method of thruster control in positioning of ships based on power control, in *Proceedings of the 4th IFAC Conference on Maneuvering and Control of Marine Craft* (MCMC '97), Brijuni, Croatia, 172–179, 1997.
14. Whitcomb, L. L. and Yoerger, D. R., Development, comparison, and preliminary experimental validation of nonlinear dynamic thruster models, *IEEE Journal of Oceanic Engineering*, 24, 4, 481–494, 1999a.
15. Whitcomb, L. L. and Yoerger, D. R., Preliminary experiments in model-based thruster control for underwater vehicle positioning, *IEEE Journal of Oceanic Engineering*, 24, 4, 495–506, 1999b.

# 4

# Computational Intelligence in Ocean Engineering

## C. H. Chen
*University of Massachusetts Dartmouth*

COMPUTATIONAL INTELLIGENCE is a generic term that includes topics in neural networks, artificial intelligence, computer vision, genetic algorithms, fuzzy logic, human-computer interaction, machine learning, intelligent robots, etc. As the "intelligence" is derived from computers or both human and machine in the case of human-machine interaction, the intelligent behavior is not that of natural intelligence. However, the computational intelligence can achieve or exceed human intelligence in many situations and thus is very useful to many ocean engineering problems. Though the application of computational intelligence has been fragmented, making a comprehensive survey difficult, there has been some success using computational intelligence in ocean engineering.

The use of computational intelligence in ocean engineering is in its infancy. The potential is enormous. A good example is the intelligent underwater vehicle that helps with search and even rescue missions in a harsh environment. We believe that more successful applications of computational intelligence in ocean engineering will emerge in the near future since the tools in computational intelligence are now greatly improved at less cost.

In this section we will present three topic areas by leading experts that will illustrate the capability of computational intelligence in ocean engineering.

The first topic involves a multivariable online intelligent autopilot design study, by R. S. Burns, R. Sutton, and P. J. Craven of the University of Plymouth. In this article Professor Burns and his colleagues present an intelligent autopilot for simultaneously controlling the dynamics of an autonomous underwater vehicle (AUV). It makes use of adaptive neural, network-based fuzzy inference system architecture.

Topic two concerns an approach to multi-robot cooperation under human supervision, by Dr. Ray Gosine and his colleagues at the Memorial University of Newfoundland. The authors make use of the discrete event system approach for accomplishing complex tasks with robots in the ocean environment.

Topic three is computer vision, by Donna M. Kocak of eMerge Interactive/HBOI and Frank M. Caimi of the Florida Institute of Technology/HBOI. The article summarizes basic principles and techniques of computer vision including specific aspects related to underwater vision. It then summarizes selected applications of computer vision found in ocean engineering references.

# 4

# Computational Intelligence in Ocean Engineering

C. H. Chen
*University of Massachusetts Dartmouth*

R. S. Burns
*University of Plymouth*

R. Sutton
*University of Plymouth*

P. J. Craven
*Racal Research, Ltd.*

R. Gosine
*Memorial University of Newfoundland*

M. Rokonuzzaman
*Memorial University of Newfoundland*

R. Hale
*Memorial University of Newfoundland*

F. Hwang
*Memorial University of Newfoundland*

J. King
*Memorial University of Newfoundland*

J. Seshadri
*Memorial University of Newfoundland*

Donna M. Kocak
*eMerge Interactive, Inc. and Harbor Branch Oceanographic Institution*

Frank M. Caimi
*Florida Institute of Technology and Harbor Branch Oceanographic Institution*

# 4.1   A Multivariable Online Intelligent Autopilot Design Study

*R. S. Burns, R. Sutton, and P. J. Craven*

This section describes the development of a multivariable online intelligent autopilot for simultaneously controlling the yaw and roll dynamics of an autonomous underwater vehicle (AUV). The six degrees of freedom nonlinear AUV model used in the design study was created using a MATLAB/Simulink environment. Simulation results are presented that demonstrate the ability of the autopilot to operate using a novel online hybrid learning rule, and to perform effectively in the presence of parameter variations within the vehicle model. Thus, the approach adopted offers an attractive alternative to more traditional control strategies.

Although remotely operated vehicles (ROVs) play an important role in the offshore industry, their operational effectiveness is limited by the tethering cable, and the reliance on and cost of some kind of support platform. Given these limitations and, in recent years, the concurrent developments in advanced control engineering theory, artificial intelligence (AI) techniques, and computation hardware for analysis, interest in the viability of employing autonomous underwater vehicles (AUVs) in operational tasks has been rekindled. Indeed, in the recently published report by the Marine Foresight Panel [1], the potential usage of AUVs was recognized.

During a typical mission scenario an AUV will experience various disturbances. These disturbances can arise from numerous sources including sea currents or variations within the AUV payload. Therefore, the ability of the autopilot system to adapt in the presence of dynamic changes is a very attractive property, and must be incorporated if a control system is to be considered as intelligent.

This section discusses online control of a realistic AUV model. In particular, the adaptive network-based fuzzy inference system (ANFIS) architecture of Jang [2] is developed for online control. A novel multivariable autopilot structure based on the co-active adaptive network-based fuzzy inference system (CANFIS) [3] is used for simultaneous online course-changing and roll-regulating control. A hybrid learning algorithm is employed in both instances, yet in an online form, to tune the fuzzy sets in the premise and consequent portions of the rule base, which collectively constitute the fuzzy controller.

The remainder of this section reviews a number of online control strategies that have been formulated for underwater vehicles. Next, the AUV simulation model used as the testbed platform for the design study is described. Consideration is then given to the proposed online scheme that utilizes an online hybrid learning rule in its execution. Finally results, discussion, and concluding remarks are given.

## Existing Online Control Strategies

The use of conventional control systems for AUVs is limited in that the hydrodynamic coefficients of a particular vehicle are not usually known until after the vehicle has been completely designed. Consequently, the repetitious design of autopilot systems can be an expensive process involving extensive testing and reconfiguration in line with changing vehicle subsystems and architectures. Subsequently, an intelligent adaptive control strategy is highly desirable to reduce design overheads and provide compensation for the time-varying disturbances and dynamics that such vehicles encounter. In recent years, neural network control schemes have been applied extensively to the problem of AUV control system design with varying degrees of success and credibility. Neural control schemes that incorporate an online learning capability facilitate adaptation of controller parameters in light of such time-varying disturbances and dynamics.

Yuh [4] provides an excellent paper on the feasibility of such an approach by applying a neural network to the design of an online underwater robotic vehicle (URV) control system. The autopilot takes the form of a discrete control law, which is subsequently employed within a continuous-time dynamic model of the URV.

Venugopal et al. [5] present an interesting review of direct and indirect neural network control schemes. Consequently, a direct neural network control strategy is reported as a simulation package based on the Ocean Voyager AUV.

Ishii et al. [6] detail a self-organizing neural-network control system (SONCS) and apply it to real-time adaptive online control of the Twin Burger AUV. The SONCS consists of an "Imaginary World" (IW) element that computes imaginary training of the controller parameters, and a "Real World" (RW) element that operates the AUV according to the control objective. Additionally, an identification network is employed as a feedforward model of the AUV plant and is required to produce state estimates for the IW's update algorithm.

More recently, hybrid neuro-fuzzy control techniques have become increasingly attractive as they can incorporate fuzzy rule-based algorithms by which the autopilot system can be initialized. Juang and Lin [7] developed and applied a self-constructing neural fuzzy inference network (SONFIN) possessing online learning capability. Based on a similar paradigm to the ANFIS of Jang [2], the parameter set of the fuzzy inference system is tuned using a neural network architecture. However, the SONFIN has more sophisticated consequent functions. Initially the consequents are set as fuzzy singletons, but through learning can be self-constructed to add elements of the more typical Takagi-Sugeno-Kang (TSK) [8] linear functions of ANFIS.

Other online approaches to the problem of control system design have been employed in the literature. For example, Corradini and Orlando [9] presented a MIMO adaptive discrete-time variable structure approach to the problem of position and orientation control of an ROV model.

## Modeling Autonomous Underwater Vehicle Dynamics

Figure 4.1 shows the complete control authority of the AUV model. However, it should be noted that for this study the upper and lower canard rudders, situated at the bow of the vehicle, are used to control the yaw dynamics (limited to ±25.2°) and roll control is achieved by use of the stern port and starboard hydroplanes (limited to ±5°). Dimensionally, the model represents an underwater vehicle that is 7 m long, approximately 1 m in diameter, and has a displacement of 3600 kg.

The equations of motion describe the dynamic behavior of the vehicle in six degrees of freedom. These equations are implemented using a nonlinear MATLAB/Simulink simulation model supplied by the Defence Evaluation and Research Agency (DERA), Sea Systems Sector, Winfrith. The model was validated against standard DERA nonlinear hydrodynamic code using tank test data and an experimentally derived set of hydrodynamic coefficients from the Southampton Oceanography Centre's AUTOSUB vehicle. In addition, the MATLAB/Simulink model structure also takes into account the dynamics of the actuators by describing them as first-order lags with appropriate limiters.

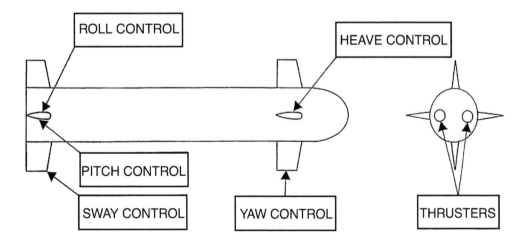

**FIGURE 4.1** The complete control authority of the AUV.

While the AUV model used in this design study is highly nonlinear and possesses severe cross-coupling components, a linearized version (see Appendix A for nomenclature) is shown in Eq. (4.1).

$$E\dot{\underline{x}} = F\underline{x} + G\underline{u} \tag{4.1}$$

where:

$$E = \begin{bmatrix} (m - Y_{\dot{v}}) & -Y_{\dot{R}} & 0 & -(Y_{\dot{P}} + mZ_G) & 0 \\ -N_{\dot{v}} & (I_Z - N_{\dot{R}}) & 0 & -N_{\dot{P}} & 0 \\ 0 & 0 & 1 & 0 & 0 \\ -(K_{\dot{v}} + mZ_G) & -K_{\dot{R}} & 0 & (I_x - K_{\dot{P}}) & 0 \\ 0 & 0 & 0 & 0 & 1 \end{bmatrix}$$

$$F = \begin{bmatrix} Y_{UV}U & (Y_{UR} - m)U & 0 & Y_{UP}U & 0 \\ N_{UV}U & N_{UR}U & 0 & N_{UP}U & 0 \\ 0 & 1 & 0 & 0 & 0 \\ K_{UV}U & (K_{UR} + mZ_G)U & 0 & K_{UP}U & -mgBG \\ 0 & 0 & 0 & 1 & 0 \end{bmatrix}$$

$$G = \begin{bmatrix} Y_{UU\delta ru}U^2 & Y_{UU\delta rl}U^2 & 0 & 0 & 1 & 1 \\ N_{UU\delta ru}U^2 & N_{UU\delta rl}U^2 & l_\phi & -l_\phi & l_\psi & -l_\psi \\ 0 & 0 & 0 & 0 & 0 & 0 \\ K_{UU\delta ru}U^2 & K_{UU\delta rl}U^2 & \gamma_\phi & -\gamma_\phi & \gamma_\psi & -\gamma_\psi \\ 0 & 0 & 0 & 0 & 0 & 0 \end{bmatrix}$$

$$\underline{u} = \begin{bmatrix} \delta_{Stern-upper} & \delta_{Stern-lower} & \delta_{Stern-port} & \delta_{Stern-starboard} & \delta_{Bow-upper} & \delta_{Bow-lower} \end{bmatrix}^T$$

## The Online Learning Scheme

By adopting the well-documented technique of "temporal backpropagation," that is backpropagation over successive time intervals, the autopilot of the AUV can be encoded as a series of Stage Adaptive Neural Networks (SANNs), as discussed in [2] and shown in Fig. 4.2. Following the success of this approach, the AUV autopilot structure herein is initially implemented as a SANN, and the simulation is then interrupted at prespecified discrete sampling points $k$ over the mission time space.

Specifically, given the state of the plant at time $t = k \times h$, where $h$ is the sampling interval width, the autopilot will generate an input to the plant based upon the modified parameter set. Implementing this procedure from $t = 0$ to $t = t_{final}$ yields the plant trajectory. This path is determined by the initial fuzzy autopilot and the output of each stage adaptive autopilot based on the ensuing parameter adaptations.

Clearly, it is necessary to make the assumptions that the delay through the controller is small, the plant is static where the next state is explicitly dependent on the last, and the required states are obtainable. Thus, by initializing the system with either the knowledge of an expert operator or, as in this example, a previously designed autopilot that performs the required task, the hybrid learning rule can be employed to recursively adapt the parameters of the stage adaptive fuzzy autopilot.

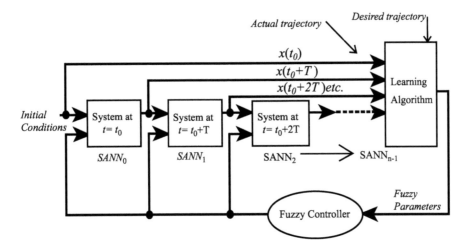

**FIGURE 4.2** Concept of a Stage Adaptive Neural Network Structure.

The cost function to be minimized can be defined as:

$$E = \sum_{k=1}^{n} \|\bar{x}(k \times h) - \bar{x}_d(k \times h)\|^2 \tag{4.2}$$

where $\bar{x}(h \times k)$ is the actual state and $\bar{x}_d(h \times k)$ is the desired trajectory at the sampling interval $t = h \times k$.

### The Hybrid Learning Rule: Online Control

The original hybrid learning rule of Jang [2] fuses two algorithms to systematically adapt the membership functions to the premise and the consequent portions simultaneously for the single input, single output case. With a time-varying system, a forgetting factor can be introduced into the sequential least squares estimator (LSE), which places heavier emphasis on more recent data pairs. This produces an algorithm that can be employed to estimate the autopilot consequent parameters online. The ability of such an algorithm to take account of more recent data pairs, when estimating the parameters of a controller, can aid in tracking a desired trajectory in the light of varying vehicle dynamics.

Introducing a matrix $W$ of forgetting factors:

$$W = \begin{bmatrix} \lambda^{m-1} & 0. & . & 0 \\ 0 & \lambda^{m-2} & . & . \\ . & . & . & 0 \\ 0 & . & 0 & 1 \end{bmatrix} \tag{4.3}$$

where $(1 < \lambda \leq 1)$, the corresponding LSE solution that minimizes the weighted error is defined by:

$$x_k = (A^T W A)^{-1} A^T W \underline{b} \tag{4.4}$$

where $k$ is the number of data pairs used in the estimate $x$. Conveniently, $x_{k+1}$ can be written as:

$$x_{k+1} = \left( \begin{bmatrix} A \\ \underline{a}^T \end{bmatrix}^T \begin{bmatrix} \lambda W & 0 \\ 0 & 1 \end{bmatrix} \begin{bmatrix} A \\ \underline{a}^T \end{bmatrix} \right)^{-1} \begin{bmatrix} A \\ \underline{a}^T \end{bmatrix}^T \begin{bmatrix} \lambda W & 0 \\ 0 & 1 \end{bmatrix} \begin{bmatrix} \underline{b} \\ b \end{bmatrix}$$

$$= (\lambda A^T W A + \underline{a}\underline{a}^T)^{-1}(\lambda A^T W \underline{b} + \underline{a}b) \tag{4.5}$$

To simplify the ensuing notation, two $n \times n$ matrices are introduced:

$$P_k = (A^T W A)^{-1} \tag{4.6}$$

$$P_{k+1} = \left( \begin{bmatrix} A \\ \underline{a}^T \end{bmatrix}^T \begin{bmatrix} \lambda W & 0 \\ 0 & 1 \end{bmatrix} \begin{bmatrix} A \\ \underline{a}^T \end{bmatrix} \right)^{-1}$$

$$= (\lambda A^T W A + \underline{a}\,\underline{a}^T)^{-1} \tag{4.7}$$

It can be shown that Eq. (4.7) can be expressed as:

$$P_{k+1} = \frac{1}{\lambda} P_k - \frac{1}{\lambda} P_k \underline{a} \left( I + \underline{a}^T \frac{1}{\lambda} P_k \underline{a} \right)^{-1} \underline{a}^T \frac{1}{\lambda} P_k$$

$$= \frac{1}{\lambda} \left( P_k - \frac{P_k \underline{a}\,\underline{a}^T P_k}{\lambda + \underline{a}^T P_k \underline{a}} \right), \tag{4.8}$$

which is the recursive least squares formula for a time varying system. Obviously, if $\lambda$ is chosen as 1 the sequential least squares formula for a time invariant system is reformulated. The sequential least squares estimator for time variant systems with multiple outputs is therefore:

$$\begin{cases} P_{k+1} = \dfrac{1}{\lambda} \left[ P_k - \dfrac{P_k \underline{a}_{k+1} \underline{a}_{k+1}^T P_k}{\lambda + \underline{a}_{k+1}^T P_k \underline{a}_{k+1}} \right] \\[3mm] x_{k+1} = x_k + P_{k+1}\, \underline{a}_{k+1} \left( \underline{b}_{k+1}^T - \underline{a}_{k+1}^T x_k \right) \end{cases} \tag{4.9}$$

The value of $\lambda$ determines the rate at which the effect of old data pairs decays. A high value of $\lambda$ ($\lambda \to 1$) produces a high rate of decay and vice versa. However, lower values of $\lambda$ can cause instability and thus the chosen value is typically taken above 0.95.

## Results and Discussion

The feasibility of the proposed SANN control scheme is assessed through the AUV simulation model. The direct learning control system was implemented as previously discussed, whereby autopilot parameter adjustment is performed at every discrete time stage. With respect to the yaw dynamics of the AUV model, the open loop time constant was calculated as approximately 1.5 sec. Consequently, the sampling interval of the SANN was set at 0.1 sec to achieve smooth overall trajectory.

### Online Yaw and Roll Results

The CANFIS technique developed in [10] is extended to one that is suitable for online control. This extension manifests itself in the form of the novel online hybrid learning rule and is applied to the CANFIS architecture via Eq. (4.9).

The 16-rule, hybrid-tuned multivariable fuzzy autopilot of [10] was used to initialize the SANN control system and thus provide a pre-tuned start point to the course-changing and roll-regulating simulation. The forgetting factor was initially set heuristically at $\lambda = 0.975$ with a step size of 5%. This produced an oscillatory response with a settling time of 20 sec and a steady-state error of 1.2%.

Further simulations, examining the influence of the forgetting factor $\lambda$ on the consequent parameter estimates, led to a more suitable choice of $\lambda = 0.99$. The responses of the AUV when employing this value are depicted in Figs. 4.3 and 4.4. This would suggest that employing a larger forgetting factor

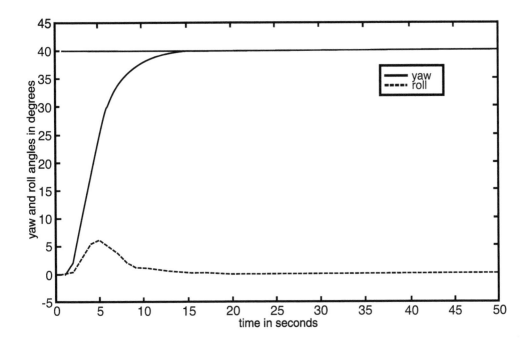

**FIGURE 4.3**  Yaw and roll responses of the AUV for a 40° course-changing maneuver using a forgetting factor of 0.99 and a step size of 5%.

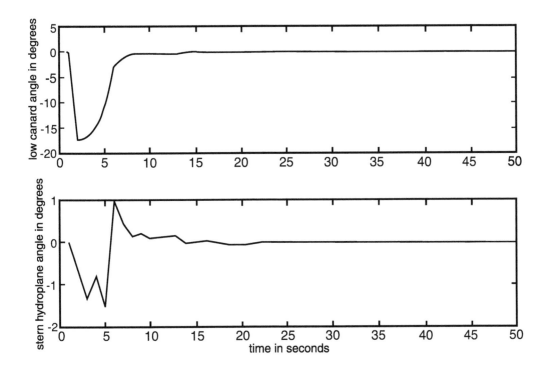

**FIGURE 4.4**  Low canard rudder and stern hydroplane responses of the AUV for a 40° course-changing maneuver using a forgetting factor of 0.99 and a step size of 5%.

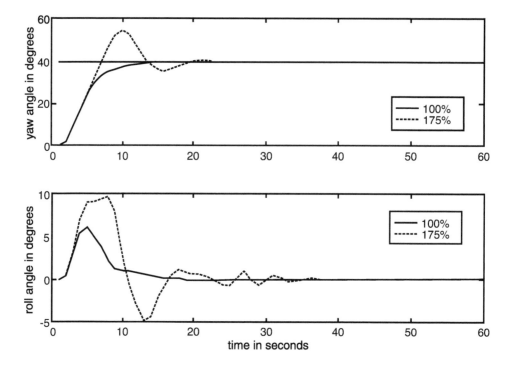

**FIGURE 4.5**    Mass variation during a 40° course change when employing the CANFIS autopilot—yaw and roll responses.

reduces the effects of new data pairs on the control system parameter estimates, as expected. This property implies that the introduction of disturbances may require a smaller forgetting factor in order that parameter estimates are updated more frequently. Nevertheless, the responses achieved when employing a forgetting factor of $\lambda = 0.99$ are superior to those attained with $\lambda = 0.975$, which illustrated that the roll cross-coupled motion was being suppressed effectively.

### Robustness Properties of the Online Autopilot

To test the robustness of the online autopilot, the mass of the vehicle was almost doubled. Figures 4.5 and 4.6 depict the yaw and roll, and low canard rudder and stern hydroplane responses of the AUV, respectively, for the nominal AUV mass and when the mass of the vehicle is increased to 175% of its nominal value. During these simulations the CANFIS online autopilot was employed with a forgetting factor of $\lambda = 0.99$ and a gradient descent step size of 5%. The AUV was initialized at a forward speed of 7.5 knots.

## Concluding Remarks

Although the AUV model used in the design study has nonlinear characteristics and severe cross-coupling terms between its principal axes, simulation results are presented that demonstrate the effectiveness of the proposed autopilot to control the yaw and roll channels. It therefore may be concluded that the approach described herein offers an attractive alternative method of designing multivariable autopilots for AUVs.

## Acknowledgments

The authors acknowledge the use of the MATLAB/Simulink simulation package kindly supplied by DERA, Sea Systems Sector, Winfrith, and thank them for their continued support.

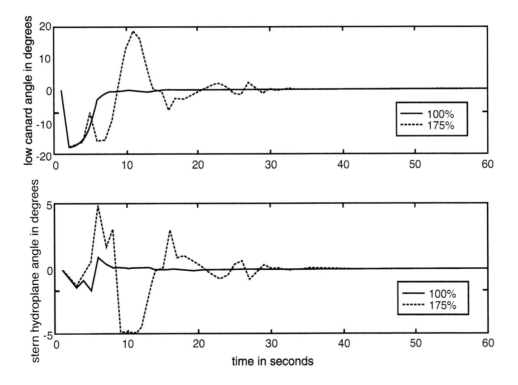

**FIGURE 4.6** Mass variation during a 40° course change when employing the online CANFIS autopilot—low canard rudder and stern hydroplane responses.

## References

1. Goodrich, D., Marine Foresight Report. Department of Trade and Industry, May, 1997.
2. Jang, J.-S. R., Self-learning fuzzy controller based on temporal back-propagation, *IEEE Transactions on Neural Networks*, 3, 714–723, 1992.
3. Mizutani, E. and Jang, J.-S. R., Co-active neuro-fuzzy modelling, *Proceedings of the International Conference on Neural Networks*, Perth, Australia, 760–765, 1995.
4. Yuh, J., A neural network controller for underwater robotic vehicles, *IEEE Journal of Oceanic Engineering*, 15, 3, 161–166, 1990.
5. Venugopal, K. P., Sudhakar, R., and Pandya, A. S., On-line learning control of autonomous underwater vehicles using feedforward neural networks, *IEEE Journal of Oceanic Engineering*, 17, 4, 1992.
6. Ishii, K., Fujii, T., and Ura, T., An on-line adaption method in a neural network based control system for AUVs, *IEEE Journal of Oceanic Engineering*, 20, 3, 1993.
7. Juang, C.-F. and Lin, C.-T., An On-line self-constructing neural fuzzy inference network and its applications, *IEEE Transactions on Fuzzy Systems*, 6, 1, 12–32, 1998.
8. Takagi, T. and Sugeno, M., Fuzzy identification of system and its application to modelling and control, *IEEE Transactions on Systems, Man and Cybernetics*, 15, 116–132, 1985.
9. Corradini, M. L. and Orlando, G., A. discrete adaptive variable-structure controller for MIMO systems and its application to an underwater ROV, *IEEE Transactions on Control Technology*, 5, 3, 349–359, 1997.
10. Craven, P. J., Intelligent Control Strategies for an Autonomous Underwater Vehicle, PhD Thesis, University of Plymouth, 1999.

## Appendix A: Nomenclature of the AUV Equation Parameters

| | |
|---|---|
| **E, F, G** | State equation matrices |
| m | Mass |
| p, r | Roll and yaw angular velocity components |
| u, v | Surge and sway linear velocity components |
| $\psi, \phi$ | Yaw and roll angles |
| $I_X, I_Z$ | Inertia components |
| K, N | Roll and yaw moments |
| Y | Y direction force component |
| B | Buoyancy force |
| G | Centre of mass |
| $K_{UP}$ | Dimensional hydrodynamic coefficients of roll |
| $N_{UU\delta ru}$ | Dimensional hydrodynamic coefficient of yaw w.r.t. canard upper |
| $\lambda_\phi, \lambda_\psi, \gamma_\phi, \gamma_\psi$ | Roll and yaw moment arm lengths |
| $\delta_{Bow-upper}$ | Input from upper canard rudder |
| $\delta_{Stern-port}$ | Input from port stern hydroplane |
| $\delta_{Stern-upper}$ | Input from upper stern rudder |

# 4.2   An Approach to Multi-Robot Cooperation under Human Supervision

*R. Gosine, M. Rokonuzzaman, R. Hale, F. Hwang, J. King, and J. Seshadri*

Technologies to perform complex underwater tasks autonomously, in order to provide for efficient and safe exploitation of offshore resources, are increasingly in demand. The recent significant advances in sensor, computer, and systems engineering will lead to the ability to complete complex underwater tasks through collaboration among multiple robots and humans in a cost effective and safe manner. In this operational scenario, a group of semiautonomous underwater robots or vehicles (e.g., remotely operated machines with limited intelligence) will perform underwater surveying, construction, inspection, and maintenance tasks under the high-level supervision of a single operator using a control station on the surface. The development of underwater robotics technology has largely been focused on underwater vehicles, onboard controls, end-effectors, surveying, and communication technologies. As these technologies mature and the problems at the interface between the machine and the application are resolved, there is a need to develop technology to provide for collision- and deadlock-free cooperation among multiple robots under high-level human supervision. This paper describes a novel approach to multi-robot cooperation under human supervision using a discrete event system approach for accomplishing complex tasks. This development is being carried out within the context of an R & D program related to the development of innovative technologies to facilitate industrial operations in harsh environments, particularly mining and offshore oil and gas. While the mining application has provided for the initial focus of this work and serves to illustrate the technology under development, it is proposed that the underlying approach will be useful for the planning and execution of underwater robot and vehicle missions for offshore oil and gas applications.

As onshore and shallow-water oil and gas deposits continue to be exploited and depleted, oil and gas companies are moving into deep water for the exploration, development, and exploitation of new deposits [1]. Safety considerations limit the ability of human divers to perform tasks in such deepwater operations, and tether management is a complex problem in utilizing underwater vehicles [2]. Limited underwater communication bandwidth significantly restricts the use of direct teleoperation of tetherless vehicles. Moreover, the requirement for continuous operator attention in direct teleoperation limits the scope of achieving optimal productivity. The unstructured nature of underwater work sites and the requirements for operator assistance in task planning preclude, at this stage, the possibility of complete automation.

It is envisioned that the next generation of underwater robots and vehicles will be equipped with intelligent end-effectors and sensors, and will be capable of performing many operations autonomously. It is also anticipated that due to the requirements of complex mechanical interactions with the environments and unpredictable nature of the environments, these machines will sometimes fail to perform elementary tasks autonomously. In order to deal with such failures, human intervention will occasionally be required. Due to the communication delay, it is not feasible to keep a human operator in the loop for continuous task monitoring. Therefore, the machine should have the capability of determining the quality of completion of each elementary task with the human operator intervening upon request. After the human intervention, the machine will resume autonomous execution of the remaining elementary tasks.

Some underwater tasks will be too complicated to be performed by a single machine alone (e.g., laying pipeline). In some cases, deployment of multiple machines in the same work site will be required to perform an elementary operation multiple times or to execute more complex operations in which different vehicles must cooperate. Example operations include the inspection of the hundreds of structural nodes of an offshore oil and gas platform, or the various coordinated operations associated with constructing a deepwater oil and gas facility. The machines need to cooperate with each other to ensure collision- and deadlock-free operation. Technology that ensures multi-vehicle coordination in a manner that is transparent to the operator will reduce the cognitive load on the operator while making it feasible to deploy multiple, cooperating vehicles under the supervision of a single operator.

Significant developments have been achieved in the area of remotely operated vehicle technology for underwater applications [3]-[7]. The main focus has been on the development of thrusters, onboard control systems, sensing systems, vehicle positioning mechanisms, and specialized grippers [8]. Much of the development has been to address problems associated with the interface (e.g., actuators, sensors) between the machine and its operating environment. There has been limited research toward devising a mechanism for multi-robot or multi-vehicle cooperation under the supervision of a single operator. A recent development in the area of task-level teleprogramming [2] to deal with communication delay does not address the issue of on-demand human intervention or cooperation among multiple vehicles.

## An Approach to Cooperation among Intelligent Vehicles and Robots

Control frameworks for cooperative mobile robots have been researched. An agent theory-based, decentralized decision-making formalism has been proposed [9]. In such a control framework, each agent shows rational and coherent behavior in order to complete a complex task cooperatively. In this application, each agent is responsible for planning and scheduling. A major problem with this approach is the difficulty in ensuring the coordination of multiple behaviors to achieve more complex behaviors. From a review of ongoing research in the area of cooperative mobile robotics, it appears that progress has been limited to the definition of the problem and the development of concepts.

A considerable scope is available for the development and integration of simulation and control technologies to facilitate cooperative behavior. Such technology would manage the sharing of the physical and human resources within an intelligent multi-robot system that can effectively deal with unplanned events in the work site and provide for input from a remote operator.

The authors have been leading the development of a novel telerobotics technology to enable an operator in a surface control room to supervise a group of semiautonomous vehicles and robots performing operations in underground mines [10]. The concept, which is based on sensorimotor augmented reality for telerobotics (SMART) [11], involves the dissociation of task specification, task execution, and task supervision. This approach also provides for a task-preserving exchange of control between teleoperated and autonomous operation of a machine. This mediation capability will enhance operational efficiency and reduce the probability of unwanted collisions.

It is assumed that the machines will fail to complete some subtasks autonomously. The disassociation of task specification and execution allows the operator to engage in independent task supervision and

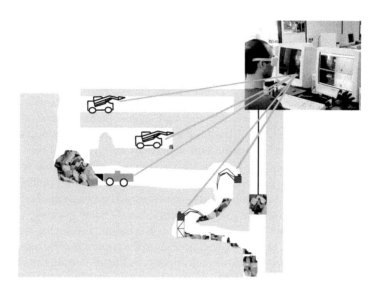

**FIGURE 4.7** Multivehicle, telerobotic underground mining scenario.

replanning for individual machines, while the other machines continue to operate autonomously. With such a system, the supervisor will intervene on demand to assist the machines in completing those subtasks. Upon termination of the human intervention, the machines will resume autonomous operation for the remaining portion of the specified task.

While the development of this concept is at an early stage, an underground mine operation based on this concept is illustrated in Fig. 4.7. In this scenario, several different semiautomated machines are required to coordinate their motions and operations in order to tunnel, blast, and move rock between an ore body and a rock-crushing unit. A human operator sits at a control console on the surface and uses the SMART concept to plan, program, and supervise the coordinated operations of the various unmanned machines.

The development of the SMART concept includes an investigation of a Petri net-based discrete event formalism [12] as the basis for the programming, simulation, and control of mobile robotic systems with provision for cooperative and flexible behavior among multiple robots and a capability for task-preserving intervention by a remote operator.

Petri net-based techniques have been developed as a means for decomposing complex tasks into simpler subtasks. With the help of this technique, these subtasks can be scheduled for execution in a particular sequence. It also provides an intuitive means for realizing synchronization among multiple tasks. Petri net-based techniques, which can be implemented as a finite state machine, provide a sound mathematical framework for task planning and scheduling. These techniques allow for off-line system simulation in order to assist the supervisor in avoiding collisions and deadlocks during the task planning phase. Petri net-based formalisms have been used successfully in planning and scheduling in a number of practical applications, including computer-aided manufacturing.

The representation of cooperative robotic tasks as a finite state machine will enable the development of an automated means for monitoring system states during execution. This run-time state monitoring allows for the detection of potential collisions and deadlocks. During system operation, the scheduling can be dynamically adjusted to avoid such situations. The provision of dynamic scheduling also has the potential to enhance the productivity of operations reducing waiting times of the machines.

## Discrete Event System (DES) Concepts Using Petri Nets

A Petri net, which is a directed bipartite graph, is used as a natural, simple, and quantitative method for modeling the behavior of a discrete event dynamic system (DES). The Petri net structure, $C$, is a four-tuple, $C = (P, T, I, O)$, where $P = \{p_1, p_2, ..., p_n\}$ is a finite set of places, and $T = \{t_1, t_2, ..., t_m\}$ is a finite

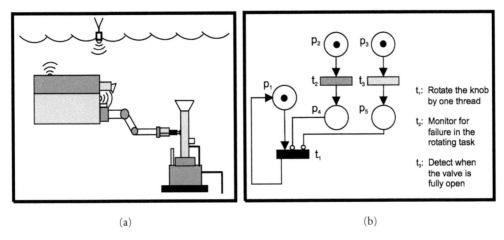

FIGURE 4.8  Underwater robot opening a valve (a) and discrete event model of valve opening task (b).

TABLE 4.1   Description of the Changes in System Behavior

| Places | Change in System Behavior Due to Presence of Token in Places |
|---|---|
| $p_1$ | Request to rotate knob one revolution. |
| $p_2$ | Request to monitor for rotation failure. |
| $p_3$ | Request to detect when valve is fully open. |
| $p_4$ | Detected a failure in rotating the knob—inhibit the rotating task. |
| $p_5$ | Detected that the valve is fully open—inhibit the rotating task. |

set of transitions. These sets of places and transitions are connected by a set of arcs mapping inputs ($I$) to outputs ($O$). An arc with only an arrow tip enables the corresponding transition once a token appears in the place. The execution of a transition is prohibited if a token appears in the place connected to this transition by an arc with a bubble at the arrow tip.

The discrete event model of the behavior of an underwater robot for a simplified task of opening a valve is shown in Fig. 4.8. After grasping the knob attached to the valve, the robot must keep rotating the knob until the value is fully opened. The robot may fail to rotate the knob, for example, as a result of seizing or corrosion and, upon detecting this failure, the robot stops trying to open the valve.

Equation (4.10) represents the different states of the execution of this valve opening task. The state of the system at the detection of a failure in the rotation task is shown in Eq. (4.11) while Eq. (4.12) represents the state after the robot has successfully opened the valve. The description of the changes in system behavior due to the presence of tokens in different places is given in Table 4.1. All processes are executed in parallel to perform this task.

$$\mu = [p_1\ p_2\ p_3\ p_4\ p_5]^T \tag{4.10}$$

$$\mu_1 = [1\ 0\ 0\ 1\ 0]^T \tag{4.11}$$

$$\mu_2 = [1\ 0\ 0\ 0\ 1]^T \tag{4.12}$$

## Use of DES to Ensure Coordination among Multiple Machines

As indicated earlier, it is proposed that multiple robots and vehicles will be required to perform some complex underwater operations, and for some operations these machines will share the same work site. An automated means of coordination among multiple machines, for example, by sharing common navigation paths without collisions, will reduce the data communication overhead and the cognitive load

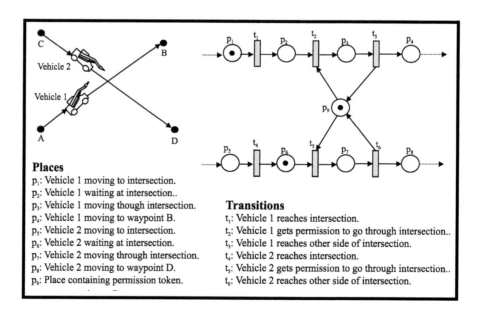

**FIGURE 4.9**    Petri net model illustrating high-level collision avoidance.

on the human operator. These issues are also of concern for cooperating underground mining robots and vehicles that are under the supervisory control of a remote operator.

The cooperative navigation task modeled as a discrete event system using a Petri net can be analyzed to ensure collision-free navigation. The utility of using a Petri net-based discrete event formalism for sharing navigation paths without causing collision has been investigated in the laboratory environment for a remote mining application. Figure 4.9 shows a Petri net model used to control the movement of two vehicles through an intersection. In order for a vehicle to enter the intersection, the "permission token" ($p_9$) must be free. Upon entering, the vehicle "holds" the permission token until it reaches the other side of the intersection, at which point the token is relinquished, giving other vehicles the opportunity to take possession of the token. In Fig. 4.9, vehicle 2 will obtain the permission token when $t_5$ fires, and will hold it while it is in the process of crossing the intersection. When vehicle 1 gets to $p_2$, it must wait until vehicle 2 has finished crossing the intersection (firing of $t_6$) before it begins to cross. In this manner, the two vehicles will never be launched into the intersection simultaneously.

## Use of DES for Seamless Human-Machine Cooperation

In order to maximize the autonomous capability of robots and vehicles and to limit the cognitive load on the human operator, the facility for task-preserving human intervention is being developed. The concept of task-preserving human intervention allows for a human operator to take direct control of one of the machines involved in a coordinated, multimachine process and to have the system resume the original autonomous task once the operator releases control of the machine. The system must be capable of determining its current state following changes made during the direct control action by the operator.

It is proposed that the discrete event framework can be used to mediate between the direct and autonomous modes of machine control.

Consider, for example, the event corresponding to a failure to rotate the knob attached to the valve. Such a failure could trigger the system to engage the human operator to intervene and take appropriate remedial action. The discrete event model of the valve opening task with the provision for task-preserving human intervention is shown in Fig. 4.10.

This concept has been tested in the laboratory environment to provide effective management of the transfer of control between the operator and the robot. Figure 4.11 illustrates a potential Petri net model

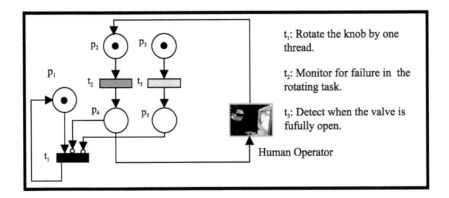

**FIGURE 4.10** Discrete event model of a valve opening task with provision of task-preserving human intervention.

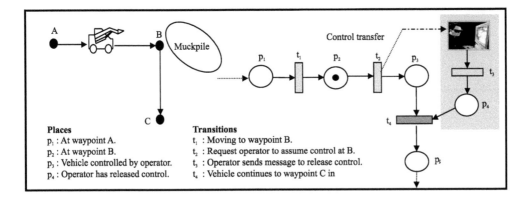

**FIGURE 4.11** Petri model illustrating potential for human-machine interaction.

for a mucking (digging) task, which incorporates task-preserving intervention by a human operator. The Petri net model manages the transfer of control between the operator and the robot. The vehicle travels autonomously from point A to B where it requests intervention for the mucking task and control is transferred to the teleoperator. When completion of mucking is signaled, control is returned to the vehicle, which continues to point C autonomously. At each point where intervention is required, a request is sent to the operator by the vehicle, thereby simplifying the task of determining which vehicles require intervention and reducing the cognitive load on the operator by continuing to allow for automatic operation of other vehicles or by temporarily suspending their operation in known machine states.

## Demonstration of Concept for Multi-Vehicle Programming and Control

The concept of using a Petri net online controller to program and control multiple vehicles was first demonstrated in a virtual mining environment created using OpenGL and illustrated in Fig. 4.12.

In order to test the concepts in a more realistic situation, however, a scaled version of a mining site was constructed and scaled models of actual construction vehicles were utilized. The laboratory test environment and the key system components are illustrated in Fig. 4.13.

A system utilizing a Petri net discrete event controller has been used to program, coordinate, control, and supervise two vehicles in a scaled surface mining environment. The system includes an operator console, including an interactive graphical workspace representation and teleoperation interface, a workspace surveillance system, including a visual vehicle tracking system and an operator interface for remote camera control, and a discrete event controller, which coordinates the sharing of the various resources.

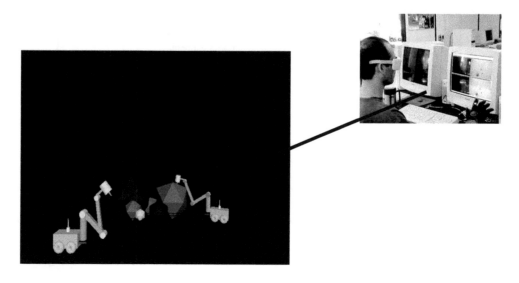

**FIGURE 4.12**    Multivehicle cooperation using Petri net simulation of virtual mining robots.

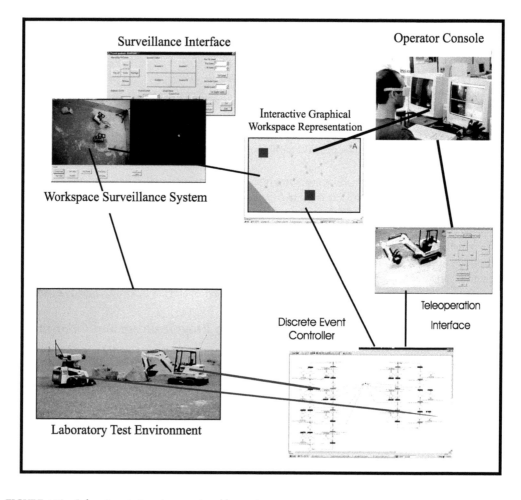

**FIGURE 4.13**    Laboratory test environment and key system components.

In the demonstration, the shared resources include the two vehicles, the workspace visual surveillance system, and the human operator.

The operator console provides the operator with a graphical representation of the work space, which will be generated and updated based on the automated scene analysis from the work space surveillance system. The console provides a means for the operator to specify way points (blue and yellow tracks on the graphical workspace representation) for each of the vehicles and the Petri net model is automatically generated "on the fly" based on these inputs. The operator console also includes a teleoperation interface that provides the remote operator with vehicle-specific views of the work site and the ability to control the vehicle function directly.

A command interpreter translates the discrete events that are generated from the graphical representation of the task (the yellow and blue paths) into low-level data and it translates low-level data into discrete events. For example, a transition firing in the discrete event controller may send the command "move vehicle 1 to point A," and the command interpreter translates "point A" into real-world coordinates for the vehicle controller. When the command interpreter receives the coordinates from the visual surveillance system (which carries out vehicle tracking) that correspond to "point A," it sends a message to the discrete event controller, "vehicle 1 at point A." The discrete event controller executes the Petri net models and sends routine commands to the command interpreter and to the operator console.

A simple task requiring the coordination of two autonomous vehicles, including task-preserving intervention by the operator, has been successfully demonstrated. Vehicle 1 is instructed to travel repeatedly between two locations in the work site along the blue path, while vehicle 2 moves back and forth along the yellow path. The paths of the vehicle cross during the task. Vehicle 1 requests operator attention at point A and resumes its autonomous operation once the operator releases control of the vehicle. While vehicle 1 is under direct control of the operator, vehicle 2 carries on its autonomous activity until it reaches a state in which it is expecting vehicle 1 to be in a required state. Vehicle 2 then suspends its activity until vehicle 1 is surrendered by the operator, resumes its autonomous operation, and reaches the desired state for the coordinated task to proceed autonomously.

## Extension of the Concept for Underwater Robots and Vehicles

The utility of using a Petri net-based discrete event formalism in the control of semiautonomous robots and vehicles has been demonstrated. A system has been implemented in order to investigate the utility of this formalism in the control of multiple, coordinated vehicles in activities requiring cooperative and flexible behavior and as a method to mediate between the autonomous and telerobotic operation of vehicles in a manner that preserves the integrity of the overall application.

It is proposed that this approach has the potential for widespread application in multi-robot or vehicle applications that require remote supervision and occasional intervention by an operator. Such applications would include the deployment of a fleet of underwater vehicles and robots for offshore operations. Consideration of this application focus is currently underway in collaboration with the oil and gas industry operating off the East Coast of Canada.

## Acknowledgments

This work is supported in part by the European Space Agency (ESA) Harsh Environments Initiative, NSERC, Petro-Canada Resources, and the IRIS Network of Center of Excellence.

## References

1. Houston, S. J. and White, J., Optimizing the value of deepwater infrastructure, *The Journal of Offshore Technology*, 7, 4, 18–22, 1999.
2. Sayers, C. P., Paul, R. P., Whitcomb, L. L., and Yoeger, D. R., Teleprogramming for subsea teleoperation using acoustic communication, *IEEE Journal of Oceanic Engineering*, 23, 1, 60–71, January 1998.

3. Brighenti, A., Parametric analysis of the configuration of autonomous underwater vehicles, *IEEE Journal of Oceanic Engineering*, 15, 3, 179–188, July 1990.
4. Cristi, R., Papoulias, P. A., and Healey, A. J., Adaptive sliding mode control of autonomous underwater vehicles in the dive plan, *IEEE Journal of Oceanic Engineering*, 15, 3, 152–160, July 1990.
5. Goheen, K. R. and Jefferys, E. R., Multivariable self-tuning autopilots for autonomous and remotely operated underwater vehicles, *IEEE Journal of Oceanic Engineering*, 15, 3, 144–151, July 1990.
6. Yuh, J., A neural net controller for underwater robotic vehicles, *IEEE Journal of Oceanic Engineering*, 15, 3, 161–166, July 1990.
7. Whitcomb, L. L. and Yoeger, D. R., Preliminary experiments in model-based thruster control for underwater vehicle positioning, *IEEE Journal of Oceanic Engineering*, 15, 3, 495–506, July 1990.
8. Lane, D. M., Davices, J. B. C., Robinson, G., O'Brien, D. J., Sneddon, J., and Elfstrom, A., The AMADEUS dextrous subsea hand: design, modeling, and sensor processing, *IEEE Journal of Oceanic Engineering*, 15, 3, 96–111, July 1990.
9. Das, S., Gonsalves, P., Kirkkorian, R., and Truszkowski, W., Multi-Agent Planning and Scheduling Environments for Enhanced Spacecraft autonomy, *The Proceedings of the Fifth International Symposium on Artificial Intelligence, Robotics and Automation in Space*, European Space Technology Center, The Netherlands, 1999, 91–98.
10. Rokonuzzaman, M., Hale, R. D., and Gosine, R. G., Modeling of Intelligent Mediation System of Sensori-Motor Augmented Reality for Telerobotics Using Discrete Event Formalism, *SPIE Conference on Mobile Robots and Autonomous Systems*, Boston, MA, September 1999 (in press).
11. Cohen, P. and Gosine, R., Sensori-Motor Augmented Reality for Telerobotics (SMART), *Proceedings of the Workshop on Harsh Environment Initiative*, European Space Technology Center, The Netherlands, 1999, 229–244.
12. Caloini, Magnani, G., and Pezze, M., A technique for designing robotic control systems based on Petri nets, *IEEE Transactions on Control Systems Technology*, 6, 1, 72–87, January 1998.

## 4.3 Computer Vision in Ocean Engineering

*Donna M. Kocak and Frank M. Caimi*

Divers, manned submersibles, and remotely operated vehicles (ROVs) rely upon the human visual system for synthesis and perception of the complex underwater visual world in order to make decisions and perform tasks. These tasks may require immediate action or delayed decision making based on collected information. In either case, there are often circumstances where it is beneficial to remove the human from the loop. For example, tasks requiring precise positioning in hostile environments could prove harmful to manned operations and may be inaccessible to ROVs requiring nearby surface support. Scientific studies resulting in the collection of vast amounts of data often become too tedious for accurate human analysis. In addition, with the advent of autonomous underwater vehicles (AUVs), all human operation is replaced with a programmed mission. Hence, automatic methods for gathering information from the surrounding environment, processing this information, and making decisions to accomplish tasks have become necessary.

Computer vision techniques provide solutions for accomplishing tasks that require little or no human intervention. Similar to the human visual system, computer vision is the *process* of extracting, characterizing, and interpreting information from images of the external world. According to Fu [1], this process can be further divided into six categories: sensing, preprocessing, segmentation, description, recognition, and interpretation. Sensing and preprocessing are low-level computer vision processes that include image formation, noise reduction, and extraction of primitive image features. Segmentation, description, and recognition are medium-level processes that extract, characterize, and label components in an image resulting from the low-level vision. Finally, interpretation is a high-level vision process that attempts to emulate cognition. These techniques have been applied to ocean engineering applications such as: search and survey, mine and ordinance detection, object recognition and tracking, image analysis and visualization, image compression, photogrammetry, reconstruction and analysis, navigational control, and image sensor fusion.

## Computer Vision Processes

This section defines the framework on which computer vision is based and provides a basic description of the processes. More detailed information can be found in various text books (e.g., [1–5]).

### Low-Level Vision

Sensing and preprocessing are low-level vision processes that do not require cognitive ability. These may be considered "automatic responses"— similar to "the sensing and adaptation process a human undertakes in trying to find a seat in a dark theater immediately after walking in during a bright afternoon. The intelligent process of finding an unoccupied space cannot begin until a suitable image is available" [1].

#### *Sensing*

Sensing is the process that produces an image. In the sea, image formation is greatly influenced by the properties of the medium. Water clarity (opacity) influenced by absorption and scattering affects the range of visibility and image quality. Typically, water clarity is measured by the numeric value of the optical attenuation constant of the medium. In water, the attenuation constant ($c$) is defined by

$$I = I_0 \exp(-cz) \tag{4.13}$$

where $I_0$ is the incident beam irradiance, $z$ is the distance over which the measurement is made, and $I$ is the measured irradiance. Prior to Eq. (4.13), a Secchi disk was used to measure water clarity [6]. When using this method, a 30-cm diameter white painted disk is lowered to a depth where it is no longer visible. This depth, $Z_d$, is called the Secchi depth and can be related directly to $c$ as follows:

$$c \sim constant/Z_d \sim 6/Z_d. \tag{4.14}$$

Although this is an approximate relation, tests in many different natural waters have confirmed its viability [7]. The constant value varies from 6 to 9 depending on the type of particle distribution in the water. The usual method for obtaining accurate measurement of the attenuation constant relies upon an instrument called a beam transmissometer. The reciprocal of $c(c^{-1})$ is referred to as the attenuation length. The attenuation length varies geographically and with depth in the ocean. Typical values are 25 to 50 m in clear water, 3 to 10 m in coastal water, and 1 m or less in estuaries. Thus, location, depth, temperature, and season can all contribute to the high degree of variability of image quality and information content. Although the Secchi technique implies a visibility of 6 attenuation lengths, conventional still or video cameras are able to produce discernable images at only 1 to 2 attenuation lengths due to limitations in dynamic range. Such systems are generally designated as "conventional" image formation systems.

The attenuation coefficient contains two parameters, $a$ and $b$, relating to the absorption and scattering properties of the medium. This relationship is defined as

$$c = a_w + b_w + a_p + b_p \tag{4.15}$$

where the subscripts $w$ and $p$ represent water and particulate components, respectively. Each of the parameters in Eq. (4.15) is spectrally dependent. The scattering coefficients increase at shorter wavelengths while the absorption coefficients depend on the characteristics of water and the suspended particulate matter. In the deep ocean, $c$ is generally minimum at a wavelength in the blue-green spectral region (460 to 490 nm). In coastal waters, the minimum is shifted into the yellow and yellow-green portion of the spectrum (530 to 560 nm) [8]. Because of this dependency, it is considered advantageous to select a specific wavelength region where the effect of absorption and scattering are minimized. This selection, however, is not always optimal since scattering can produce visual artifacts that occlude image features, while absorption reduces the available light flux. Both processes are nearly independent spectrally. Nevertheless, laser illumination near the spectral minimum of the attenuation constant is characteristically used when imaging is required at greater than 2 attenuation lengths ("extended range" imaging). In contrast to conventional image formation methods using a camera, extended range imaging methods using laser illumination tend to reduce image artifacts resulting from scattered light generated by artificial illumination sources.

**Conventional Image Formation**—Conventional imaging uses a camera to map the spatial reflectance properties of a three-dimensional (3-D) scene to a two-dimensional (2-D) representation. Cameras provide high resolution and sensitivity, and fast sampling rates (30 frames per second) permitting real-time performance, rich visual cues such as texture, shading, and surface markings useful for identification, and passive sensing ideal for covert operations. Cameras are also available as standard equipment on most underwater vehicles. The primary image impairments observed when using these systems are the contrast reduction and the loss of illuminance at many attenuation lengths. Contrast reduction results as the background illumination $E_s$ due to scatter increases in comparison to the scene illumination $E_d$ at the camera lens, as shown in Fig. 4.14. As the contrast is reduced the dynamic range requirement of the camera increases. In Fig. 4.15, the brightly illuminated regions are produced by scattered light from two illumination sources. The target is illuminated from forward-scattered light that is present over a limited spatial portion of the entire image area. Bands of different reflectivity on the target appear with varying contrast in the reproduced image. If the scattered field was not present, image impairments would be governed solely by the exponentially decreasing value of $E_s$. At seven attenuation lengths the illuminance is reduced by a factor of $10^{-5}$ in comparison to the value at one attenuation length. Consequently, low light level cameras, such as SIT, ISIT, ICCD, and I²CCD, are of great utility in undersea imaging applications. Dynamic range remains an important issue in the use of low light level cameras and artificial illuminators.

As shown in Fig. 4.14, any imaging system using artificial illumination will generate a scene illumination that is reduced in comparison to the backscattered illumination as the range is increased, resulting in a reduced signal to noise (S/N) ratio. Every effort is made to reduce the backscattered illumination by the placement of lights and the reduction of the angular field of illumination. The objective is to reduce the overlapping fields of view between the camera and the light source. Generally, the light-scattering process is of the same order of magnitude as the absorption process, thereby blurring detail and directing scattered photons back into the camera at a time of arrival prior to that of the reflected, unscattered

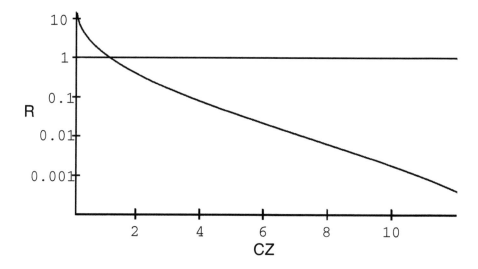

**FIGURE 4.14**   The ratio $R$ of $E_d$ to $E_s$ vs. attenuation length $cz$ for $c/b = 3$.

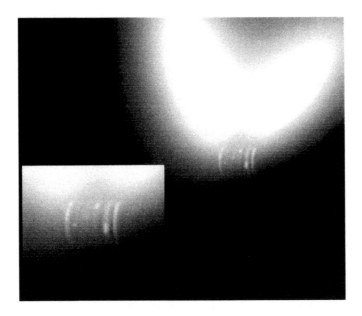

**FIGURE 4.15**   Typical underwater image at greater than two attenuation lengths. (Insert highlights target details.)

photons. The relative degree of scatter in the forward or backward directions is described by the asymmetry factor $g$ defined by [9]

$$g = \int_0^{2\pi} \int_0^{\pi} \cos\theta [P(\theta)/4\pi] d\Omega, \tag{4.16}$$

where $P(\theta)$ is the scalar phase function and $\theta$ is the angle with respect to the direction of propagation. The scalar phase function is related to the volume scattering coefficient $b$ and volume scattering function $\beta$ by

$$P(\theta) = 4\pi\beta(\theta)/b \tag{4.17}$$

with $b$ being defined as,

$$b = 2\pi \int_0^{2\pi} \beta(\theta)\sin(\theta)d\theta. \tag{4.18}$$

The asymmetry factor is zero for isotropic scattering, near minus one for highly backscattering media, and near one for highly forward scattering media like seawater. The relative value of $b$ in relation to $a$ is spectrally variant and for pure water is at maximum near the absorption minimum at approximately 475 nm giving water its deep blue color. The volume scattering function is naturally of great importance in determining the apparent contrast of in-water images. Its form is approximately characterized for pure water over a limited angular range [10] and is very nearly that expected for Rayleigh scattering

$$\beta(\theta) = \beta(\theta)_{90} \left[\frac{\lambda_0}{\lambda}\right]^{4.32} [1 + 0.835\cos^2(\theta)] \tag{4.19}$$

In seawater, the function is considerably forward peaked, but still a minimum at 90 degrees. More refined scattering functions applicable to imaging analysis can be found in [11]. At near forward angles (milliradians or less) the scattering is described by refractive index variations existing in the medium (turbulence) and may be expressed as a root mean square (rms) variation in refractive index from which is estimated an expected modulation transfer function [12].

Conventional imaging can be used in conjunction with image gradient techniques (i.e., shape from shading, shape from texture, etc.) to infer 3-D data from single 2-D images. The resulting image is often referred to as 2½-D. Optical flow (shape from motion), stereo vision, and active vision techniques can also be employed to produce 3-D range and intensity (voxel) images using more than one conventional, 2-D view. Implementation details are derived from techniques found in medium-level vision.

**Extended Range Image Formation**—Methods for extended range imaging utilize range gating, synchronous scanning, and polarization discrimination of specialized light sources. For implementation, these schemes usually require the use of lasers to produce narrow illumination beams with high power in the spectral regions associated with maximum seawater transparency. Initial attempts at producing advanced systems using these techniques were impeded by technological limitations associated with laser energy, size, efficiency, and pulse repetition rate, as well as positional stability of the imaging platform.

**Time-Gated Imaging**—Time-gated systems typically operate by scanning a laser beam in a point-by-point fashion over the field of view (FOV). A single burst of photons is produced over each element of the scan with a physical length in space that is much smaller than the beam spread at the object. The physical burst length must also be smaller than any difference in photon path length that could originate from scatter along the direct path from laser to object and back that would be at sufficient magnitude to compete with detection of the unscattered return beam. An example of the operation and geometric configuration of such a system is shown in Fig. 4.16.

In operation, an optical pulse is produced by the laser and is broadcast into space at an angle determined by the instantaneous position of the scan optics. The pulse is directed toward the object and is spread in time and spatial extent as it propagates. The photon packet strikes the object and is reflected according to the reflection and scattering characteristics of the object surface. Most natural undersea objects produce diffuse reflection patterns resulting in a nearly spherical spreading loss of reflected light. The detector remains gated "off" until the photons are about to arrive at the detector aperture, thus eliminating higher intensity signal contributions from scattered photons that may arrive from nearby scattering centers. Once the detector is gated "on," a time history of the pulse is taken, and an estimate of its magnitude is

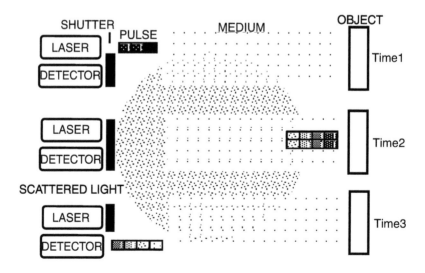

**FIGURE 4.16**   Time-gating sequence for rejection of scattered light.

recorded for display as a pixel intensity at the assumed scan location. The process is continued sequentially until the desired number of scan elements is recorded.

In water, light travels at nearly 4 ns/m so that for most applications laser pulses as short as 1 to 5 ns are suitable. Adequate pulse energies are required to overcome absorption and scattering losses over long round-trip path lengths and to overcome detector noise for a given detector aperture. Advances in solid-state laser technology have improved the efficiency and reduced the size of blue-green lasers such that handheld units are feasible, but pulse energy and efficiency is still not adequate for some applications. If the range gate is wide in comparison to the pulse, the backscattered radiance, $B_b$, is given as the product of $\beta(\theta = \pi)$, the range gate width $dz$, the irradiance at the object $E$, and $e^{-\alpha}$, while the target radiance is the product of the target reflectance $\rho$, $E$, and $e^{-\alpha}$. Thus, the contrast, $c$, is independent of range and is proportional to $\rho$ and inversely proportional to $\beta(\pi)$ and $dz$ under ideal conditions of detection.

In order to "screen" scattered light from the detector, a shutter or gating device is used—often a gated photomultiplier or image intensifier tube—due to its fast response and high sensitivity. The return from the detector is quantized in amplitude for each time increment allowing a 2-D reflectance image or 3-D surface map to be constructed. The 3-D information can be derived from time-of-flight information for each pulse burst. However, depth resolution at close range requires extreme temporal resolution—10 ps$^{-1}$ for approximately 1 mm.

For any time-gated system, laser pulse rates, output energy per pulse, receiver aperture, and detector noise floor govern performance, but definite range performance improvements are possible. For instance, imaging has been demonstrated at greater than 5 to 6 attenuation lengths, in comparison to 1 to 2$^{+}$ attenuation lengths for conventional camera systems.

An extension of the time-gated method is the *LIDAR* method, where time-of-flight information is used to obtain range estimates. These systems are typically used in airborne remote sensing and some undersea military applications. Pulse repetition rates suitable for video frame rate imaging are desirable for many applications involving relative motion between the detection platform and the scene.

Alternative time gating methods use pulsed lasers to illuminate the entire scene, or portions of it, at one brief window in time and produce extended range reflectance (2-D) images. In these systems, the requirement for a rapid pulse repetition rate is relaxed, but the required energy per pulse is increased. Detectors commonly used with this illumination method include gated intensified cameras, such as ISIT or ICCD types. An advantage of this approach is that standard low light cameras are readily adapted without the need for special purpose scanners or detectors.

**Field-Limited Imaging**—Another method for the reduction of scattered light effects on the image involves the synchronous scanning of a narrow laser beam with a narrow field-of-view detector. In this case, the scattered light returned to the detector is reduced by minimizing the overlap between the illuminated volume and the detector field of view, and placing the overlap region closer to the object in the scene, as shown in Fig. 4.17. This process is analogous to following the light from a laser pointer with a telescope, and assumes some knowledge about the distance to the surface.

**Structured Illumination Imaging**—*Line scan systems* producing 3-D data have been developed using laser, structured illumination. These systems operate in either a "fly-by" or "scan mode" and rely upon spatial disparity of the laser projector and sensor to determine range by triangulation. Fly-by systems typically use a single laser stripe projection to scan broadside to the vehicle motion in order to produce a series of one-dimensional (1-D) range maps that can be analyzed as the scan crosses the object under inspection. By contrast, scan mode systems employ a high resolution 2-D laser scanner in addition to a high resolution position sensitive detector, such as a 1-D CCD array, to detect the laser reflection at every point of the scan. A 2-D range map is produced, but at a slow rate (sometimes taking minutes).

*Synchronous scan (SS) systems* operate by minimizing the common volume occupied by the laser illumination and the detector field of view. One type of laser line synchronous scan (LLSS) system uses multiple-wavelength excitation and detection. With this strategy, fluorescence and color imaging can be accomplished at increased optical depths (4 to 6 attenuation lengths). Other systems incorporate a "push broom" imaging concept using pulsed lasers and high-speed detectors capable of simultaneously estimating

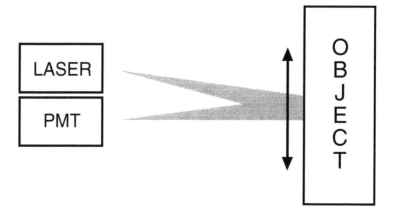

**FIGURE 4.17**   Synchronous scan geometry.

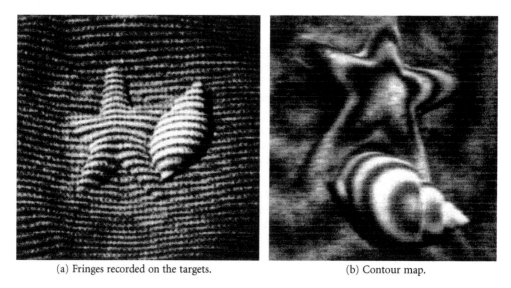

(a) Fringes recorded on the targets.                    (b) Contour map.

**FIGURE 4.18**   Video Moiré input and output images. (*MTS Journal*, with permission.)

time of flight along a complete line of illumination. This approach can allow longer distance images to be produced that include the range coordinate as well as the x-y reflectance image (voxels).

*Spatial quasi-coherent interferometric systems* utilize spatially modulated illumination (SMI) to generate real-time 3-D surface contours (Moiré) of remote objects with variable resolution and a zoom capability. Variable spatial frequency vertical interference fringes are projected onto a target and are viewed with a video camera. As shown in Fig. 4.18a, the 3-D surface shape of the target distorts the fringes [13]. A reflectance image taken with a standard camera would be featureless by comparison. Through processing that includes a reference signal from a second video camera, Moiré depth contours can be derived as shown in Fig. 4.18b. Targets of varying size and at varying ranges, can be illuminated with structural illumination sized to the detail of the depth information desired. In addition to the real-time depth contour display, the successive capture of a series of target images with shifted structural illumination allows the off-line generation of an accurate 3-D model of the target surface. The form of the desired output pattern must be carefully chosen, with some considerations including immunity to misalignment, and whether the recognition system is to be autonomous, computer aided, or human. By combining arbitrary projection, arbitrary filtering, and optical processing, it is possible to produce a target recognition system that is immune to target range, rotation, and alignment.

**Coherent Imaging**—*Holographic systems* record the phase relationships between a spatially and temporally coherent reference beam and an object beam. They are advantageously applied to the imaging of particles in small volumes of water because of the large depth of field available in comparison to conventional microscopic imaging systems. At large optical distances, holographic methods have failed to meet expectations due to the destruction of the coherent wavefront by turbulence, refractive index variability, and scattering.

*Temporal quasi-coherent interferometric systems* use modulated waveforms to acquire 3-D information. These systems are less sensitive to range information than holographic systems, and offer variable resolution depending upon the waveform modulation frequency or phase.

**Acoustic Imaging**—Scanning sonar, either mechanical or by array electronic scanning, is used for forward imaging applications. Narrow-beam mechanical scan, phase comparison, side-scan, synthetic aperture, and multibeam are just a few sonar types. Usually, narrow-beam sonar has a thin beam of 1 to 2 degrees in the horizontal direction and scans over a multisecond period. Continuous transmission frequency modulation (CTFM) is a subset of this category. Phase comparison sonar uses the phase information to determine bearing on two or more wide beams at a time. Data rate is improved over mechanical scan systems, but bearing resolution is proportional to SNR. Side-scan sonar uses a line array to produce a narrow (1°) horizontal beam and a wide (30 to 70°) vertical beam. Side-scan operates by observing the interruption of the acoustic reverberation caused by an object in the beam. Images are difficult to interpret for the untrained observer. Multibeam sonar either steers a multiplicity of single beams or duplicates angular sectors to acquire a complete image in the time it takes one pulse to travel to the target and back. Thus, the information rate is high. Many different configurations are possible but are beyond the scope of this discussion. Synthetic aperture techniques rely upon coherently summing many returns from a sonar system as it passes a target area. The resolution of the system is increased by the synthetic size of the aperture that is formed by many data records put together to make a much larger time record. Angular resolution from diffraction theory is related inversely to the aperture width [14].

### Preprocessing

Image formation methods can produce noisy and imperfect images as a result of faulty detectors, inadequate or nonuniform lighting, blur due to motion or out-of-focus optics, undesirable viewpoint, noise due to the properties of the medium, poor SNR, etc. Although higher quality images can be produced if the problem is corrected prior to acquisition, spatial-domain or frequency-domain preprocessing techniques can be applied to reduce noise and enhance details. For brevity, these techniques will be discussed as applied to conventional 2-D (intensity) images; however, similar methods can be applied to $2\frac{1}{2}$-D and 3-D range and intensity images.

**Spatial-Domain Methods**—Spatial-domain methods operate on the image pixels directly. Several methods use pixel neighborhood operations and convolution masks, and can be represented as

$$g(x, y) = f[I(x, y)] \qquad (4.20)$$

where $I(x, y)$ is the input image, $f$ is the operator on $I$ defined over some neighborhood $(x, y)$, and $g(x, y)$ is the resulting (output) image. Smoothing operators such as *averaging and median filtering* remove high-frequency variability introduced by sampling, quantization, and transmission disturbances during image acquisition. Smoothing convolves a (weighted) mask symmetrically around each pixel to compute an average of the neighboring pixels and then uses this average as the center pixel value in the smoothed image. Median filtering ranks pixels in a neighborhood according to brightness and selects the median value for the center pixel. This is a nonlinear, lossy operation, but it is particularly useful for removing random noise and replacing missing pixels.

Morphological operators, such as the *dilate operator*, can be used to broaden the gradients in an image by blurring features. This has the advantage of removing small pixel groups (or noise specks) and smoothening jagged edges of larger pixel groups. The *erode operator*, on the other hand, reduces high-intensity pixels in a neighborhood yielding sharper image features. Multiple applications and various combinations of these operations can produce specialized filtering effects.

Edge information can be obtained from an image using *gradient* and *Laplacian operators*. These operators are high-pass filters that remove low frequencies corresponding to gradual overall variations in brightness and enhance high frequencies corresponding to points, lines, and edges. The gradient of an image $f(x, y)$ at location $(x, y)$ is defined as the 2-D vector

$$G[f(x, y)] = \begin{bmatrix} G_x \\ G_y \end{bmatrix} = \begin{bmatrix} \dfrac{\partial f}{\partial x} \\ \dfrac{\partial f}{\partial y} \end{bmatrix}. \tag{4.21}$$

Commonly, edge detection uses the magnitude of the gradient over a neighborhood in the image defined by the Sobel or Kirsch masks [3]. The Laplacian is a second-order derivative operator defined as

$$L[f(x, y)] = \frac{\partial^2 f}{\partial x^2} + \frac{\partial^2 f}{\partial y^2}. \tag{4.22}$$

Because the second derivative is highly sensitive to noise, this operator is not commonly used for edge detection and instead is used to determine whether a given pixel is on the dark or light side of an edge.

*Histogram equalization* can be used to increase contrast in images potentially affording enhanced visual quality. For each of the possible intensity levels in an image, $j = 0, 1, \ldots, (l-1)$, the count of the number of pixels in an image at intensity $j$ can be stored in a histogram. Variable $l$ is the number of discrete intensity levels, equal to 256 for 8-bit resolution. A plot of the histogram (intensity vs. count) reveals information about the image—peaks correspond to the more common intensity levels and valleys correspond to the less common intensity levels. In sparse images, where a uniform background is dominant, peaks represent the background and valleys represent features. Images with little contrast may contain empty regions on either side of the histogram, indicating that the full range of intensity levels is not being used. Histogram equalization expands the contrast by spreading image pixel values over the entire available intensity range. For each intensity value $k$ in the original image, the histogram function $H$ is calculated as

$$H(k) = n_k, \tag{4.23}$$

where $n_k$ is the number of pixels in an image with an intensity value equal to $k$ and $\Sigma n_k = n$, the total number of pixels in the image. Local histogram operations can be applied to selected areas of an image. This can be used to reveal subtle differences in intensity within an area; however, it can distort relationships between brightness of regions that are often important in the medium-level vision processes.

**Frequency-Domain Methods**—Frequency-domain methods operate on complex pixels resulting from taking the frequency-based transform of an image. As with the spatial-domain methods, these preprocessing methods can be used to remove certain types of noise, apply large convolution masks, enhance periodic features, and locate defined features in images. They can also be used to measure images to determine periodicity or preferred orientation. The most common method is the *Fourier transform* defined by

$$F(u) = \frac{1}{N} \sum_{x=0}^{N-1} f(x) e^{-i2\pi ux/N}, \tag{4.24}$$

where $N$ is the number of sampled points along the function $f(x)$ (assumed to be uniformly spaced), $x$ is a real variable representing either time or distance in one direction across an image, $f$ is a continuous and well-behaved function, $i$ is $\sqrt{-1}$, and $e\text{-}2\pi ux/N = \cos(2\pi ux) - i\sin(2\pi ux)$ (Euler's formula). Similarly, the reverse Fourier transform is defined by

$$f(x) = \frac{1}{N}\sum_{u=0}^{N-1} F(u)e^{i2\pi ux/N}. \tag{4.25}$$

The advantage of using frequency-domain methods is the improved efficiency over spatial-domain convolution methods that require multiple accesses to each pixel in the image. Furthermore, frequency-domain transforms are invariant to translation in amplitude and can be made invariant to rotation and scale changes with proper preprocessing.

## Medium-Level Vision

Segmentation, description, and recognition are considered medium-level vision functions. These processes extract, characterize, and identify objects in an image resulting from low-level vision. Although the human visual process incorporates a certain degree of intelligence when performing these processes, computer vision techniques can be applied to yield similar results. Some methods incorporate *a priori* information or "learning" techniques as a form of intelligence.

### Segmentation

Segmentation is the process of extracting objects from an image. Techniques are based on finding similarities or discontinuities in either static or sequential scenes. When multiple objects are present in a scene, they can overlap in the image and cause improper segmentation. For example, two or more objects may be merged into one region or a single object may be bifurcated into two or more regions. Segmentation results depend on the robustness of the algorithm. Algorithms using 3-D information have advantages over their 2-D counterparts in situations where objects are partially occluded. Unless the separation in range between objects is very small (less than a preset threshold value), merging and splitting of objects will not occur.

**Similarity Methods**—Similarity methods group pixels based on common features. One method, *thresholding*, uses predetermined pixel intensities as the criteria for grouping. This can be based on two-level (binary) or multilevel thresholding criteria where the threshold values are selected using histogram information, boundary characteristics, or multiple variables including color, texture, and pixel location. Methods such as *region growing* and *splitting* and *merging* use region information for grouping. Pixel aggregation, for example, starts with a "seed" point defining a region and expands out to neighboring pixels depending on whether or not the pixels satisfy predetermined threshold criteria. Another method uses "*image differencing*," or motion between images, to discriminate objects—a technique commonly used by humans and animals.

**Discontinuity Methods**—Objects in a scene are often separated from each other and the background by discontinuities such as edges, lines, and points. Using *edge information* provided by low-level vision, local analysis methods compare strength and direction of the gradient operator at the edge pixels and link those with similarities representing a single object. Global methods, such as the *Hough Transform* or *graph-theoretic techniques*, use the entire image to link object boundaries. The Hough Transform, for example, links edge pixels that conform to a predefined shape of the object.

*Shape from shading* provides another discontinuity method for segmenting object shape from a single image. This technique is best applied to locate image objects whose surface properties exhibit uniform reflectance and little or no texture detail. Images containing surface detail are better treated with an alternative method, *shape from texture*. The reflectance map of the image is defined by $R(p, q) = \cos\theta_i$, where $p$ and $q$ are the first derivatives of the range $z$ with respect to pixel location $(x, y)$ and $\theta_i$ is the angle between the surface normal and the direction toward the light source. Therefore, the reflectance map is actually a measure of the curvature of the surface. The reflectance map is equal to the intensity image map for spherical objects, however, in general, shape must be computed from the intensity gradient in the image. One-dimensional contours can be obtained from knowledge of the image intensity gradient and from solution of several differential equations [15] to obtain the whole surface.

Another method incorporating physical properties uses an *active contour model*, or *snake*, to perform segmentation [16]. A snake is an energy-minimizing spline whose position is defined parametrically by

$$v(s) = (x(s), y(s)),$$ (4.26)

where *s* is the contour length and (*x*, *y*) is the pixel coordinate. The snake uses physics-based methods to locate an object, feature, or contour by a dynamic minimization process using the energy functional:

$$E_{snake} = \int_0^1 E_{snake}(v(s))ds$$ (4.27)

or

$$E_{snake} = \int_0^1 \{E_{int}(v(s)) + E_{image}(v(s)) + E_{con}(v(s))\}ds.$$ (4.28)

Three types of energies influence the snake: internal spline energy ($E_{int}$), image energy ($E_{image}$), and external constraint energy ($E_{con}$). The internal spline energy, represented as

$$E_{int} = (\alpha(s)|v_s(s)|^2 + \beta(s)|v_{ss}(s)|^2)/2,$$ (4.29)

controls the "elasticity" of the snake by the first-order term and the "rigidity" of the snake by the second-order term. The weights $\alpha(s)$ and $\beta(s)$ control the importance of these terms. For example, setting $\beta(s) = 0$ removes the rigid component and causes the snake to develop a corner. The image energy, represented as

$$E_{image} = w_{line}E_{line} + w_{edge}E_{edge} + w_{term}R_{term},$$ (4.30)

is a weighted combination of line ($E_{line}$), edge ($E_{edge}$), and termination ($E_{term}$) components. The simplest example of the line functional is the image (I) pixel intensity,

$$E_{line} = I(x, y).$$ (4.31)

Depending on the sign of $w_{line}$, snakes will be attracted to either light or dark lines. Next is the edge functional which can be represented by any edge-finding algorithm; for example, setting

$$E_{edge} = -|\nabla I(x, y)|^2$$ (4.32)

causes snakes to be attracted to contours with large image gradients (areas of high contrasts like edges). The termination functional allows the snake to find terminations of line segments and corners represented by

$$E_{term} = \frac{\partial \theta}{\partial n_1}.$$ (4.33)

Energy functionals incorporate prior knowledge of the features being sought, which include pixel intensity, edge, and termination descriptors. When analyzing sequential images, snakes can be used to combine segmentation and tracking into a single step.

Time-varying information can also be used to segment an image. *Optical flow (shape from motion)* is the apparent motion of the brightness pattern associated with the brightness of objects in an image sequence. It assumes that in the absence of other scene events, the brightness of an object surface remains relatively constant as the surface moves relative to the viewer. This is the so-called "brightness constancy assumption" [15]. Object surfaces are identified by observing locations where the optical flow changes rapidly. A "relaxed brightness constancy assumption" is best used in underwater imaging where brightness changes can be modeled for changes in illumination resulting from relative motion.

## Description

Features of each segmented object are identified and stored as descriptors for use in the recognition process. The descriptors must provide enough information to uniquely identify an object, notwithstanding size, location, and orientation. Three types of descriptors are boundary, regional, and 3-D.

**Boundary Descriptors**—The shape of an object can sometimes be used to uniquely define the object. *Chain codes, signatures, polynomial curves, shape numbers, moment invariants*, and the *Fourier boundary descriptors* are all methods of describing object shapes and boundaries [1]. In cases when shape alone cannot uniquely define an object, regional descriptors can be used.

**Regional Descriptors**—Regional descriptors define geometrical shape and surface characteristics of objects. Simple information such as *area, perimeter, compactness, curvature*, and the *length of the major* and *minor axes* can be used [1]. When scaling distortions are introduced in the image, other surface features such as *color* and *texture* can be used. Texture information can be represented statistically where the surface is defined as being coarse, smooth, etc., or structurally where the surface is defined as an arrangement of primitive features. *Granulometric features* are pattern descriptors based on mathematical morphology that contain shape and texture information [17].

*Fourier* and *wavelet transforms* are other types of regional descriptors [19]. When the Fourier transform is used as a regional descriptor, it must be separately applied to different regions of the image. Unlike the Fourier transform, wavelet transforms produce time and frequency localization of the input image. For example, the 1-D Fourier transform maps a 1-D function of a continuous variable into a 1-D sequence of coefficients, whereas the wavelet transform provides a 2-D array of coefficients. "A wavelet representation is much like a musical score where the location of the notes tells where the tones occur and what their frequencies are" [18]. There are many types of wavelets and each can be advantageous for a particular image analysis application. Like the Fourier transform, wavelets may be considered a class of filters for preprocessing as well.

**3-D Descriptors**—Range images can be described using 3-D descriptors. *Planar patches, gradients, line* and *junction labeling, generalized cones, ellipsoids*, and *superquadrics* provide a means of describing 3-D data [20]. Extensions of frequency-domain analysis techniques are applicable to description of three and higher dimensional data sets. 3-D descriptors are more complex and require more processing time than conventional, 2-D representations.

## Recognition

Once the object has been segmented from an image and its features described by a set of descriptors, a label can be assigned for classification. This process is typically performed based on matching using either decision-theoretic or structural approaches. Difficulties in matching can result from variations in object size and orientation due to the viewing perspective, and shape distortions in nonrigid, deformable objects.

**Decision-Theoretic Methods**—Decision-theoretic methods use *decision functions* to match the descriptors of an object to a predefined class. *Statistical classifier* techniques base the decision on a statistical comparison of the descriptors. Another approach bases the decision on using *correlation coefficients*. Position and intensity of the object, for example, can be correlated to those of the object class.

If the resulting correlation is greater than a preset threshold, then a match is made. This method assumes that intensity has been normalized and that both objects maintain a consistent size and orientation.

*Vector quantization (VQ) classifiers* provide classification of input image data in vector form [19]. A variety of methods including ad hoc techniques have been devised for creation of a set of basis vectors. The criteria for association of an input vector to a particular class (codebook) can be determined by minimizing a cost function such as Euclidean distance or entropy.

**Structural Methods**—Unlike decision-theoretic methods, structural methods use geometrical relationships for recognition. Objects are broken into "pattern primitives" specifying length, direction, and order. *Matching methods,* such as shape numbers and string matching or *syntactic methods* that contain rules governing the way the primitives are interconnected, can be used for recognition.

Other structural approaches compare objects with *3-D models* [20]. Because a 3-D model can be (automatically) scaled and rotated to approximate the real-world object, problems associated with variances in the viewing perspective are alleviated. With the addition of constraints based on dynamic rules, physics-based 3-D models can be developed to control deformable modeling primitives and create new modeling operators that transform shapes into other shapes. These methods can efficiently represent the behavior of complex objects, motions, and materials.

### High-Level Vision

High-level vision involves interpreting an image so that a decision can be made to accomplish a task. This is the goal of a computer vision system and the solution to autonomous operation in an unconstrained working environment. Some of the previous classification techniques can be extended over multiple data sets and over data from multiple sensors (image fusion) to provide high-level decision theoretic approaches toward automated tasking. Interactive control and decision-making functions are often included in these high-level algorithms that are typically implemented on special-purpose, high-speed processors.

## Ocean Engineering Applications

Many applications exist for computer vision techniques at all levels in the undersea environment. Of the many techniques that have been implemented, a few examples are presented.

### Search and Survey

The high image acquisition rate available with laser line scan (LLS) or LIDAR systems is well suited to coverage of broad areas of the seafloor, which is often necessary in search and survey missions. The long-range viewing capability of these systems is also beneficial for detection, classification, and localization (DCL) in poor visibility conditions. Automated processing of large databases produced in these searches is highly desirable for identification of candidate search objects.

LLS imaging systems have been developed for rapid, broad area search and survey [21, 22]. One of these systems was instrumental in the search and recovery efforts for the TWA Flight 800 wreckage off the coast of New York [21]. This search mission covered 110,000,000 square feet of seafloor at a rate of 300,000 square feet per hour. The imaging system was towed at a height of approximately 15 feet, yielding 2-D images with resolution of less than a quarter of an inch. Over 700 individual targets were identified in poor visibility conditions. The imaged debris was precisely located at the time of the scan by a DGPS system. This information along with a description and the date and time of its discovery was recorded to a Geographic Information System (GIS) database for information management and archival.

While not an underwater application, an infrared camera has been developed for maritime imaging that allows viewing of small targets that may be unclear on a radar screen or invisible with conventional cameras due to fog or darkness. The Advanced Maritime Infrared Imaging System (AMIRIS) [23] utilizes DSP-based preprocessing techniques on raw infrared image data to produce visually enhanced representations. The system also combines the video and infrared images permitting display of low thermal contrast targets that may be seen in the video but not in the infrared.

## Mine and Ordnance Detection

Although sonar and electromagnetic signature characterization is useful for the in-water, long-range detection and localization of mines, identification is often better served using alternate sensor means. The potential for high resolution and for observation of optical signatures at multiple wavelengths offers some advantage for detection as well as identification for laser-based image formation systems over other methods. Aerial reconnaissance using high power laser imaging systems is of great interest due to the potential high speed, wide area search and independence from natural lighting artifacts. Range limitations from through-water sensing are the primary drawbacks.

Programs exist for characterizing both subsurface and aerial imaging techniques using advanced image formation and sensor fusion methods [24]. Exploitation of spectral characteristics of the environment in both active and passive imaging modes have been extensively investigated, as well as different system configurations. The LLS systems are adaptable to multiple wavelength operation [25]. For in-water use, quasi-coherent image formation methods are being used to allow contouring of features with LLS-type imaging systems [26]. Other advanced systems are being developed with line-imaged, streak tube detectors operated in a "push broom" fashion [27]. Range or altitude is obtained over the line image as it is swept forward by vehicle motion.

Detection of mines using aerial reconnaissance methods is particularly dependent upon refraction (scattering) at the air-sea interface for systems operating in the visible spectral region. Wave slope, as well as the dynamic nature of the sea surface present particularly difficult problems if images are to be formed. The problem is exacerbated when few pixels cover the target dimensions. Analytical techniques for mine detection and observation via transmarine boundary layer (t-MBL) imaging are based on vision-based techniques for multi-look image formation. Two basic methods are commonly employed. One method is the *restore-before-detect* approach where the entire image is approximately reconstructed using *a priori* knowledge of the surface topography. The second approach employs a *detect-before-restore* where only the estimated target pixels are reconstructed after detection of possible targets. The latter is more subject to environmental problems such as wind-driven chop breakup of the target. Examples of algorithmic approaches for restoration of t-MBL images can be found in [28, 29].

## Object Recognition and Tracking

Object recognition and tracking are important to numerous applications including unobtrusive identification or monitoring, inspection, vehicle control, or accurate computation of correspondences through time-varying sequences. Applications may be commercial, military, or scientific, and consequently have a variety of specialized requirements for false alarm rate, accuracy, computational speed, complexity, and power consumption.

Two scientific studies use automated techniques to identify plankton in conventional video images. In one system, active contour models (snakes) track bioluminescent plankton and identify the organism based on the kinetics of their emission [30, 31]. Video images are preprocessed using the dilate operator to blur features, thus providing longer-range attraction for the snakes, and using the erode operator to sharpen features in a copy of the original image. The copy is subtracted from the dilated image to produce an edge image. Snakes placed on this image seek the outlines of each emission and track them through sequential ISIT video frames, allowing features such as duration, size, peak intensity, rate of change, and location to be recorded and used in the species identification. A decision-theoretic approach, based on certainty theory [32], is used to classify the plankton with nearly expert level accuracy. In all, this automated approach allows researchers to process the overwhelming amount of data needed to characterize the spatial and temporal relationships of the plankton. Methods such as these can be generalized to other video analysis tasks in underwater imagery.

In another system, large numbers of plankton images detected by a towed underwater video system are classified by pattern recognition [33]. Invariant moments features and Fourier boundary descriptors are combined with grayscale morphological granulometries to form a feature vector containing shape and texture information. Recognition is achieved using a decision-theoretic approach based on learning vector quantization (LVQ) classifiers to sort the identified plankton into one of six categories. The LVQ uses a parallel-training algorithm implemented on neural network architecture to optimize codebook vectors.

Neural network structures are particularly useful pattern analysis tools "because they provide an interpretational 'glue' or framework which links together a variety of methods and viewpoints, thus constituting a generic form of pattern analysis methods" [34]. Another application uses neural network classifiers to assort buried objects from acoustic image data [35].

An automated recognition system employed for visual inspection of pipelines receives video input from a CCD camera mounted on a ROV and processes the data in real time to locate the pipe's profile [36]. A vertical edge detection algorithm is applied to the image where regions of interest are selected and a contour extraction algorithm is used to locate the pipe. A least-squares algorithm approximates the straight lines forming the pipe's profile. These approximations can be used as tracking features to direct vehicle navigation in automated inspection [37].

## Image Analysis and Visualization

In an attempt to learn more about the ocean and the life contained therein, scientific research often involves monitoring various biological processes and events. Special-purpose imaging systems are often developed to capture and analyze these events.

The distribution and characterization of marine snow aggregates at a particular location in the ocean is one scientific area of interest. Due to the fragility of marine snow aggregates, noncontact optical techniques are necessary for determining particle shape, size, and distribution through the water column [38, 39]. However, the use of conventional methods to photograph an optically defined volume of water have resulted in size ambiguity, reduced accuracy of the results due to noise, and extensive post-processing prior to data analysis [40]. One system that has been developed to record marine snow is the Shadowed Image Particle Profiling and Evaluation Recorder (SIPPER) [41]. SIPPER uses two high-speed line scan cameras and two collimated, laser-generated light sheets to record each particle from two orthogonal directions. Particles are imaged within a sampling tube at a distance of 96 mm and at a sample rate of 7500 lines/second; resolution of the two cameras is 4096 and 2048 pixels per line, respectively. To ensure proper scaling of the images, particle velocity through the tube is measured. Once the data is collected, 3-D reconstruction methods can be applied to combine the 2-D images, thereby producing a visual representation of the desired volumetric data.

Instantaneous 3-D images of marine snow aggregates have been recorded using a 3-D laser scanner [42, 43]. The scanner produces accurate, high-resolution (200 × 200 pixel) surface maps at a nominal distance of 7 cm and at a rate of 20 images per second. Filtering and segmentation methods are used to extract particle range, surface area, volume (estimated), and direction and speed of motion from the collected images. The example in Fig. 4.19 illustrates a pseudo-color rendering of segmented marine snow particles, where color is selected based on the range coordinate (purple is farthest from the detector and red is closest). The relative size and shape estimate of each particle is used to construct the rendered version of the 3-D data. The particles shown ranged in size from 0.5 to 2.0 mm. Using this information, the density of the aggregates can be statistically estimated over a larger area and correlated with other local information to determine the importance of these particles as a food source to bottom-dwelling fish.

A similar 3-D laser scanner was developed to accurately measure fish size, shape, motion, and population density in a given volume of water [43]. An example sequence was recorded to track the motion of fish in an aquarium. The 200 × 200 pixel images were recorded at 20 frames per second at a standoff distance of 1 m. The acquisition rate was selected to minimize motion blur due to fish and platform movement. Sixty frames (3 sec) of the sequence were analyzed to determine the fish coordinates and velocity vectors. Four frames, at times 0.35, 1.15, 1.75, and 2.75, are shown in Fig. 4.20, and the 3-D motion paths for each fish throughout the sequence are shown in Fig. 4.21.

## Image Compression

Real-time display and processing of images and the transmission of images over network-based and low-bandwidth communication channels in ocean applications has increased the need for image compression methods. Many of the standards in use suffer from various impairments that can be observed at high magnification and low contrast; these include block-like appearance, noise sensitivity, and artifact production.

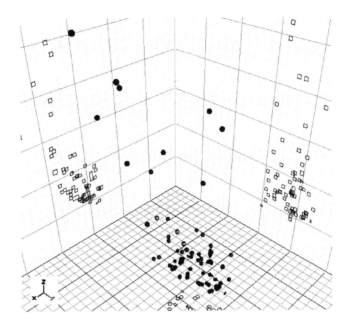

**FIGURE 4.19**   Render of segmented marine snow range map recorded by a 3-D laser scanner.

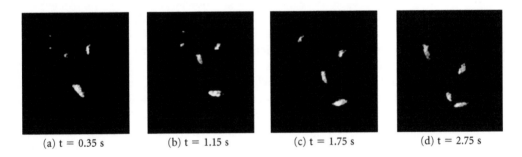

(a) t = 0.35 s          (b) t = 1.15 s          (c) t = 1.75 s          (d) t = 2.75 s

**FIGURE 4.20**   Selected range maps from an image sequence recording fish motion. (International Ocean Systems Design, with permission.)

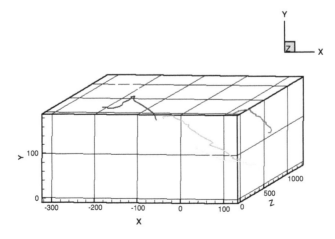

**FIGURE 4.21**   3-D motion paths of the scanned fish.

Research in this area is emphasizing higher compression ratios to achieve lower transmission bandwidth, fidelity of transmission for both machine or man in-the-loop applications, and automated recognition and feature identification within the compressed data format (e.g., [44]).

Tailoring existing methods or designing more advanced compressive strategies for specific underwater imaging applications allows lower bandwidth acoustic transmission of images with minimal degradation. A system using JPEG and EPIC standard data compression algorithms has been developed to transmit LLS, still camera, and sonar images over an acoustic modem at rates up to 10k bps at ranges exceeding 2000 m [45]. Input image sizes range from 64 kb for low-end, and 8-bit CCD images to 8 MB for 16-bit electronic still camera or LLS images. Without compression, an 8 MB image would take just under two hours for transmission over a 10k bps link. Typically, the amount of compression can be adjusted with algorithmic parameters and is determined by how much degradation the application will allow. For example, little or no degradation is allowable for data archiving, whereas only an approximation of shape is needed for some object classification. For the purpose of this application, some loss was allowed as long as man-made objects on the ocean seafloor could be uniquely identified. Experimental results showed that a JPEG compression ratio (CR) of 10:1 preserved all features, while a CR of 60:1 caused severe blocking effects prohibiting identification. However, even with a small, 10:1 reduction, an 8 MB image would take only a little more than ten minutes for transmission over a 10k bps acoustic link. The EPIC algorithm showed similar results, achieving a slightly higher compression ratio but with less reliability due in the presence of transmission bit errors. Since JPEG operates on small data blocks that contain synchronization and restart markers, this method is less sensitive to bit errors in the data link. Furthermore, JPEG utilizes little memory, is a standard (portable) format, compresses images of variable size, and allows for immediate viewing of low resolution images without decompression. JPEG, however, is meant for human visualization. Quantization levels introduced by this algorithm at high CRs pose problems for automated analysis. In the case of transmitting format-specific data (e.g., sonar), *a priori* knowledge of the data types, formats, and resolutions can be used with adaptive filtering techniques to achieve higher CRs over standard lossless methods.

In a similar application, a custom algorithm called BLAST (Blurring, Local Averaging, Sharpening, and Thresholding) was developed specifically for compressing underwater images [46–48]. Compression is accomplished via space-variant quantization of small, locally averaged neighborhoods. Advantages of this method include a more uniform spatial error distribution, local operations allowing efficient sequential processing or very fast processing on parallel architectures, and efficient decompression requiring only lookup table and convolution operations. Additionally, using an elliptical encoding block (EBLAST) yields improved performance. Elliptical blocks reduce disruption of oblique straight lines with increasing block size, distribute more information near the periphery of an encoding block to its neighboring blocks, and spatially interleave encoding block information during decompression—contributing to enhanced compressibility and a more visually attractive image. CRs up to 280:1 have been achieved using this algorithm with good results. Figure 4.22 compares EPIC, BLAST, and EBLAST image compression results of a typical underwater image. Aerial images of underwater targets are shown in Fig. 4.23 using the BLAST algorithm at 125:1 compression ratio.

## Photogrammetry

Measurement of object size, area, volume, and shape, as well as distance between objects, object density, and movements are common requirements in scientific and industrial offshore applications. Photogrammetric systems are useful for making these measurements. One such system produces graphic overlays indicating range, area of coverage, and other information on video images as they are displayed [13]. The system uses one of two different laser configurations. In the simplest configuration, two parallel laser projections are aligned in a direction along the optical axis of a video camera. A third beam is oriented along an angle with respect to the parallel beams and is adjusted to be within the desired field of view at the maximum viewing distance. The video image containing the laser reflections is captured and a simple thresholding algorithm is used to segment the lasers from the background. Since frame-to-frame variation in laser position is small, position information from the previous frame serves as a starting

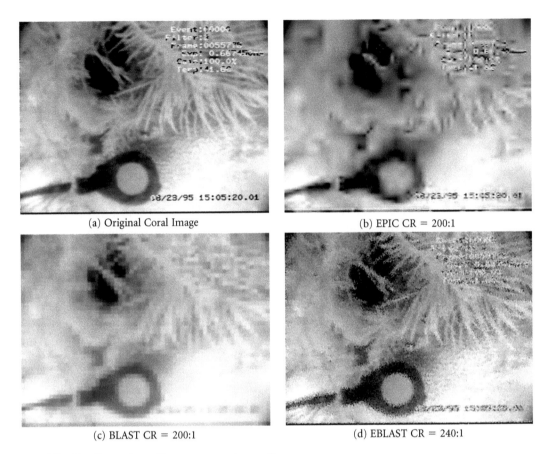

(a) Original Coral Image

(b) EPIC CR = 200:1

(c) BLAST CR = 200:1

(d) EBLAST CR = 240:1

**FIGURE 4.22** Comparison of image compression results.

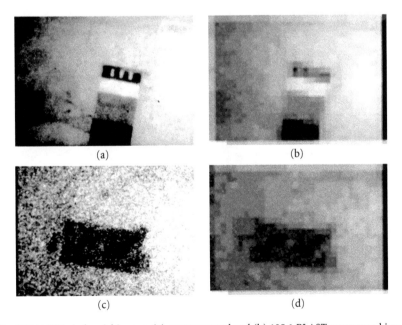

(a)

(b)

(c)

(d)

**FIGURE 4.23** 256 × 180 pixel aerial images: (a) contrast panel and (b) 125:1 BLAST compressed image, (c) target and (d) 125:1 BLAST compressed image. Source images provided by NSWC Coastal Systems Station.

point to minimize search time for segmentation. With a constant magnification video camera lens, the distance to a planar surface or point is obtained from the relative position of the reflections produced by the angled beam with respect to one of the parallel beams. The two parallel beams establish a size reference from which the dimensions of the field of view can be established.

In another configuration, four (or more) laser projections are arranged parallel to the video camera axis [49]. For a fixed camera magnification, the parallel configuration establishes the location and tilt of a planar reflecting surface. Additional laser projections improve the accuracy of the estimates and can be useful for obtaining the same information when a zoom lens is used with the video camera. Real-time video systems providing spatially variant grid map overlays [50, 51] have been demonstrated in the laboratory, and initial studies have been completed for benthic or other surface imaging applications [52].

## Reconstruction and Analysis

Images of an object or scene recorded at different locations are often combined to form a single 2-D or 3-D representation. This representation can be used to assist humans or to augment algorithm performance in local search and exploration efforts, object localization and recognition, and navigation. 3-D representations can additionally be used in the construction of virtual reality worlds. The formation process covering a broad work space is referred to as mosaicking or scene reconstruction, whereas representation of a single object is referred to as object reconstruction.

Mosaics of underwater images are often formed using video images [53–56]. A stereo-video approach produces both 2-D and 3-D mosaics [57]. The disparate images from each camera allow the derivation of object shape and size information. Figure 4.24a shows the left and right video images taken as the system was moved across the scene in clear water. Figure 4.24b shows the 2-D mosaic produced from one camera view, and Fig. 4.24c shows the 3-D mosaic reconstruction. Reconstruction in turbid water is often inadequate without the addition of other visual cues (i.e., shape from shading [58, 59], shape from motion [60], etc.) or additional extended range image data.

Extended range, 3-D acoustic images are used in another reconstruction application, as shown in Figs 4.25(a-e) [61]. In this technique, the raw 3-D acoustic data is preprocessed to reduce noise introduced

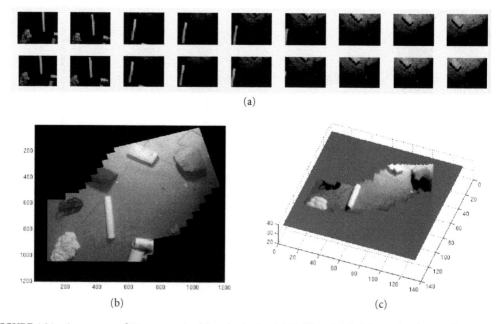

(a)

(b)                                                                    (c)

**FIGURE 4.24**  A sequence of 10 stereo pairs (a) and calculated 2-D (b) and 3-D (c) mosaics. (From Khamene, A. and Negahdaripour, S., *Proc. Oceans '99*, CD, Seattle, WA, September 1999. With permission from IEEE.)

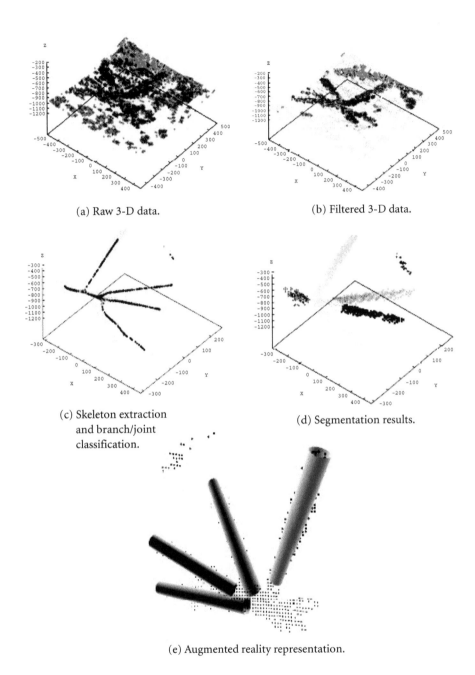

(a) Raw 3-D data.

(b) Filtered 3-D data.

(c) Skeleton extraction
and branch/joint
classification.

(d) Segmentation results.

(e) Augmented reality representation.

**FIGURE 4.25** Object reconstruction using 3-D acoustic images (a-e). (From Giannitrapani, R., Trucco, A, and Murino, V., *Proc. Oceans '99*, CD, Seattle, WA, September 1999. With permission from IEEE.)

by the environment and image formation process (Figs. 4.25a-b). As the first step in segmentation, an erosion operation is combined with a pixel-wise "shape" classification (straight-line or branched line) to produce a 3-D skeletal representation (Fig. 4.25c). Using this as a template, a connected component algorithm is applied to all image pixels to produce the segmented result (Fig. 4.25d). The geometric shape of the segmented regions are analyzed and classified as either "pipe-like" or "not-pipe-like" elements, and then synthetic representations are generated (Fig. 4.25e).

## Navigational Control

Automated methods for piloting underwater vehicles can be useful to the scientific, commercial, and military community. Periodic sampling, inspection and repair of subsea structures, tracking specific features, maintaining a position or specific distance from a desired surface, and maneuvering into docking platforms are among the many tasks that can be automated. For example, optical techniques have been established for ROV station keeping [62, 63] and ROV/AUV navigation [64, 65], where motion of the vehicle is computed based on visual cues from the texture of the seafloor. Other optical methods for ROV station keeping derive motion directly from the image sequences (motion-by-video) [54, 66, 67]. In one such application [54], a 2-D image mosaic of the operational area is constructed and updated during navigation. Frame-to-frame motion estimation is used in comparison to the mosaic map for position estimation of the vehicle. Video data captured at $128 \times 120$ pixel resolution is compressed using block averaging to $64 \times 60$ pixels to reduce overhead in the algorithm and thereby achieve real-time processing.

Acoustic methods have also been developed and tested for AUV navigation and docking. In one application [68], fast DSP processors are used to optimize performance allowing multiple AUVs to communicate with each other and to reference moorings during survey operations.

## Image Sensor Fusion

The notion of sensor fusion holds various interpretations. In its simplest interpretation, multiple sensors of the same type can be used simultaneously to provide redundant sensing. If data from one sensor appears aberrant or if the sensor fails altogether, that sensor can be ignored and data from one or more other sensors can be used instead. Another interpretation is the idea of equipping an undersea vehicle with various, disparate sensors in order to collect a variety of information. For example, the combination of acoustic, optical, electromagnetic, chemical, and GIS data is required for many undersea vehicle missions. The main objective is to combine the data in an effective manner for either human or machine interpretation. "Data mining" and vector quantization methods [19] provide a framework for storing and retrieving such information. Perhaps the most useful interpretation for computer vision applications is to use disparate sensors, with overlapping fields of view, to look at the same information—with the idea that if environmental conditions prohibit or degrade the performance of one type of sensor, the other will not be affected and can be used instead. In conditions favoring the use of both, the data can be compared and filtered accordingly (averaged, etc.). This method is complementary and robust. As an example, an underwater vehicle may be equipped with two cameras configured for stereo imaging and a 3-D laser line scanner for collecting range information. In favorable conditions, range data from both systems can be combined to provide accurate, 3-D range data. However, in turbid conditions or cases when a longer viewing distance is needed, just the extended range data can be used. An extension of this, and perhaps the most complete interpretation, includes additional cues to further enhance the desired information or provide other useful information. In the previous example, image gradient information (e.g., surface texture) can be added to assist in deriving the range or to provide additional information that may be required for recognition or interpretation. In each of these cases, methods are needed for detecting failures, combining the data quickly and intelligently, and configuring the systems such that there is no interference between sensors. In the latter case, it may be desirable to perform laser scans at a rate less than 30 fps so that the camera has unobstructed (no laser) views. Real-time sensor fusion techniques combining sonar and video information can be found in [69].

In some applications it may be necessary to integrate diverse data types into a pictorial display where image attributes such as color (hue), intensity, coordinate location in (x, y, z) space, pixel density, texture, and pattern can be used to represent multidimensional data. Advancements in this field are highly application dependent, but are commonly observed in virtual reality presentations where the observer's location and orientation in space govern the images presented to the human visual system. Generation of virtual environments derived from actual image data may also be considered a form of sensor fusion. Once the environment is generated, algorithms can be implemented for path planning and object detection and avoidance. Developments in creating virtual environments are predicated upon large image databases, rapid processing and high-resolution display.

# References

1. Fu, K. S., Gonzalez, R. C., and Lee, C. S. G., *Robotics: Control, Sensing, Vision, and Intelligence*, McGraw-Hill, New York, 1987.
2. Marr, D., *Vision*, W. H. Freeman and Co., San Francisco, CA, 1982.
3. Ballard, D. H. and Brown, C. M., *Computer Vision*, Prentice-Hall, Englewood Cliffs, NJ, 1982.
4. Weeks, A. R., *Fundamentals of Electronic Image Processing*, SPIE/IEEE Press, 1996.
5. Russ, J. C., *The Image Processing Handbook*, 3rd ed., CRC Press, Boca Raton, FL, 1999.
6. Tyler, J. E., The Secchi Disc, *Limn. and Ocean.*, 13, 1, 1, 1968.
7. Gordon, H. R. and Wouters, A. W., Some relationships between Secchi depth and inherent optical properties of natural waters, *Applied Optics*, 17, 21, 3341, 1978.
8. Prieur, L. and Sathyendranath, S., An optical classification of coastal and oceanic waters based on the specific spectral absorption curves of phytoplankton pigments, dissolved organic matter, and other particulate materials, *Limn. and Ocean.*, 26, 4, 671, 1981.
9. Jerlov, N. G., *Marine Optics*, Elsevier Oceanography Series, Elsevier, Amsterdam, 1976.
10. Morel, A., Optical properties of pure water and seawater, *Opt. Aspects of Oceano.*, (N. Jerlov and E. S. Nielsen, eds.), 1, Academic Press, New York, 1974.
11. Fournier, G. R. and Jonasz, M., Computer-based underwater imaging analysis, *Proc. SPIE, Airborne and In-Water Underwater Imaging*, 3761, 62, Denver, CO, July, 1999.
12. Hodara, H., Refractive index variations in seawater, *AGARD Lecture Series 61 on Optics in the Sea*, 2.2–1, Neuilly Sur Seine, France, 1973.
13. Caimi, F. M., Blatt, J. H., Grossman, B. G., Smith, D., Hooker, J., Kocak, D. M., and Gonzalez, F., Advanced underwater laser systems for ranging, size estimation, and profiling, *MTS Journal*, 27, 1, 31, Spring, 1993.
14. Karp, S., Gagliardi, R. M., Moran, S. E., and Stotts, L. B., *Optical Channels*, Plenum Press, New York, 1988.
15. Horn, B., *Robot Vision*, MIT Press/McGraw-Hill, New York, 1986.
16. Kass, M., Witkin, A., and Terzopoulos, D., Snakes: Active contour models, *Proc. 1st ICCV*, 259, 1987.
17. Matheron, G., *Random Sets and Integral Geometry*, John Wiley & Sons, New York, 1975.
18. Burrus, C. S., Gopinath, R. A., and Guo, H., *Introduction to Wavelets and Wavelet Transforms*, Prentice-Hall, Englewood Cliffs, NJ, 1998.
19. Key, G., Schmalz, M. S., and Caimi, F. M., Performance analysis of tabular nearest neighbor encoding algorithm for joint compression and ATR over compressed imagery, *Proc. SPIE Mathematics of Data/Image Coding, Compression, and Encryption II*, 3814, 127, Denver, CO, July, 1999.
20. Barr, A. and Witkin, A., Topics in Physically-Based Modeling, ACM SIGGRAPH '89, Course #30, Boston, MA, July, 1989.
21. Saade, E. J. and Carey, D., Laser line scan operations during the TWA Flight 800 search effort, *Proc. Oceans '96*, CD, Fort Lauderdale, FL, September, 1996.
22. Leatham, J. and Coles, B. W., Use of laser sensors for search & survey, *Proc. UI '93*, 171, January, 1993.
23. AMIRIS Product Design Specification, p/n 105419, Rev. B, eMerge Vision Inc., July 8, 1997.
24. Strand, M. P., Underwater electro-optical system for mine identification, *Proc. SPIE Detection Technologies for Mines and Minelike Targets*, 2496, 487, June, 1995.
25. Strand, M. P., Coles, B. W., Nevis, A. J., and Regan, R. F., Laser line-scan fluorescence and multi-spectral imaging of coral reef environments, *Proc. SPIE Ocean Optics XIII*, 2963, 790, February, 1997.
26. Mullen, L., Contarino, M., Laux, A., Concannon, B., Davis, J., Strand, M., and Coles, B. W., Modulated laser line scanner for enhanced underwater imaging, *Proc. SPIE Airborne and In-Water Underwater Imaging*, 3761, 2, Denver, CO, July, 1999.
27. McLean, J. W., High resolution 3-D underwater imaging, *Proc. SPIE Airborne and In-Water Underwater Imaging*, 3761, 10, Denver, CO, July, 1999.

28. Schmalz, M., Reconstruction of submerged targets from multiple airborne images. 1. Theory and error analysis, *Proc. SPIE Airborne and In-Water Underwater Imaging*, 3761, 83, Denver, CO, July, 1999.

29. Schmalz, M., Reconstruction of submerged targets from multiple airborne images. 2. Algorithms and simulation results, *Proc. SPIE Airborne and In-Water Underwater Imaging*, 3761, 95, Denver, CO, July, 1999.

30. Kocak, D. M., Lobo, N. D. V., and Widder, E. A., Computer vision techniques for quantifying, tracking and identifying bioluminescent plankton, *IEEE J. Oceanic Eng.*, 24, 1, 81, January, 1999.

31. Kocak, D. M., Quantifying, tracking, and identifying bioluminescent plankton using active contour models, Thesis (M.S.), University of Central Florida, Orlando, 1997.

32. Lugar, G. F. and Stubblefield, W. A., *Artificial Intelligence and the Design of Expert Systems*, Benjamin/Cummings, Redwood City, CA, 1989.

33. Tang, X. and Stewart, W. K., Plankton image classification using novel parallel-training learning vector quantization network, *Proc. Oceans '96*, 3, 1227, Fort Lauderdale, FL, September, 1996.

34. Lowe, D., What have neural networks to offer statistical pattern processing?, *Proc. SPIE Adaptive Signal Processing*, 1565, 460, Great Malvern, Worcesterhire, UK.

35. Granara, M., Pescetto, A., Repetto, F., Tacconi, G., and Trucco, A., Statistical and neural techniques to buried object detection and classification, *Proc. Oceans '98*, 3, 1269, Nice, France, September, 1998.

36. Conte, G., Zanoli, S., Perdon, A. M., Tascini, G., and Zingaretti, P., Automatic analysis of visual data in submarine pipeline, *Proc. Oceans '96*, 1213, Fort Lauderdale, FL, September, 1996.

37. Zanoli, S. M. and Zingaretti, P., Underwater imaging system to support ROV guidance, *Proc. Oceans '98*, 1, 56, Nice, France, September, 1998.

38. Honjo, S., Doherty, K. W., Agrawal, Y. C., and Asper, V. L., Direct optical assessment of large amorphous aggregates (marine snow) in the deep ocean, *Deep-Sea Research*, 31, 67, 1984.

39. Carder, K. L., Costello, D., and Steward, R. G., State-of-the-art instrumentation for measuring ocean aggregates, ONR Aggregates Dynamics in the Sea Workshop Report, *American Institute of Biological Sciences*, 131, 1986.

40. Carder, K. L. and Costello, D. K., Optical effects of large particles, *Ocean Optics*, Oxford Monographs on Geology and Geophysics No. 25, Oxford University Press, Oxford, 243, 1994.

41. Samson, S., Langebrake, L., Lembke, C., and Patten, J., Design and initial results of high-resolution Shadowed Image Particle Profiling and Evaluation Recorder, *Proc. Oceans '99*, CD, Seattle, WA, September, 1999.

42. Caimi, F. M., Kocak, D. M., and Asper, V. L., Developments in laser-line scanned undersea surface mapping and image analysis systems for scientific applications, *Proc. Oceans '96*, Supp., 75, Fort Lauderdale, FL, September, 1996.

43. Kocak, D. M. and Caimi, F. M., Surface metrology & 3-D imaging using laser line scanners, *Int. Ocean Systems Design*, 3, 4, 4, July/August, 1999.

44. Negahdaripour, S., Xu, X., Khamene, A., and Awan, Z., 3-D motion and depth estimation from sea-floor images for mosaic-based station-keeping and navigation of ROVs/AUVs, *Proc. Workshop AUV*, Cambridge, MA, August, 1998.

45. Eastwood, R. L., Freitag, L. E., and Catipovic, J. A., Compression techniques for improving underwater acoustic transmission of images and data, *Proc. Oceans '96*, CD, Fort Lauderdale, FL, September, 1996.

46. Schmalz, M. S., Ritter, G. X., and Caimi, F. M., Performance evaluation of data compression transforms for underwater imaging and object recognition, *Proc. Oceans '97*, 1075, Halifax, Nova Scotia, October, 1997.

47. Caimi, F. M., Kocak, D. M., Schmalz, M. S., and Ritter, G. X., Comparison and development of compression algorithms for AUV telemetry: recent advancements, *Proc. MTS Ocean Com. Conf.*, 2, 1139, Baltimore, MD, November, 1998.

48. Kocak, D. M. and Caimi, F. M., DSP hardware implementation of transform-based compression algorithm for AUV telemetry: recent advancements, *Proc. Oceans '98*, 3, 1624, Nice, France, September, 1998.

49. Caimi, F. M. and Tusting, R. F., A parallel beam array size estimation system for underwater use, HBOI Internal Report, 1988.

50. Wakefield, W. W. and Genin, A., The use of a Canadian (perspective) grid in deep-sea photography, *Deep-Sea Research*, 34, 3, 469, 1987.

51. Tusting, R. F. and Davis, D. L., Laser systems and structured illumination for quantitative undersea imaging, *MTS Journal*, 26, 4, 5, 1992.

52. Devin, M. G., A laser-based method to quantify the species diversity of the deep-sea benthic megafauna, Thesis (M.Sc.), Florida Institute of Technology, Melbourne, FL, 1998.

53. Marks, R. L., Rock, S. M., and Lee, M. J., Real-time video mosaicking of the ocean floor, *IEEE J. Oceanic Eng.*, 20, 3, 229, July, 1995.

54. Negahdaripour, S. and Xu, X., Direct motion estimation from sea floor images for navigation and video mosaicking, *Proc. Oceans '97*, CD, Halifax, Nova Scotia, October, 1997.

55. Singh, H., Howland, J., and Yoerger, D., Quantitative photomosaicking of underwater imagery, *Proc. Oceans '98*, 1, 263, Nice, France, September, 1998.

56. Gracias, N. and Santos-Victor, J., Automatic mosaic creation of the ocean floor, *Proc. Oceans '98*, 1, 257, Nice, France, September, 1998.

57. Khamene, A. and Negahdaripour, S., Building 3-D elevation maps of sea-floor scenes from underwater stereo images, *Proc. Oceans '99*, CD, Seattle, WA, September, 1999.

58. Zhang, S. and Negahdaripour, S., Recovery of 3D depth map from image shading for underwater applications, *Proc. Oceans '97*, 1, 618, Halifax, Nova Scotia, October, 1997.

59. Yu, C. H. and Negahdaripour, S., Orientation and distance recovery of Lambertian planar surfaces in light-attenuating media from optical images, *J. Opt. Soc. Am. A*, 8, 1, 217, January, 1991.

60. Negahdaripour, S., Revised interpretation of optical flow; integration of radiometric and geometric cues in optical flow, *IEEE Trans. PAMI*, 20, 9, September, 1998.

61. Giannitrapani, R., Trucco, A., and Murino, V., Segmentation of underwater 3D acoustical images for augmented and virtual reality applications, *Proc. Oceans '99*, CD, Seattle, WA, September, 1999.

62. Negahdaripour, S. and Yu, C. H., Passive optical sensing for near-bottom stationkeeping, *Proc. Oceans '90*, Washington, DC, September, 1990.

63. Marks, R. L., Wang, H. H., Lee, M. J., and Rock, S. M., Automatic visual station keeping of an underwater robot, *Proc. Oceans '94*, Brest, France, 1994.

64. Huster, A., Fleischer, S. D., and Rock, S. M., Demonstration of a vision-based dead-reckoning system for navigation of an underwater vehicle, *Proc. Oceans '98*, 1, 326, Nice, France, September, 1998.

65. Branca, A., Stella, E., and Distante, A., Autonomous navigation of underwater vehicles, *Proc. Oceans '98*, 1, 61, Nice, France, September, 1998.

66. Jin, L., Xu, X., Negahdaripour, S., Tsukamoto, C., and Yuh, J., A real-time based stationkeeping system for underwater robotics applications, *Proc. Oceans '96*, 3, 1076, Fort Lauderdale, FL, September, 1996.

67. Xu, X., Negahdaripour, S., Automatic optical station keeping and navigation of an ROV; sea trial experiments, *Proc. Oceans '99*, CD, Seattle, WA, September, 1999.

68. Singh, H., Catipovic, J., Eastwood, R., Freitag, L., Henriksen, H., Hover, F., Yoerger, D., Bellingham, J. G., and Moran, B. A., An integrated approach to multiple AUV communications, navigation and docking, *Proc. Oceans '96*, 1, 59, Fort Lauderdale, FL, September, 1996.

69. Lagstad, P. and Auran, P. G., Real time sensor fusion for autonomous underwater imaging in 3D, *Proc. Oceans '96*, 3, 1330, Fort Lauderdale, FL, September, 1996.

# 5

# Fiber Optics in Oceanographic Applications

Frank M. Caimi
*Florida Institute of Technology*

Syed H. Murshid
*Florida Institute of Technology*

Barry G. Grossman
*Florida Institute of Technology*

Clay Kirkendall
*Naval Research Laboratory*

Tony Dandridge
*Naval Research Laboratory*

George Wilkins

## 5.1 Overview of Fiber Optics in Oceanographic Applications

*Frank M. Caimi*

The transmission of signals over fiber optic media has now become a trend of major economic importance. This is due mainly to the many advantages offered over conventional transmission methods using metallic conductors. An ability to carry high bandwidth signals and to eliminate electromagnetic interference in high signal environments are perhaps the most significant. In underwater applications, the use of fiber optics for long-haul telemetry in transoceanic telephone transmission is well-known; however, fiber optics is also of great importance in sensing, as well as in vehicle control and communications. Much of the theory relating to fiber optic transmission was developed decades ago, and recent advancements have been primarily in cabling, networking, sensor development, and specialized connector and splice designs. With these advancements have come reductions in component costs, and resulting commercial opportunities.

The success of the technology derives from developments in the early 1970s when fiber transmission losses less than 20db/km were achieved by Corning. At that time semiconductor diode sources were in developmental stages providing lifetimes insufficient for telemetry use. Within 10 years, these problems were overcome and sources with lifetimes greater than $10^6$ hours became available.

Initially, single fibers were difficult to excite and bundled single fibers were proposed to alleviate the problems associated with low brightness, poorly collimated sources. As technology improved, better-collimated, high power sources were developed to improve power transfer to the active or "core" region of an individual fiber. This trend eventually provided for the cost effective use of large core (62.5 Mm diameter) multimode fibers and eventually small core (<10 micrometer) single-mode fibers for long haul, low loss signal transmission. The advantage of single-mode fibers is their low loss (<1 db/km) and high bandwidth capability—100 Ghz-km in comparison to 1 Ghz-km for multimode fibers.

Today, fiber bundles are used for power delivery rather than communications; e.g., for medical treatment and cauterization, area lighting, and a host of industrial applications. Single fibers are now used for short- and long-haul communications including transoceanic telemetry, Internet connectivity, Remotely Operated Vehicle control and video linkage, chemical, acoustic, and EM sensing, pressure hull health assessment, etc. The applications have become so varied that fibers are available in a variety of core sizes, compositions, and constructions, each with specific advantages and requirements. Knowledge of the basic theory of fiber optic waveguides is necessary for understanding the potential uses, limitations, and day-to-day handling and preparation considerations.

## Physical Basis for Optical Fiber Transmission

Light propagation in optical fibers is essentially a manifestation of the familiar Snell's Law for a cylindrical geometry. Willebrord Snell (1591–1626) was the first to explain the "Law of Refraction," which requires that a photon striking a surface at an angle $\theta_1$ with respect to the surface normal imposes a specific requirement on the trajectory of the photon leaving the surface. The law was first published by French philosopher Descartes (1596–1650), who became aware of Snell's work. Snell's law may be derived from the momentum-matching condition for photons at a dielectric interface, as follows:

$$p_1 = h\upsilon\sqrt{\mu\varepsilon_1}\sin(\phi_1) = p_2 = h\upsilon\sqrt{\mu\varepsilon_2}\sin(\phi_2)$$

here,

$$\sqrt{\varepsilon_1} \equiv n_1, \quad \sqrt{\varepsilon_2} \equiv n_2$$
$$n_1\sin(\phi_1) = n_2\sin(\phi_2)$$

(5.1)

Here, the refractive indices in both media are $n_1$ and $n_2$ corresponding to the dielectric constants $\varepsilon_1$ and $\varepsilon_2$, and momentum $p_1$ and $p_2$ in medium 1 and 2, respectively. The quantity $\sqrt{\mu\varepsilon_1}$ is the reciprocal of the energy propagation speed in medium 1, indicating that the speed of propagation is reduced to less than the speed of light (i.e., $c/n_1$) when $n_1$ is greater than 1. In the case where the refractive index in medium 1 is greater than that of medium 2, we have a condition where the right-hand side of Eq. (5.1) is a maximum when $\phi_2$ is 90°. This condition is shown in Fig. 5.1 and suggests that $\phi_1$ reaches a maximum angle beyond which light no longer propagates in medium 2. This so-called "critical angle" is obtained from Eq. (5.1):

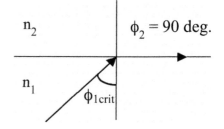

**FIGURE 5.1** Critical Angle Refraction at a dielectric interface.

$$\sin(\phi_{1\,crit}) = \frac{n_2}{n_1} \le 1$$

(5.2)

It is preferable to express the angle in terms of the complementary angle $\theta_1 = 90 - \phi_1$, so that the light propagation angle is measured with respect to the plane of the interface. Light is therefore totally reflected back into medium 2 at angles $0 \leq \theta_1 \leq \cos^{-1} \frac{n_2}{n_1}$. This condition of "total internal reflection" is a common experience for swimmers looking up toward the sky while underwater. At sunset, the sun appears at about a 48° angle from vertical. At angles beyond 48° from vertical, the swimmer sees reflections from within the pool. The critical angle for this case is given by $\phi_{1\text{crit}} = \sin^{-1}(1/1.33) = 48°$, since the refractive index of water is 1.33. In fibers, the medium with the larger index is typically glass with $n = 1.5$.

## Fiber Construction: Step Index and Graded Index Types

In optical fibers, the geometry is cylindrical rather than planar as shown in Fig. 5.2. The fiber is composed of a center high refractive index region or "core" and a surrounding lesser refractive index region or "cladding." The fiber is also coated with additional layers for protection of the cladding/core structures, which, if glass, are particularly sunsceptible to mechanical damage and to water degradation. The situation is as shown in Fig. 5.3. A primary buffer covering the glass cladding is used as a first layer of mechanical protection. The first buffer is usually a UV-cured acrylate coating of nominal 62.5-$\mu$m thickness. The coating is chosen for rapid speed of application since the "drawing speed" in the manufacture of the fiber is a primary cost consideration in production. The outer or secondary buffer is sometimes a gel filling within a loose tube or a "tight buffer" consisting of a tightly bound elastomeric coating chosen for water impermeability, durability, and bonding strength. In some specialty fibers, the buffer coating is metal, such as gold. This fiber construction can be particularly useful for through-hole penetrations where a metal-glass seal is required.

The simplest fiber construction is the *step index* type shown in Fig. 5.2, having uniform refractive index within the core and cladding regions. The refractive index of the cladding ($n_{\text{clad}}$) is only slightly smaller than that of the core ($n_c$) in most communications fibers, suggesting the definition of the fiber refractive index profile height parameter $\Delta$.

$$\Delta \equiv \frac{n_c^2 - n_{\text{clad}}^2}{n_c^2} \equiv \sin^2\{\theta_{\text{crit}}\} \approx \frac{n_c - n_{\text{clad}}}{n_c}, \quad (\Delta \ll 1) \tag{5.3}$$

In this case, when $\Delta$ is much smaller than one, the critical angle is given simply as

$$\theta_c \cong \sqrt{2\Delta} \tag{5.4}$$

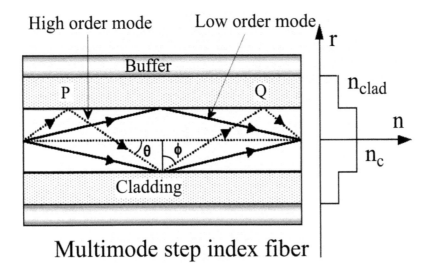

FIGURE 5.2   Cylindrical geometry for optical fiber (step index profile).

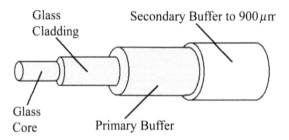

**FIGURE 5.3**   Common fiber construction showing buffer coatings.

Improvements in communications fiber performance can be obtained, in comparison to the step index construction, if the core refractive index is engineered to follow a prescribed (power law) profile according to

$$n_c^2(r) = n_{clad}^2 \{1 - \Delta(r/a)^q\} \tag{5.5}$$

where $r$ is the radial distance from the fiber axis, $a$ is the radial extent of the core, and $q$ is a power law profile parameter. When $q \to \infty$, the profile is essentially a *step index type*, and when $q = 2$, the profile is parabolic. Fibers constructed with finite values of $q$ are members of a class called *graded-index profile fibers*.

The profile affects the propagation of light along different paths within the core. Paths may be at a particular angle $\theta$, and are therefore of different length as light traverses the length of the fiber. Since the speed of propagation is related to the refractive index from Eq. (5.1), photons traveling different paths will arrive at the opposite end of the fiber at different times. The net effect of this *temporal dispersion* is to distort pulses sent along the fiber. In signal processing terminology, the fiber exhibits a *delay spread* or *finite impulse response* that directly affects the *coherence bandwidth* [1] of the fiber optic communication system. The effect of selecting a particular refractive index distribution or profile within the core region is to alter the dispersion characteristics of the fiber. In the case where $q = 2$, the dispersion is greatly reduced in comparison to the step profile fiber.

It can be shown that a hyperbolic secant core index profile defined by

$$n^2(r) = n_c^2 \operatorname{sech}^2 \left\{ \sqrt{2\Delta}\, \frac{r}{a} \right\} \tag{5.6}$$

allows a ray transit time that is equal for all ray trajectories* or modes within the core. It hypothetically produces no temporal difference for rays traveling at different angles with respect to the core longitudinal axis. This profile acts essentially as a graded index lens reducing the optical path length for photons traveling at steep angles. This can be deduced from the following considerations. For any graded index profile, the rays may propagate in a periodic manner according to the equation of motion

$$n(r) \cos \theta_z(r) = n(r = 0) \cos \theta_z(r = 0) \equiv \bar{\beta}$$

$$\text{or} \qquad z(r) = \bar{\beta} \int_0^r \frac{dr}{[n_c^2 - \bar{\beta}^2]^{1/2}} \tag{5.7}$$

This equation is a restatement of Snell's Law for a graded refractory medium having radial distribution. Equation (5.7) is derived easily from the Eikonal Equation, which is obtained directly from Fermat's variational principle $\delta \int (\vec{n} \cdot d\vec{s}) = 0$ [4]. Accordingly, the refractive index and ray trajectory angle at

---

*It is strictly not valid, however, for rays not passing through the fiber meridional axis.

the core longitudinal axis completely determine the ray angle at every point $r$. Stated another way, *the product of the refractive index and the cosine of the propagation angle is a constant ($\bar{\beta}$) independent of position along the ray path.* Unlike the step index fiber, a graded index refractive index profile as defined by Eq. (5.5) will produce a ray path that is quasi-sinusoidal with a maximum radial penetration distance depending on the value of the refractive index and angle $\theta$ at the point $r = 0$. The refractive index at this point, the so-called turning point, is equal to $\bar{\beta}$:

$$n(r_{\max}) = n(r = 0)\cos\theta_z(r = 0) \equiv \bar{\beta} \tag{5.8}$$

This means that rays traveling at steep angles to the fiber axis will travel a greater distance but at a greater speed at increasing radial distance due to the reduced refractive index in those regions. The choice of the power law index $q$ is essential in obtaining the lowest optical path (time or phase) difference between paths (modes). It can be shown that the parabolic index gradient ($q = 2$) is nearly optimal for minimizing modal dispersion in a power law profile, graded index, fiber.

## Numerical Aperture: Light Collection Ability

The ability of the fiber to accept or collect light is a direct consequence of the critical angle relationship since light is accepted for

$$0 \leq \theta \leq \theta_{\text{crit}} = \cos^{-1}\frac{n_{\text{clad}}}{n_c}\sqrt{2\Delta} \tag{5.9}$$

A measure of the light collecting ability is obtained from the numerical aperture *NA*—defined as the sine of the acceptance angle outside the fiber (usually in free space). The expression for *NA* is easily expressed in terms of $\Delta$

$$NA = \sin\theta_{\text{air}} = n_c\sin(\theta_{\text{crit}}) = n_c\{1 - \cos^2(\theta_{\text{crit}})\}^{1/2} = n_c\left\{1 - \frac{n_{\text{clad}}^2}{n_{\text{nc}}^2}\right\}^{1/2} \tag{5.10}$$

$$= n_c\sqrt{2\Delta} = \sqrt{n_c^2 - n_{\text{clad}}^2}$$

$$\text{here,} \quad \frac{\sin^2(\theta_c)}{2} = \Delta$$

Numerical apertures between 0.1 and 0.5 or more are found in commercially made fibers, depending on the fiber geometry and the materials chosen for the fiber core and cladding. Typical values for glass fibers are 0.15 to 0.2 and for plastic are 0.4 to 0.5. The numerical aperture in *graded index* fibers is considered as a local value; that is, it varies according to the radial distance from the fiber longitudinal axis.

The numerical aperture defines the angular acceptance angle of the fiber for guided waves; however, the core diameter can determine the amount of optical flux (number of photons captured) from an finite source. Therefore, larger core fibers are preferred for large or extended sources (such as LEDs) if significant power is to be coupled into the fiber to overcome losses associated with long runs.

## Modal Structure

Certain photon trajectory angles are allowed for total internal reflection according to geometric "ray theory" given by Eqs. (5.2) and (5.4). Photons propagating within these limits are designated as "bound." In terms of electromagnetic wave theory, waves propagating at different angles with respect to the fiber axis within these limits are called "bound modes." Although a distinct similarity exists between the photon trajectory in *ray theory* and the propagation vector from *wave theory*, the former initially appears to allow an infinite number of angles to propagate, while the latter allows only specific waves at distinct propagation angles to transit the fiber [2]. Remarkably though, the ray theory approach can produce a description of the discrete values allowed for $\theta$ if phase shifts along the path are considered. Although this approach for derivation of the allowed modes is beyond the scope of this discussion, it is worthwhile to illustrate the method. For the step index fiber in Fig. 5.2, consider the path from P to Q for a meridional ray—a ray that repeatedly passes through the fiber longitudinal axis. The phase of the light following this path is

$$\Phi = k_0 n_c \frac{2a}{\sin\theta} - \varphi_c \tag{5.11}$$

where $k_0$ is the free space propagation constant, $2a/\sin(\theta)$ is the ray path length, and $\varphi$ is the phase associated with reflection from the core/cladding interface. The latter can be obtained from the Fresnel reflection coefficients and Eq. (5.1) [3]. This phase shift must be equal to that obtained from the path length along the waveguide axis, i.e., $(2a)\cot\theta$, multiplied by the phase shift per unit length $k_{0\,eff}$, i.e.,

$$\Phi_z = [2a\cot\theta]k_{eff} = [2a\cot\theta]\frac{2\pi n_c}{\lambda_{Z\,effective}} = [2a\cot\theta]\frac{2\pi(n_c\cos\theta)}{\lambda_0} \tag{5.12}$$

The allowed modes are obtained by setting Eq. (5.11) equal to Eq. (5.12) and solving for integer multiples of $2\pi$ phase difference. The latter quantity in parentheses ($n_c\cos\theta$) is called the *effective refractive index* ($\bar{n}$), *ray invariant* ($\bar{\beta}$), or *normalized propagation constant* ($\beta/k$). The *waveguide propagation constant* ($\beta = k_0 n_c\cos\theta$) is found in the last factor of Eq. (5.12). Both quantities take on discrete values determined by the solution of the resulting transcendental equation subject to the following limitation

$$k_0 n_{clad} \le k_0 n_c\cos\theta_j \le k_0 n_c$$
$$\text{i.e.,} \quad k_0 n_{clad} \le \beta_j \le k_0 n_c \tag{5.13}$$

which is just a restatement of Eq. (5.9). The index $j$ designates specific values associated with each allowed mode. The complete solution for a cylindrical waveguide is somewhat more complex than Eqs. (5.11) or (5.12) suggest. The geometry of a cylindrical fiber allows more than just the rays along the meridian, as assumed. In fact, rays may also travel in a helical fashion along the fiber axis. These are designated as "skew" rays and account for some of the unique characteristics of cylindrical waveguides. A complete description and derivation can be found elsewhere [3].

What is perhaps most significant is that the value of the propagation constant $\beta$ is bounded to within a small range of values for the case when the cladding and core refractive indices are closely matched. This is a desirable condition for obtaining high bandwidth operation, as the modal dispersion, or time delay between modes, is minimized. The time required for each mode to travel the fiber length is $z\bar{\beta}_j/k_0 c$.

A great amount of effort has been expended in solving the wave equations for various fiber geometries, index profiles, and perturbations in order to find the lowest loss construction, resistance to losses in jacket and cladding materials, loss tolerance to bending, etc.

**Multimode Fibers**

It should be evident from Eq. (5.13) that the number of modes propagating in the core is determined by the difference between the $n_c$ and $n_{clad}$. A limited number of modes are spaced in that interval based on a finite number of solutions to the transcendental modal equation. So, it should be desirable to minimize this difference, i.e., to minimize $\Delta$, in order to minimize the time delay between modes. This is exactly the case, but the modal solutions also depend upon the ratio of the core size to the free space wavelength. This dependency is usually specified by the *V number*, *waveguide parameter*, or *normalized frequency* defined by

$$V \equiv \frac{2\pi a}{\lambda_0}\{n_c^2 - n_{clad}^2\}^{1/2} = (k_0 n_c a)(2\Delta)^{1/2} \tag{5.14}$$

It turns out that the value for $V$ reasonably describes the fiber optical properties in terms of just a few physical parameters. For instance, the number of bound modes is given for the power law profile [Eq. (5.5)] as

$$N_b = \frac{V^2}{2}\left[\frac{q}{q+2}\right] = (kn_c a)^2 \Delta, \qquad V \gg 1 \qquad (5.15)$$

The number of bound modes allowed for a standard communications fiber ($a = 62.5/2 \ \mu m$, $q = 2$, $\Delta = 0.001$, $n_c \sim 1.5$, and $\lambda = 1.3 \ \mu m$) implies $V = 13$ and $N_b \sim 42$. The step-index fiber contains double this number. Clearly, the number of modes is influenced predominantly by the waveguide parameter $V$, since it predominantly specifies the fiber construction and wavelength of operation. To minimize the distortion of a transmitted pulse whose energy is distributed across all modes, it is desirable to restrict core radius $a$ and the fractional refractive index difference $\Delta$ to obtain a minimum number of modes, i.e., one single mode. This is possible for $V < 2.4$ in *step index* fibers and the resulting mode is designated the $HE_{11}$ mode. "*HE*" stands for "hybrid electric" meaning that the field has both an electric and magnetic $z$-axis (fiber axis) field component. This is in contrast to a TE or TM wave that would have zero $E_z$ or $H_z$ field components, respectively, in the direction of propagation. The existence of these longitudinal field components is a direct consequence of the transverse gradient of the refractive index in the fiber core/cladding region producing a nonzero term for $\nabla_t \ln n$ in the longitudinal field solutions obtained from Maxwell's equations [2].

A description of the magnetic and electric fields associated with each mode can be obtained from a solution of Maxwell's equations. The general form for the axial field components is [2]:

$$\begin{bmatrix} E_z \\ H_z \end{bmatrix} = \begin{bmatrix} A\dfrac{J_n(U\frac{r}{a})}{J_n(U)}\left(\begin{matrix}\cos n\varphi_{n \ even} \\ \sin n\varphi_{n \ odd}\end{matrix}\right)e^{-\beta z} \\[3ex] B\dfrac{J_n(U\frac{r}{a})}{J_n(U)}\left(\begin{matrix}-\sin n\varphi_{n \ even} \\ \cos n\varphi_{n \ odd}\end{matrix}\right)e^{-\beta z} \end{bmatrix} \quad (r < a) \qquad (5.16)$$

where the core parameter is defined as

$$U \equiv a(k_0^2 n_c^2 - \beta^2)^{1/2}, \qquad (5.17)$$

and $n$ is the order of the Bessel function of the first kind J, and $\varphi$ is the azimuthal coordinate in the fiber cross-section. A similar expression can be obtained for the fields in the cladding region:

$$\begin{bmatrix} E_z \\ H_z \end{bmatrix} = \begin{bmatrix} A\dfrac{K_n(W\frac{r}{a})}{K_n(W)}\left(\begin{matrix}\cos n\varphi_{n \ even} \\ \sin n\varphi_{n \ odd}\end{matrix}\right)e^{-\beta z} \\[3ex] B\dfrac{K_n(W\frac{r}{a})}{K_n(U)}\left(\begin{matrix}-\sin n\varphi_{n \ even} \\ \cos n\varphi_{n \ odd}\end{matrix}\right)e^{-\beta z} \end{bmatrix} \quad (r > a) \qquad (5.18)$$

where the cladding parameter $W \equiv a(\beta^2 - k_0^2 n_{clad}^2)^{1/2}$, and $n$ is the order of the Modified Bessel function of the 2nd kind. Within the core, the fields are oscillatory in the radial dimension according to the behavior of the J-type Bessel functions and the value of $U$. Outside the core region, the waves decay exponentially with $r$ for $W > 0$. The transverse field components are obtained from Eqs. (5.17) and (5.18) by direct application of Maxwell's equations. The values of $U$ and $W$ are found from solutions of the eigenvalue equations obtained by field matching at the core/cladding boundary. The eigen functions are quite complex for the step index fiber, and approximate solutions are available for the "weak guidance" case ($\Delta \ll 1$). A variational formulation for the eigenfuctions has also been found for the graded index fiber [6]:

$$\frac{U}{(1 + 2/q)^{1/2}} J_{n+1}\left[\frac{U}{(1 + 2/q)^{1/2}}\right] \Big/ J_n\left[\frac{U}{(1 + 2/q)^{1/2}}\right] = W \frac{K_{n+1}(W)}{K_n(W)} - \frac{W^2}{q+2}\left[1 - \frac{k_{n-1}(W)K_{n+1}(W)}{K_n^2(W)}\right] \quad (5.19)$$

Since the Bessel functions of the first kind are similar to harmonic functions, they exhibit oscillatory behavior. Solutions of Eq. (5.19) therefore determine the modes, which are indexed according to the number of azimuthal variations $n$, and the $m$-radial variations associated with the J-type Bessel functions. The roots of Eq. (5.19) are designated $\beta_{nm}$ since $U$ and $W$ depend only on $\beta$ for a given waveguide parameter $V$; that is, the waveguide construction and wavelength of operation (defined by $V$) uniquely specify the allowed modal distribution and propagation constants $\beta_{nm}$. The relationship between $U$, $W$, and $V$ is given by:

$$V^2 = U^2 + W^2 \quad (5.20)$$

as can be easily verified from Eqs. (5.14), (5.17), and (5.18). The propagation constant $\beta_{nm}$ is found from the definition of $U$ [Eq. (5.17)] and $V$ [Eq. (5.14)]

$$\beta_{n,m} = k_0 n_c\left(1 - \frac{2\Delta U_{n,m}^2}{V^2}\right)^{1/2} \quad (5.21)$$

For this reason, the modal description of a particular fiber waveguide is often shown on a $\beta/k$ (normalized propagation constant) vs. $V$ graph for specific values of $U$ corresponding to a particular mode. The waveguide parameter or "normalized frequency" specifies the wavelength while $U$ takes on specific values according to solutions of the eigenvalue Eq. (5.19).

The cut-off region for a given mode, i.e., the frequency below which (wavelength above which) the mode ceases to propagate in the core region, occurs when $W \to 0$ (i.e., $U = V = V_c$). According to Eq. (5.19), this occurs for the second mode ($LP_{11}$) when

$$V_c \cong 2.405\left(\frac{q+2}{q}\right)^{1/2} \quad (5.22)$$

To ensure single mode operation of a power law, graded index fiber, $V$ must be less than this value. Also at cutoff, the propagation constant $\beta_{nm}$ is equal to $k_0 n_{cl}$, i.e., the phase velocity becomes equal to that of a plane wave propagating in the cladding.

The exact mode descriptions are found from the particular allowed structure of the electric and magnetic fields in the fiber. For $n = 0$, $TE_{0m}$ and $TM_{0m}$ modes are the standard designations depending on whether the z-component of the electric or magnetic field is zero [see Eq. (5.16)]. For TE, $e_x = e_z = h_y = 0$, and for TM, $h_x = h_z = e_y = 0$. These modes correspond to "ray paths" along the meridian, i.e., passing through the fiber longitudinal axis and having two distinct linear polarization states for the transverse field. The EH and HE modes are comprised of "skew rays" that rotate azimuthally as they travel along the fiber axis. As such, these modes are hybrids of the TE and TM modes, since the polarization state is rotating as the wave propagates. For $n = 1$, either $HE_{1m}$ or $EH_{1m}$ are the designated nomenclatures [8], the former being predominantly transverse magnetic. When $\Delta \ll 1$, the z-component of the field becomes small in comparison to the transverse fields and the waves approximate Transverse Electromagnetic (TEM)

**TABLE 5.1**  Cut-off Wavelength for Single Mode Operation vs. Wavelength and Δ

| Wavelength (nm) | Δ (n ~ 1.46) | Fiber Core Diameter (μm) |
|---|---|---|
| 850 | 0.002 | 7 |
| 1300 | 0.002 | 10.7 |
| 1500 | 0.002 | 12.3 |
| 850 | 0.001 | 4.9 |
| 1300 | 0.001 | 7.5 |

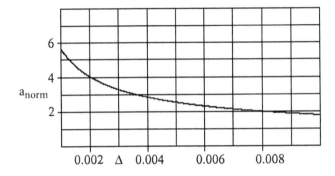

**FIGURE 5.4**  Normalized core radius ($a/\lambda$) vs. index profile height Δ for single mode operation ($V < 2.4$). (Single mode operation for all values below curve).

waves or "plane waves" that occur in free space. Under these approximations, the waves can be considered as linearly polarized waves and are sometimes designated LP (linear polarized) modes. Modal field descriptions are complex, requiring solution of the vector wave equation, and are treated in detail elsewhere [2, 5(Chapter 3)].

### Single Mode Fibers

Single mode operation of a *step index fiber* is achieved by setting $V = V_c = 2.4$ in Eq. (5.14). This dependency is plotted in Fig. 5.4 where the normalized core radius ($a/\lambda$) is plotted against the core/cladding refractive index fraction over the range $0.001 < \Delta < 0.01$. The curve shows points where $V = 2.4$. Points below the curve correspond to smaller $V$-numbers. Clearly, there are a large number of values possible for $a$ and the indices of refraction for the core and cladding. If losses are to be reduced, however, it can be shown that the majority of the energy flow in the fiber is within the core region when the $V$-number is near maximum for a given propagating mode. For a single mode, step index fiber, this means that losses associated with energy in the cladding (e.g., due to microbends) are minimized if a $V$-number near or just below 2.4 is used.

The fiber will support single mode operation at greater than the cutoff wavelength defined by

$$\lambda \geq \lambda_c = \frac{2\pi a n_c \sqrt{2\Delta}}{V_c} = \frac{2\pi a n_c \sqrt{2\Delta}}{2.405}_{\text{step index}} \tag{5.23}$$

At lesser wavelengths, the fiber will support additional "higher-order" modes.

A computation of the fiber diameter at cutoff is shown in Table 5.1. Because the core size is quite small, it is difficult to couple to single mode fibers and to make joints that are mechanically aligned and stable.

### Coupling and Alignment Loss

Three types of coupling loss are readily defined: (1) lateral shift, (2) axial tilt, and (3) $z$-axis separation. A lateral shift of a single mode fiber by an amount $s$ over a fraction of the core radius $a$ produces a

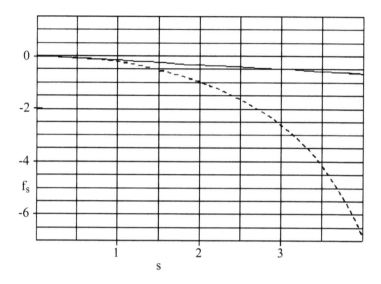

**FIGURE 5.5**  Coupling loss $f$(dB) for lateral misalignment $s$. Dashed: single mode 9-micron diameter core from Eq. (5.24), Solid: multimode graded index ($q = 2$) 50-micron diameter core from Eq. (5.25).

fractional loss in flux given by [9]

$$f_s = 0.5\left[\frac{sWJ_0(U)}{aJ_1(U)}\right]^2 = (s/a)^2 \text{ for } V \geq 1.5$$

$$f_s(dB) = 10\log_{10}[1 - (s/a)^2]$$

$$\tag{5.24}$$

The loss for a power law profile, multimode fiber is somewhat more demanding relative to the core size

$$f_s^{\text{multimode}} = \frac{2s(q + 2)}{a\pi(q + 1)m} \tag{5.25}$$

A comparison of single mode and parabolic profile multimode losses are shown in Fig. 5.5.

A tilt by an amount $\delta$ between two single mode fiber axes produces a fractional loss approximated by [9]

$$f_\partial = (n_c k\sin(\delta))^2\left[\frac{aJ_1(U)}{WJ_0(U)}\right]^2 = (0.7n_c ka\sin(\delta))^2 = \frac{V^2}{4\Delta}\sin^2\delta, \quad V > 1.5$$

$$f_\delta(dB) = 10\log_{10}\left(1 - \frac{V^2}{4\Delta}\sin^2\delta\right), \quad V > 1.5$$

$$\tag{5.26}$$

For $V = 2$, the loss is 2.7% for a tilt angle of 0.3 degrees and $\Delta = 0.001$.

The fractional loss in multimode power law profile fibers with uniformly excited modes is

$$f_\delta^{\text{multimode}} = \frac{(\sin\delta)\Gamma(2/q + 2)}{(2\pi\Delta)^{1/2}\Gamma(2/q + 2)} \tag{5.27}$$

where the Gamma function is indicated as a function of the power law index $q$. A somewhat simpler and more accurate expression is given for small misalignments of step index fibers

$$f_\delta^{\text{multimode}} = \frac{0.72\sin^2(\delta)}{\Delta} \tag{5.28}$$

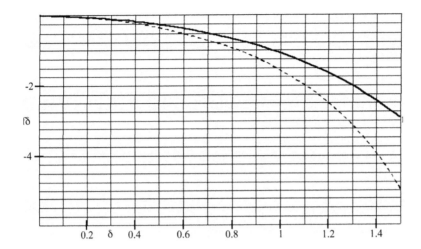

**FIGURE 5.6** Coupling loss $f_\delta$ (dB) due to angular misalignment $\delta$ (degrees) of: (a) dashed line: single mode fibers ($V = 2$, $\Delta = 0.001$) according to Eq. (5.26), and (b) solid line: multimode fiber ($\Delta = 0.001$) according to Eq. (5.28).

From Eq. (5.28), a tilt angle of 7% of the numerical aperture produces a loss of 0.72%. Parabolic index fibers ($q = 2$) are more complex, and for the same 7% angular misalignment, exhibit a loss of 2.3% or 0.1dB [5]. Angular misalignmnet loss for single mode and multimode fibers is shown vs. $\Delta$ in Fig. 5.6.

Longitudinal or $z$-axis misalignment for a step index fiber is estimated by

$$f_z^{\text{multimode}} = \frac{z_{gap}\sqrt{2\Delta}}{4a} \tag{5.29}$$

This expression is much less demanding than that for lateral displacement [Eq. (5.25)] by a factor of $8/\pi(2\Delta)^{1/2}$. A 50 $\mu$m core fiber with NA = 0.25 will show a splice loss of 0.1db for an offset of 2.5 $\mu$m and 1 degree angular misalignment. Additional losses of 0.3 db due to reflection are added to the above if air separates the fiber ends. Index matching fluid may be used in a splice connection or, alternately, fusion slicing may be used to reduce this effect.

## Modal Noise

For years, ROV operators have observed modal noise in amplitude-modulated, video fiber optic links. It occurs whenever optical power travels through mode selective devices, such as multimode fibers, connectors, or winch slip rings. The noise originates from separate spectral lines in a quasi-coherent source that split into multiple waveguide modes (angles of propagation) in the fiber simultaneously. At the end of the fiber these modes have different phase relationships and interfere with each other causing speckle patterns in the output beam. (Note that adjacent longitudinal modes of the laser do not interfere with each other since they are separated by greater than 50 GHz.) Graded index fibers show more speckle patterns than step index fibers due to the existence of fewer modes and lower multimode dispersion. Single mode fibers near cutoff may exhibit some modal noise characteristics.

The speckle pattern (if stationary) would not contribute to amplitude or intensity variations at the detector, but whenever connectors, bends, or temperature changes are introduced, the speckle pattern is altered and statistical variations, resulting in intensity noise, are introduced. There are several ways to reduce modal noise: (1) use lasers with many longitudinal modes or use of LEDs, (2) use single mode fibers, (3) use large area detectors, (4) avoid the use of connectors within a coherence length of source end of the fiber, and (5) use high quality connectors. Suggestion 1, as well as 3, will necessarily reduce the bandwidth of the communications system.

Single mode fibers can also be plagued with a form of modal noise due to polarization effects. Since the polarization of the lowest order mode is free to change with the smallest environmental perturbations, polarization-sensitive devices such as beam splitters or diffraction gratings, will contribute to noise in the system. Polarization-maintaining fiber may be required to reduce this effect [7].

## Cladding Modes

Any optical power coupled into the fiber core outside the numerical aperture cone is not bound to the fiber and escapes into the cladding. Because the cladding may be overcoated with a buffer of generally higher refractive index, the power escapes into the buffer and can be subject to scattering and absorption. Cladding modes generally dissipate within a short distance ($\sim$10s m), however a cladding mode stripper or filter can remove them closer to the source. They are observed as light being emitted at greater angles than specified by the numerical aperture of the fiber.

## Mode Stripping

Cladding modes can propagate for a short distance in fibers, perhaps as much as 100 m. If it is necessary to remove them in a shorter segment of fiber, a cladding mode stripper can be employed. The stripper is formed by removing all buffer and coatings in a short section of fiber (several centimeters). A chemical or mechanical method can be used. The fiber section is then immersed in a fluid of greater refractive index than the cladding. Glycerin, microscope immersion oil, and some polymers can be used.

## Mode Filters

A mode filter or mixer is used to create an equilibrium mode distribution in multimode fibers that is independent of light launching conditions. Graded index fibers with a 50-$\mu$m core diameter typically allow about 500 core modes, each subject to different attenuation and propagation characteristics. Losses are generally greater for the higher order modes. After some distance, the higher order modes dissipate by conversion to lower order modes (mode mixing) or by attenuation as a result of bending losses, fiber ellipticity, etc. Generally the mode distribution becomes stable after several kilometers. The mode filter is used to create the equilibrium mode distribution without resorting to long fiber lengths and can be used to measure fiber attenuation over short lengths. Because the mode distribution is effectively reduced, the fiber exhibits an *effective numerical aperture*. For a graded index fiber of 50/125 constructions this may be about 0.11.

The simplest mode filter consists of a fiber wound on a mandrel. In this device, the higher order modes are converted to cladding modes due to the fiber curvature. A loss of several db can be expected. The filter diameter and number of turns are determined experimentally to produce the same modal distribution seen at the end of a long section of the same type of fiber. Coupled, short segments of fibers with different core sizes may also be used for mode filtering.

## Dispersion

Dispersion refers to the variation of the delay characteristics of an optical fiber. The delay characteristics change with excitation frequency (free-space wavelength) and with modal distribution. Dispersion results from several factors: multiple modes propagating within the waveguide simultaneously (modal dispersion), changes in the refractive index of the glass or plastic with frequency (material dispersion), and changes of propagation constant associated with variations in $V$ and (consequently) $U$ with frequency (waveguide dispersion).

Modal dispersion is predominant in multimode fibers and limits operational bandwidth due to increasing signal delay depending on the fiber length. Bandwidth length products of gigaHertz-kilometer are possible in well-designed systems. Manufacturers typically quote the bandwidth capability of a fiber as a distance-length product or in terms of the pulse dispersion, with units of nanoseconds per kilometer-nm, where the dispersion is normalized to the bandwidth of the source and the fiber length, i.e., $D = z^{-1}(dt/d\lambda)$. In some instances a bandwidth curve is given for different run lengths.

*Material* and *waveguide dispersion* are dominant sources of dispersion in single-mode fibers, and impose severe restrictions on the choice of wavelength and signal bandwidth for long runs. *Waveguide* dispersion is associated with changes in the waveguide propagation constant with wavelength and for single mode fiber is about 2ps/km-nm at 1300 nm. Material dispersion is associated with changes of the refractive index of the core and cladding materials with wavelength. Modal dispersion results from temporal delays between different propagating modes and is of primary concern in multimode fibers.

## Modal Dispersion

Simple expressions can be derived using ray theory for modal (intramodal) dispersion in step index fibers. The pulse spread, $t_d$ is found by subtracting the maximum delay time, $t_{max}$, from the minimum delay time, $t_{min}$, for a fiber of length $z$.

$$t_{max} = \frac{n_c}{c\cos\theta_c}z, \qquad t_{min} = \frac{n_c}{c}z$$

$$t_d = t_{max} - t_{min} = \frac{n_c}{c}z\left(\frac{n_c}{n_{clad}} - 1\right) \cong \frac{n_c}{c}\Delta z \tag{5.30}$$

Small profile parameters $\Delta$ are preferred for minimum pulse spread.

A similar expression can be obtained for power law profile, graded index fibers.

$$t(\bar{\beta}) = \frac{n_c}{c}\frac{z}{q+2}\left[\frac{qn_c}{\bar{\beta}} + \frac{2\bar{\beta}}{n_c}\right] \tag{5.31}$$

The minimum time delay is found by differentiating Eq. (5.30) with respect to $q$, yielding $q_{optimum}$ and the pulse spread.

$$q_{optimum} = 2\sqrt{1 - 2\Delta}$$

$$t_d \cong \frac{n_c}{c}\frac{\Delta^2}{8}z \tag{5.32}$$

The pulse spread is $\Delta/8$ times less than the step index fiber and the optimum profile is approximately parabolic ($q = 2$) for small values of the profile parameter. The spread is highly sensitive to the value of $q$ and can increase by an order of magnitude for variations about $q_{optimum}$ equal to $\Delta$. These expressions are also dependent upon the material dispersion, i.e. the variation of refractive index of the core and cladding with wavelength. Similar to Eq. (5.31),

$$t(\bar{\beta}) = \frac{n_a}{c}\frac{z}{q+2}\left[\frac{(p+q)n_c}{\bar{\beta}} + \frac{(2-p)\bar{\beta}}{n_c}\right] \tag{5.33}$$

where, $\quad n_a \equiv n_c(\lambda) - \lambda\dfrac{dn_c(\lambda)}{d\lambda} \quad$ and $\quad p \equiv \dfrac{n_c(\lambda)}{n_a}\dfrac{\lambda}{\Delta(\lambda)}\dfrac{d\Delta(\lambda)}{d\lambda}$

The minimum transit time is

$$t(\bar{\beta}_{min}) = \frac{n_z}{c}\frac{2z}{q+2}\sqrt{(p+q)(2-p)} \tag{5.34}$$

In this case the optimum profile is shifted from $q \sim 1.98$ to $q \sim 2.2$ for $p = -0.14$ according to

$$q_{optimum} = \sqrt{4(1 - 2\Delta)} - p[1 + \sqrt{1 - 2\Delta}] \tag{5.35}$$

Under these conditions, the value for the pulse spread $t_d$ from Eq. (5.32) is reduced by the factor $n_a/n_c$. The bandwidth $B$ is approximately the reciprocal of the pulse spread, i.e.,

$$B \approx \frac{8c}{n_a \Delta^2} z^{-1} \quad \text{or}$$

$$Bz \approx \frac{8c}{n_a \Delta^2}$$

(5.36)

From Eq. (5.36), length-bandwidth products $Bz$ of approximately 16 Ghz-km are possible for $\Delta = 0.01$. Smaller profile heights can produce dramatic increases in bandwidth according to theory.

## Material Dispersion

The material dispersion results in a pulse spread given simply by the derivative of the group delay $\tau_m$ with respect to wavelength.

$$t_{dm} = \frac{d}{d\lambda}\tau_m = \frac{d}{d\lambda}\left(\frac{n_a}{c}z\right) = \left\{-\frac{\lambda}{c}\frac{d^2 n}{d\lambda^2}\right\}z\delta\lambda$$

(5.37)

where $\delta\lambda$ is the spectral width of the source. The term in brackets is the material dispersion, which ranges from $-120$ ps/nm-km at 800 nm, to zero at 1270 nm for pure silica. This zero dispersion wavelength can be increased somewhat by doping with germanium. An example of operation at a region of higher dispersion is an LED with an 800 nm peak wavelength and a spectral width of 40 nm. The resulting pulse spread is 4.4ns/km.

## Waveguide Dispersion

The waveguide dispersion is due to variation of the propagation constant $\beta$ with wavelength. The group delay $\tau_{wg}$ is given by

$$\tau_{wg} = \frac{z}{c}\frac{d\beta}{dk} = \frac{zn_{\text{clad}}}{c}\left[1 + \Delta\left\{b - \frac{2bJ_n^2(Ua)}{J_{n+1}^2(Ua)J_{n-1}^2(Ua)}\right\}\right]$$

$$b \equiv 1 - \left(\frac{Ua}{V}\right)$$

(5.38)

The pulse spread may be found by taking the derivative of Eq. (5.38) with respect to wavelength.

$$t_{dw} = \frac{-zn_{\text{clad}}\Delta}{c}(\delta\lambda)V\frac{d}{dV}\left\{b - \frac{2bJ_n^2(Ua)}{J_{n+1}^2(Ua)J_{n-1}^2(Ua)}\right\}$$

(5.39)

Plots of the term in brackets and its derivative vs. $V$ are available elsewhere [10, 11].

For $V$ greater than 2 but less than 2.4, the expression for the pulse spread per unit length for $\Delta = 0.01$ and $n_{\text{clad}} = 1.5$ becomes:

$$\frac{t_{dw}}{z} = -0.003\frac{\delta\lambda}{c\lambda}$$

(5.40)

This compares to $-0.02\frac{\delta\lambda}{c\lambda}$ for the material pulse spread per unit length at 900 nm. At 1300 nm however, the waveguide dispersion is dominant at about $-2$ ps/nm-km and this is the major bandwidth-limiting factor for single mode fibers.

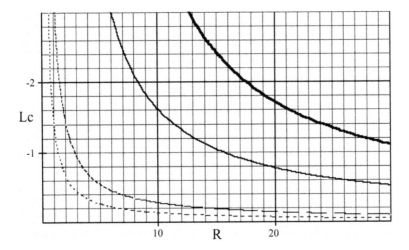

**FIGURE 5.7** Curvature loss $L_c$ (dB) vs. radius of curvature R (cm) for multimode fibers (50 μm dia., λ = 1.3 μm). Bold: Δ = 0.001, parabolic index $q$ = 2; Solid: Δ = 0.001, step index; Dashed: Δ = 0.01, parabolic index $q$ = 2; Dotted: Δ = 0.001, step index, Δ = 0.01.

## Fiber Losses and Signal Attenuation

Signal attenuation may be due to many factors, including absorption and scattering in the fiber core, nonuniformities in the core/cladding boundary, microbending due to buffer, jacket, or cable deformations, macrobending and fiber curvature loss, poor coupling from source to fiber and/or in fiber connectors or splices, mechanical damage from exceeding the tensile stress/strain limit, cross-sectional size perturbations due to lateral strain, microcracking due to water intrusion, etc.

### Curvature Losses

An optical fiber that is subject to a constant radius of curvature experiences losses over very short lengths. The losses increase dramatically above a certain critical radius of curvature [5]. Assuming the modes in a fiber are equally distributed, the curvature loss $L_c$ in decibels is given by

$$L_c(dB) = 10\log\left(1 - \frac{q+2}{2q\Delta}\left[\frac{2a}{R} + \left(\frac{3\lambda}{4\pi n_c^2 R}\right)^{2/3}\right]\right) \tag{5.41}$$

The loss versus R for several values of $q$ is shown in Fig. 5.7. Single mode losses are determined by the mode field diameter, i.e., how tightly the field is confined to the core. As the wavelength in increased, the V-number decreases and the wave field diameter increases, thereby increasing losses due to bending.

### Mechanical Damage, Water Intrusion, and Cracking

It is well-known that moisture breaks the silicon bonds of stressed fibers and lowers the failure stress level. It is therefore important to use a fiber that is buffered with stress-relieving coatings that are impervious to water penetration. In air, a maximum strain limit is about 4%. By contrast in liquid nitrogen, the breaking strain is nearly 14% due to absence of water.

### Environmental Effects

**Temperature**—The refractive index of glass is roughly between $10^{-7}$ and $10^{-8}$/K over a large temperature range. Geometric changes in the fiber also occur with temperature due to the thermal expansion coefficient of bulk glass $\sim 10^{-6}$/K.

**Hydrostatic Pressure**—Hydrostatic pressure changes the refractive index of bulk glass on the order of $10^{-7}$/atm. Volumetric changes of $10^{-8}$/atm are also observed. The changes are generally so small that it is only in interferometric sensor applications that they are significant.

**Tension**—Tension has the effect of shrinking the cross-section of the optical fiber as well as increasing its length. These effects provide the basis of operation for many types of fiber-optic sensors.

### References

1. Proakis, J. G., *Digital Communications,* 2nd ed., McGraw-Hill, New York, 1989.
2. Snyder, A. W. and Love, J. D., *Optical Waveguide Theory,* Chapman and Hall, New York, 1983.
3. Buckman, A. B., *Guided Wave Photonics,* Harcourt Brace Jovanovich, Orlando, FL, 1992.
4. Born, M. and Wolf, E., *Principles of Optics,* 4th ed. Pergamon Press, New York, 1970.
5. Miller, S. E., and Chynoweth, A. G., *Optical Fiber Telecommunications,* Academic Press, New York, 1979.
6. Olshansky, R. and Okashi, T., Analysis of wave propagation in optical fibers having core with α-power refractive index distribution and uniform cladding, *IEEE Trans. Microwave Theory Tech,* MTT-24, 416–421, 1976.
7. Epworth, Modal noise-causes and cures, *Laser Focus Magazine,* Sept. 1981, 109–115.
8. Snitzer, E., Cylindrical dielectric waveguide modes, *J. Opt. Soc. Am.,* 52, 491–498.
9. Cook, J. S., et al., Effect of misalignment of coupling efficiency of single mode fiber butt joints, *Bell Sys. Tech J.,* 52, 139–1448.
10. Keiser, G., *Optical Fiber Communications,* 2nd edition, McGraw-Hill, New York, 1991.
11. Gloge, D., Weakly Guiding Fibers, *App. Opt.,* 10, 2252–2258, 1977.

## 5.2  Basics of Fiber-Optic Strain Sensors

*Syed H. Murshid and Barry G. Grossman*

Fiber optic sensor systems offer many potential advantages over the conventional strain-sensing schemes. The nearly inert nature of the optical fiber coupled with its very high sensitivity make it ideal for use in difficult and challenging applications, such as those encountered in the oceanographic environment [1]. The major advantages [2] of fiber optic sensors as compared to conventional electronic sensors for seaboard applications include:

- Small size
- Light weight
- High sensitivity
- Corrosion resistance
- High electrical resistance
- Low power consumption
- Optical and electrical multiplexing
- Safe in hazardous and explosive environments
- Immunity to electromagnetic interference
- Immunity to radio frequency interference
- Remote, real-time operation
- Low cost

Typically, sensing instrumentation measures parameters induced or changed by the measurand. This is true for sensors based on piezoelectric, capacitive, resistive, and most other types of phenomena. Similarly, fiber optic sensors utilize measurand-induced changes in light waves propagating in glass or plastic fibers. These changes can be in light intensity, phase, polarization, and spectral distribution or in propagation mode. The change in the sensor output with the measurand can either be calculated or determined experimentally. Over the last ten years, fiberoptic sensors have been built to demonstrate

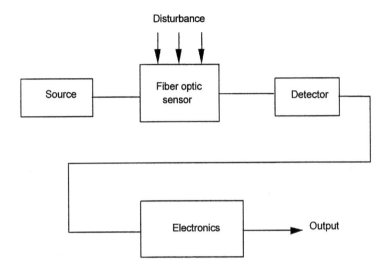

**FIGURE 5.8**  Basic building blocks of fiber optic sensor systems.

measurement of most parameters, including strain, temperature, force, vibration, pH, salinity, and electromagnetic and acoustic signals. The geometrical versatility of the fiber in combination with multiple sensor configurations results in a transducer system that can be tailored to any real-world application. Most high performance fiber optic sensors for underwater applications involve coupling and detection of measurand-induced strain on the fiber. Hence accurate strain measurement is the key to fiber optic sensor systems.

## Overview of Fiber Optic Strain Sensors

Fiber optic sensor systems consist of a light source, some optical components that generally include optical fiber, couplers, polarizers, filters etc., and a detector with appropriate signal processing to detect and analyze the final output. Fig. 5.8 shows the basic building blocks of the fiber optic sensor system [3]. There are many different methods of classification for these sensors, but generally they are categorized into two types. These two categories [4] are the intensity-modulated or non-interferometric sensors and the phase-modulated or interferometric sensors. In intensity modulation, the measurand causes a change in the amplitude of the optical power in the fiber. Some examples of intensity-modulated fiber optic sensors include microbend, macrobend, evanescent mode, and fluorescence sensors. These sensors are generally simple to make, involve low cost, and have adequate sensitivity and resolution for a large number of applications. However, the sensitivity of intensity-modulated sensors is significantly lower than that of phase-modulated sensors.

## Intensity-Modulated Sensors

The intensity at the output of an amplitude-modulated sensor is proportional to the measurand. In general, the output of such a sensor can be written as

$$I_{out} = I_{in} \cdot f(\text{Measurand}). \tag{5.42}$$

Usually, the sensor is characterized by an attenuation coefficient that is a function of the measurand. The typical response of an amplitude-modulated sensor is shown in Fig. 5.9. It contains three regions, two nonlinear regions including the cut-off and saturation regions, and one linear region in which the sensor is usually designed to operate.

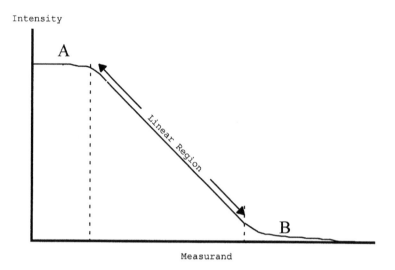

**FIGURE 5.9**   Response of amplitude-modulated sensor.

**Region A:** Region A depicts a stage where the system is yet to detect any perturbations on the sensor. In this region, either the detector is saturated or the measurand has not sufficiently affected the sensor. Generally there is a minimum perturbation threshold that must occur before any change in light intensity can be detected.

**Linear Region:** An intensity-based or amplitude-modulated sensor is mostly designed to operate in this region. It is typically referred to as the active region of the sensor. The slope of the linear region is a function of the sensitivity of the sensor. Steeper slopes can be achieved with more sensitive sensors. Small slopes are desirable for a wider dynamic range of measurement.

**Region B:** In region B, the system does not respond to the measurand. Generally this occurs when the measurand causes such an extensive attenuation of light that the sensor output intensity can no longer be detected.

The output of the intensity-modulated sensor only depends on amplitude of the detected signal. The state of polarization and coherence of the light source is insignificant. Hence, an incoherent source is generally preferred as it is less expensive and eliminates the effects of interference that may be indistinguishable from the measurand. A suitable reference can also be added to compensate for drift in the source and other undesired environmental effects. A number of intensity-modulated sensors are available, but common examples include the attenuation gap sensors, the microbend sensors, and the macrobend sensors.

## Attenuation Gap Sensor

The attenuation gap sensor is constructed by aligning the ends of two fibers in close proximity. Light propagating through one of the fibers is transmitted out of the fiber, through the gap between the fibers, and into the second fiber where it continues to propagate. The amount of light that is coupled into the second fiber depends on the intensity of light entering the gap, the transmissivity of the material in the gap, and the alignment of the fibers across the gap. Illustrated in Fig. 5.10 are two possible architectures [47] for this type of sensor.

The gap sensor can also operate as a single-ended sensor if a mirror is placed at the end of the output fiber. This architecture requires more components but is twice as sensitive as the transmission mode gap sensor because the light traverses the gap twice. Light launched into the fiber passes through a coupler, through the sensing gap, and into another fiber terminated with a reflective end. The reflective end reflects the light back through the sensing gap and into the coupler where some of the reflected light is coupled into the fiber leading to a detector. The single-ended configuration is twice as sensitive, but may lose too much light traversing the sensor gap a second time.

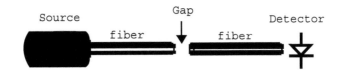

Transmission Mode

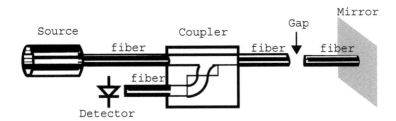

Reflection Mode

**FIGURE 5.10**  Attenuation gap sensor architectures.

The measurand affects the amount of light coupled into the second fiber across the gap. Either the alignment of the two fibers or the transmission coefficient of the material in the gap can modulate the amount of light coupled into the second fiber.

For example, a pH sensor could be made by mixing a pH-sensitive dye in a solution residing in the gap between the two fibers. As the pH changes, the dye also changes its optical characteristics and starts attenuating more or less light. The change in attenuation can therefore be detected and an assessment of the pH can be made.

A gap sensor can also be made into a colorimeter, by using a broadband source. A colored dye or material can be placed in the gap affecting the spectral content of the transmitted light. The detector could determine the color content of the gap by analyzing which wavelengths have been attenuated and which have been transmitted.

Any measurand modulating the amount of light coupled across the gap can be characterized by an attenuation coefficient, $\alpha$. In general, the attenuation coefficient can be a function of wavelength and/or environmental factors. The intensity within the gap sensor can be expressed mathematically with the attenuation coefficient as

$$I(z) \;=\; CI_0 e^{-\alpha z} \tag{5.43}$$

Here, the constant $C$ accounts for the amount of light that couples into the second fiber due to the geometry of the diverging light from the first fiber.

The coupling constant, $C$, is derived from Fig. 5.11 for the case where the fibers are perfectly aligned. The fraction of light coupled into the second fiber is approximated to be the square of the ratio of the cone radius entering the core of the second fiber to the total cone radius of the diverging light from the first fiber. That is,

$$\frac{P_{\text{coupled}}}{P_{\text{total}}} \;=\; \left[\frac{a_2}{R}\right]^2 . \tag{5.44}$$

By simple trigonometry, $R$ can be found to be

$$R \;=\; L \cdot \tan \theta_m + a_1 \tag{5.45}$$

where $\theta_m$ is the acceptance angle.

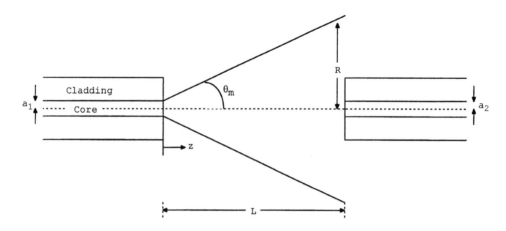

**FIGURE 5.11**   Light inside gap sensor.

Therefore, the power ratio becomes

$$\frac{P_{coupled}}{P_{total}} = \left[\frac{a_2}{a_1 + L\tan\theta_m}\right]^2.$$   (5.46)

Applying this ratio to the intensity equation, the final form of the output intensity of the gap sensor is obtained.

$$I = I_0 e^{-\alpha L}\left(\frac{a_2}{a_1 + L\tan\theta_m}\right)^2 = I_0 e^{-\alpha L}\left\{\frac{a_2}{a_1 + L\tan[\sin^{-1}(NA)]}\right\}^2$$   (5.47)

   Source coherence and light polarization are insignificant because the gap sensor only measures intensity and does not depend on the phase or state of polarization. Incoherent source is generally desirable so that interference effects do not disrupt the system. As with all intensity sensors, source drift can be a problem if a reference is not used. Fluctuations in source intensity are indistinguishable from attenuation due to the gap sensor.

**Microbend Sensor**

The microbend loss causes the light propagating in the fiber to decrease. This loss occurs due to redistribution of energy between the core mode and the core modes and the cladding modes, i.e., the guided, higher order core modes are coupled to the weakly guided cladding modes (radiative modes) in multimode fibers. This mode coupling occurs when the fiber geometry is changed by environmental perturbations that induce physical periodic bends in the fiber. These small changes in geometry cause lower order core modes to be coupled to the higher order core modes and, at the same time, the higher order core modes to radiative modes. The microbend effect can be explained with the help of mode coupling theory.

   The ratio of index of refraction for optical fibers $n(r)$ is generally described by the following equation [5]:

$$n^2(r) = n_0^2\left[1 - 2\Delta\left(\frac{r}{a}\right)^\alpha\right]$$   (5.48)

where $n_0$ is the index of refraction at the center of the core, $r$ is the radius outward from the core center, $a$ is the core radius, $\alpha$ is the index profile (i.e., step or graded), and $\Delta$, the relative index difference between

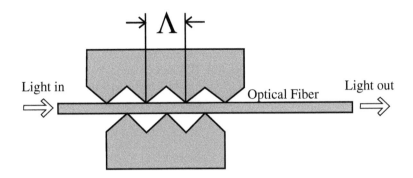

**FIGURE 5.12**   Spatial wavelength of external microbending deformations.

the core and cladding, is calculated by:

$$\Delta = \frac{n_0^2 - n_{cl}^2}{n_0^2} = \frac{(NA)^2}{2n_0^2} \qquad (5.49)$$

where $n_0$ is the index of refraction at the center of the core, $n_{cl}$ is the cladding index, and $NA$ is the numerical aperture. For modes to be exactly coupled together, the external deformations must have spatial wavelength, $\Lambda$ (Fig. 5.12). This spatial wavelength, $\Lambda$, is related to successive modes by its respective propagation constant, $\beta$:

$$\beta' - \beta = \Delta\beta = \frac{2\pi}{\Lambda} \qquad (5.50)$$

The spacing between mode groups $(\Delta\beta_m)$ in terms of the propagation constants with principal mode numbers $m$ and $m + 1$ is given as:

$$\Delta\beta_m = \frac{2}{a}\left(\frac{\alpha\Delta}{\alpha + 2}\right)^{1/2}\left(\frac{m}{M}\right)^{(\alpha - 2)/(\alpha + 2)} \qquad (5.51)$$

where $a$ is the core radius, $\alpha$ is the index profile, $M$ is the number of modal groups guided by the fiber, and $M^2$ is the total number of modes guided by the fiber. The total number of modes guided by the fiber is:

$$M^2 = \left(\frac{\alpha}{\alpha + 2}\right)(n_0 ka)^2\Delta \qquad (5.52)$$

where $n_0$ is the index of refraction at the center of the core, $a$ is the core radius, $\alpha$ is the index profile, and $k = 2\pi/\lambda$ is the free-space propagation constant for the wavelength of light used.

The spatial wavelength derivation is based on the difference between these mode groups in fibers with specific index profiles. The two profile cases of fibers used are step index ($\alpha = \infty$) and graded index ($\alpha = 2$). Step index means that the index of refraction is constant throughout the core region. For the step index case ($\alpha = \infty$), Eq. (5.51) simplifies to

$$\Delta\beta_m = \beta_{m+1} - \beta_m = \frac{2\sqrt{\Delta}}{a}\left(\frac{m}{M}\right) \qquad (5.53)$$

where $m$ is the mode number, $M$ is the number of modal groups guided by the fiber, $a$ is the core radius, and $\Delta$ is the relative index difference between the core and cladding. In a step index fiber, the spacing between the modes is based on the mode number and therefore varies for all of the modes. This means that only those modes with spacing equivalent to the external deformations will be coupled among themselves, while the rest of the modes will not be coupled at the same time. From Eqs. (5.50) and (5.53) it can be seen that lower order modes are coupled with large spatial wavelengths while high-order modes are coupled with smaller spatial wavelengths. Maximum microbend loss occurs when the highest order mode couples to the first radiative mode, or $m = M$. The result after combining Eqs. (5.50) and (5.53) with $m = M$ yield, Eq. (5.54), the spatial wavelength for optimum sensitivity [6] in a step index fiber

$$\Lambda_{SI} = \frac{\sqrt{2}\,\pi a n_0}{NA} \tag{5.54}$$

where $a$ is the core radius, $n_0$ is the index of refraction at the center of the core, and $NA$ is the numerical aperture of the fiber. A graded index fiber has an index of refraction that parabolically decreases in value. It has its highest refractive index at the center of the core, which decreases radially, and the minimum value of refractive index is at the cladding. For the graded index case ($\alpha = 2$), Eq. (5.51) simplifies to

$$\beta_{m+1} - \beta_m = \frac{\sqrt{2\Delta}}{a} \tag{5.55}$$

where $a$ is the core radius and $\Delta$ is the relative index difference between the core and cladding. Unlike the step index case, the formula for the graded index case has no dependence on the mode number. Because the index profile for the graded index fiber is parabolic, the space between all of the modes is the same. Therefore, all of the modes will couple at the same spatial wavelength, providing a much higher transfer of modal power from the core into the radiative modes. Combining Eqs. (5.50) and (5.55) results in Eq. (5.56), the spatial wavelength for optimum sensitivity for graded index fibers given by

$$\Lambda_{GI} = \frac{2\,\pi a n_0}{NA} \tag{5.56}$$

Therefore, the output intensity of the light guided within the fiber decreases with more and more external perturbations because of greater mode coupling. This effect of intensity modulation of light in optical fibers is utilized in microbend sensors. The measurand periodically deforms the sensing fiber. As a result the light intensity received at the detector changes. This modulation of intensity has a close to linear correlation to the measurand. A typical construction of the microbend sensor is shown in Fig. 5.12. The signal-carrying fiber is placed between two deformers. The measurand, such as force or the weight of a vehicle, acts on one or both of the deformers. This results in perturbations in the fiber geometry. The amount of perturbations has a direct relationship to the measurand. Therefore the output intensity has one-to-one correspondence to the measurand. Figure 5.13 shows the typical intensity response of the microbend sensor to applied load.

The microbend sensor was one of the first fiber optic sensors to be developed. It was first proposed and demonstrated in 1980 [7, 8]. The fiber optic sensor systems program of the Navy drove early interest in microbend sensors, mainly directed toward hydrophone applications. Since then, over 100 different studies have appeared in the literature. The tremendous interest in microbend sensors is a result of the many unique advantages it offers. Some of these include

- Mechanical and optical efficiency.
- Low parts count and simple construction.
- Higher reliability at low cost.

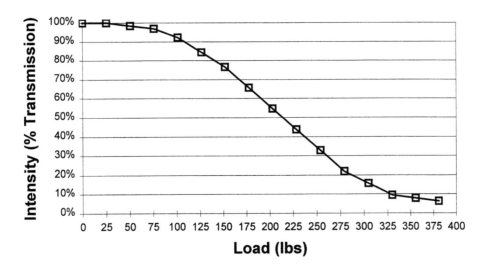

**FIGURE 5.13** Intensity response of the microbend sensor to applied load.

- Automatic sensor health monitoring, hence fail-safe operation.
- No mechanical coupling of fiber to other components needed, hence it is immune to differential thermal expansion.
- Ability to operate in hostile, high temperature, and explosive environments.
- Ease of readout and minimal signal processing.

The microbend sensor also offers all the advantages of fiber optic sensors including light weight, immunity to RFI and EMI, ability to survive harsh process conditions and electrical insulation, etc. A number of fiber optic sensors based on the principles of microbending have been developed for both oceanographic and land-based applications. They include load and acceleration sensors [9], axle detection, counting, and classification systems [10], weigh-in-motion sensors [11], pressure sensors [12], pore water pressure sensors [13], and shock sensors [14], etc.

## Macrobend Sensor

Macrobend sensor operation depends on the attenuation of light caused by sharp bends in the optical fiber. This attenuation owes to critical angle violations induced by external bends in the fiber. System architectures for this type of sensor are identical to the gap sensor. These architectures are shown in Fig. 5.14. The single ended configuration is twice as sensitive because the light traverses the sensor twice, but requires more components.

Critical angle violations cause microbend effect. When light propagating through a fiber encounters a sharp bend, some modes reach the core-cladding interface at angles that are less than the critical angle. These modes eventually become cladding modes and are quickly attenuated. Figure 5.15 graphically depicts critical angle violations at bends.

It is interesting to note that a critical radius [15] of curvature exists, above which losses are negligible. The critical radius is given by

$$R_C \cong \frac{3\,n_{\text{cladding}}\lambda_0}{4\,\pi \cdot NA_0^3} \qquad (5.57)$$

where *NA* is the numerical aperture of the fiber, *n* denotes the refractive index, and $\lambda_0$ is the wavelength of the light source.

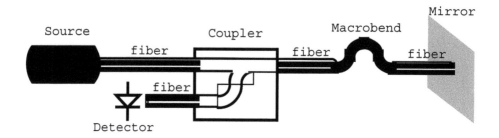

FIGURE 5.14   Macrobend sensor architectures.

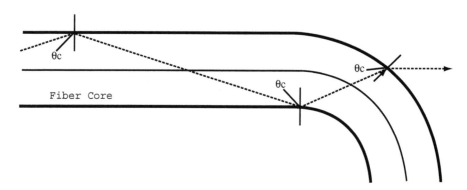

FIGURE 5.15   Critical angle violation.

As with most losses, macrobending loss can be mathematically expressed as an attenuation coefficient, $\alpha_{\text{bends}}$. The output intensity can therefore be written as

$$I(z) = I_0 \cdot e^{-\alpha_{\text{bends}}z}. \tag{5.58}$$

The macrobend attenuation coefficient can be written as

$$\alpha_{\text{bends}} = A_1 e^{-A_2 r} \tag{5.59}$$

where $A_1$ and $A_2$ are constants depending in the physical parameters of the fiber. These coefficients are determined experimentally.

## Bragg Grating

The architecture of a Bragg grating system is similar to most intensity sensor systems. It can operate in both transmission or reflection mode. These configurations are shown in Fig. 5.16.

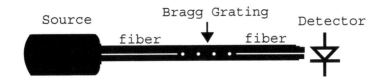

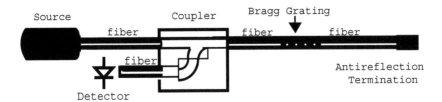

**FIGURE 5.16**   Bragg grating architectures.

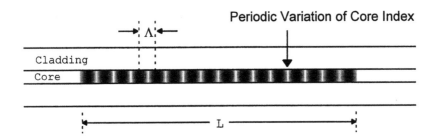

**FIGURE 5.17**   Fiber Bragg grating.

A Bragg grating is simply a standard fiber that at some point begins to have a small periodic change in the core index. The distance between equivalent index variations is called the spatial periodicity ($\Lambda$), while the total distance covered by all the variations is the gauge length ($L$) as shown in Fig. 5.17.

The Bragg grating acts like an optical filter by reflecting resonant wavelengths. A band of wavelengths centered on a specific wavelength is reflected back toward the source from the grating. This specific wavelength is called the Bragg wavelength, is a function of the index grating, and is given by the following equation

$$\lambda_{\text{Bragg}} = \frac{2 \cdot \Lambda \cdot n_{\text{avg}}}{\ell} \tag{5.60}$$

where $\Lambda$ is the grating periodicity, $n_{\text{avg}}$ is the average or overall effective index of the grating, and $\ell$ is the order of the grating (Usually $\ell = 1$). All other wavelengths are transmitted through the grating. This produces a band pass effect at the reflection end. At the transmission end, the source spectrum minus the band centered around the Bragg wavelength is received. Typically, a Bragg grating has a bandwidth of around 0.1 nm and can have reflection coefficients at the Bragg wavelength close to 100%.

The Bragg grating is suitable for fiber optic sensing applications, as the average index of the grating is sensitive to the environment. It should be noted that the average index of refraction for all optical fibers is environmentally sensitive, and this also holds for the index profile of a Bragg grating residing in a fiber. As a result, the Bragg wavelength is also environmentally sensitive. Hence, perturbations imposed on the grating will cause a shift in the Bragg wavelength that can be detected.

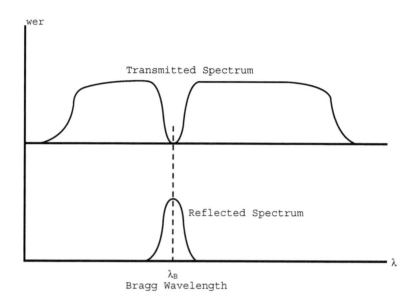

**FIGURE 5.18**    Bragg grating output spectra.

Many desirable features are offered by Bragg grating sensors. Since the grating only affects wavelengths near the Bragg wavelength for each individual grating, these sensors can easily be multiplexed. Multiple gratings with different periodicities can be placed along the same fiber. These can operate independent of each other. The Bragg grating can be made inexpensively when extreme precision is not required. Also, when used as a single-ended sensor, it inherently reflects at the Bragg wavelength, hence it is not necessary to add a reflective coating at the end of the fiber. Bragg grating sensors are mechanically strong as the sensor is protected inside the fiber.

On the negative side, Bragg gratings cannot be used for high temperature applications as the index variation begins to diminish with temperature. However, a temperature sensor can be made based on this phenomenon. Generally, spectral signal processing is required for these sensors, which can be expensive. One solution to this problem is to place an optical filter with a sharp roll-off at the equilibrium Bragg wavelength. As the measurand varies around some equilibrium point, a linearly increasing or decreasing signal can be generated.

The typical output of a transmission mode Bragg grating will look very different from that in reflection mode. Ideally, the sum of the two outputs will equal the source light. The spectra of these outputs are shown in Fig. 5.18.

The source for the Bragg grating sensor needs to be broadband relative to the specific Bragg wavelength. Otherwise the Bragg wavelength will be undetectable. Generally, white light sources are used for Bragg grating sensors.

## Interferometric Fiber Optic Sensors

Interferometric sensors usually consist of single mode fibers with coherent sources. The perturbation caused by the measurand modulates light by changing the phase or the state of polarization. The output of most interferometric sensors including the Mach-Zehnder, Michelson, Polarimetric, and Extrinsic Fabry-Perot (EFPI) sensors exhibit a sinusoidal change in intensity with relative phase difference. The phase angle or the induced phase shift of the output is related to the measurand. The sensitivity of the sensor to the measurand is measured in terms of the induced phase shift of the reflected light [16]. The phase shift $\varphi$ for light propagating in a single mode fiber interferometric sensor of gauge length $L$, with refractive index $n_1$ can be written as

$$\varphi = \frac{2\pi L n_1}{\lambda_0}. \tag{5.61}$$

The force $F$ will predominantly change the length $L$. The induced change, $\Delta L$, depends on Young's modulus, $E$, of the material. Mathematically, it can be represented as

$$\Delta L = \frac{FL}{AE} \tag{5.62}$$

where $A$ is the cross-sectional area of the fiber. The Young's modulus for quartz glass is $2 \times 10^{11}$ *Pa*. Hence the resultant change of phase $\Delta\varphi$ due to an axial force $F$ will be

$$\Delta\varphi = \frac{2\pi n_1 LF}{\lambda_0 AE} \tag{5.63}$$

Similarly, if the same fiber sensor is subjected to a uniform radial pressure $P$, then the gauge length $L$ of the sensor will also increase due to the Poisson effect. If $\xi$ is the Poisson's ratio of the material ($\xi = 0.2$ for quartz glass), then the increase in length can be represented [17] as

$$\Delta L = \frac{2\xi PL}{E} \tag{5.64}$$

and the change in phase [18] of the sensor becomes

$$\Delta\varphi = \frac{4\pi n_1 \xi LP}{\lambda_0 E} \tag{5.65}$$

If the temperature changes by an amount $\Delta T$, then both the length and refractive index of the fiber sensor will change. The change in length $\Delta L$ is represented as

$$\Delta\Lambda = \alpha\Delta T, \tag{5.66}$$

where $\alpha$ is the coefficient of thermal expansion of the material. For silica glass the value of $\alpha = 5 \times 10^{-6}$ *m.K*$^{-1}$. and the coefficient of thermal changes in the index of refraction is $\frac{dn}{dT} = 7 \times 10^{-6}$ *m.K*$^{-1}$. Therefore, the total change in phase can be represented as

$$\Delta\varphi = \frac{2\pi L}{\lambda_0}\left[\frac{dn}{dT} + n_1 \alpha\Delta T\right]. \tag{5.67}$$

The original and the phase-shifted signals interfere with each other while propagating back and the photo detector produces a corresponding electrical output that is sinusoidal in nature. Therefore force, pressure, and temperature can be measured directly. To measure other parameters such as shock and magnetic field, a transducer is generally incorporated to convert the desired field into a parameter that the sensor can measure. For example, a magnetostrictive material can be attached to the fiber to change the length of the fiber in response to the external magnetic field. Similarly, a coating that incorporates magneto-strictive material can be used to cover the fiber to convert the changing magnetic field into change in axial pressure. Problems can be encountered due to nonlinearity of response of the transducer and changes in ambient conditions like change in environmental temperature or pressure. Therefore, maintaining the stability of the fiber sensor arms without masking the effect of the desired field is important.

## Overview of Interferometric Optical Strain Measurement

Interferometric fiber optic strain gauge measuring axial strain was demonstrated as early as 1978 [19]. Since then, most of the effort is focused on detecting strain along the uniaxial direction [20, 21, 22]. Some of the most common interferometric fiber optic sensors used for measurement of strain include Mach-Zehnder, Michelson, Polarimetric, Few-mode, and Fabry-Perot interferometers, mostly because of their high sensitivity. Different combinations of sensors and techniques have been tried for different applications.

## Classification of Interferometric Fiber Optic Sensors

Many different ways of classifying interferometric fiber optic sensors are available including single-fiber-path, dual-fiber-path, and intrinsic and extrinsic sensors. In intrinsic fiber optic sensors, the sensing takes place within the fiber, whereas in extrinsic optical fiber sensors, the sensing takes place in a region outside the fiber. In applications where structural degradation due to external components is undesirable, fiber optic sensors are classified as two-fiber-path or single-fiber-path fiber optic sensors. The former includes Mach-Zehnder and Michelson interferometric sensors while the latter includes Polarimetric, Two-mode, Fabry-Perot, and Bragg grating types of interferometric sensors.

### *Two-Fiber Interferometric Fiber Optic Sensors*

Over the years, the most popular fiber interferometers [23] have included the Mach-Zehnder and Michelson interferometers, which are optical fiber implementations of classical optical interferometers. The Mach Zehnder [24] and Michelson have been widely used for studying strain [25]. Despite the high sensitivity and a good dynamic range of these sensors, difficulties occur in multiplexing sensors of this type due to the large number of fiber optic cables and components. This makes them unsuitable for most practical applications. This is especially true for measuring internal strain fields because proper isolation of the reference arm is critical for both Mach-Zehnder and Michelson interferometers. The two-fiber-path arrangement is less suitable for many applications due to lower common mode rejection, greater intrusiveness, and the need for preserving the phase at the interface of structure. Because of their extremely high sensitivity, isolating the sensing region from undesired environmental perturbations is critical. The Mach-Zehnder and Michelson sensors are both two-fiber sensors. This means that their output is determined by the difference in the optical path length (phase) between the sensing fiber and the reference fiber beams. The biggest disadvantage of these sensors is that measurement accuracy depends not only on the reference fiber being totally isolated from measurand changes, but also variations in temperature and other environmental conditions that can effect the output signal. Also, since the two interconnection fibers between the sensor head and the measurement/drive electronics are part of the interferometer, they can induce significant error into measurements. With extra care in design of the sensor system these errors can be reduced in the laboratory, but not enough to be generally employed for practical sensing applications.

### *Single-Fiber Interferometric Fiber Optic Sensors*

Sensors designed to measure strain or strain-related parameters including the fiber optic sensors should meet the following criteria [26].

- Intrinsic nature for maximum stability.
- Point sensing configuration with high lead insensitivity.
- Sensitivity only to the measurand.
- Well behaved, with repeatable response.
- All fiber construction for greater operational stability.
- Linearity of response.
- Single-fiber construction for better common-mode rejection.
- Single-ended design for simplicity in installation and connection.
- High sensitivity with adequate operating range.
- Insensitivity to environmental perturbations.

- Minimum degradation of the host structure.
- Interrupt immune with capability for absolute measurement.
- Ease of manufacture, mass production, and multiplexibility.

In lieu of the above criteria, especially when minimum degradation due to the presence of the sensor is an important criterion, a single-fiber sensor system is preferred. Although many fiber optic sensors have been suggested and used for strain and strain-related process measurement [27], only the Polarimetric [28], the two-mode [29], the Fabry-Perot, and the Bragg grating sensors have proved to be suitable for such measurements. It should be noted that the Bragg grating sensors are sometimes treated as interferometric sensors due to their high sensitivity and nature of output. Selection of an individual sensor depends on the process to be sensed and the compliance of the sensor with the application requirements. Table 5.2 compares these four types of single-fiber sensors.

The two-mode or the modal domain sensors operate on the phenomenon of mode pattern phase modulation that is caused by an external disturbance. Detailed theory and applications are discussed in a number of references [30, 31, 32]. The axial sensitivity of the two-mode fiber optic sensor is better than the polarimetric sensor [33]. However, the two-mode fiber optic sensor displayed very low sensitivity to twisting, compression, and bending [34]. Although the two-mode fiber optic sensor is a single-fiber sensor, the offset splicing requirement makes fabrication difficult. Therefore, specialized elliptical core high birefringence (HiBi) fiber with two-mode operation at a desired optical wavelength must be obtained, which makes it expensive.

The fiber optic version of the polarimetric sensor utilizes birefringence in a polarization maintaining a single-mode fiber. A two-dimensional transverse strain measurement, using an induced birefringence polarimetric sensor, can be examined and the help of an external force can create initial optical eigen axes where these axes rotate due to the large beat length [35]. Hence, this low-induced birefringence fiber is not suitable for practical applications. On the contrary, two well-defined optical eigen axes already exist in high birefringence fiber. The difference in the phase retardation between the two optical axes in the polarimetric sensor changes due to the external disturbance. The high transverse sensitivity [36] of the polarimetric sensor allows the HiBi fiber to be used for detecting transverse strain. In ship-based structures where large gauge lengths are appropriate, the lower sensitivity of the polarimetric fiber optic sensor can be acceptable.

The Bragg grating and the intrinsic and extrinsic Fabry-Perot interferometric fiber optic sensors are preferable in terms of multiplexing potential, sensitivity, and spatial resolution. These sensors have the capability of point sensing and have spectrally dependent outputs that minimize problems associated with variable losses from connectors and fiber leads. The major advantage of the extrinsic Fabry-Perot interferometers (EFPI) over the intrinsic Fabry-Perot fiber sensor is its immunity to problems associated with the state of polarization and sensitivity to only the axial strains [37].

**Common Interferometers**

This section briefly describes five commonly encountered fiber optic interferometric sensors.

**TABLE 5.2**   Comparison of Single-Fiber Sensor Systems

|  | Fabry-Perot | Bragg Grating | Two-Mode | Polarimetric |
|---|---|---|---|---|
| Localized | Yes | Yes | Nearly | Nearly |
| Sensitive to temperature | No | Yes | Yes | Yes |
| Sensitivity to strain direction | Difficult | Yes | Difficult | Difficult |
| Linearity of response | Good | Excellent | Good | Good |
| Single-ended operation | Yes | Yes | Difficult | Difficult |
| Sensitivity and range | Excellent | Excellent | Poor | Poor |
| Absolute measurement | Possible | Yes | Possible | Possible |
| Multiplexing within structures | Yes | Yes | Difficult | Difficult |
| Ease of production | Yes | Yes | Costly | Costly |

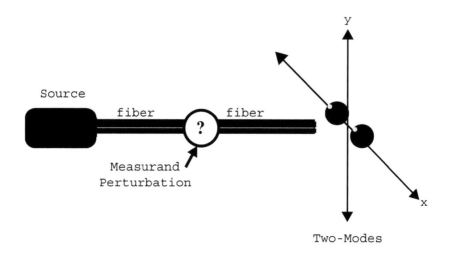

**FIGURE 5.19**   Two-mode sensor architecture.

### Two-Mode Interferometer

The purpose of the two-mode sensor is to provide a means of having both the signal and reference arms of an interferometric system in the same fiber. A typical configuration is shown in Fig. 5.19.

The configuration shown in Fig. 5.19 has the advantage of eliminating phase delay-caused non-measurand perturbations that are encountered when the reference arm is located at even a slightly different physical location. The two-mode sensor, however, is less sensitive (on the order of 30 dB less sensitive) because both modes propagating in the fiber are affected by the measurand in the same way but by different amounts. By using two fibers for the signal and reference arms, the reference arm may be kept 100% isolated from the measurand. Measurand isolation of the two-mode sensor is more easily achieved because most non-measurand perturbations are automatically compensated for in the sensor. The two-mode sensor is also stronger and more easily managed because it only has one fiber.

The two-mode sensor is constructed by operating a single-mode fiber at a slightly smaller source wavelength. Ideally, this increases the $V$-number (normalized frequency) to a value of approximately 2.5 to 3.8, where only two modes will propagate. These are the $LP_{01}$ and $LP_{11}$ modes.

Three types of perturbations will produce a phase change between the two propagating modes. They are changes in the $x$-polarized propagation constant (i.e., asymmetric strain), changes in the $y$-polarized propagation constant (asymmetric strain), and axial strain.

### Polarimetric Interferometer

The polarimetric sensor operates on the interference of different polarization modes. It has the advantage of using the same fiber for both the signal and reference arms of the system. A phase difference is induced by perturbations that affect the two polarization modes differently. The architecture of a polarimetric sensor system is shown in Fig. 5.20.

Given the phase difference between the two orthogonal polarization modes, $\beta$, Table 5.3 summarizes the type of source polarization.

Light enters the system with some polarization. After the first polarizer, only the polarization components along the two eigen axes are coupled into the fiber. The light propagates through the fiber, which also acts as an integrating sensor. Any perturbations that affect the relative phase velocity between the two polarization modes in the fiber produce a phase delay. At the second polarizer, the polarization components along a common angle, $\alpha$, of the two polarizations exiting the fiber are combined and interfere. An interference pattern is produced as the relative phase velocities are modulated as a function of the measurand.

**TABLE 5.3**  Summary of Source Polarizations

| Linear polarization: | $\beta = 0°$ |
|---|---|
| Circular polarization: | $\beta = 90°$ |
| Elliptical polarization: | else |

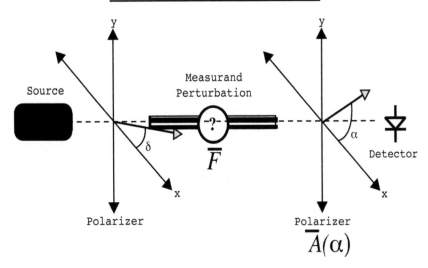

**FIGURE 5.20**  Figure 5.20 Polarimetric sensor architecture.

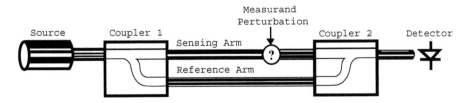

**FIGURE 5.21**  Mach-Zehnder architecture.

In general, for maximum visibility, both interfering beams should be of the same magnitude. Ideally, the source and input polarizer will produce light with equal intensity, in both polarization modes. Also, at the second polarizer, the two polarization modes should be recombined equally with an angle of 45° between them. In short, $\delta = 45°$ and $\alpha = 45°$ for maximum fringe visibility.

The polarimetric interferometer has the advantage of having both the signal and reference arm in the same fiber. It utilizes a polarization-maintaining fiber where the two allowed polarization modes act as the signal and reference beams. Since both arms are in the same fiber, many undesired perturbations will affect the two polarization modes equally. Perturbations producing asymmetric stress and strain will affect the relative phase velocities and produce an interference pattern.

Sources with long coherence lengths are not as crucial for the polarimetric sensor as with other interferometers. Since the signal and reference beams reside in the same fiber, the phase difference remains much shorter than in typical two-fiber interferometers. Proper consideration of the polarization of the source is important and the source must distribute light equally between the two polarization modes of the system, otherwise the visibility of the sensor will be lower.

*Mach-Zehnder Interferometer*
The simplest of the fiber optic interferometric sensors is the Mach-Zehnder. Its architecture is shown in Fig. 5.21. A laser light source usually provides the coherent light for the Mach-Zehnder interferometer. Not shown in the diagram is the mechanism used to couple the laser light in the fiber.

This usually consists of a lens and possibly a polarizer. A coupler is used to split the light into two separate fibers, which in turn are recoupled at the output of the system where the split beams recombine and interfere. Depending on the optical path difference of the two arms of the system, the recombined light will exhibit constructive interference, destructive interference, or somewhere in between. As the optical path difference changes due to external perturbations, the detected light will produce an interference fringe pattern and oscillate between bright and dark fringes due to constructive and destructive interference.

Usually one of the two fibers in the system is kept isolated from the environmental effects being measured and is used as a reference. This fiber is called the reference arm. The other fiber is exposed to the measurand and is called the signal, or sensing arm. It is important to keep the reference arm exposed to all other environmental perturbations except for the measurand. Otherwise, it will be difficult to decipher the measurand from other perturbations imposed on the signal fiber. The two arms of the system recombine through a coupler. A detector may be placed at the output fiber of the coupler.

External perturbations include any environmental effects on the fiber that will change the index of refraction, which will affect the optical path length of the signal arm. These include effects such as temperature, pressure, strain, etc. For the Mach-Zehnder interferometer, one output fringe is produced whenever the optical path length of the signal beam (relative to the reference beam), changes an amount equal to the wavelength of source.

### Michelson Interferometer

The Michelson interferometric sensor has the advantage of being able to operate as a single-ended sensor and, as a result, is more sensitive. The Michelson architecture, however, requires more components and less light can be detected. In Fig. 5.22, a laser light source provides coherent light for the Michelson interferometer. The figure does not show the mechanism that is used to couple the laser light into the fiber. This usually consists of a lens and possibly a polarizer.

A coupler is used to split the light into two separate fibers. Like the Mach-Zehnder interferometer, one arm is the signal arm while the other is the reference arm. At the end of each of these arms is a mirror, or a reflective coating on the end of the fiber. The light in the two arms is reflected back to the coupler and therefore transverses the fiber arms twice. The light reflected back down the arms is recombined at the coupler. A detector is placed at the source side of the coupler to receive a portion of the recombined light. As with the Mach-Zehnder, the phase delay of the signal light depends on the optical path difference between the signal and reference arms.

Like the Mach-Zehnder interferometers, the reference arm of the Michelson interferometer is also kept isolated from the measurand. As light propagates through the signal arm twice, the Michelson interferometer is twice as sensitive as the Mach-Zehnder interferometer. If each of these were exposed to the same perturbation, the optical path difference experienced by the Michelson interferometer would be twice that experienced by the Mach-Zehnder. Therefore, the Michelson sensor will produce two fringes for each fringe produced by the Mach-Zehnder sensor for the same perturbation.

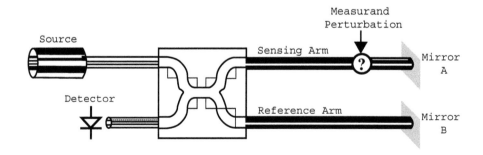

**FIGURE 5.22**   Architecture of Michelson interferometer.

**TABLE 5.4**  Trades Study of Popular Interferometric Fiber Optic sensors

| Interferometer | Advantages | Disadvantages |
|---|---|---|
| Mach-Zehnder | Very sensitive | Integrating sensor, lead sensitive |
| Michelson | Very sensitive | Integrating sensor, lead sensitive |
| Fabry-Perot | Robust, sensitive, point sensor with insensitive leads | Less sensitive than others |

### The Extrinsic Fabry-Perot Sensor

The first successful operation of the quadrature phase-shifted fiber optic EFPI sensor for detection of amplitude and relative polarity of dynamically varying strain was demonstrated in 1991 [38]. Ever since, the EFPI sensors have been widely used to monitor damage and fatigue of materials and structures [39, 40, 41, 42]. The major advantages of the extrinsic fiber optic Fabry-Perot interferometric sensors over the intrinsic versions are minimization of polarization-related problems and sensitivity to axial strains alone [43]. Most of the EFPI sensors involved in these projects were attached to the surface of the sample. The accuracy of the EFPI sensor for measuring strain depends on controllable parameters including the construction process, the efficiency of light coupling between the air gap and the accuracy of measurement of the gauge length of the sensor. The former is important because better coupling between the fiber ends results in better signal/noise ratio and higher dynamic range. The latter is more important than the former because of the sensor's high sensitivity. The sensitivity of an EFPI sensor is a function of the gauge length and the wavelength of the optical source. The gauge length is generally determined during the fabrication of the sensor and it plays an important role in the accuracy of strain measurement. The theory of operation of extrinsic Fabry-Perot fiber optic sensor is straightforward.

The Fabry-Perot reflection sensors use only one interconnection fiber and since it is not part of the interferometer, it does not induce errors from external perturbations. To be accurate, all the fibers included in almost every interferometric scheme including the Fabry-Perot interferometric sensors are part of the sensor. But in case of the Fabry-Perot interferometer, both the reference and sensing signals travel almost exactly the same path length, except for the air gap, hence the effects of any external perturbations are experienced equally by both the reference and sensing signals. As a result, the signals cancel each other. The only remaining critical sensing parameter is the axial spacing or the air gap between the two fiber end faces. Therefore, the Fabry-Perot sensor has a unique point-sensing geometry of only a few millimeters that makes it best suited for strain and shock sensing and a host of other applications. It is less sensitive than the Mach-Zehender and Michelson interferometers. This lower sensitivity of the Extrinsic Fabry-Perot interferometer can be attributed to its sensitivity to change in the air gap of the Fabry-Perot cavity alone. As a result, it is only sensitive to axial strain and strain in other directions does not affect the output of the extrinsic Fabry-Perot interferometer. Table 5.4 summarizes the trade-offs between the three major fiber optic interferometric sensors, i.e., the fiber optic versions of the Mach-Zehender, Michelson, and the Fabry Perot interferometric sensors.

The output intensity variation of an EFPI sensor is sinusoidal and the amplitude envelope of the output intensity varies with strain. In addition, it is the only sensor that is sensitive to axial strain alone.

The construction of a typical EFPI sensor is shown in Fig. 5.23. The fabrication process is simple. A single-mode and a multimode fiber are carefully cleaved and cleaned to produce flat surfaces. The two fibers are then inserted into a hollow core glass tube, which is used to align and hold the two fibers together. Small dots of epoxy are generally used to bond the fibers to the glass tube. Accurate remeasurement of the gauge length is important for the sake of accuracy. The gauge length for this type of EFPI sensor is defined as the distance between the inner edges of the two-epoxy regions inside the hollow core silica tube.

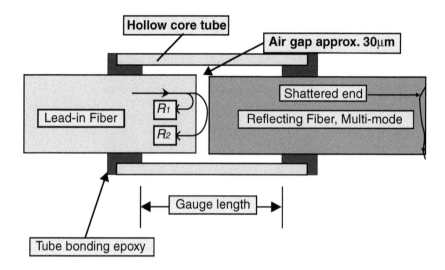

**FIGURE 5.23** Typical construction of extrinsic Fabry-Perot interferomete.

**The Strain-Fringe Relationship of the EFPI Sensor**

The EFPI sensor is one of the most versatile and robust interferometric sensors. Typically it does not require any calibration. The number of optical interference fringes, as seen by the detector, is physically related to the displacement of the air gap ($\Delta L$) and the wavelength of the light ($\lambda$). This relationship can be represented as:

$$m = \frac{2\Delta L}{\lambda} \tag{5.68}$$

The strain-fringe relationship can be given:

$$m = \frac{2L}{\lambda}\varepsilon_x \tag{5.69}$$

The axial strain, $\varepsilon_x$, is defined as the change in length per unit length, ($\frac{\Delta L}{L}$). In the above equation, $L$ is the gauge length of the EFPI sensor. From the above equation, it can be seen that one fringe corresponds to a displacement of 0.3164 $\mu$m of the air gap if a Helium-Neon (He-Ne) laser source is used, as the He-Ne has an operating wavelength of 632.8 nm. For an EFPI sensor having a gauge length of one centimeter, the resolution is 31.64 $\mu$-strains per fringe.

The Fig. 5.24 shows the comparison of strain obtained by electrical strain gauge to that obtained from the EFPI optical strain gauge. The EFPI sensor was embedded in a small cube of epoxy. The dimensions of the cube were 35 $\times$ 35 $\times$ 35 mm, while the gauge length of the sensor was 3.9 mm. The epoxy was subjected to static loading under a loading machine.

The EFPI strain sensor shows excellent linear response and correlates well with the theory and experimental results. This type of sensor is not limited to the measurement of static strain only. Murshid et al. [44] have successfully used the EFPI sensor to measure dynamic strain and dynamic processes such as shock, including a state-of-the-art shock measurement system that can measure shocks at a resonant frequency in excess of 200 kHz.

## Fiber Optic Rotation Sensors

Most oceanographic applications of fiber optic sensors primarily involve strain sensing, but use of the Sagnac [45, 46] effect is also popular and it is commonly applied for short-term rotation measurements. High resolution, low cost, light weight, and near zero startup delay are the key features of fiber optic

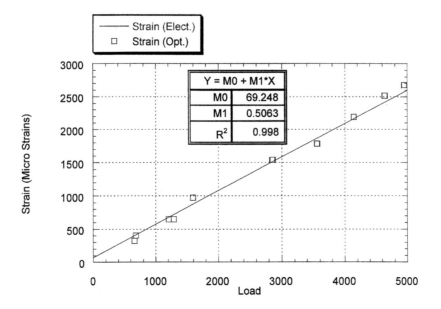

| | Y = M0 + M1*X |
|---|---|
| M0 | 69.248 |
| M1 | 0.5063 |
| R² | 0.998 |

**FIGURE 5.24** Comparison of the EFPI output to the electrical strain gauge.

gyroscopes. These gyros tend to have higher drift over time, hence they are utilized for short-term naval missions. Different approaches have been suggested (including low coherence sources), with varied success to overcome the long-term drift associated with these sensors. The measurement of rotation is of considerable interest for undersea applications, especially in the design of navigation systems. Undersea vehicles requiring autonomous operation at depth must rely upon inertial sensing to obtain positional estimates with reference to a starting location, since GPS data may not be generally available. Under these conditions, an inertial navigation system is employed to keep track of rotational and translational acceleration, from which velocity and position can be obtained by integration. Inertial Navigation Systems (INS) incorporating mechanical gyroscopes and accelerometers have been in use since the 1940s, but the quest for greater accuracy, lower cost, and higher performance has led to interest in optical rotation rate sensors or "gyros." Optical gyroscopes use an interferometric means of sensing rotation based on the Sagnac effect and, if properly designed, can provide much higher accuracy (lower drift rate) than their mechanical counterparts. Historically, the best performing optical gyros utilize a ring-laser configured interferometer of the largest diameter possible. More recently, the "fiber optic gyro" or "FOG" has been proposed to extend the accuracy and lower the cost of ring-laser gyros (RLGs). The fiber gyro *in theory* has all the benefits of the RLG, but with smaller diameter. Problems associated with source coherence, optical scattering, and stability have limited the performance of the FOG to short-range navigation use. Nevertheless, the benefits of no moving parts, no warm up, and low cost have allowed FOGs to be implemented in missions of increasing duration where extreme navigational accuracy is not required.

Fiberoptic rotation sensors rely on the Sagnac effect to produce an optical path difference between counter-rotating light beams proportional to the rotation rate. The goal of the optical system design is to produce two light beams of equal intensity such that one rotates in the clockwise direction while the other propagates in the counterclockwise direction. The rotation rate is sensed measuring the interference between the two coherent beams at an output port. When a vessel containing this optical structure is radially stationary, the optical transit time is the same for both beams. When the plane containing the optical ring rotates in one specific direction, the beam traveling in that specific direction has to cover a slightly longer path as compared to the beam traveling in the opposite direction, before the two beams interfere with each other. This causes a difference in the propagation time taken by the two counter-propagating optical beams. The difference in phases due to propagation delay is detected and processed in order to obtain information about rotation. This can be done with a very high degree of resolution and accuracy.

Consider a fiber rotation sensor wound in a coil of $N$ turns of cross-sectional area $A$. It can be shown that the optical path difference produced between counter rotating beams is given by:

$$\Delta L = 4AN\Omega/c$$

where $\Omega$ is the rotation rate and c is the speed of light in free space. For a sensor mounted on a platform, rotating at 0.015 degrees per hour and having an area of 100 cm squared with only turn of fiber, it can be shown that $\Delta L = 10^{-15}$ cm—a very small number. Therefore, many turns of fiber are generally used to increase the delay for a given rate of rotation in order to improve sensitivity.

The optical gyro has been in use for some time. It provides low drift rates and circular error performance (CEP) when used as part of an inertial navigation system. The performance of the Sagnac interferometer improves with the area of the ring per loop. Hence, the fiber gyro is becoming more practical for high performance applications.

## References

1. Dandridge, H., and Cogwell, G., Fiber optic sensors for Navy applications, *IEEE Lightwave Communications Systems* (LCS), 79–89, 1991.
2. Giallorenzi, T. G., Bucaro, J. A., Dandridge, A., Sigel G. H., Cole, J. H., Rashleigh, S. C. and Priest, R. G., Optical Fiber Sensor Technology, *IEEE Journal of Quantum Electronics*, QE-18, 4, April 1982.
3. Powers, J. P., *An Introduction to Fiber Optic Systems, The Aksen Associates Series in Electrical and Computer Engineering*, 1993.
4. De Paula, R. and Moore, E., Fiber Optic Sensors Overview, SPIE Vol. 566, Fiber Optic and Laser Sensors III, 1985, 2–11.
5. Berthold III, J. W., Historical Review of Microbend Fiber Optic Sensors, *Journal of Lightwave Technology*, 13, 7, July 1995.
6. Lagacos, N., et al., Microbend Fiber Optic Sensor, *Appl. Opt.*, 26, 2171–2180, 1987.
7. Fields, J. N., Attenuation of a parabolic index fiber with periodic bends, *Appl. Phys. Lett.*, 36, 799–801, 1980.
8. Fields, J. N., et al., Pressure Sensor, *J. Acoust. Soc. Am.*, 67, 816–818, 1980.
9. Lagakos, N. and Bucaro, J. A., Optimizing fiber optic microbend sensor, *Proc. SPIE*, 718, 12–20, 1987.
10. Grossman, B. and Cosentino, P., Development and Implementation of Fiber Optic Vehicle Detection and Counter System, Technical Report No. FL/DOT/RMC/06650-726, January 1996.
11. Grossman, B. and Cosentino, P., Fiber Optic Sensors for Measurement of Weigh-in-Motion, Requested paper at the NATDAC '96 Meeting in Albuquerque, NM.
12. Grossman, B., Cosentino, P., Kalajian, E., Kumar, G., Doi, S., and Vergese, J., Fiber Optic Pore Pressure Sensor Development, referred and published in *Proc. of the Transportation Research Board*, 1994.
13. Grossman, B., Cosentino, P., Doi, S., Kumar, G., and Vergese, J., Fiberoptic Pore Pressure Sensor for Civil Engineering Applications, *Proc. of SPIE Fiberoptic Sensors and Smart Structures Conference*, September, 1993, Boston, MA.
14. Miers, D. R., Raj, D., and Berthold, J. W., Design and characterization of fiber optic accelerometers, *Proc. SPIE*, 838, 314–317, 1988.
15. Powers, J. P., *An Introduction to Fiber Optic Systems*, Asken Associates, Homewood, IL, 1993, Chap. 3.
16. Caimi, F. and Murshid, S., Oceanographic equipment and instruments, *Wiley Encyclopedia of Electrical and Electronics Engineering*, Vol. 15, John Wiley & Sons, New York, 1999, 81–93.
17. Wilson, J. and Hawkes, J., *Optoelectronics: An Introduction*, Prentice-Hall, New York, 1989.
18. Powers, J. P., *An Introduction to Fiber Optic Systems*, Asken Associates, Homewood, IL, 1993.
19. Butter, C. and Hocker, G., Fiber optic strain gauge, *Applied Optics*, 17, 18, 2867–2869, September 1978.
20. Costandi, W., A Fiber Optic Sensor for the Simultaneous Detection of Strain and Temperature, Masters Thesis, Florida Institute of Technology, Melbourne, FL, 1992.

21. Murphy, K., Michnel, F., Claus, R., Tran, T., and Miller, S., Optical fiber sensors for measurement of strain and acoustic waves, *Proc. SPIE*, 1918, 110–120, 1993.

22. Farhadiroushan, M. and Giles, I., High birefringence optical fiber pressure sensor, *SPIE*, 949, 162–169, Fiber Optics 1988.

23. Udd, E. ed., *Fiber Optic Sensors: An Introduction for Engineers and Scientists*, Wiley Interscience, New York, 1991.

24. Watanabe, S. and Cahill, R., An overview of Mach-Zehnder interferometric sensors, *SPIE*, 566, Fiber Optic and Laser Sensor III, 16–21.

25. Jackson, D. and Jones, J., Fiber optic sensors, *Opt.*, 33, 1469–1503.

26. Measures, R. M., Fiber Optic Strain Sensing, in *Fiber Optic Smart Structures*, John Wiley & Sons, New York, 1995.

27. Sienkiewicz, F. and Shukla, A., Evaluation of a fiber optic sensor for strain measurement and an application to contact mechanics, *Experimental Techniques*, July/August 1994.

28. Costandi, W., A Fiber Optic Sensor for the Simultaneous Detection of Strain and Temperature, Master's Thesis, Florida Institute of Technology, Melbourne, FL, 1992.

29. Alavie, A., An Interrupt Immune Two-Mode, Single-Ended Localized Fiber Optic Sensor, Ph.D. Dissertation, Florida Institute of Technology, Melbourne, FL, 1992.

30. Blake, J., Huang, S., Kim, B. and Shaw, H., Strain effects on highly elliptical core two-mode fibers, *Optics Letters*, 12, 9, September 1987.

31. Wang, A., Wang, Z., Vengsarker, A., and Claus, R. Two-mode elliptical-core fiber sensors for measurement of strain and temperature, *SPIE*, 1584, 294–303, 1991.

32. Erikhoff, W., Temperature sensing by mode-mode interference in birefringent optical fibers, *Optics Letters* 6, 204–206, 1981.

33. Costandi, W., A Fiber Optic Sensor for the Simultaneous Detection of Strain and Temperature, Master's Thesis, Florida Institute of Technology, Melbourne, FL, 1992.

34. Huang, S., Blake, J., and Kim, B., Perturbation effects on mode propagation in highly elliptical core two-mode fibers, *Journal of Light Wave Technology*, 8, 1, 23–33, January 1990.

35. Calero, J., Wu, S., Pope, C., Chuang, S., and Murther, J., Theory and experiments in Birefringent optical fibers embedded in concrete structures, *Journal of Light Wave Technology*, 12, 6, 1081–1091, June 1994.

36. Sirkis, J., Unified approach to phase-strain-temperature models for smart structure interferometric optical fiber sensor, *Optical Engineering*, 32, 4, 2867–2869, April 1993.

37. Erikhoff, W., Temperature sensing by mode-mode interference in Birefringent optical fibers, *Optics Letters*, 6, 204–206, 1981.

38. Murphy, K. A., Gunther, M. F., Vengsarker, A. M., and Claus, R. O., Quadrature phase-shifted, extrinsic Fabry-Perot optical fiber sensors, *Optics Letters*, 16, 4, 273–275, February 1991.

39. Murphy, K. A., Gunther, M. F., Vengsarker, A. M., and Claus, R. O., Fabry-Perot fiber optic sensors in full-scale fatigue testing on an F-15 aircraft, *Proceedings of SPIE, The International Society for Optical Engineering*, 1588, 134–142, 1991.

40. Carman G. P., Murphy, K., Schmidt, C. A., and Elmore, J., Extrinsic Fabry-Perot interferometric sensor survivability during mechanical fatigue cycling, *SEM Spring Conference on Exp. Mech.*, Dearborn, MI, June 1993, 1079–1087.

41. Carman, G. P. and Mitrovic, M., Health monitoring techniques for composite materials employing thermal parameters and fiber optic sensors, *SPIE*, 2191, 244–256.

42. Murphy, K. A., Gunther, M. F., Tran, T. A., and Claus, R. O., Plastic deformation analysis in metal using absolute optical strain sensor, *SPIE*, 2191, 308–313.

43. Murphy, K. A., Gunther, F. M., Plante, A. J., Vengsarkar, A. M., and Claus, R. O., Low profile fibers for embedded smart structures applications, *Proceedings of SPIE, The International Society for Optical Engineering*, 1588, 2–13, 1991.

44. Murshid, S. and Grossman, B., Fiber optic Fabry-Perot interferometric sensors for shock measurement, Invited Paper, 44th International Instrumentation Symposium, Aerospace Industries and Test Measurements Division, May 1998, Reno, NV.

45. Bergh, R. A., Lefevre, H., and Shaw, H., An overview of fiber optic gyroscopes, *J. Lightwave Technology,* LT-2, 2, 91–107, 1984.

46. Ezekiel, S. and Arditty, H., Eds., Fiber optic rotation sensors and related technologies, Springer series in optical sciences, Springer-Verlag, Berlin, 1982.

47. Rumpf R. C., Fiber Optic Temperature Sensor, M.S. Thesis in Electrical Engineering, Florida Institute of Technology, Melbourne, FL, December, 1997.

# 5.3   Fiber Optic Acoustic Sensors

*Clay Kirkendall and Tony Dandridge*

Fiber optic sensor systems have received considerable interest over the last few decades for use in undersea sensing applications [1]. Fiber optic sensors provide high sensitivity and dynamic range, wide bandwidth, are easily multiplexed, and are suitable for a wide range of sensing applications. In addition, fiber optic sensors are typically completely passive, which is a particular advantage in undersea applications as no electrical power is required at the *sensor* or array level.

Fiber optic sensing as a whole is a varied subject with many different approaches to sensing a variety of different fields. The methods used to couple the desired field to the fiber, as well as the parameter of light which is exploited to make the sensor, have been optimized and refined for each sensing application [2, 3]. Underwater acoustic sensing is no exception and while no single sensor design or multiplexing approach has been selected for all cases, the field has been considerably narrowed. The dynamic range requirement in underwater acoustic sensing allowing for the detection of small pressure changes in the presence of large hydrostatic pressures severely limit the optical interrogation approaches. Almost without exception, high performance fiber optic underwater sensor systems are based on interferometric detection of optical phase shifts resulting from strain coupled into the optical fiber. Despite the limitation on optical interrogation approaches, the field of fiber optic underwater sensing is rather diverse with multiple interferometric configurations and sensor multiplexing schemes seeing use in practical applications. As will be seen, fiber optic transducer design is also very diverse. The mechanical properties of optical fiber provide for a tremendous flexibility in the geometric configuration of the transducer. The mechanical properties of the transducer can be optimized for sensitivity, directionality, bandwidth, etc. with minimal impact from the sensing fiber.

Each interferometric configuration and optical multiplexing scheme has its own set of advantages and disadvantages that can be optimized for a particular application. The following sections will focus on the options for the main building blocks of an interferometric fiber optic underwater sensor system as well as on some of the issues associated with fiber optic sensing. The chapter will start with a brief overview of interferometric sensing, interferometric configurations, and the effects and issues associated with polarization. Then optical multiplexing schemes that have seen widespread use in underwater sensing will be described. Once the groundwork of the optical subsystem has been described, the design of the actual fiber optic sensor transducers will be discussed along with trade-offs and system performance implications. Finally, several fiber optic undersea sensing applications are described along with some results from deployed systems.

## Fundamentals

Practically all fiber optic underwater sensor systems are based on the interferometric measurement of strain induced in the optical fiber by the field to be measured. This is an intrinsic sensing configuration that compares the relative phase of light that has passed through a section of fiber that has been strained through exposure to the measurand field with light that has passed through an unexposed reference path, and allows extremely small strains to be detected. For example, interferometric systems have been built and deployed that can detect one ten-millionth of a radian ($5.7 \times 10^{-6}$ degrees), which corresponds to a strain in 1 meter of fiber of approximately $10^{-14}$. Mach-Zehnder and Michelson interferometers are the

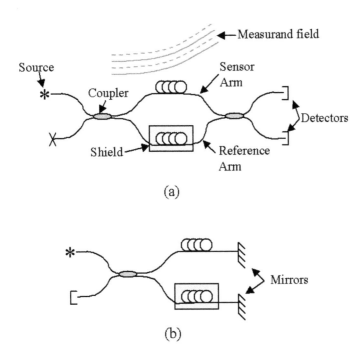

FIGURE 5.25   Mach-Zehnder (a) and Michelson (b) fiber optic interferometer configurations.

most common interferometer configurations used in deployed systems and are shown schematically in Fig. 5.25 along with some of the terms applied to different parts of the interferometer. In the Michelson configuration the light passes through the sensing and reference arms twice (once each way), doubling the optical phase shift. For identical length sensing fiber, the scale factor of the Michelson interferometer is twice as large as the Mach-Zehnder interferometer. In applications where the transmissive properties of the Mach-Zehnder interferometer are not required, the Michelson configuration is usually favored due to the scale factor improvement and lower cost (single coupler).

The total optical phase (in radians) of light traveling through a fiber is given by:

$$\phi = nkL = \frac{2\pi nL}{\lambda} \tag{5.70}$$

where $n$ = the index of refraction, $k$ = optical wavenumber ($2\pi/\lambda$), and $L$ = the physical path length of the section of fiber. Modulation of the index of refraction, wavenumber, or path length of the fiber results in a change in the total phase of the light. The differential phase shift due to a change in the physical path length $L$ is given by:

$$\Delta\phi(L) = \frac{2\pi n\Delta L}{\lambda} \tag{5.71}$$

The most direct way to detect this phase change is to use an interferometer to compare it to light from the same source that is unperturbed. It can be shown that the output intensity of an interferometer with an input signal $\phi_s \cos \omega_s t$ introduced into one arm is given by [1, 5]:

$$I = \frac{I_o}{2}[1 + V\cos(\phi_e + \phi_s \sin\omega_s t)]$$

$$= \frac{I_o}{2}[1 + V(\cos\phi_e \cos(\phi_s \sin\omega_s t) - \sin\phi_e \sin(\phi_s \sin\omega_s t))] \tag{5.72}$$

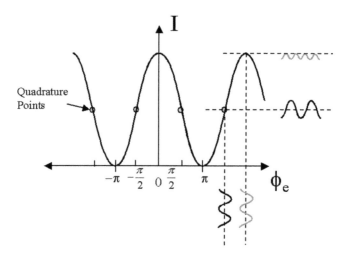

**FIGURE 5.26**   Interferometric response showing signal fading and frequency doubling with small signal input as a function of interferometer phase bias.

where $\phi_e$ is an environmental phase shift and $V$ represents the visibility of the interferometer. The visibility is determined by characteristics of the interferometer such as coupler-splitting ratios, optical loss, polarization effects, and the coherence of the optical source. As can be seen, the output intensity varies with the cosine of the phase shift that contains the desired signal. This is a nonlinear function that will result in interferometric signal fading and frequency doubling, as shown in Fig. 5.26, depending on the value of $\phi_e$. To overcome interferometric signal fading, either the interferometer must be actively locked in quadrature or another means of obtaining quadrature signals must be used such as phase or frequency modulation [4]. The interferometer can be linearized by locking it in quadrature [6], i.e., by forcing $\phi_e = \pm \pi/2$, which reduces Eq. (5.72) to:

$$I = \frac{I_o}{2}[1 + V \sin(\phi_s \sin\omega_s t)] \Rightarrow \frac{I_o}{2}[1 + V\phi_s \sin\omega_s t]$$

for small $\phi_s$. This approach requires feedback to each sensor and is generally not practical for an array of sensors. Phase or frequency carriers can be used to overcome interferometric signal fading and effectively linearize the interferometric response. The approach used most often in deployed systems to date introduces a high frequency phase modulation on the interferometric output and is commonly referred to as phase generated carrier (PGC) [4]. Expanding Eq. (5.72) to include phase carrier $\beta \sin \omega_\beta t$ gives:

$$I = \frac{I_o}{2}[1 + V \cos(\phi_e + \phi_s \sin\omega_s t + \beta \sin\omega_\beta t)]$$

Expanding the $\cos(\sin(...))$ terms using Bessel functions results in:

$$I = \frac{I_o}{2}[1 + V(J_0(\beta)\cos\Theta - 2J_1(\beta)\sin\omega_\beta t \sin\Theta + 2J_2(\beta)\sin 2\omega_\beta t\cos\Theta$$
$$-2J_3(\beta)\sin 3\omega_\beta t\sin\Theta + \cdots)]$$

where $\Theta = \phi_e + \phi_s \sin \omega_s t$. The amplitude of the signal at $\omega_\beta$ is proportional to the sine of the desired signal, while the amplitude of the signal at $2\omega_\beta$ is proportional to the cosine of the desired signal. Downconverting the signal at $\omega_\beta$ and $2\omega_\beta$ gives the sine and cosine of the desired signal phase, as shown schematically in Fig. 5.27. If $\beta$ is chosen such that $J_1(\beta) = J_2(\beta)$, then the amplitude of the sine and cosine terms will be balanced. Converting from the rectangular coordinate system to polar coordinates by taking the arctangent of $\sin\{\Theta\}/\cos\{\Theta\}$ or using a differentiate and cross-multiply demodulator [2]

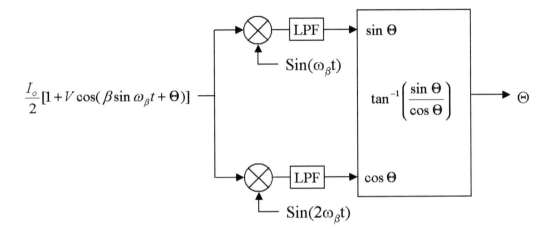

$$\frac{I_o}{2}[1 + V\cos(\beta\sin\omega_\beta t + \Theta)]$$

**FIGURE 5.27**  Phase-generated carrier down-conversion and demodulation.

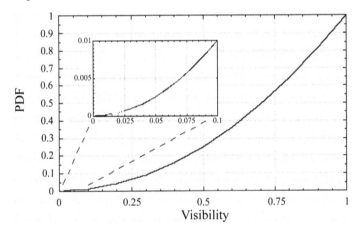

**FIGURE 5.28**  PDF of the visibility of a fiber optic interferometer with random polarization states.

recovers the phase of the interferometer, which is directly proportional to the input signal (i.e., strain in the fiber). Other methods for demodulating interferometric sensors that are gaining acceptance include straight heterodyne approaches where a frequency carrier is used with FM demodulation and hybrid approaches where optimized phase and frequency carrier amplitudes are used, minimizing the demodulation electronics [7].

## Polarization

One factor in the interferometric visibility, V in Eq. (5.70) is the state of polarization (SOP) in the two arms of the interferometer. If the SOP of the optical signal from the two interferometer paths are not aligned when recombined, they will not perfectly interfere and polarization-induced fading (PIF) of the optical signal will result [8]. If the SOP of the two beams happen to be orthogonal to each other, there will be no interference, i.e., $V \Rightarrow 0$ in Eq. (5.72), and as a result no signal will be recovered.

Approaches to addressing PIF include active input polarization control [9], the use of Faraday-rotator mirrors [10], and polarization diversity receivers (PDR) [11]. If nothing is done to address PIF, the SOP for the two beams can be assumed to be randomly distributed and the probability distribution function (PDF) (or cumulative distribution function) for the visibility can be calculated [12]. Figure 5.28 shows the PDF for an individual sensor with no polarization control. From Fig. 5.28, the probability that the

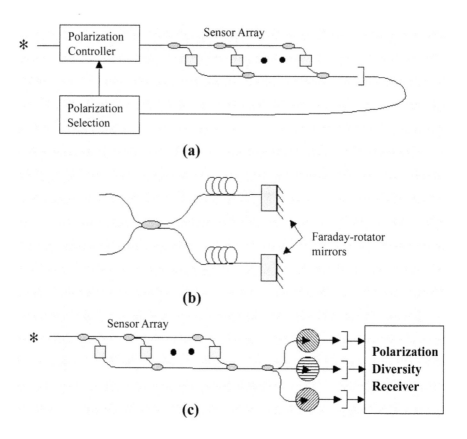

**FIGURE 5.29**   Three approaches to overcoming or eliminating polarization-induced fading: input polarization control (a); Faraday-rotator mirrors (b); tri-cell mask PDR (c).

PIF is greater than 50% is just 25%, i.e., statistically the optical interference signal will be less than half strength, one-quarter of the time. The probability that the optical signal is less than quarter strength is just 1/16. The probability of a complete signal fade is unlikely, but it can happen. In many applications even a small probability of a complete signal fade, which corresponds to a sensor dropout, is unacceptable and some method to prevent a complete signal fade must be used.

By controlling the input SOP, one can always find an input SOP that will perfectly align the SOP of the two paths at the output of a single interferometer, to provide optimum interference [9]. This approach has been extended to arrays of sensors, shown in Fig. 5.29(a) [13–15], where it has been shown that with input polarization control an input SOP can always be found to guarantee the worst PIF for the array will be less than $\sin(\pi/2N)$, where $N$ equals the number of sensors. For example, the maximum polarization-induced signal fade for 4-, 8-, and 16-element sensor arrays with input polarization control is 38, 20, and 10%, respectively. This approach has not seen much use due to the restriction on array size and the difficulty in tracking the optimum input SOP. The use of Faraday-rotator mirrors as shown in Fig. 5.29(b), which guarantees perfect SOP alignment in most applications [10], has only seen limited use primarily due to the cost of the components.

A common approach to overcoming PIF in practical systems is the use of a polarization diversity receiver (PDR) as shown in Fig. 5.29(c) [11]. For the tri-cell PDR, the input optical signal is split into three paths and each one is interfered across a polarizer at a different orientation. Optimum performance is achieved when the polarizers are 60° apart, which guarantees that the worst signal fade on one of the three polarizers is just over 50%. By continuously selecting the largest of the three outputs for demodulation, the PDF for the tri-cell PDR can be predicted as shown in Fig. 5.30. The tri-cell mask uses the minimum number of polarizations to prevent a complete signal fade. A PDR with two masks reduces

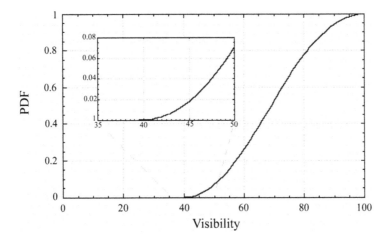

**FIGURE 5.30** Visibility PDF for the tri-cell mask PDR.

the probability of a complete fade compared to no polarization control, but can still fade completely. Using more than three masks will reduce the maximum signal fade, but it will also add more optical attenuation and require more demodulation electronics. It should be noted here that although the tri-cell mask approach protects against complete signal fades, the added attenuation from the signal splitting and polarizers must be accounted for when comparing approaches. For comparison purposes, the tri-cell mask approach guarantees a return from each sensor that is typically 7 to 10 dB down relative to the receiver input level. Conversely, if nothing was done to address PIF, then the probability, from Fig. 5.28, that any individual sensor has a greater than 10 dB fade is just 1% (of course, this is per sensor; for an array, the probability that at least one sensor has faded is quite a bit larger). The PDR does provide a guaranteed optical power budget that allows system designers to allocate their optical power and design a system that will meet requirements for all polarizations.

## Multiplexing

To realize the cost advantages of fiber optic sensor systems, some form of sensor multiplexing must be performed to amortize the opto-electronics cost over multiple sensors. Fortunately, fiber optics lends itself to multiplexing and there are a large number of approaches to multiplexing fiber optic sensors [2, 3]. In practice, only a few of these approaches have seen widespread use in underwater systems and those will be the focus in this section.

Figure 5.31(a) shows a 2 × 2 PGC frequency division multiplexed (FDM) sensor array utilizing two sources to interrogate two sensors each. The outputs of the sensors are then recombined in such a way that each return fiber carries one output from each source. If different phase carrier frequencies are used for each source, $\omega_1$ and $\omega_2$ in Fig. 5.31(a), then the output from the sensors on each return fiber can be separated using frequency selection in the receiver. The concept can be expanded to a generic N * M FDM matrix as shown in Fig. 5.31(b). The multiplexing penalty with the FDM architecture is the optical splitting loss, 1/( N * M), and the increased shot noise on the detector due to N return signals per detector. FDM is a high performance multiplexing approach with continuous interrogation allowing for wide bandwidth and high dynamic range.

Time division multiplexing (TDM) is typically achieved using a pulsed source to interrogate an array of interferometric sensors with different optical delays built into their respective paths. Figure 5.32(a) shows a forward coupled ladder architecture where each successive sensor has an added delay T. These fiber delays typically consist of a few tens of meters of optical fiber with negligible loss. A serial TDM architecture is shown in Fig. 5.32(b). Here, the fiber used to make the sensor is also used for the delay. This constrains the sensor placement and timing, but the sensors and architecture are considerably simpler with significantly

**TABLE 5.5**  Trade-Offs and Performance Issues

| Multiplexing | Sensor Configuration | # of Fibers | # of Sensors | Sensors per Fiber | Noise | Comments |
|---|---|---|---|---|---|---|
| FDM | Both | N+M | N * M (50–100) | M (4–16) | Low | Continuous interrogation Wide bandwidth High dynamic range |
| TDM | Transmissive Reflectrometric | 1 1 | 10–100 8–32 | 10–100 8–32 | Med. | Low cost Limited bandwidth High multiplexing gain |
| Hybrid WDM/TDM | Transmissive Reflectrometric | 1 1 | 40–1600 32–512 | 40–1600 32–512 | Med. to Low | Low cost Limited bandwidth Highest multiplexing gains |

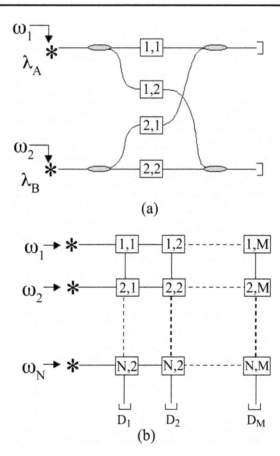

**FIGURE 5.31**  Two-by-two FDM array (a) and a generalized N-by-M FDM array (b).

fewer components, which lowers cost. The couplers in Fig. 5.32(b) can be replaced with partial reflectors, further simplifying the architecture, but cross-talk due to multiple reflections must be taken into account. The multiplexing penalty with TDM is the optical splitting loss that goes as $1/N^2$ ($N$ = the number of sensors) and aliased noise due to the inherent undersampling of the optical return [16].

Hybrid approaches utilizing both wavelength division multiplexing (WDM) and TDM offer the highest channel counts per fiber. With the proliferation of optical amplifiers, hybrid WDM/TDM sensor systems with channel counts of 200 to 1000 sensors per fiber and standoffs/array spans of several hundred kilometers are becoming practical. Table 5.5 presents some of the trade-off and performance issues associated with the multiplexing schemes presented above.

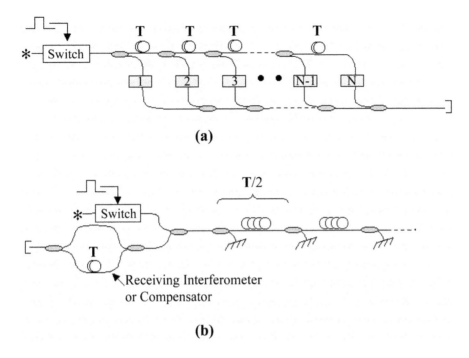

(a)

(b)

**FIGURE 5.32** TDM forward-coupled ladder array (a) and serial array (b).

## Hydrophone Design

The development of fiber optic hydrophones has been ongoing for the last 25 years. As stated previously, because of the requirement of high dynamic range and sensitivity (often determined by the fact that they have to detect very small pressure changes in the presence of large hydrostatic pressures), interferometric techniques have shown themselves to be the only viable approach. The fundamental approach is to measure acoustically induced strain in a length of optical fiber. In the early 1980s at the Naval Research Laboratory (NRL), a considerable effort was expended developing mircobend sensors (i.e., amplitude-based sensing) for this application. However, the requirement of detecting the required pressure signals (i.e., acoustic), which are typically ~200 dB smaller than the ambient hydrostatic pressure, proved too great a technical challenge and the approach was abandoned. Throughout the 1980s and 1990s, numerous groups tried other approaches, some employing diaphragms with Fabry-Perot interferometric approaches, others using Bragg gratings as the sensing elements. To date none of these approaches have come close to achieving the performance of the fiber interferometer systems, which alone appear able to meet the stringent specifications required for hydrophone systems.

The first work on fiber hydrophones dates from 1977 with work at NRL and TRW. This work used simple coils of optical fiber with only the primary coating on the fiber. The interferometer was formed with bulk optical components and only the coil of fiber was immersed in the water. In these devices the acoustic field interacted with the silica optical fiber, which is relatively incompressible. Consequently, the strain induced in the fiber was small and these devices were very unresponsive. At this point, it is necessary to introduce two terms that are important in defining the characteristics of fiber hydrophones. First is the hydrophone's *responsivity*, which is defined as the optical phase shift generated by a change in pressure, in this case by the acoustic field ($d\phi/dP$), typically measured in dB re radian/$\mu$Pascal (Rad/$\mu$Pa). As this parameter is directly proportional to the length of sensing fiber in the hydrophone, it is useful to introduce the term *normalized acoustic responsivity* (NAR), which normalizes the hydrophone reponsivity by the length of sensing fiber (i.e., $((d\phi/\phi)/dP)$). The units of this term are inverse $\mu$Pa with typical measurements in dB re 1/$\mu$Pa. From Eqs. (5.70) and (5.71), $d\phi/\phi = \Delta L/L = \zeta\varepsilon_f$, where $\varepsilon_f$ is the strain in the fiber and $\zeta$ is the strain optic correction factor (~0.78 for silica fiber) [1]. These early demonstrations of "bare"

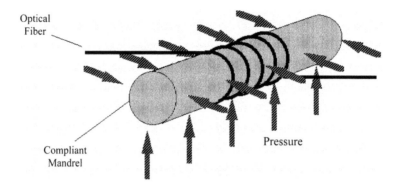

**FIGURE 5.33**   Fiber optic mandrel hydrophone.

fiber hydrophones led to hydrophones with normalized acoustic responsivities of $\sim -340$ dB re $(1/\mu\text{Pa})$. With typical lengths of $\sim 30$ meters of sensing fiber, hydrophone responsivities of approximately $-170$ dB re rad/$\mu$Pa are achieved. Using Eq. (5.70), the smallest acoustic field that can be resolved with this sensor is determined by the optical noise floor of the system, $\phi_{\min}/\sqrt{\text{Hz}}$, and the hydrophone responsivity. As discussed previously, the noise floor of a fiber optic sensor system is determined by a number of parameters. If one is interested in interrogating a single sensor, an optical noise floor in the acoustic band of $\sim -120$ dB re rad/Hz is achievable. Even with the bare fiber sensor described, a noise floor close to a quiet ocean environment (below 1 kHz) can be achieved.

In early hydrophone development it became clear that a greater responsivity to the acoustic field was desirable. Two approaches were investigated in the early 1980s: (a) coating the fiber with an acoustically compliant material, and (b) wrapping the fiber around an acoustically compliant structure (a mandrel). A number of different materials were investigated. In the coated fiber case, the material had to be compressible (low bulk modulus) while having the ability to couple the acoustically induced strains to the "stiff" fiber (requiring a high Young's modulus of the coating material). In the cylindrical mandrel case, shown in Fig. 5.33, the responsivity is determined by the compressibility of the mandrel material assuming that the fiber wrap follows the mandrel. The responsivity of this structure is given by $1/(3 * B)$ where B is the Bulk modulus. With materials such as nylon, the NAR is approximately $-325$ dB, which results in hydrophone responsivities of $-155$ to $-160$ dB re rad/$\mu$Pa. To avoid low-frequency resonances, plastics with high Young's modulii were used.

While this type of sensor, interrogated with a low noise electro-optic system behaved well in static environments (in 1983 NRL demonstrated a nylon mandrel based fiber hydrophone at sea that demonstrated a $\sim 10$ dB below sea state zero noise), similar design exhibited poor noise performance in the more dynamic towed array environment. This towed array test (also conducted in 1983) demonstrated that incorrectly mounted, a plastic mandrel based hydrophone could become very sensitive to acceleration, which can lead to poor noise performance in dynamic environments. Testing this type of sensing element for towed arrays continued at NRL until 1987, where by suitable mounting they achieved noise performance within 1 to 2 dB of the best conventional designs.

In this time frame, the solid plastic mandrel was superceded by thin-walled, air-backed metal hydrophones shown schematically in Fig. 5.34. In this design, pioneered independently by Goodyear, Litton, and NRL, the radial strain of the composite thin-wall cylinder is [17]:

$$\varepsilon_{\text{thin}} = -\frac{Pr}{Yt}\left(1 - \frac{v}{2}\right)$$

where $r$ is the cylinder radius, $Y$ the Young's modulus of the composite wall material (including the fiber and adhesive), $t$ the wall thickness, and $v$ is the composite Poisson's ratio. This expression is based on thin wall theory and is generally considered accurate for $r/t > 10$. For the first time, hydrophones could

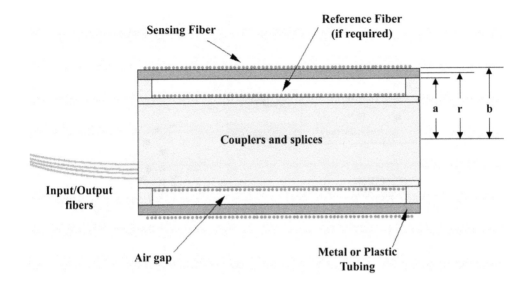

**FIGURE 5.34**  Fiber optic air-backed mandrel hydrophone design.

be designed to optimize performance for specific operating hydrostatic pressure requirements (typically by modifying the wall thickness), rather than relying on the material properties of plastics. Typical responsivities were increased to ~−140 dB re rad/$\mu$Pa for operating pressures up to ~1000 psi using wall materials such as aluminum or brass. A number of successful demonstrations were carried out using this design of fiber hydrophone including a 48-channel towed array in 1990 (built by Litton under contract to NRL), a 49-channel planar array, as well as a fixed surveillance array built and tested by NRL in the Arctic in 1990 and 1992. One issue noted with this hydrophone design was that for small diameter mandrels (required for certain towed array and planar array applications), very thin walled tubing was required to maintain responsivity. This required machining, which not only led to increased cost, but to nonuniformity of the wall, which sometimes led to premature buckling of the tube under high hydrostatic pressures.

To overcome deficiencies in the air-backed, thin-walled metal fiber hydrophone design, NRL began work to utilize air-backed, polycarbonate tubing based sensors for the high performance sensor arrays in the Arctic. Because of the lower Young's modulus of the polycorbonate wall material, greater wall thicknesses are possible. Modeling the transducer as a thick-walled cylindrical shell results in a composite radial strain of [18, 19]:

$$\varepsilon_{thick} = -\frac{P[b^2(1-2v)+a^2(1+v)]}{Y(b^2-a^2)}$$

where $a$ is the inner radius and $b$ is the outer radius of the cylinder. This equation is accurate for $r/t > 10$. The use of plastic tubing had the added benefit that these designs allowed an approximate 10 dB increase in responsivity for the same operating pressure. Hydrophones of this design were tested by NRL in the Arctic in 1992. Fiber hydrophone responsivities for towed array and planar array applications are typically −130 dB re rad/$\mu$Pa, while for bottom-mounted arrays, −120 dB re rad/$\mu$Pa are routinely achieved. For shallower water operation, NRL has demonstrated hydrophones with responsivities of ~−110 dB re rad/$\mu$Pa. Currently designs of fiber hydrophones for Navy applications use air-backed plastic mandrels. Successful demonstrations of this technology include a number of bottom-mounted surveillance arrays (both vertical line arrays (1994) and horizontal line arrays (1996)). Arrays of this design are now in the preproduction phase of development. A number of very successful towed array tests have been conducted for thin-line array applications using hydrophones of this design. The fiber optic towed array is in the

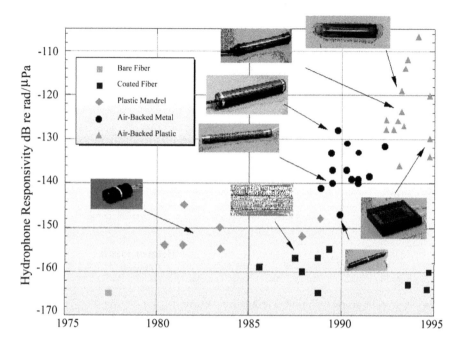

**FIGURE 5.35**   Evolution of fiber optic hydrophone responsivities.

process of being transitioned to production. The current next generation planar array (now in production) for the Virginia class attack submarine utilizes air-backed polycarbonate mandrels. The evolution of fiber optic hydrophone design and responsivities is shown in Fig. 5.35. The trend to higher responsivities is not expected to continue since the strain in the fiber resulting from the hydrostatic pressure is near the material limits. The responsivity could only be significantly increased by dramatically limiting the depth capability.

## Applications

### Towed Arrays

Probably the major enabling advances in fiber optic sensor technology have been in the areas of packaging and multiplexing. For example, in the towed array area (in which an array of hydrophones is towed behind a ship for military or geophysical applications), by 1983 NRL had tested fiber optic hydrophones in a towed array at sea, which in the laboratory environment provided the required 10 dB below sea state zero threshold detection. These hydrophones were interrogated by the same basic phase generated carrier approach used today with current fiber optic sensor systems; they used 2 × 2 fiber couplers, semiconductor diode lasers, and employed noise-reduction circuitry. However, these hydrophones in the towed environment produced noise levels considerably in excess of their piezoceramic counterparts; the problem was associated with the fiber optic hydrophone's response to extraneous noise fields. The generic approach to fiberoptic towed arrays is shown in Fig. 5.36.

Today's towed array fiber optic hydrophones are ~25 dB intrinsically more responsive to acoustic fields, but are 50 dB less responsive to extraneous noise fields. This improvement comes directly as a result of packaging the fiber optic interferometer. The laser sources, optical components, and demodulators used in the U.S. Navy's December 1990 tow test of the 48-channel All Optical Towed Array (AOTA), a joint NRL/Naval Underwater Warfare Center (NUWC) program, were similar to those used in 1983. However, this multiplexed system demonstrated state-of-the-art flow noise performance at the array level as well as the hydrophone level, meeting or exceeding the best conventional piezoceramic towed arrays. Often the packaging innovations are very specific to particular applications and as such may be of limited

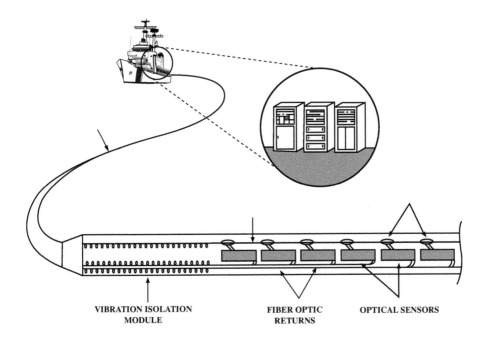

**FIGURE 5.36**   Fiber optic towed array.

interest to a general fiber sensor audience. One final part of the AOTA program was the "at sea" demonstration of fiber optic sensors for array depth, temperature, roll, pitch, and heading. The Navy's development of fiber optic sensors for towed arrays is continuing, especially in the area of thin-line towed arrays where the technology is transitioning for a projected production date of 2004.

### Fixed Arrays

Reducing the size and weight of hydrophone arrays to make them easier to transport and deploy has received considerable attention. In the case of a vertical line array, as the array elements become smaller and the cable becomes thinner, there is less drag by the water currents on the array and it will stay vertical with less flotation and less inherent cable strength. In 1994, NRL deployed a 16-element vertical line acoustic array using frequency division multiplexing interrogation. This 16-element vertical array is 55 meters in length and is interrogated over a 10 km tether cable. The riser cable to which the hydrophones are connected is a tapered cable containing nine optical fibers at its base, and only the Kevlar strength member and the fairing at the top. The riser cable is 3 mm in diameter at the base, excluding the haired fairing. It is constructed by binding nine optical fibers, each buffered to 0.9 mm with Hytrel, and the strength member together in a weaving process, which also forms the fairing. The hydrophones are connected to optical fiber breakouts, which were made during the cable manufacturing process from the riser cable. The hydrophones, breakout couplers, and splices are potted in a hydrodynamically shaped fin, in which the leading edge contains the riser cable. The array is deployed at a distance from the tending ship (or shore) and interrogated optically over the tether cable. There are no electronic components between the tending ship and the vertical line array. All of the hydrophones had a flat frequency response between DC and 20 kHz (the system bandwidth for all 16 channels) and had the same responsivity to within 0.5 dB, the responsivity did not vary with temperature or hydrostatic pressure (up to 2900 psi). This system was tested in a Fjord in Norway and was collocated with the 8-element MARS magnetometer array (discussed below). The configuration and a sample of spectral data collected over a period of time are shown in Figs. 5.37 and 5.38.

For many systems, however, the number of fibers required in the FDM matrix approach is unattractive, especially for systems where there is a very long distance between the sources/receivers/electronics and the passive array, such that a cable with a large fiber count is very expensive. For this type of system, TDM

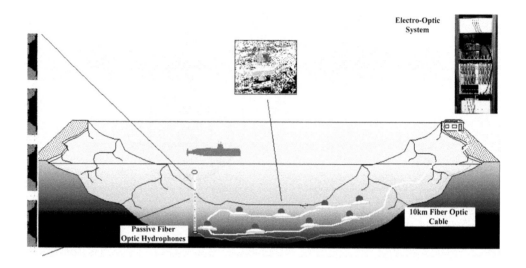

**FIGURE 5.37**   Fiber optic magnetic sensor array and vertical line hydrophone array.

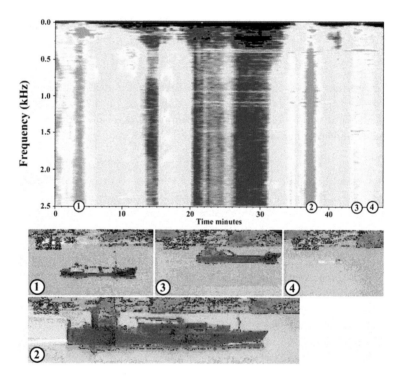

**FIGURE 5.38**   Frequency-time plot of one hydrophone in the vertical line array showing the signal from different targets.

multiplexing augmented with WDM multiplexing becomes extremely attractive. In 1996, NRL/NRAD deployed a hybrid WDM/TDM array capable of interrogating 128 fiber optic sensors over a single fiber [19, 20]. In the demonstration two wavelengths were used to interrogate two half-populated (to reduce cost) TDM arrays for a total of 64 hydrophones. This demonstration included an optical amplifier telemetry module to increase the standoff distance to shore. A system schematic is show in Fig. 5.39. The source

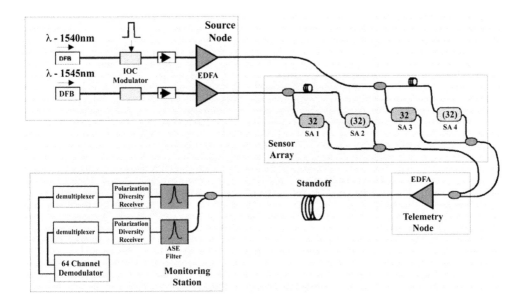

**FIGURE 5.39** WDM/TDM fiber optic bottom-mounted array system. Source node is capable of interrogating 128 fiber optic sensors, but array is only half-populated.

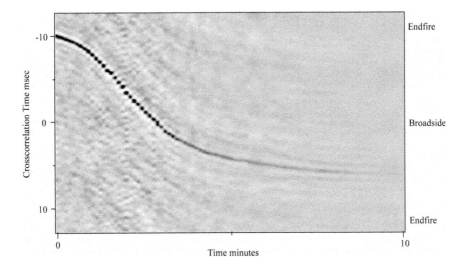

**FIGURE 5.40** Cross-section between two sensors of a passing vessel.

electro-optics and telemetry nodes were battery powered and the array was totally passive. Figure 5.40 shows the cross-correlation between two channels as a target passes by the array.

Unlike conventional sensors, which typically have band-limiting filters built into the sensor, fiber optic hydrophones are wide bandwidth devices operating from DC to tens of kilohertz. This allows for the recovery of the surface wave action, which is well below the frequency range of interest, and is shown in Fig. 5.41.

For most applications, the detection threshold performance of 10 dB below sea state zero (SS0) and light shipping level, demonstrated by most fiber optic hydrophone systems, is satisfactory. However, some applications require threshold detection considerably better than SS0. One such application area is that of acoustic measurements in the Arctic, shown in Fig. 5.42. The electro-optic system consisted of a Nd:YAG diode pumped laser (1.3 $\mu$m operation), commercially available cable (2 km in length), low

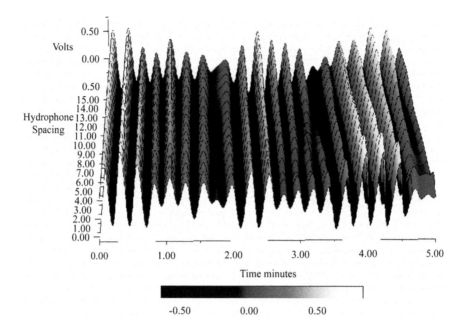

**FIGURE 5.41**   Reconstruction of surface wave-crossing, bottom-mounted array.

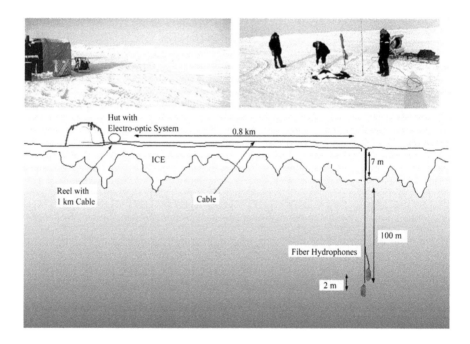

**FIGURE 5.42**   Arctic acoustic test.

noise demodulators, and noise reduction circuitry developed at NRL. In April 1990, two fiber optic hydrophones were used to measure the ambient acoustic noise under shore-fast ice at the mouth of Independence Fjord in the vicinity of Kap Eiler Rasmussen, Greenland. The system operated without any difficulty for the entire nine-day period of the experiment. During the experiment extremely quiet data was obtained with acoustic noise levels at 500 Hz corresponding to 26 dB below SS0 and a system noise floor of 33 dB below SS0.

An array consisting of six fiber optic hydrophones and two diagnostic channels was tested under pack ice in the vicinity of 87.45°N 14°W in the spring of 1992. The primary purposes of this test were to demonstrate the operation of an array of fiber optic hydrophones in this environment as well as to improve the low-frequency noise performance of the previous fiber optic hydrophone system tested in the Arctic. An additional smaller array was interrogated simultaneously with the 6-element array. The hydrophone elements in the 6-element array were placed 40 m apart, and the first hydrophone element was located 40 m below the ice surface. The array was deployed through a hole in the ice located 47 meters from the shelter, and was connected to the interrogation system in the shelter with a fiber optic cable.

A single laser (Lightwave Electronics Model 120) was used to interrogate all eight fiber optic hydrophones, as well as three diagnostic channels. The coupling ratios of the array breakouts were chosen such that an equal amount of optical power was present at each device. Individual returns from each hydrophone were brought back through the fiber optic cable and detected and demodulated inside the shelter. The demodulation technique was the standard phase-generated carrier and a subtraction technique was used for laser noise reduction. Threshold detection of ~30 dB below sea state zero with improved low frequency performance was demonstrated with this system. The entire experiment was operated on battery power. The development of the high responsivity air-backed plastic hydrophones led to the ability to interrogate larger numbers of hydrophones with TDM interrogation.

## Planar Arrays

Theoretically, in a shot noise-limited system, the most efficient time and frequency division multiplexing approaches have similar performance for architectures without severe cross-talk. Yet, at the time the determination had to be made as to the multiplexing approach for the Light-Weight Planar Array (LWPA) program (a submarine-mounted acoustic array), the maturity of the frequency division approach led to its choice for this multielement (~50 elements) demonstration effort. Figure 5.43 shows the concept of LWPA. LWPA, which is a joint NUWC/NRL Advanced Technology Demonstration of a 56-channel fiber optic array (consisting of 49 planar fiber optic hydrophones and 7 diagnostic channels), is designed to meet a stringent set of Navy specifications. The array was tested for acceleration sensitivity and reflection gain on a full-scale submarine hull section at Seneca Lake, NY in January and February of 1993. The same array was taken to Lake Pend Oreille and flow-noise tested on the KAMLOOPS buoyant submarine model in April of 1993. Litton Guidance and Control Systems and Martin Marietta constructed the array under contract to the Naval Undersea Warfare Center (NUWC). The 56 channels were frequency division multiplexed and demodulated with phase-generated carrier techniques. The fiber optic hydrophone array was capable of unattended operation under battery power since the KAMLOOPS model was not

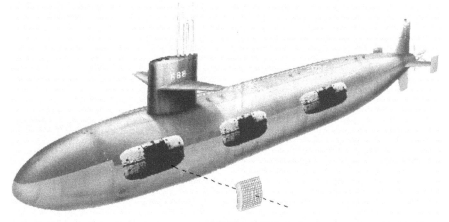

Planar Array of Hydrophones

**FIGURE 5.43**  Planar hull array concept.

manned during testing. The LWPA array met or exceeded all of the relevant performance requirements. This system is now in production for the Virginia class attack submarine.

## Trends

Recent advances in WDM and optical amplifier technology are impacting fiber optic sensor system (FOSS) architectures and capabilities. Ten years ago, WDM was not practical due to the lack of availability of sources and multiplexing components. Today, mainly due to the telecommunications industry, WDM technology is fairly mature with multiple sources for lasers and multiplexing components. Similarly, the capability to optically amplify the signal in the fiber has had a profound effect on FOSS architectures. Optical amplifiers work on the average, not peak, signal power and as a result can amplify low duty cycle pulses to very high powers quite efficiently. This overcomes the high optical loss in large TDM arrays enabling TDM technology. In the past, the span (source to detector) of a FOSS was limited by the available power of the source. Now, optical amplifier repeaters allow for spans of several hundred kilometers, dramatically increasing the capabilities. These advances are changing the direction of fiber optic sensor technology.

## Non-Acoustic Sensing

Although acoustic applications have received the most interest in underwater fiber optic sensing, a number of non-acoustic applications have been investigated. In the early 1990s, magnetic field sensor arrays were investigated for military applications and more recently acceleration and displacement sensors have been receiving considerable interest. As previously discussed, fiber optic interferometers are highly sensitive instruments for measuring strain induced in an optical fiber. As such they are suitable for a wide variety of sensing applications limited only by the requirement that a transducer can be devised to convert the desired measurand field to a strain or displacement.

Intrinsic fiber optic magnetic field sensors have been fabricated using magnetostrictive materials to convert magnetic fields to strain [22]. Bonding a section of optical fiber to the magnetostrictive material couples the induced strain into the fiber where it can be detected using interferometric approaches. In the 1 mHz to 10 Hz frequency range (the range of interest) thermal drifts in the interferometer severely limit the performance of the sensor. One way to achieve the desired performance is to use the nonlinear response of the magnetostrictive material to up-convert the low frequency magnetic information by means of a magnetic dither field applied to the sensor [22]. In 1993, an array of 24 fiber optic magnetometers configured as eight vector magnetometers was deployed and operated continuously for one year in a fjord in Norway [23, 24] (see Fig. 5.37). Figure 5.44(a) shows the track of a target passing the array and Fig. 5.44(b) shows the response from the sensor at the closest point of approach. This application

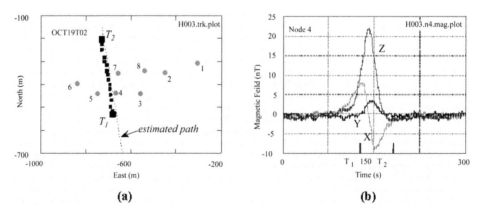

(a)                                                                (b)

**FIGURE 5.44**  Track of a target over the MARS system (a) and the signals from the vector magnetometer at the closest point of approach (b). The circles in (a) represent the vector magnetometer locations.

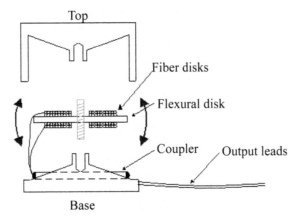

Top

Fiber disks

Flexural disk

Coupler    Output leads

Base

**FIGURE 5.45**    Schematic of an exploded flexural disk fiber optic accelerometer package.

significantly differs from previously discussed applications in two ways: a single interferometer was used to multiplex multiple sensors, and active feedback was used to bias the interferometer and magnetic sensors. Three magnetic sensors and one node telemetry channel were multiplexed inside each interferometer using frequency division multiplexing (operating on the nonlinear response of the magnetostrictive material). By up-converting the magnetic signal to a lower noise operating point of the interferometer, a minimum detectable phase of 0.1 $\mu$rad/$\sqrt{Hz}$ was achieved. This is 10 to 20 dB lower than any other deployed array. One advantage of this multiplexing approach is that the cost of the Faraday-rotator mirrors used to overcome PIF can be amortized over multiple sensors making their use more affordable. The requirement for a powered node removes one of the main advantages of fiber optics in undersea applications, a totally passive array. Optically powering the dither electronics could alleviate this disadvantage.

The advantages of fiber optic undersea sensing have more recently drawn attention in oil exploration and oil well management applications. For these applications there is significant interest for acceleration and displacement (geophone) sensors in addition to conventional hydrophones. A number of different approaches have been developed for fiber optic accelerometer and displacement transducers [25, 26]. One approach is the use of a flexural disk, as shown in Fig. 5.45, which converts the displacement of the fiber disk, due to acceleration, to a strain in the fiber [27, 28]. This same basic approach can be used to detect displacement by selecting the proof mass and flexural disk properties such that the frequency of interest is above the mass-spring resonance of the structure. The Navy has been involved in the development of both acceleration sensors and displacement sensors for acoustic detection, primarily for hull arrays. These sensors have seen little field testing to date (one 16-channel system employing a mix of both displacement and acceleration sensors was performed by NRL in 1996) but several major systems are in progress. Recently, there have been significant advances in the performance of fiber accelerometers with noise figures of $-150$ dB re g/$\sqrt{Hz}$ being realizable over a wide bandwidth (10 kHz).

## References

1. Giallorenzi, T. G., Bucaro, J. A., Dandridge, A., Siegel, G. H., Cole, J. H., Rashleigh, S. C., and Priest, R. G., Optical fiber sensor technology, *IEEE J. Quant. Electron.*, 18, 626–665, 1982.
2. Kersey, A. D., Dandridge, A., and Tveten, A. B., Overview of multiplexing techniques for interferometric fiber sensors, *SPIE Proc. Fiber Optic & Laser Sensors V*, 838, 184–193, 1987.
3. Kersey, A. D. and Udd, E., Eds., *Fiber Optic Sensors: An Introduction for Engineers and Scientists*, John Wiley & Sons, New York, 1991, chap. 11.

4. Dandridge, A., Tveten, A. B., and Giallorenzi, T. G., Homodyne demodulation scheme for fiber optic sensors using phase-generated carrier, *IEEE JQE*, 18, 10, 1982.

5. Dandridge, A. and Udd, E., Eds., *Fiber Optic Sensors: An Introduction for Engineers and Scientists*, John Wiley & Sons, New York, 1991, chap. 10.

6. Dandridge, A. and Tveten, A. B., Phase compensation in interferometric fiber optic sensors, *Optics Letters*, 7, 6, 1982.

7. Bush, I., Cekorich, A., and Kirkendall, C. K., Multi-channel interferometric demodulator, *Proc. of 3rd Pacific Northwest Fiber Optic Workshop*, 1997.

8. Stowe, D. W., Moore, D. R., and Priest, R. G., Polarization fading in fiber interferometric sensors, *IEEE J. Quantum Electron.*, 18, 1644, 1982.

9. Kersey, A. D., Marrone, M. J., Dandridge, A., and Tveten, A. B., Optimization and stabilization of visibility in interferometric fiber optic sensors using input-polarization control, *JLT*, 6, 10, 1599, 1988.

10. Kersey, A. D., Marrone, M. J., and Davis, M., Polarisation-insensitive fibre optic Michelson interferometer, *Elec. Lett.*, 27, 6, 518–19, 1991.

11. Frigo, N. J., Dandridge, A., and Tveten, A. B., Technique for elimination of polarization fading in fiber interferometers, *Electron. Lett.*, 20, 319, 1984.

12. Kersey, A. D., Marrone, M. J., and Davis, M., Statistical modelling of polarisation induced fading in interferometric fibre sensors, *Elec. Lett.*, 27, 6, 481–483, 1991.

13. Kersey, A. D., Marrone, M. J., and Dandridge, A., Experimental investigation of polarisation-induced fading in interferometric fibre sensor arrays, *Elec. Lett.*, 27, 7, 562–3, 1991.

14. Boger, Y. S. and Tur, M., Polarization-induced visibility limits in interferometric fiber optic sensor arrays, *Elec. Lett.*, 27, 8, 622–3, 1991.

15. Tur, M., Boger, Y. S., and Shaw, H. J., Polarization-induced fading in fiber optic sensor arrays, *J. Lightwave Technol.*, 13, 7, 1269–76, 1995.

16. Kirkendall, C. K., Kersey, A. D., Dandridge, A., Marrone, M. J., and Davis, A. R., Sensitivity Limitations Due to Aliased High Frequency Phase Noise in High Channel-Count TDM Interferometric Arrays, OFS-11, 1996.

17. Love, A. E. H., *Mathematical Theory of Elasticity*, 4th ed., Cambridge University Press, Cambridge, U.K., 1934.

18. Timoshenko, S. P. and Goodier, J. N., *Theory of Elasticity*, McGraw-Hill, New York, 1936.

19. McDearmon, G. F., Theoretical analysis of a push-pull fiber optic hydrophone, *J. Lightwave Tech.*, 5, 647, 1987.

20. Davis, A. R., Kirkendall, C. K., and Dandridge, A., 64-channel all optical deployable acoustic array, *Optical Fiber Sensor Conference*, Williamsburg, VA, October 1997.

21. Kirkendall, C. K., Davis, A. R., Dandridge, A., and Kersey, A. B., 64-Channel all-optical deployable array, *NRL Review*, 63–65, 1997.

22. Bucholtz, F. and Udd, E., Eds., *Fiber Optic Sensors: An Introduction for Engineers and Scientists*, John Wiley & Sons, New York, 1991, chap. 12.

23. Bucholtz, F., et al., Multichannel fiber-optic magnetometer system for undersea measurements, *IEEE J. Lightwave Technol.*, 13, 7, 1385–1395, 1995.

24. Bucholtz, F., et al., Undersea performance of eight-element fibre optic magnetometer array, *Electron. Lett.*, 30, 23, 1974–1975, 1994.

25. Vohra, S. T., Danver, B., Tveten, A., and Dandridge, A., Fiber optic interferometric accelerometers, *AIP Conf. Proc. 368, Acoustic Velocity Sensors: Design, Performance, and Applications*, 285, 1995.

26. Brown, D. A., Fiber optic accelerometers and seismometers, *AIP Conf. Proc. 368, Acoustic Velocity Sensors: Design, Performance, and Applications*, 260–273, 1995.

27. Vohar, S. T., Danver, B., Tveten, A., and Dandridge, A., High performance fiber optic accelerometers, *Elec. Lett.*, 33, 1997.

28. Nau, G. M., Danver, B. A., Tveten, A. B., Dandridge, A., and Vohra, S. T., Fiber optic interferometric displacement sensors for vibration measurements, *OFS Proc.*, October 1997.

# 5.4 Fiber Optic Telemetry in Ocean Cable Systems

## *George Wilkins*

The past two decades have seen major improvements in the strength, ruggedness, and attenuation of deep-sea armored coaxial cables. Some of these gains have been due to more efficient geometries, and to the development and application of precise cable analysis techniques. Another contributor has been the availability of armor steels with higher strengths at little or no sacrifice of flexibility.

During these same years, even greater advances occurred in the performance and reliability of fiber optic cables. These advances have been so striking that fiber optics is beginnng to replace electrical telemetry in many areas of undersea telemetry. This replacement should be especially rapid in systems where performance is constrained by some mixture of:

a.  A need for more bandwidth than conventional telemetry cables can supply without compromising system constraints on the cable's diameter, volume, weight, and handling system.[1]

b.  A need, beyond the capability of conventional telemetry, for the cable and handling system to be very much smaller, lighter, and more transportable.

c.  An anticipated growth of system data requirements that will require a major expansion of telemetry bandwidth, but with no increase in cable diameter. Cable diameter must become essentially independent of bandwidth.

d.  A requirement for long cable runs (many 10s of km) without a repeater. For example, if the cable cannot be used to transmit electrical power, then transmitters and receivers must be located (only) at the ends of the telemetry system.

e.  A constraint that telemetry system cost (including handling and deployment) must be as low as possible. For example, the data link may be expendable.

This section discusses recent developments that have brought armored coax telemetry to a high level of maturity. It goes on to describe contributions that fiber optics can make in exceeding these levels, as well as special guidelines and hazards involved in the use of this new technology. Cable design examples will be presented and explained. Finally, in a topic where knowledge and guidelines are still embryonic, this section will discuss the influences that fiberoptic telemetry can and should have on the cable handling system.

## The Starting Point—The Deep-Sea Armored Coax

Experience has taught that the development of deep-sea optical cables has been an evolutionary sequence rather than revolutionary. We have known the capabilities and the limits of those conventional cable technologies from which the development began. The development objectives have generally been known. Finally, we have been able to maintain a relatively clear view of the path that best connected these two end points. This is a reasonable definition of continuous evolution. In a revolution, progress tends to be sudden and often discontinuous with the loss of many participants.

### The Navy/Scripps Armored Coax

The armored coaxial cable sketched in Fig. 5.46 should be familiar, since it has been in common use as a deep-sea oceanographic tether since about 1970. It is often known as the Navy or Scripps deep tow cable, and has its most probable origin in the 1960s as an armored lift cable for elevators and mine shafts. This cable, which is still in general use, has a number of serious design flaws.[2]

---

[1]The classic reason usually given for (often reluctantly) choosing fiber optics is the need for telemetry bandwidth. Other constraints, of the types described above, can convert this initial timidity to enthusiasm.

[2]Ref. [1] describes the response of this cable to accelerated flexure over 35.5-cm-diameter sheaves at 6350-kg tension.

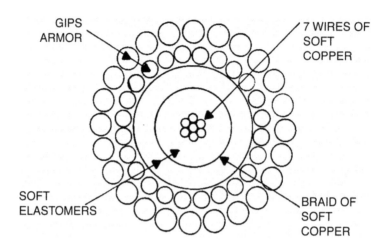

**FIGURE 5.46**   Navy/Scripps deep-sea coax circa 1972.

1. The soft copper wires of the center conductor are too large and stiff, and the conductor becomes inelastic at a very low value of tensile strain. When the cable is subjected to severe tensile or flexure cycling, these wires are likely to "ratchet." That is, each stress cycle can cause the center conductor to suffer another increment of permanent strain. Finally, the conductor becomes so much longer than the surrounding cable structure that it Z-kinks into the dielectric spacer. At this point, the conductor may either break or its component wires may force their way through the insulation to short against the shield conductor.
2. The shield conductor is fabricated as a soft copper braid and tends to break up rapidly during loaded flexure.
3. The elastomeric materials that form the dielectric spacer and the coax jacket are soft and can easily be punctured by the metal shards resulting from defects 1 and 2.
4. The contrahelical steel armor package is initially stress balanced but, because of the larger outside wires, has a 2/1 torque mismatch. As a result, the cable suffers very high rotation under load when one end is unconstrained. This unbalances stress in the two armor helices and results in a 30% loss of strength. (The "one end free" mode is common in deep oceanographic systems where the instrumentation package is normally fitted with a swivel.)

While this strength loss is serious, the lack of torque balance is of even greater concern. If the package strikes the sea floor and unloads the cable, then the considerable rotational energy stored in the lowest cable section is suddenly released. The resulting reaction can lift the now-slack cable high enough to form a loop. When the cable is again loaded, this loop can tighten into a hockle, or worse, with catastrophic results.

### The UNOLS Armored Coaxial Cable

During the summer of 1983, the Woods Hole Oceanographic Institution (WHOI) was asked by the National Science Foundation to negotiate a large purchase of deep-sea armored cables for the U.S. National Oceanographic Laboratories (UNOLS). In response, WHOI hosted a meeting of UNOLS representatives to determine what the nature and design of these cables should be. The author was working at WHOI that summer and was asked to attend.

The UNOLS group quickly agreed on the critical issues. First, a single cable design should be adopted so that deep-sea tethered systems could be readily transferred among the UNOLS laboratories. Second, the diameter of this standard tether should be 17.3 mm (0.680"); a "best" fit to the oceanographic winches used by most laboratories. Third, the new design should be a major improvement over the current deep-sea tether cable.

A fourth point was accepted in principle. The new tether cable should be a "last generation" version of the conventional coax. That is, the next generation deep-sea cable should include fiber optic telemetry.

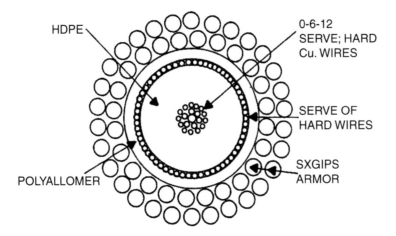

HDPE

0-6-12
SERVE; HARD
Cu. WIRES

SERVE OF
HARD WIRES

SXGIPS
ARMOR

POLYALLOMER

**FIGURE 5.47**   UNOLS deep-sea coax circa 1983.

As a goal, the new armored coax should also be compatible in size, strength, power transfer, and handling characteristics with this new electro-optical (E-O) tether.

The new armored coax, Fig. 5.47 was built by The Rochester Corporation (TRC) in late 1983. It is similar to one formally proposed to the Navy by the author in 1973 and built by TRC for the Naval Oceanographic Office in 1982. It differs from these primarily in the use of work-hardened copper[3] and ultra-high-strength steels. Key features of the 1983 design include:

1.  The center conductor was changed to a 12-around-6 structure of half-hard copper wires, formed as a unidirectional helix around a monofilament. This increased the tensile compliance of the conductor during load or flexure cycling.
2.  To increase its penetration resistance, the coax dielectric was fabricated from a high-density polyethylene.
3.  To eliminate crossovers and self-abrasion, the braided-shield conductor was replaced with a served shield of work-hardened copper wires. A conductor of this type is likely to exhibit radiation leakage and cross-talk, since the interwire openings behave somewhat like the Taylor's slits in an optical diffraction experiment. To correct for this, a copper-backed-polyester tape was wrapped around the copper wires to establish short circuits across the slits. This technique was used successfully in 1974 during the development of the Kevlar-armored tether cable for the Navy's Remote Unmanned Work System [2, 3].
4.  The armor wires were a special galvanized steel that combined excellent flexure performance with an ultimate tensile strength of 22,000 kg/cm² (315,000 psi). A common wire diameter was used to simplify shipboard armor repairs.

### Comparison of the New and Old Coax Designs

Complete torque balancing of the new cable design would have required unbalancing the tensile stresses in the two armor layers and, therefore, would have resulted in a severe reduction of cable strength. As a compromise, a slight residual torque was allowed. The resulting rotation caused barely 1% loss in cable strength. More important, the energy stored in cable rotation was so small that, for reasonable payload in-water weights, the sudden loss of tension during any bottom impact could never cause the cable to lift high enough to form a loop or hockle. This had been the first design goal.

---

[3]Copper wires are usually work hardened during the drawing process. But the cable manufacturer, often responding to a customer's specification, is likely to purchase the wires in a soft annealed temper. To satisfy this requirement, the hardened (i.e., elastic) wires must be heated and annealed to return them to a dead soft condition. Instead of paying a cost penalty for tempered wires, the customer may actually pay a premium to eliminate this desirable feature.

**TABLE 5.6**    Performances of Two Deep-Sea Armored Coaxes

| Cable Parameter | 1972 Navy/SIO | 1983 UNOLS |
|---|---|---|
| Diameter (mm) | 17.3 | 17.3 |
| Strength (kg) | | |
| Ends fixed | 17,000 | 17,800 |
| One end free | 11,900 | 17,300 |
| Weight (kg/km) | | |
| In air | 1070 | 1020 |
| In water | 820 | 795 |
| Free length (m) | | |
| Ends fixed | 20,800 | 22,300 |
| One end free | 14,500 | 21,800 |
| Payload (kg)[a] | | |
| Ends fixed | 1880 | 2350 |
| One end free | None | 2150 |

[a]For operations to 6000 m, with the lower cable end free to rotate. The system's static strength/weight safety factor is 2.5.

Table 5.6 compares the performance of the new armored coax to that of its Navy/Scripps predecessor. Note that for the both-ends-fixed operating mode, there was very little increase in cable ultimate strength. This was because much of the additional strength that could have been gained by selection of SXGIPS steel armor was reinvested to reduce the RF attenuation of the cable's (now larger) coax core. (Attenuation in the UNOLS cable was approximately 67 dB at 5 MHz through a 6000-meter length).

The second major design goal had been to reduce the loss of tensile strength that normally results when one cable end is free to rotate. This is a much more general operational situation. As Table 5.6 shows, the new armored coax loses little strength for this condition. In fact, it is about 45% stronger than the Navy/Scripps design.

The payoff of this low tension/rotation sensitivity is made clearer if we examine its impact on the tether's ability to support deep-sea weights. Because of its weight and low strength, the Navy/Scripps cable had an inherent (i.e., self-weight) safety factor less than 2.5 during deployment (one end free) to 6000 m. Adding the weight of an instrumentation package at this depth forced the cable static safety factor to fall even farther short of this reasonable constraint.

The Navy/Scripps cable could carry a 1500-kg (in-water weight) package to a depth of 6000 meters only if its static safety factor was reduced to 1.85. The low torque and rotation of the UNOLS design gave it a static safety factor of 2.76 for this same payload and depth.

Successful operation of the UNOLS cable was reported by both Canadian and U.S. oceanographic laboratories, including Scripps and WHOI. The cable was reported to be well behaved, with little performance degradation during tensile or flexure cycling. Cable torque and rotation were reportedly so low during deep-tow operations that no differences were observed when the payload swivel was switched in and out of the system [4].

The UNOLS armored coax cable was also used successfully by WHOI's Deep Submergence Laboratory during the search for RMS TITANIC in 1985. In its role as the primary instrumentation and support tether for the deep-sea search system ARGO, this "last generation coax" telemetered the first video photographs of the wreckage of that ship.

### New Nonlinear Analysis Techniques

In these newer designs, both performance and reliability have been aided immeasurably by the introduction of computerized analysis techniques [5, 6, 7]. These recognize nonlinearities in the cable geometry and allow cable performance to be precisely balanced in both stress and torque. The results are precise, with accuracy and reliability that fit well with normal manufacturing tolerances.

New versions of these analysis programs can analyze the effects of bending on cable strength and even support a limited insight into lifetime and failure modes if the cable is subjected to loaded flexure. Two of these analysis programs are available in a condensed version, which can be run on a personal computer [8].

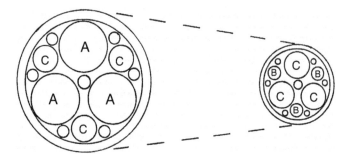

**FIGURE 5.48** Effect of converting coaxes (a) into optical fibers (b).

### An Insight into Fiber Optic Telemetry

To reduce the UNOLS cable's RF attenuation, it was necessary to increase the cross-section of its coaxial core. If it had not been for this constraint, the use of ultra-high-strength steels in the new armor package would have increased cable strength to more than 23,000 kg (50,000 lb).

But bandwidth in the armored coax is inescapably entwined with the core's cross-section and, through it, with the cable's diameter and strength. The low conductor resistances and thick dielectrics demanded by telemetry constraints leave little room for optimization of the cable's power or strength functions.

The tyranny of these interactions served as an important stimulus for UNOLS's desire to convert from coaxial to fiber optic telemetry in the next generation deep-sea tether. That decision is rewarded by several advantages offered by fiber optics.

**Advantage #1**—It is almost axiomatic that one optical fiber can provide an enormous increase in telemetry bandwidth. If single-mode fiber optic technology is used, data rates of at least 1000 megabits/second can be achieved over an 8- to 10-km cable run. When system telemetry constraints require it, a single optical fiber can simultaneously send data in one direction and command signals (at a different wavelength) in the opposite direction. Even if we discount such high data rates by 20 times in order to convert from digital data to an equivalent analog, a single tiny (0.125-mm diameter) optical fiber can support uplink telemetry at data rates that correspond to at least 10 channels of high resolution television.

**Advantage #2**—This bandwidth can be almost independent of cable diameter. In many designs, the tiny optical fiber can be placed into inconsequential crannies of the cable cross-section or can even be used in roles normally relegated to void fillers.

This characteristic is illustrated in Fig. 5.48 which shows the effects of replacing three coaxial subcable units (a) with three optical fibers (b) in a hypothetical cable core. The three power conductors are left unchanged. Before the switch, these power conductors were relatively inconspicuous and, in fact, were used as void fillers to firm up and round out the cross-section of the core. After the conversion, they became the dominant elements in the core—and the void-filling role in the now much smaller cable core has been assumed by the optical fibers. Any further diameter reduction must be obtained through a rethinking of the cable's power and strength functions.

**Advantage #3**—In most E-O cables, the power and telemetry functions will be nearly independent, so much so that they can be separately optimized. This is a near-revolutionary departure from the coax design approach, where optimization of one function (generally telemetry because it is so critical) is likely to ride roughshod over subordinate system requirements.

For example, the conductors and dielectric insulation in an "acceptable" telemetry coax can usually transfer far more power than the system needs. In response to this freedom, the power subsystem is often allowed to grow, i.e., to consume the excess power and, ultimately, to demand it.

## The Nature of Fiber Optic Communications

An optical fiber can be considered as a simple quartz waveguide, which traps light rays and constrains them to propagate within a very small and extremely transparent rod. Fiber attenuation (Fig. 5.49) can be less than

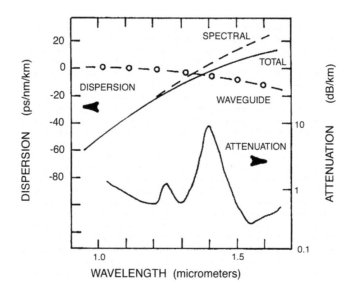

**FIGURE 5.49**   Attenuation and dispersion spectra for fused silica optical fibers.

0.1 dB/km, and telemetry over 250-km continuous fiber lengths has been demonstrated in the laboratory. Bandwidths, i.e., bandwidth-length products, of 200,000 Mb-km/sec have also been achieved [9]. As the figure shows, bandwidth is normally greatest at 1.3 micrometers, where total dispersion passes through zero.

On the bandwidth/attenuation scale noted above, fiber optic communication through the cable length needed to support deep-sea operations seems almost trivial. For example, a single 10-km optical fiber, operated at a standard T4 rate of 270 MB/sec, can easily support four or five digital TV channels. Yet such a system's bandwidth-length product is only 1% of demonstrated capability.

## Physical Properties of Optical Fibers

Physically, an optical fiber is a solid fused silica rod, with a normal diameter of 0.125 mm (0.005"). For protection and isolation from external stresses, this rod is coated with one or more layers of plastic. Typically, silicone rubber is used as the primary coating, but UV-cured acrylates are becoming popular in this role. The secondary coating is usually an extruded elastomer (e.g., hytrel or nylon).

In a short gauge, the fiber's ultimate tensile strength can be almost 50,000 kg/cm$^2$ (700,000 psi). This is equivalent to an ultimate strain of about 7%, far greater than the breaking strains of cabling steels or the lightweight Kevlar or Spectra fibers.

But in a practical telemetry system, the optical fiber must be used in continuous lengths of several kilometers. In such lengths, an optical fiber acts very much like a chain—a chain that, somewhere, must have a weakest link. It is the physical strength of this weakest link that will determine the practical tensile strength (or ultimate strain) of the optical fiber.

Typical optical fibers for deep-sea tether cables can be specified to survive a "proof" stress of about 10,500 kg/cm$^2$ (150,000 psi). This is equivalent to a tensile strain of 1.5%. In this test, the fiber is passed between two sheaves so that it experiences a tensile stress with duration of about 1 sec. Any fiber section weaker than the proof stress will fail.

At a relatively small cost premium, the proof strain can be increased to about 2.0%. But even the lower of these strain limits is greater than the elastic limit of cable armoring steels, although it is still much less than the 2.4% failure strain of Kevlar-49.

## Light Propagation in Optical Fibers

Figure 5.50a sketches the cross-section of an optical fiber. Light propagation along the fiber will take place in a central core (the shaded zone). The sketch shows a ray of light moving at incidence angle $\theta_0$, from a medium of refractive index $n_0$ into a core of index $n_1$. If the external medium is air, then $n_0 = 1$.

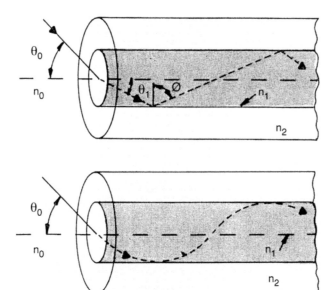

**FIGURE 5.50** (a) Step index fiber. (b) Graded index fiber.

Refraction at the core/air interface reduces the external angle of incidence to a value $\theta_1$. The two angles are related by the Law of Sines.

$$n_0 \sin \theta_0 = n_1 \sin \theta_1 \tag{5.73}$$

To trap the light ray inside the fiber, the optical core is surrounded by a transparent cladding,[4] which has a lower index of refraction ($n_2 < n_1$). The largest value of $\theta_0$ that can propagate within the core belongs to a ray that suffers total internal reflection at the core/cladding interface. This occurs when $\theta_0$ is equal to "Brewster's angle," defined by the relationship:

$$\sin \theta_0 < (n_2/n_1)\sin(90°) = (n_2/n_1) \tag{5.74}$$

$$= \cos \theta_1 \tag{5.75}$$

$$= \sqrt{1 - \sin^2 \theta_1} \tag{5.76}$$

Using Eqs. (5.74) and (5.76),

$$\sin \theta_1 = \sqrt{1 - (n_2^2/n_1^2)} \tag{5.77}$$

Combine Eqs. (5.73) and (5.77) with $n_0 = 1$ to solve for the external incidence angle.

$$\sin \theta_0 = n_1 \sqrt{1 - (n_2^2/n_1^2)} \tag{5.78}$$

$$= n_1 \sqrt{(n_1^2 - n_2^2)/n_1^2} \tag{5.79}$$

---

[4]The optical fiber's core and cladding are normally ultrapure silica and one of these components is doped to achieve the desired values of $n_1$ and $n_2$. The fiber may contain a second silica cladding that provides physical protection. Finally, the fiber is protected by one or more plastic jackets or "buffers" to a diameter of 0.25 to 1.0 mm.

The fiber's core and cladding indices of refraction seldom differ by more than one percent, so the quantity $(n_1 + n_2)$ is usually set equal to $2n_1$. Also, it is customary to define an index difference term $\Delta = (n_1 - n_2)/n_1$. Equation (5.79) becomes

$$\sin \theta_0 = n_1 \sqrt{\frac{(n_1 + n_2)(n_1 - n_2)}{n_1}\frac{}{n_1}} \qquad (5.80)$$

$$= n_1 \sqrt{2\Delta} \qquad (5.81)$$

$$= NA \qquad (5.82)$$

This last term is the "numerical aperture," the maximum external incidence angle at which light can be "launched" into an optical fiber, that is, trapped within the fiber and propagated by it. The square of NA is proportional to the solid angle defined by this launch cone, so that fiber NA is equivalent to the "speed" of a camera lens.

In Fig. 5.50a, fiber refractive index was discontinuous at the core/cladding interface. This type of behavior defines a "step index" fiber. Many other index profiles are possible. In almost all of them, the fiber core's index of refraction varies with radius according to the general constraint that

$$n = n_1[1 - \Delta(r/a)^a] \qquad (5.83)$$

where

$\quad n$ = the index of refraction on and along the fiber axis.
$\quad r$ = the radial distance from that axis.
$\quad a$ = the radius at which (by definition) the optical core ends and the cladding begins.
$\quad \alpha$ = a parameter that defines the shape of the transition between core and cladding.

For a step index (SI) fiber, $\alpha = \infty$, although few manufacturers can manufacture a fiber with a value greater than about 50. The optical behavior of a step index fiber is shown in Fig. 5.50a.

When $\alpha = 2$, the index of refraction of the fiber core changes from $n_1$ to $n_2$ as a parabolic function of fiber radius between $r = 0$ and $r = a$. Figure 5.50b shows how a light ray moves along the core of this "graded index" or "GI" fiber.

If all other parameters are identical, the graded index optical fiber can have a much greater bandwidth-length product than a fiber with a step index. The reason becomes apparent when the light paths in Fig. 5.50a and 5.50b are compared.

SI The optical power in a digital data 'bit' contains light rays at all angles less than $\mathrm{Sin}^{-1}(NA)$. Those rays that enter the fiber at large angles of incidence must travel a greater total distance and, therefore, move more slowly along the fiber axis than the lower-angle rays. Due to this speed difference, the bit's initial square shape begins to broaden and round off. Ultimately, the bit interferes with adjacent bits. Bit contrast is reduced, and a bandwidth-length limit has been reached.

GI Here, the core's refractive index decreases from $n_1$ on the fiber axis to $n_2$ at the core/cladding "interface." As it enters a region of smaller index, the light ray must move faster. This helps high-angle light rays keep up with their paraxial counterparts. In fact, if $a$ is precisely equal to 2, then all light rays of the same wavelength propagate along the core at the same speed. This increases the bandwidth-length product as much as 50 times compared to the step index fiber.

### Fiber Optic Propagation Modes

It can be shown by wave theory that light propagating in an optical fiber is quantized. That is, the fiber can accept only certain discrete angles of incidence (i.e., angular modes). Also, the number of angular modes "$N$" that can be accepted by a fiber is finite.

$$N = \frac{\alpha.\Delta}{2 + \alpha}\left[\frac{2 \pi n_1}{\lambda}\right]^2 \qquad (5.84)$$

**TABLE 5.7**   Typical Optical Fiber Parameters

| Parameter | Step | GI | S-M #1 | S-M #2 |
|---|---|---|---|---|
| $\lambda$ ($\mu$m) | 1.3 | 1.3 | 1.3 | 1.3 |
| $a$ ($\mu$m) | 25 | 25 | 2.0 | 4.0 |
| $n_1$ | 1.5 | 1.5 | 1.5 | 1.5 |
| $\alpha$ | 40 | 2 | 40 | 40 |
| $\Delta$ | 1% | 1% | 1.38% | 0.34% |
| $N$ | 328 | 164 | 1 | 1 |

**TABLE 5.8**   Power/Bandwidth Comparison of Three Optical Fiber Types

| Fiber Type | Bandwidth-Length (Mbit-Km/sec) | Launch Power (relative) |
|---|---|---|
| Step index | 20 | 80 |
| Graded index | 1,000 | 40 |
| Single-mode | 15,000 | 1 |
| Single-mode[a] | >200,000 | 1 |

[a] When the signal is launched by an ultra-monochromatic (e.g., distributed-feedback) single-mode laser.

where the parameters $a$, $\Delta$, $n_1$, and $\alpha$ have already been defined, and $\lambda$ is the wavelength. Note that, because of its high $a$ value, the SI optical fiber contains twice as many propagation modes as the graded index fiber.

## Single-Mode Fibers

If $\Delta$ or $a$ are decreased then the number of modes that can be propagated by an optical fiber must also decrease. In the limit, the fiber propagates only a single mode. This limit is reached [12] when the fiber satisfies the inequality:

$$V = (2\pi n_1 a / \lambda)\sqrt{2\Delta} < 2.405 \qquad (5.85)$$

Table 5.7 shows typical fiber parameters for step index, graded index, and single mode fibers. Note that the single-mode fiber can have quite different values of index difference and core radius, so long as they satisfy Eq. (5.85).

Each fiber type has advantages and disadvantages. The large core diameters of the step index and graded index fibers allow them to accept greater launch powers from injection laser diodes (ILDs) and light-emitting diodes (LEDs). But the single-mode has a bandwidth-length product far greater than any obtainable with the large core fibers. These properties are shown in Table 5.8.

### Fiber Bending and Microbending

In designing for the use of optical fibers in cables, it is extremely important, in fact essential, to recognize that the optical fiber cannot be treated like a copper conductor. For example, the fiber will be absolutely unforgiving of small-radius bends. The effect of such a bend will be to convert the highest order propagation modes into modes that leave the fiber.

This loss is immediate and can be quite large. It occurs even if the fiber completes only a very small arc of such a bend. After the initial loss, the curved fiber will continue to exhibit a steady-state radiation loss that is caused by:

1. A repopulation of the highest order modes because the bend is also perturbing the fiber's more tightly bound modes.
2. The immediate stripping of the new "highest order" residents out of the fiber.

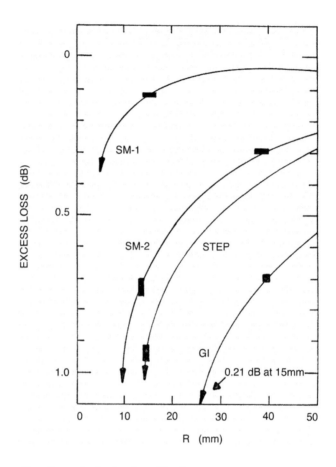

**FIGURE 5.51**  Effects of bending and microbends on fiber loss.

The fraction of total propagated energy that will be immediately lost is approximately equal to [11, 13]

$$\text{Loss} = 1 - \frac{\alpha + 2}{\alpha}\left[\frac{1}{2\Delta}\right]\left[\frac{0.002a}{R} + \frac{0.003\lambda}{4\pi n_1(1-\Delta)R}\right]^{2/3} \tag{5.86}$$

Here, "R" is the radius of curvature of the fiber axis in millimeters. Figure 5.51 shows how typical optical fibers respond to uniform bending at various radii of curvature. The four fibers plotted in the figure are identical to the ones described in Table 5.7.

Because of its high bandwidth and large core, the GI fiber has become quite popular in undersea tether and tow cables. As Fig. 5.51 shows, it is also extremely sensitive to even moderate radii of curvature. Part of this sensitivity, a fault shared with the step index fiber, is a concentration of optical power at higher (i.e., more loosely bound) modes. Another reason (this time not shared) is that the high order modes in the GI fiber have a closer angular spacing than those in the step index fiber. As a result, more modes, and therefore more optical energy, can be perturbed by a given fiber bend.

Notice the low bending responses of the two single-mode fibers. Note also how that response is sharply reduced if core radius *a* is decreased while increasing index difference Δ. The freedom to make such changes in a single-mode fiber is limited by the need to satisfy the constraint imposed by Eq. 5.85.

Figure 5.51 also shows the effect of "microbends," a much more "dangerous" form of fiber curvature. In a typical microbend, the linearity of the fiber axis is perturbed by some imperfection in the local environment. As a result, the fiber axis suffers a highly localized bending, which combines small amplitude with a small spatial period. Typical microbending conditions will have a spatial period comparable to the fiber's buffered diameter.

This is, unfortunately, exactly the condition that can excite resonant scattering (i.e., leakage) from the fiber's more loosely bound propagation modes. Energy in these modes will be converted into unguided radiation modes.

Under these conditions, a single microbend can cause an immediate attenuation loss of approximately [14]:

$$\text{Loss(dB)} = \left[\frac{H^2}{\Delta^3}\right]\left[\frac{a^4}{b^6}\right]\left[\frac{KE'}{E}\right]^{3/2} \qquad (5.87)^5$$

where the following definitions apply:

$H$ = displacement induced by the microbend ($\mu$m).
$b$ = radius of the fiber optical cladding ($\mu$m), with a value generally less than the physical fiber radius.
$E'/E$ = a ratio of the fiber buffer's tensile modulus to the modulus of the fiber.
$K$ = a constant with a value near unity.

Figure 5.51 includes a semiquantitative model to illustrate how a relatively gentle microbend can cause a very large local attenuation increase. The model assumes that an already-curved fiber is subjected to a perturbation that reduces its radius of curvature by 1 mm. At a bend radius of 40 mm, additional optical attenuation is almost negligible. If the microbend occurs at 15 mm, then the resulting loss is quite significant. This is especially true if one considers that, in a poorly designed or badly built cable, such microbends can occur with a frequency of 100s/km. In the author's experience, one such cable had an excess attenuation of 1400 dB/km.

This bending/microbending scenario is not academic. It is, in fact, a reasonable description of what actually happens in a cable when a microbend perturbs a fiber that has been assembled in a relatively small-radius helix. Causes of microbending range from roughened surfaces of adjacent components, to imperfections in the protective fiber buffer, to localized Z-kinking caused by yielding of metal conductors. Several methods can be employed to reduce the impact of fiber bending and microbending:

1. Use single-mode optical fibers. In addition to microbend protection, these fibers will increase system bandwidth by one or two orders of magnitude (compared to a GI fiber).
2. Increase the radius of fiber curvature in the cable. As a rule of thumb, this radius should never be less than 50 mm.
3. Float the fiber within a hydrostatic cable environment, so that it is not forced to make high-pressure contacts with adjacent surfaces.
4. Make these adjacent surfaces as smooth as possible. For example, a conductor neighbor should contain small wires, and the specification for assembly of its jacket should require a pressure extrusion.
   The combination of large wires and a tubed extrusion can allow the memory of these wires to carry through to the outer surface of the insulation. If the fiber is forced to conform to this surface, its axis may be repeatedly bent so that it acts like a grating. Extremely high excess losses can result.
5. Increase the thickness of the fiber buffer to isolate the fiber from external anisotropic (i.e., bending) forces.
6. Encase the optical fiber in a hard protective metal tube. This is an extreme version of the technique normally used to protect the fiber—an initial coating with a soft plastic buffer to form a quasi-hydrostatic environment, followed by a secondary annulus of a harder plastic.

---

[5]Murata [15] describes an equation with a similar form, except that $b^6$ is replaced by $b^4d^2$, where $d$ is the physical radius of the optical fiber. See also Kao [16].

For certain applications, this sixth approach is attractive, since multiple roles can be assigned to a protective metal tube. In addition to its primary anti-microbending assignment, the metal tube can seal the fiber inside a hydrostatic environment. It can serve as a dedicated conductor, and as a major strength element in the cable. For selected applications, the fiber, metal tube, and (optional) jacket can serve as the entire cable.

Previous papers [17, 18] have reported success with techniques to form multi-kilometer lengths of metal tube as hermetic containers around one or more optical fibers. The application of this approach to undersea cables has also been discussed [18–20].

## The Use of Optical Fibers in Ocean Cables

The shotgun marriage of power and telemetry functions in an electromechanical coax cable allows a limited set of options in choosing the geometry of that cable. Conductor resistances and dielectric thicknesses can be varied, but the basic shape of the cable remains quite static. Figures 5.46 and 5.47 represent a fair statement of the limits to the design freedom one has in the conventional E/M coax.

This is not so in the E-O cable. Here, the ability to separate power and telemetry functions allows a much greater freedom to choose from among distinctly different cable geometries. Three of these configurations, representing designs with 1, 2 and 3 electrical conductors, are sketched in Fig. 5.52. The operating characteristics of each of these types will be discussed in some detail.

One generic type of E-O cable design is not treated here. That is the "hybrid" or "straddle" approach, a design that attempts to keep one foot in the new optical technology and the other foot solidly planted in electromechanical technology. A typical member of this cable design family will contain several types and sizes of power conductors. It will contain one or more coax subcables, and perhaps a twisted-pair data cable or two. Somewhere within this layout, it will try to accommodate "*N*" (usually a large number to ensure "safety") optical fibers.

This section intends to present a rational set of recommendations for the proper use of fiber optics in undersea cables. The hybrid cable design violates almost every one of these recommendations.

### The N-Conductor E-O Cable

**3-Conductors**—While this cable design (Fig. 5.52) is intended for use with 3-phase power transfer, it allows considerable operating freedom for the power system. The design can operate with a cable-return power circuit (unbalanced line). It is even possible to parallel the conductors and operate the system with a seawater return.

The optical fibers in the 3-conductor cable design are overjacketed to a size that rounds out the core's cylindrical cross-section. This allows them to ride gently in the channels defined by the several conductors.

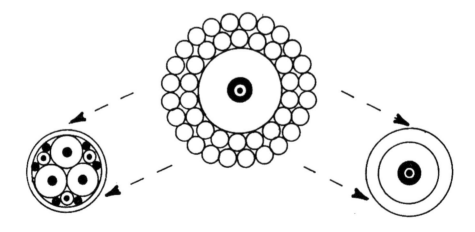

**FIGURE 5.52** Alternative core designs for electro-optical cables.

It will be shown that this design is relatively inefficient in trading off power transfer versus strength. At the same time, it gives excellent protection and strain relief to the electrical and optical conductors.

Some 3-conductor designs incorporate a strength member (e.g., Kevlar-49) within the fiber overjacket. Some designers avoid this approach for good reason. The main value of the Kevlar is to make the cable company's task easier, so that more back tension can be applied to the fiber unit as it is served into the conductor structure. Once the cable has been built, the higher tensile modulus of the strengthened fiber unit, driven by the fact that cable strain is determined by cable load acting on the cable armor, causes the fiber unit to experience relatively high tension in the E-O core. This forces the unit to press against local conductors while attempting to migrate to the cable center. Under such high tensile and bearing stresses, the Kevlar yarns can also act as local microbending centers that increase fiber optical attenuation.[6]

**2-Conductors**—The right-hand sketch in Fig. 5.52 shows a coax-like structure, with two electrical conductors and one or more optical fibers. The associated ability to provide in-cable circuit return for a power system can be a critical advantage. For example, this design might be used for an E-O cable that must be dipped into the sea from a helicopter.

But the cable-return capability is achieved at the cost of lower power-versus-strength efficiency, since the requirement for two layers of insulation reduces the cross-section that can be dedicated to load-bearing structure. This coaxial E-O cable design is not recommended if diameter, strength, and weight are critical system elements, unless the use of a cable circuit return is a categorical imperative.

**1-Conductor**—In both the 1- and 2-conductor designs, the optical fiber(s) are contained within and protected by a closed and welded metal tube. In the single-conductor design approach to be discussed here, this tube also serves as the cable's only electrical conductor, so that the power circuit must be completed by a seawater return.

Figure 5.53 shows the centers of three design options for this conductor. The shaded region in the centers of designs (1) and (2) is a void-filling gel that provides the optical fibers with a hydrostatic environment. In design (3), the metal tube is given a thin elastomeric jacket. This layer serves as a bedding layer for the (work-hardened) copper conductors added to reduce and tune cable resistance.

In the limit, any of these designs might contain only one optical fiber, and that fiber might operate in a full-duplex mode. For example, telemetered data can move in one direction at one wavelength while command signals flow in the other direction at a second wavelength.

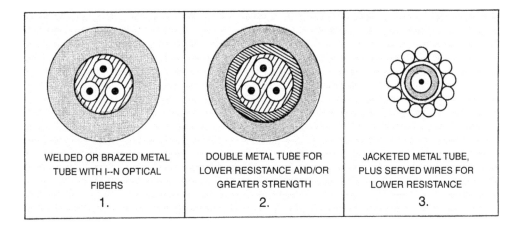

| WELDED OR BRAZED METAL TUBE WITH I--N OPTICAL FIBERS | DOUBLE METAL TUBE FOR LOWER RESISTANCE AND/OR GREATER STRENGTH | JACKETED METAL TUBE, PLUS SERVED WIRES FOR LOWER RESISTANCE |
|---|---|---|
| 1. | 2. | 3. |

**FIGURE 5.53**   Center-conductor options for metal-tubed optical fibers.

---

[6]If strength is required in the buffered fiber, it should be obtained with materials that offer stiffness in axial compression, e.g., fine steel wires or glass filaments.

The attenuation of the optical fiber in the formed and welded tube will generally be less than the level claimed by the fiber manufacturer. (This means only that those fiber stresses that can increase fiber attenuation are less inside the tube than they are on a "zero-stress" measurement spool.)

The single-conductor cable design operates with a seawater circuit return, an approach that is sometimes challenged as a critical weakness of the single-conductor concept. Some scientists have been known to defend grounding the shield conductor to the armor of their deep-sea coax cables with the argument that circuit return is through the armor. At best, less than 1% of system current will flow through the armor. The remainder will pass through the sea, through the painted hull of the ship and, by the most devious means, back to the system ground. It would be to operate the system with a dedicated, efficient (and known) seawater return!

A valid question might be: "What are the physical limits to a seawater return circuit?" Europe offers the most compelling operational experience with this type of circuit. For example, the Skagerrak Sea power cable delivered 250 mw between Norway and Denmark over a 130-km single-conductor circuit with a seawater return. Until a second cable was installed in the late 1970s, to take advantage of the fact that the cable's resistance was slightly lower than that of the ocean, this power system operated uneventfully with a seawater circuit return of 1000 A [21].

In the 1- and 2-conductor designs in Fig. 5.52, the metal tube lies parallel to the cable axis. This means that helical geometry cannot be used to provide the E-O conductor with strain relief. The conductor must furnish whatever tensile compliance is needed to avoid permanent strain at very high cable tension.

Figure 5.54[7] demonstrates typical stress/strain behavior for one alloy commonly used to build ultraminiature versions of the metal-tubed optical cable. The tube is almost completely elastic for strains as

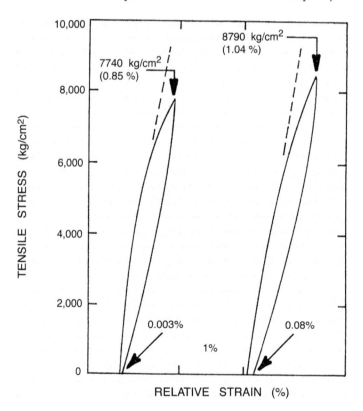

**FIGURE 5.54**   Olin type 638 bronze alloy spring temper 10% conductivity.

---

[7]Two U.S. companies are able to fabricate metal-tubed optical fibers in lengths of 10+ km. They are Olin Corporation (New Haven, CT) and KT Armor Tech (San Diego, CA).

great as 0.85%. For an alloy with 50% conductivity (e.g., Olin 195), the elastic strain limit is 0.7% [17]. Both of these strain-limiting levels are higher than the elastic limits of conventional "compliant" stranded copper conductors.

## Designs for Electro-Optical Undersea Cable

No single E-O cable design can be "optimum" for all system constraints and for all operating conditions. Design choices must still depend on the often painful resolution of conflicting requirements for cable strength, power transfer, and diameter. This is not unusual—in fact, it is the norm in the design of tether cables. What makes the E-O cable unique is the fact that system bandwidth no longer plays a critical role in that design conflict. It is instructive to follow the path of a typical E-O tether design [22, 23]:

1. The design analysis begins with a cable diameter constraint, to ensure that the cable can operate from existing handling systems. (This analysis evaluates cable performance at 7.94-, 9.53-, and 17.3-mm diameters (5-, 6-, and 11/16-in.)

2. A second constraint sets a minimum value for the power that must be delivered to system instrumentation. This limit is best expressed as a power-length product (P-L), i.e., as a power P delivered to a payload through cable length L.

3. The P-L specification may be accompanied by a specified value of supply voltage or cable current. This constraint is often imposed for no obvious reason and can have a very serious impact on cable strength/diameter trade-offs. This interaction will be discussed later.

4. Other parameters that can affect the choice of geometry for the E-O core include the number of fibers, requirements for strain relief, or even special constraints placed on the electrical power system. For example:

    a. The need for optical and electronic simplicity at the ends of the cable might veto the use of full duplex telemetry, forcing the cable to contain two or more optical fibers.

    b. The need for additional tensile strain relief might force the selection of a 3-conductor cable core because of its helical structure.

    c. A propulsion motor in the instrumentation package might demand 3-phase power.

5. With these parameters and constraints in hand, a (somewhat) arbitrary initial value can be chosen for system supply voltage. This choice allows the cable electrical resistance to be calculated, leading to design values for conductor dimensions, for insulation thickness and, finally, for the cross-section and overall diameter of the E-O core. As a special constraint to minimize electrical pinholes, the dielectric might be specified to have a thickness greater than some minimum value.

6. The cable and core diameters define the annulus available for load-bearing armor so that, assuming reasonable values were selected for the armor coverage and helix angles, we can calculate cable strength, weight, payload capability, and "free length."

7. If the conflict between strength and diameter is critical, then the supply voltage should be "optimized." That is, supply voltage will be adjusted until a value is found for which cable strength (or strength/weight) is a maximum.

8. Finally, the approximate (i.e., linear) cable design can be refined. Such tuning involves the use of nonlinear analysis techniques to adjust the conductor and armor helix angles for simultaneous stress and torque balance [5–8].

## Power-Diameter-Strength Optimization of E-O Cables

The sensitivity of component diameters to supply voltage is shown in Fig. 5.55 for the 1-conductor cable design. Note that the metal tube's thickness is limited to 27% of tube diameter. This constraint recognizes that the tube wall may buckle in the forming die if the relative thickness exceeds this limit.

Also, a 1.0-mm minimum thickness constraint has been imposed on the cable insulation to avoid pinhole effects. This constraint affects only the low-voltage side of the curve, where a thinner insulation will satisfy the constraint that dielectric stress be equal to 1970 V/mm (50 V/mil).

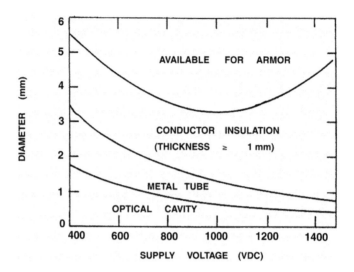

**FIGURE 5.55**    Component diameters vs. supply voltage for P = 1000 Watts, L = 8 km, T = tube thickness/diameter ratio = 0.27.

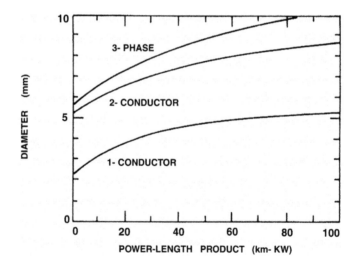

**FIGURE 5.56**    Growth of E-O core diameter with power-length product.

The diameter of the E-O core must also grow as the system's P-L product is increased. This effect is shown in Fig. 5.56 for all three cable geometries. Each curve represents a family of "optimum" design solutions, i.e., the smallest jacket or dielectric diameter possible for a given P-L product. A 50% voltage drop in the cable is assumed for all analyses except the one plotted in Fig. 5.60.

Note that the diameters of the 2- and 3-conductor cores are much more sensitive to the value of P-L, since the multiconductor cores must use more insulation to achieve the required voltage protection and physical isolation. In addition, the 3-phase cable core is inefficiently filled.

Figure 5.57 shows the effects of P-L growth on cable strength for all three conductor designs and for a cable diameter of 7.94 mm (5/16″). Figures 5.58 and 5.59 present similar data for cable diameters of 9.53 mm (3/8″) and 17.3 mm (0.680″).

In the smaller cables, it is clear that the multiconductor designs have severely limited abilities to supply both power and strength. A somewhat smaller penalty is exacted at the largest diameter (17.3 mm),

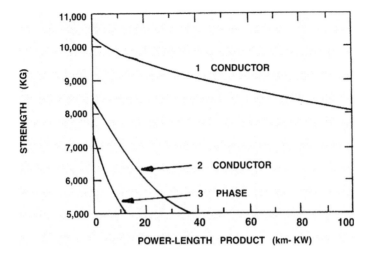

**FIGURE 5.57**  Power-length product vs. strength. Cable O.D. = 7.94 mm.

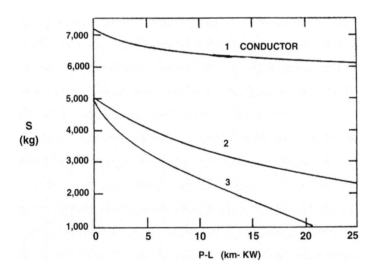

**FIGURE 5.58**  Variation of strength with power-length product. Cable O.D. = 9.53 mm.

since that cable has sufficient cross-section to absorb moderate P-L values without too great a theft of space from the load-bearing armor.

## System Trade-Offs and "Off-Optimum" Cable Designs

Figure 5.60 gives another example of the trade-offs that become possible once telemetry bandwidth has been removed from the design conflict. Here, contours of constant cable strength are plotted in voltage-current space. The maximum achievable strength for this 7.94-mm-diameter, single-conductor cable is also shown. The cable design satisfies all constraints listed in Fig. 5.55.

The performance contours in Fig. 5.60 could as easily have been drawn in terms of cable payload capability, or cable weight, or cable strength/weight ratio. Whatever parameter is most critical to the system can be plotted.

The insights offered by contour plots of this type can be of considerable value in helping to resolve design conflicts among such system parameters as cable diameter and strength, operating voltage, and

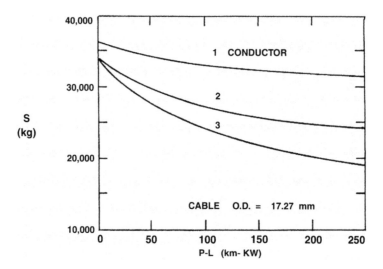

**FIGURE 5.59**   Power-length product vs. strength.

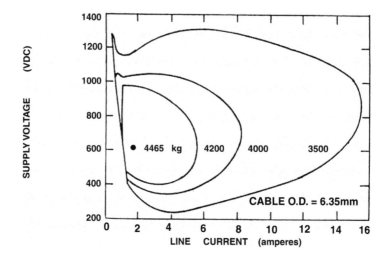

**FIGURE 5.60**   Contours of constant cable strength in voltage/current plane for: delivered power = 500 Watts; tube conductivity = 50%; cable length = 7 km; tube T/O.D. ratio = 0.27.

line current. At the very minimum, this insight can force system conflicts and compromises to take place on more rational grounds, i.e., with more light and less heat.

### The Next Generation

In the past, the author has (somewhat reluctantly) recommended the 3-phase, 3-fiber cable geometry for the "next generation" deep-sea E-O tether cable. A typical cross-section for this cable is shown in Fig. 5.61. The primary reasons for selecting this geometry were:

1. Its larger diameter (17.3 mm) gives some measure of relief from the normal inefficiency of this design in trading off strength versus power-length product.
2. The use of three optical fibers simplifies the telemetry system. One fiber can be dedicated to uplink data, one to downlink commands, and one kept in reserve for expansion or for use as a spare.

**TABLE 5.9**    Four Options For Deep-Sea Electro-Optical Tether Cable

| Performance Parameter | 17.3-mm Diameter 3-Conductor Cable | | 7.9-mm Diameter 1-Conductor Cable | |
|---|---|---|---|---|
| | 10,000 W | 20,000 W | 1000 W | 5000 W |
| Optimum voltage | 864 V | 1026 V | 1148 V | 1397 V |
| Strength (kg) | 23,470 | 20,240 | 6600 | 5790 |
| Weight (kg/km) | | | | |
| In air | 1135 | 1032 | 276 | 260 |
| In water | 927 | 818 | 233 | 216 |
| Free length (km) | 25.3 | 24.8 | 28.3 | 26.8 |
| Payload (kg)[a] | 3830 | 3190 | 1240 | 1020 |

[a]For operations to 6000 m, with strength/weight safety factor of 2.5.

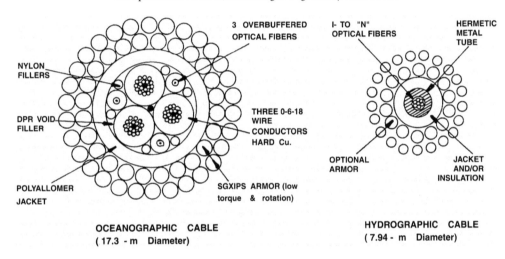

FIGURE 5.61  Options for deep-sea electro-optical cables.

3. The helical core geometry gives additional strain relief for both fibers and electrical conductors.
4. The cable geometry is a relatively close match with the well tested and successful design of an earlier E-O tow cable that was also built for operation under very high stress conditions [18].

At the same time, it is hard to ignore the many advantages of the single-conductor E-O cable. It may well be that the "next" generation should see two basic types of deep-sea tethers:

**Oceanographic:**  A heavy duty cable will be similar to the unit sketched in Fig. 5.61, and will operate from the standard oceanographic winch. It will support massive instrumentation packages, supplying them with at least 10 kW through a 10-km length.

**Hydrographic:**  A smaller cable (probably 7.94 mm) will operate from a standard hydrographic winch. It will deploy moderate-sized packages to abyssal depths and supply them with a few kilowatts of power. It will probably contain one optical fiber-operated full duplex, but can incorporate at least three fibers within a slightly larger conductor tube.

Table 5.9 shows how these two designs might perform in deep ocean operations. The two examples given for each design differ only in the amount of power they export (through a common 8000-m length).

## Handling Systems for E-O Cables

The handling characteristics of the armored deep-sea E-O tether should be much like those of conventional armored coaxes or 3-phase cables. In either case, the cable's internal geometry will closely resemble that of its electromechanical counterpart. The armor packages are (deliberately) as similar as possible. Differences in cable handling should occur primarily as responses to new operational capabilities offered by the fiber, or to new requirements that the fiber brings with it.

## Optical Slip Rings

At the shipboard end of an E-O cable, uplink telemetry will be received as very low-level optical signals. As in conventional Elm systems, these signals must be continuously fed from the rotating cable storage drum through some kind of slip ring to a stationary receiving point. The problem is more difficult (but not impossible) if the cable contains more than one optical fiber. Choices for handling the data at the winch include:

1. Pass the optical signals directly through the slip ring. But the signals will be quite weak, as much as 30 dB (1000 times) weaker than the power levels launched into the fiber at the other end of the cable. Any oil, dirt, or particulate matter within the slip ring can have disastrous effects on data transmission especially on bit error rate, which with fiber optics can readily be $10^{-9}$.
2. Receive (and transmit) the optical signals at a repeater located in a splash-proof housing within the core of the storage reel. The power level required by each optical channel is no more than 1 to 3 watts, so that the repeater can easily be powered for missions of days with conventional batteries. Signal transfer options from this point include:
   a. Multiplex the data at a high bit-rate, then transmit a composite signal through the slip joint.
   b. Accomplish the same purpose with wavelength division multiplexing (WDM). That is, transmit each signal at a discrete wavelength. This approach involves combining a spectral multiplexer (two signals sent in one direction) with a spectral duplexer (two signals sent in opposite directions). Optical signals can be sent through both ends of the cable reel axle.
   c. As a third alternative, use angular multiplexing. Let each half of the rotary joint be a blown up version of the step index fiber in Fig. 5.50a.[8] Light injected in one end of the joint at a given incidence angle will exit the other end at the same angle. Different data channels pass through the joint at different angles.
   d. As a near-term fallback, optical digital data can be converted to an electrical data stream, and telemetry can be fed through a conventional coaxial slip joint. If this is done, it is vital that the telemetry system remain digital, i.e., that data precision not be lost through premature conversion to an analog format.

## Fiber Flexure Performance

Although its absolute strength is low, the optical fiber is not the most sensitive element in the cable. It is much like a strand in a spider web—weak but very tough.

As long as the optical fiber is neither stressed to a level more than (about) 50% of its proof strain nor assaulted by other components, its performance under loaded flexure should be at least as good as that of any other cable element.

## Fiber Pressure/Temperature Response

Optical fiber attenuation is almost completely insensitive to (at least) the $-50°C$ to $+75°C$ temperature range. Any attenuation response to deep-sea pressures, to topside temperatures, or to handling system-bearing pressure will actually be due to other causes. Likely culprits include imperfections in the fiber's plastic buffer, and microbending induced in the fiber by adjacent cable components. These should have been eliminated (or at least observed) during cable manufacture and initial testing.

---

[8]We might use a section from the preform from which a fiber is drawn, since it is a scaled up (about 200 times) model of the optical fiber.

## E-O Cable Operating Stresses and Strains

It is important to remember that the optical fiber is not at all like its copper wire ancestor. For example, the copper wire can be, in fact, normally is, fully yielded and deformed during the production and cabling of an electrical conductor.

A braided copper conductor is a good example. In a good braid, each wire repeatedly makes small-radius, fully yielded crossovers of its companion wires. This does not increase wire resistance, since electrons can easily blunder around such sharp turns. Just one of these crossovers would either break an optical fiber or would induce sufficient attenuation to consume most of the system's optical power margin.

As a general rule, the optical fiber should not be subjected to high tensile loadings for long periods of time.

1. During a deployment of hours to days, fiber tensile strain should be limited to no more than 30% of the proof strain to which it was originally subjected. Typical values might be 1.5% proof strain and 0.45% deployment strain. Any strains due to fiber curvature must be counted. Some would argue that the percent-of-proof-strain limit level should be 20%.
2. During storage periods of weeks to years, the fiber's strain level should be no greater than (about) 0.10%. A related precaution is given below.

## Storage Conditions for E-O Cables

In conventional cables, high operating loads can permanently strain the copper wires. This is not necessarily bad as long as the cable is never allowed to become completely slack. If all cable tension is released, drawback of the elastic armor may cause the copper conductors to Z-kink.[9]

**Solution #1**     Using monofilament core wires and work-hardened wires, make the copper conductors as elastic and compliant as possible.

**Solution #2**     If the cable has been loaded to such a point that Z-kinking is likely, then it must never again be allowed to see low or zero tension.

The first solution also applies for the E-O cable, since it is no more than a common sense design technique for all highly stressed cables that contain materials subject to low-strain yielding.

The second solution is easy to apply to the E/M cable, since long-term tensile loads are unlikely to damage a cable component. In fact, it is common practice to operate such cables directly off a winch, so that the inner wraps see full deployment stresses and strains throughout their time in storage.

This method of operation would probably be fatal to an E-O cable. Without re-reeling, the fiber spends its operating life suffering the highest stresses it has ever sustained during deployment. The most likely result of such storage is that the optical fibers will creep to failure, and failure will occur on the storage reel, since that is where the cable spends most of its life.

## Winches for E-O Cables

In deep ocean operations, the E-O cable should never be operated directly off a storage winch. The conventional traction winch should work well with an optical or electro-optical cable. It also solves the storage tension problem, since selection of on-reel storage tension can be independent of operating conditions.

For smaller E-O cables (such as the hydrographic unit described previously), serious consideration should be given to use of the linear or "caterpillar" winch. The traction that this winch can deliver is proportional to the diameter of the cable. But the pulling force required is proportional to the cable's cross-section, i.e., to the square of its diameter. Therefore, the winch becomes more efficient as the cable becomes smaller.

Another advantage of such a winch is that the cable need not be bent during the transition from high to low tension. In very small cables, this leads to a handling system concept in which the linear winch is

---

[9]The Japanese name for this phenomenon is "zaku-tsu."

rotated so that the pulling axis is parallel to the axis of the loaded cable. The approach can essentially eliminate cable degradation due to scuffing and loaded flexure.

## References

1. Gibson, P. T., et al., Experimental investigation of electromechanical cable, *Proceedings of 8th Annual MTS Conference,* Washington, D.C., September 1972.
2. Wilkins, G. A., et al., Lightweight cables for deep tethered vehicles, *Proceedings of The MTS-IEEE Oceans '75 Symposium,* San Diego, CA, September 1975.
3. Wilkins, G. A., et al., Production and performance of a Kevlar-armored deepsea cable, *Proceedings of the MTS-IEEE Oceans '76 Conference,* Paper 9-A, San Diego, CA, September 1976.
4. Koelsch, D., Deep Submergence Laboratory, Woods Hole Oceanographic Institution (personal communication).
5. Knapp, R. H., Nonlinear analysis of a helically-armored cable with nonuniform mechanical properties, *Proceedings of MTS-IEEE OCEAN '74 Symposium,* 155, September 1974.
6. Nowak, G., Computer designs of electromechanical cables for ocean applications, *Proceedings of MTS-IEEE OCEAN '74 Symposium,* 293–305, September 1975.
7. Knapp, R. H., Torque and stress balanced design of helically armored cables, *Trans. ASME Journal of Engineering of Industry,* 103, 61–66, February 1981.
8. Knapp, R. H., Mechanical Engineering Department, University of Hawaii. Also, Philip Gibson, Tension Member Technology, Huntington Beach, CA, (personal communication).
9. Frisch, D. A. and Ranner, P. J., Unrepeatered submarine systems, *Proceedings of International Conference on Optical Fiber Submarine Telecommunication Systems,* 77–83, Paris, February 1986.
10. Marcatili, E. A., Objectives of early fibers: Evolution of fiber types, in *Optical Fiber Telecommunications,* Miller, S. E. and Chynoweth, A. G., Eds., Academic Press, New York, 1979.
11. Marcuse, D., et al., Guiding properties of fibers, in *Optical Fiber Telecommunications,* Miller, S.E. and Chynoweth, A. G., Eds., Academic Press, New York, 1979.
12. Snitzer, E., Cylindrical dielectric waveguide modes, *J. Optical Society of America,* 51, 491–498, 1961.
13. Gloge, D., Bending loss in multimode fibers with graded and ungraded core index, *Appl. Optics,* 11, 2506–2512, 1972.
14. Olshansky, R., Distortion losses in cabled optical fibers, *Appl. Optics,* 14, 20–21, 1975.
15. Murata, H., et al., Optimum design for optical fiber used in optical cable system, *4th ECOC Conf.,* Geneva, September 1978.
16. Kao, C. K., et al., Fiber cable technology, J. of Liqhtwave Technology, LT-2, 4, 479–488, 1984.
17. Smith, W., et al., Metallic encapsulation of optical fibers, presented at *MTS Cable/Connector Workshop,* Houston, TX, January 1984.
18. Wilkins, G. A., Fiber optics in a high-stress (undersea) environment, presented at *Third Int. Symposium on Offshore Mechanics and Arctic Engineering (OMAE),* New Orleans, LA, 12–16 February 1984.
19. Wilkins, G. A., A miniaturized, transoceanic, fiber optic communications cable, Proceedings FOC, San Francisco, CA, September 1981.
20. Wilkins, G. A., How small can an electro-optical, transoceanic cable be? *Proceedings of Int. Telemetry Conference,* San Diego, CA, 267–280, October 1981.
21. Hauge, O., et al., The Skagerrak HVDC cables, *Proceedings of Int. Conference on Large High Voltage Electric Cables,* Paper No. 21-05, Paris, September 1978.
22. Wilkins, G. A., The many dimensions of fiber optics in undersea systems, *Proceedings of OCEAN SPACE '85,* Tokyo, 729–737, June 1985.
23. Wilkins, G. A., Tradeoffs in optimization of deepsea E-O cables, *Proceedings of OMAE (ASME),* Tokyo, April 1986.

# 6

# Current Measurement
# Technology

Albert J. Williams 3rd
*Woods Hole Oceanographic*
*Institution*

## 6.1   Current Measurement Technology

### Principles of Current Measurement

Fluid motion in the ocean or other large bodies of water can be separated into three classes of motion. The first is mean transport, bearing on the physical oceanographic question, "Where does the water go?" Adequate sampling in space is generally a critical issue in capturing the flux of water and other scalar advected quantities such as heat and salt or sediment in suspension. Often the significant flow is weak and the threshold of measurement or the linearity near zero flow in sampling instrumentation is crucial. More often, the mean flow is buried in large zero mean flows such as wave motion, vibration of the sensor, or strong turbulent fluctuations. In these cases, linearity of the sensor over a large dynamic range is important.

The second class of motion that drives current measurement is wave motion. Waves alone do not transport fluid (except a very little bit from Stokes drift) but are reversing flows requiring special capabilities of mechanical sensors. An issue in measuring waves is separation of the directions and the frequency of the wave components. Current meters are not the usual instruments that determine this, but they can do so if they are three-axis sensors or provide additional information such as pressure.

The third class of fluid motion is turbulence, in which there is neither net transport nor closed orbits of particle motion. Turbulence is chaotic. The scale of turbulent fluctuations is small, extending from the generation scale, generally meters, to the dissipation scale, generally millimeters. The challenge for current measurement here is the size of the sensing volume, the rate of sampling, and the vector resolution of direction.

Traditional current measurement addressed only the first class of motion. Stream gauging with weirs does not require velocity sensors at all. But such techniques for flow measurement are not adaptable to open bodies of water. For these, point measurements of current were developed. Parallel developments for wind measurements resulted in anemometers, but meteorologists and aviators displayed even greater interest in the determination of wind direction. Most of the patents recorded from 1910 to 1935 under the category of measurement of flow were for wind vanes. Direction of flow has been equally important for oceanographers and technology has been as limited by the determination of direction as it has been for speed. Prior to 1980, most current meters separated the measurement of speed and direction. Since then, measurement of vector flow components has become more common.

If one wants to know where the water goes, why not mark the fluid with a tracer or a drogue and see where the tracer or drogue goes? This is the principle of Lagrangian measurement of flow. By contrast, the measurement of flow at fixed points in space provides a Eulerian description of the flow field. Modern fluid mechanics is a field theory and Eulerian measurements fit the models naturally. Lagrangian drifters or float trajectories are both easier to comprehend and harder to include in models. At the limit of a large number of Lagrangian trajectories and an equally large number of Eulerian flow measurements, their two descriptions can be interchanged. Sparse observations, however, are not equivalent, although each can reveal processes of transport.

## Mechanical Sensors

In streams where the direction of flow is generally known, propeller-type current meters, such as the Price flow meter, are used [1]. In open water, the Savonius rotor and vane were most common until 1990 and are still used in moored arrays without surface buoys. The Savonius rotor makes a full revolution when moved a constant length through still water or equivalently when the water in which it is immersed has moved past it by that distance. There is no directional preference although it is assumed that it senses only the horizontal component of flow [2]. Bearing friction sets a threshold for motion detection, typically 1 to 3 cm/s. Above the threshold, the revolutions are linear in water displacement. The Geodyne [3] and later the Vector Averaging Current Meter (VACM) [4] current meters added a compass and vane to the Savonius rotor to resolve flow direction. In the Geodyne meter, the compass and vane were read every preset number of revolutions of the rotor (counted with magnetic reed switch closures inside the housing from a magnet on the rotor) and recorded at that instant. The VACM read the compass and vane every eighth revolution of the rotor and resolved the motion into east and north components that were accumulated and stored at preset intervals in time. This reduced the spurious errors produced by spot readings of direction in the Geodyne instruments when they were subjected to wave motion or mooring strumming.

A very successful use of the Savonius rotor has been in the Recording Current Meter (RCM) instruments of Aanderaa [5]. The housing with the Savonius rotor is mounted on a vane to point into the current and swivel on the mooring. This permits a single reading of the compass to resolve the flow direction and eliminates the magnetic vane follower coupling of the rotor/vane instruments. If the excursions of the water in any direction are greater than the response length of the vane, about one meter in the RCM4 and about 1/2 meter in the VACM, the vane will follow the direction of motion reasonably well. If the excursion length is approximately equal to the response length of the vane, the behavior is poor, and if the excursion length is less than the response length of the vane, the sensed direction of flow can even be backwards. Use of current meters with Savonius rotors is now generally restricted to moorings with subsurface buoyancy so that they are not subject to errors caused by wave pumping and their use is thus restricted to depths below the wave zone. Strumming of the mooring where the flow is very strong can also present problems for these instruments (and others as well).

In 1980, the rotor and vane originally used in the Vector Measuring Current Meter (VMCM) [6] were replaced by two orthogonal propellers, or fans. The response length of the fans is only a few centimeters so it performs well, even in waves, and is less subject to error from mooring motion. This is the mechanical current meter most used with surface moorings where meteorological data are required as well as surface currents. In a special application, a third axis of flow sensing was added for measurements of Langmuir cells with their downwelling flows. In the year 2000, it is difficult to buy a complete deep-sea current meter with mechanical velocity sensors. Yet many are in the inventory of oceanographic agencies and institutions and will be used as long as electronic retrofits can modernize them [7].

## Acoustic Sensors

Mechanical friction in a current meter makes it impossible to measure very low velocities (deep-sea boundary layers for example) and moving parts are vulnerable to fouling by fishing line and marine growth. For these reasons, a sensor with no moving parts is attractive to users. Acoustic, electromagnetic, and light scattering sensors have all been used for current measurement. Acoustic sensors are presently dominating the commercial current meter field.

Acoustic travel-time current sensors depend on transmission of sound between fixed transducers to measure the component of flow along the transducer axis. The technique is more than 50 years old but has given rise to commercial current meters only since 1975. A burst of sound is typically transmitted by a pair of opposed piezoceramic transducers and the two groups of ultrasonic waves travel in opposite directions through the water at the speed of sound in the moving fluid to arrive at the opposite transducer, now a receiver [8]. The acoustic burst traveling downstream arrives first, or its phase is advanced with respect to the burst traveling upstream. The time difference or the phase difference is linearly proportional to the component of velocity along the acoustic axis and inversely proportional to the speed of sound squared. The accuracy and sensitivity of the measurement is limited more by the flow-obstructing characteristics of the transducer supports than the electronic timing or phase measurement circuitry. There is no friction to limit low velocity sensitivity, although the lack of a distinctive effect at zero velocity requires that the zero point be determined by calibration. Two- and three-axis variations of the acoustic travel-time current meter are available from several manufacturers [9]. This class of instrument is now called a point sensor since there are profiling acoustic current meters available as well.

Doppler current sensors scatter light or sound from inhomogeneities in the water and determine the velocity of the water by the change in frequency of the scattered light or sound. The component of velocity along the acoustic axis of a monostatic Doppler sensor or along the angle bisector of a bistatic acoustic or light-scattering instrument is measured. This directionality makes the Doppler current meter a vector velocimeter. Acoustic Doppler Velocimeters (ADVs) are typically bistatic point sensors. This means that the sound is transmitted along one axis and received by a separate transducer along another axis. In the typical three-axis ADV, the three receive axes cross at a point and the three Doppler signals define a vector velocity at that point [10]. This volume can be as small as 1 cm in diameter and is physically remote from the transducer supports to give a minimally disturbed sensing volume. Scattering is produced by mineral particles, living or dead marine organisms, or by micro bubbles, all of which are very effective scatterers at the frequencies of 5 to 10 MHz used in these instruments. For lower frequency ADVs or profiling instruments, thermal inhomogeneities are effective scatterers as well. Concentration of scatterers can vary by about 5 orders of magnitude, but some scatterers are required and this limits the application of ADVs to regions near boundaries or near the surface where scatterers can be expected. The sensitivity (mm/s) of an ADV is sufficient to study turbulence where its small measurement volume and high sample rate provide the required spectral response in frequency and wave number. An important class of acoustic Doppler current meter is the Acoustic Doppler Current Profiler (ADCP) where monostatic measurements as much as 1000 meters from the transducer permit velocity components to be measured at each point along the acoustic path. Mounted on a ship looking downward, a profile of velocity from the surface to the bottom can be obtained in shallow water where the bottom Doppler return gives the ship's speed [11]. Mounted on the bottom looking upward, the ADCP can measure the velocity profile almost to the surface. The scattering volumes are not the same for the separate beams of the ADCP since the same transducer is used for transmitting and receiving (monostatic) and the beams are inclined to the vertical to provide different horizontal components. Far from the transducers the beams are spread apart. The ADCP range depends upon frequency since absorption of sound increases with frequency. Long-range ADCPs use frequencies below 100 kHz where individual pings (the transmission of a short burst of sound) scatter from inhomogeneities in a volume a meter or more in diameter. The Doppler shift in frequency from a single ping does not amount to many cycles at this frequency and the resolution in velocity from one ping can range from 0.2 to 0.5 m/s. But averaging can reduce the uncertainty and in most cases the averaging can be assumed to be unbiased so that the resolution can be improved at the expense of sample rate. As the range of the profile is increased, a second cause of reduction in sample rate is the need to allow the first ping to die away before sending the next ping. Ambiguity about where the scattering came from could result if more than one ping was received at a time. The trade off in the range-resolution-sample rate product was hard to beat until 1990 when coded pulses were employed to permit multiple pings to be in the water at once. The broadband ADCP technique [12] increased the RRS (range-resolution-sample rate) product fiftyfold. ADCPs from several manufacturers dominated the meter market in the year 2000.

Correlation Sonar is an acoustic technique for measuring current that is still experimental. Originally proposed and tested for a bottom-referenced speed log on ships, the inhomogeneities in transmission of the sound through the water moving across the acoustic beam create a speckle pattern or scintillation in the plane of the receiver normal to the beam [13]. The phase delay where the correlation is maximum between two receivers separated in the along-current direction gives the velocity component of the inhomogeneities in the direction of the receiver displacement. The benefit of this technique is that it averages across a river or strait where total transport is needed while the instrumentation can be on the shore.

## Other Sensors

While mechanical sensors were the principal types in use until 1980 and acoustic sensors are the principal types being manufactured in 2000, there are others that have been in use since 1970 and are still important today. Electromagnetic sensors, Laser Doppler Velocimeters, and radar backscatter sensors are some of these.

EM, or electromagnetic velocity sensors, depend on the voltage (emf) generated by motion of a conductor in a magnetic field. The water serves as the moving conductor while the magnetic field is either the earth's magnetic field or is generated in the sensor. The Marsh-McBirney family of sensors [14] and the S4 current meter by InterOcean [15] are spherical bodies with a coil inside and electrodes on the skin. By reversing the magnetic field and synchronously detecting the small voltages measured at the electrodes, the relatively large but slowly varying surface potential of the electrodes can be rejected and the current-induced voltage amplified and measured. Using a single coil and two pairs of electrodes in the plane of the coil, two components of velocity can be measured. A second coil in a plane perpendicular to the first with a third pair of electrodes in the plane of the second coil but not in the plane of the first provides a three-axis sensor of velocity. This vector measurement is attractive and made these sensors some of the first alternatives to mechanical sensors around 1970. They have no moving parts, they can be made small, and they are fast. There is a trade-off between sensitivity and power since the magnetic field is generated with an electric current, which requires power, but the signal is proportional to the magnetic field. The magnetic field is also concentrated near the surface of the sensor where the fluid velocity is most affected by the skin of the sensor. In fact, there is a boundary layer hysteresis in reversing flow that is troublesome unless the electrodes are spaced a small distance out or the surface has artificial roughness added with sharp ribs.

The geomagnetic electrokinetograph, or GEK, had scientific use in the 1950s and its principle is in use in several families of current meter in 2000. This principle is that conductive fluid moving in the earth's magnetic field generates an emf in the direction mutually perpendicular to the magnetic field and the flow. In a broad current such as in the ocean, the emf drives an electric current that returns through slower moving water or conductive seafloor and the resistive voltage drop due to this current can be measured by a sensor advected with the flow. The sensor measures the difference between its own velocity and the conductivity-weighted average of velocity to the bottom. Free fall instruments including an expendable electromagnetic current profiler (XCP, XEMVP) [16] are used to study deep currents and low frequency internal waves.

The Laser Doppler Velocimeter (LDV) has been an important laboratory tool for flume studies of flow over topography and around bodies since 1970. As in wind tunnels, the flow is seeded with scatterers to enhance the signal strength. The benefits of small measurement volume and high sample rates are attractive for turbulence studies and the LDV technique has been exploited for deep-sea and coastal measurements. In the deep sea, without seeding, the signal strength is low and the sample rates are less than in the lab. More particles and micro-bubbles in the near-shore environment give a better signal and greater sample rate [17], but LDV has not seen the acceptance of ADV. The smaller sample volume and the shorter wavelength of the LDV limit the number and size of optimal scatterers so that the signal it produces is more intermittent than with the ADV. At very high turbidity, where there are sufficient scatterers, there is a light attenuation problem that can be overcome only at increased laser power. But LDV offers the highest spatial resolution and, with sufficient scatterers, the highest sample rate for turbulence studies.

Electromagnetic propagation in the radio spectrum can be used to measure current. Radar is Doppler shifted by the velocity of the waves that cause sea clutter in radar presentations. This Doppler shift can be used to determine the current at the surface if the radar wavelength is tuned to the waves that are returning the radar signal. By using VHF frequencies for the radar, Bragg scattering can be obtained from gravity waves with wavelengths of 10 to 40 meters, waves that have a well-known deep-water propagation speed. In the absence of current, there should be two Doppler shifts in the radar return from these waves, one for waves approaching the radar and the other for waves receding from the radar. If the Doppler shift is displaced from what is expected, the water on which the waves are propagating can be assumed to be moving. This is the basis of several commercial systems such as CODAR [18] and OSCR [19]. An over-the-horizon military system [20] has provided surface current maps from a distance of 1000 km. When two radar stations have coverage of a coastal region with different viewing angles, a vector current map can be constructed within the overlap region. The propagation speed of gravity waves in shear currents is related to their wave length. By using two frequencies in the radar, a current shear estimate can be made. The wavelength of the gravity wave determines the depth at which a current affects the propagation speed of the wave. Shallow current affects short wavelength gravity waves; deeper currents affect long wavelength waves. In shallow water, the wave speed is not that of deep water but related to the depth rather than the wavelength. So the radar can be used also as a depth sounder where the separation of the two Doppler shifts is a measure of the shoaling effect. So far, the radar current measurement has not been exploited from satellite observations of radar backscatter, although such use is possible in principle. It has seen its greatest use in the littoral region where the radars can be set up on the shore.

## Lagrangian Floats

Global coverage with Lagrangian drifters is a solution to the problem of measuring current everywhere that is inexpensive compared to deploying and servicing moorings everywhere. A program to do this has been proposed and named ARGO [21]. The sparseness of Lagrangian observations that causes problems for certain studies when only a few floats are deployed is then no longer such a concern, at least on the global scale. In a sense, the proposed seeding of the ocean with 1000 to 3000 floats is like seeding a flow for differential particle imaging velocimetry (DPIV) [22] where the velocity field can be reconstructed from the displacement of each Lagrangian particle over a short time interval.

Concerns with Lagrangian measurements of flow center on what flow the drifter measures. Drogued surface buoys have been used to measure the Gulf Stream and Kuroshio; warm and cold core Gulf Stream rings; and the western drift that carries warm water across the Atlantic to northern Europe. These floats were tracked by satellite. They had ropes or drogues suspended beneath them to follow the water rather than the wind. Subsurface floats, called Swallow floats [23], are less compressible than water, and move with the current at their ballasted depth. But a Swallow float is a compromise between isobaric and isopycnal. Being less compressible than water, it tends to seek a constant depth even if the water sinks or rises. This makes it less successful at tagging a parcel of water for many studies than if it tracked the density and was isopycnal. Increasing the compressibility of the float to match that of its target water mass makes it a better water tag, but complicates its ballasting and depth control. An important development has been depth control based upon temperature sensing. Bobbers that make daily excursions around their target isotherm and readjust their displacement track their target water better than isobaric floats [24]. These have been tracked acoustically from low-frequency timed transmissions from the float or by satellite, as in the case of Autonomous Lagrangian Circulation Experiment (ALACE) drifters that surface periodically and are tracked by satellite.

## Merged Observations

Eulerian and Lagrangian formulations are the classic ways to view current measurement. Autonomous Underwater Vehicle (AUV) supported velocity sensors are going to require us to consider merged observations that are neither the classic fixed in space nor the tagged particle measurements. AUVs move at velocities that are only a few multiples of the current that they might measure, so they are making velocity observations from a series of positions and cannot be considered true Eulerian measurements, even when the AUV velocity is subtracted from the measurement. Yet they are certainly moving with respect to the water and are not a passive particle so they cannot be viewed as Lagrangian. Depending on the question that is asked of the sensor, they can be assumed to approximate Eulerian measurements from an array of positions. But their observations are certainly biased if internal waves or deep convection are the subject of study. Classifiers for target processes can be constructed that take the AUV speed as a parameter for separation [25]. Ocean wanderers carrying velocity sensors may join our more traditional moored and free-drifting instruments, so these concerns must be addressed.

## Summary

Measurement of current is inherently more complex than temperature or density because current is a vector rather than a scalar quantity. Stability of the sensor and earth coordinate direction are important. Motion of water in the ocean is a fundamental feature that underlies transport and mixing of heat, salt, chemical species, and suspended particles. At the small scale, turbulence is responsible for friction and stress that dissipates waves and erodes sediment. Waves transport energy but are fluid motions that can be measured as current, even though there is little net transport of water. Observations of current have progressed from point measurements to profiles and to surface maps. Arrays of sensors to reveal structures in the velocity field can use large numbers of an inexpensive mooring [26] or can combine surface and profiling instruments for presenting 3-D processes. Global coverage will require a new type of array, possibly an array of Lagrangian floats or self-powered gliders [27] that will provide current information from their differential motion between transmissions or from current sensors on their moving bodies.

## References

1. Kenney, B. C., An experimental study of vane coupling in Price current meters, in *Proceedings of the IEEE Fifth Working Conference on Current Measurement,* IEEE Catalog Number 95CH35734, 1995, 10–14.
2. Saunders, P. M., Overspeeding of a Savonius rotor, *Deep-Sea Res.,* 27, 755–759, 1980.
3. Richardson, W. S., Stimson, P. B., and Wilkens, C. H., Current measurements from moored buoys, *Deep-Sea Res.,* 10, 369–388, 1963.
4. Halpern, D., Pillsbury, R. D., and Smith, R. L., An intercomparison of three current meters operated in shallow water, *Deep-Sea Res.,* 21, 489–497, 1974.
5. Aanderaa, I. R., A recording and telemetering instrument, Technical Report, Fixed Buoy Project, NATO Subcommittee for Oceanographic Research, vol. 16, 1964.
6. Weller, R. A. and Davis, R. E., A vector measuring current meter, *Deep-Sea Res.,* 27A, 565–582, 1980.
7. Strahle, W. J. and Martini, M. A., Extending and expanding the life of older current meters, in *Proceedings of the IEEE Fifth Working Conference on Current Measurement,* IEEE Catalog Number 95CH35734, 5–9, 1995.
8. Williams, A. J. III, Tochko, J. S., Koehler, R. L., Grant, W. D., Gross, T. E., and Dunn, C. V. R., Measurement of turbulence in the oceanic bottom boundary layer with an acoustic current meter array, *J. Atmos. and Oceanic Tech.,* 4, 312–327, 1987.
9. Kun, A. and Fougere, A., A new low-cost acoustic current meter design, in *Proceedings of the IEEE Sixth Working Conference on Current Measurement,* IEEE Catalog Number 99CH36331, 150–154, 1999.
10. Lohrmann, A., Cabrera, R., Gelfenbaum, G., and Haines, J., Direct measurements of Reynolds stress with an acoustic Doppler velocimeter, in *Proceedings of the IEEE Fifth Working Conference on Current Measurement,* IEEE Catalog Number 95CH35734, 205–210, 1995.
11. Edelhauser, M., Rowe, F., and Kelly, F., Long-range current profiling from moving vessels, in *Proceedings of the IEEE Sixth Working Conference on Current Measurement,* IEEE Catalog Number 99CH36331, 287–294, 1999.
12. Brumley, B. H., Cabrera, R. G., Deines, K. L., and Terray, E. A., Performance of a broadband acoustic Doppler current profiler, *IEEE J. of Oceanic Engineering,* 16, 402–407, 1991.

13. Clifford, S. P. and Farmer, D. M., Space-time analysis of acoustic scintillations in ocean current sensing, in *Proceedings of the IEEE Third Working Conference on Current Measurement*, IEEE Catalog Number 86CH2305-1, 78–81, 1986.

14. Cacchione, D. A. and Drake, D. E., A new instrument system to investigate sediment dynamics on continental shelves, *Marine Geology*, 30, 299–312, 1979.

15. Trageser, J. H. and Elwany, H., The S4DW, an integrated solution to directional wave measurements, in *Proceedings of the IEEE Fourth Working Conference on Current Measurement*, IEEE Catalog Number 90CH2861-3, 154–168, 1990.

16. Sanford, T. B., D'Asaro, E. A., Kunze, E., Dunlap, J. H., Drever, R. G., Kennelly, M. A., Prater, M. D., and Horgan, M. S., An XCP user's guide and reference manual, Tech. Rep. APL-UW TR 9309, Applied Physics Laboratory, University of Washington, Seattle, WA, 59 pp., 1993.

17. Trowbridge, J. H. and Agrawal, Y. C., Glimpses of a wave boundary layer, *J. Geophys. Res.*, 100, 20729–20743, 1995.

18. Barrick, D. E. and Lipa, B. J., Using antenna patterns to improve the quality of SeaSonde HF radar surface current maps, in *Proceedings of the IEEE Sixth Working Conference on Current Measurement*, IEEE Catalog Number 99CH36331, 5–8, 1999.

19. Shay, L. K., Graber, H. C., Ross, D. B., and Chapman, R. D., Mesoscale ocean surface current structure detected by high-frequency radar, *J. of Atmos. and Oceanic Tech.*, 12, 881–900, 1995.

20. Georges, T. M., and Harlan, J. A., The case for building a current-mapping over-the-horizon radar, in *Proceedings of the IEEE Sixth Working Conference on Current Measurement*, IEEE Catalog Number 99CH36331, 14–18, 1999.

21. Argo Science Team 1998. On the design and implementation of ARGO. An initial plan for a global array of profiling floats, ICPO Report (21), Godae Report (5), Bureau of Meteorology, Melbourne Australia.

22. Bertuccioli, L., Roth, G. I., Katz, J., and Osborn, T. R., A submersible particle image velocimetry system for turbulence measurements in the bottom boundary layer, *J. Atmos. and Oceanic Tech.*, 16, 1635–1646, 1999.

23. Davis, R. E., Webb, D. C., Regier, L. A., and Dufour, J., The autonomous Lagrangian circulation explorer (ALACE), *J. Atmos. and Oceanic Tech.*, 9, 264–285, 1992.

24. Price, J. F., Bobber floats measure currents' vertical component in the Subduction Experiment, *Oceanus*, 39, 1, 26, 1996.

25. Zhang, Y., Spectral feature classification of oceanographic processes using an autonomous underwater vehicle, Ph.D. thesis, Massachusetts Institute of Technology and Woods Hole Oceanographic Institution, Woods Hole, MA, 2000.

26. Williams, A. J.3rd, Benthic "weather" current meter array, in *Proceedings of the IEEE Sixth Working Conference on Current Measurement*, IEEE Catalog Number 99CH36331, 236–240, 1999.

27. Curtin, T. B., Bellingham, J. G., Catipovic, J., and Webb, D., Autonomous oceanographic sampling networks, *Oceanography*, 6(3), 86–94, 1993.

# Index